agi

the agi

source book

for geographic

information

systems

1996

association for geographic information

ited by
David R. Green
Department of Geography
University of Aberdeen

David Rix
MVM Consultants

Chris Corbin
Corbins Consultancy

published by
**the Association for
Geographic Information**

*marketed and
distributed by*
Taylor & Francis

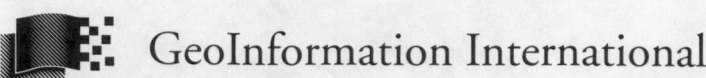

GeoInformation International

Discover our new products

New: *International GIS Dictionary*   **Launch Date: November 1995**

International GIS Dictionary includes GIS terms from all over the world and also from related disciplines, such as Remote Sensing. Edited by top GIS experts, Karen Kemp and Rachael McDonnell, the *International GIS Dictionary* provides an invaluable source for professionals and students. ISBN 1 899761 19 5

New: *GeoCube v1.5*   **Launch Date: December 1995**

A colourful electronic GIS-based course for the professional and educational market. GeoCube v1.5 offers a comprehensive introduction into the concepts of GIS. It will bring the concepts of GIS alive - with the aid of text, illustrations and animations! ISBN 1899761 32 2

Announcing a new journal: *Transactions in GIS*   **Launch Date: January 1996**

This new Journal, *Transactions in GIS*, will provide a forum for the exchange of ideas, techniques, approaches and experiences in the rapidly growing international field of Geographical Information Systems (GIS). *Transactions in GIS* will include review and research articles, book reviews, comments, forthcoming events and abstracts. *Transactions in GIS* will be an invaluable tool for those in the research and development side of GIS as well as those in planning and decision making. ISSN 1361 1682

 NEW: Census Users' Handbook

Edited by Stan Openshaw

Paper 1 899761 06 3
Price £19.99 net

 NEW: GIS for Business and Service Planning

Edited by Paul Longley, GrahamClarke

Paper 1 899761 07 1
Price £19.99 net

 Understanding GIS: The ARC/INFO® Method; ESRI

UNIX and Open VMS ™- Version 7
Paper 1 899761 04 7   Price £23.99 net

PC Version
Paper 0 582 21433 5   Price £21.99 net

 NEW: GIS for Business: Discovering the Missing Piece in Your Business Strategy
GeoInformation International

Paper 1 899761 05 5
Price £24.99 net

 Postcodes: The New Geography

J F Raper, D W Rhind, J W Shepard

Cased 0 582 09270 1
Price £29.99 net

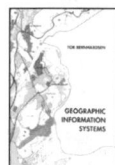 Geographic Information Systems

Tor Bernhardsen

Paper 82 991928 3 8   Price £19.99 net

 Geographical Information Systems: Principles and Applications
Edited by D J Magurie, M F Goodchild, D W Rhind

Cased 0 582 05661 6
Price £148.00 net

 ARC Macro Language: Developing ARC/INFO® Menus and Macros with AML™
ESRI

Paper 1 899761 54 3
Price £26.99 net

 Bringing Geographical Information Systems Into Business

David J. Grimshaw

Paper 0 582 22549 3   Price £17.99 net

For further details or a complimentary book catalogue please contact: Heather Burkinshaw, GeoInformation International, 307 Cambridge Science Park, Milton Road, Cambridge, CB4 4ZD, UK. Tel: +44 1223 423020 Fax: +44 1223 425787

D
910.285
AGI

D
910.6041
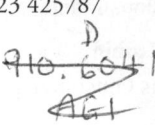

contents

trade directories 1996 115

guide to the use of the directories 117

miscellaneous reference 1996 315

foreword

This is the seventh edition of the Association for Geographic Information's source book, and the second in its revamped format. The 1995 edition considerably increased the number of directory entries. This new edition adds still further to the list and some 400 organisations are now included.

All the information in the detailed directories has been provided by the organisations themselves from a questionnaire distributed in early summer 1995. No charge has been made for the basic entries in the books, although some have chosen to expand on their entry by taking further space. Although the editors have done their best to ensure that the information included is accurate, no responsibility can be taken for any errors that may be included. We would of course be grateful to receive notification of any errors or omissions.

As before, the directories are supplemented by a series of articles on business aspects of GIS, by some of the country's leading experts. There is also a substantial amount of general reference material.

The AGI's thanks are due to the three editors, David Green, David Rix and Chris Corbin for the effort they have put in to compiling the book.

We do hope that you will find the book useful in your day to day work. Your comments and criticism would be welcome.

Article editors:	David R. Green and David Rix
Directory editor:	Chris Corbin
Publisher:	Shaun Leslie
Advertising sales:	MJ Marketing
Design:	Michael O'Reilly Creative Consultancy
Typesetting:	Thameslink Ltd
Printing:	Page Bros, Norwich
Price:	£29.50
ISBN:	1 874059 20 9
	0-7484-0496-1

Marketed and distributed by Taylor & Francis, Rankine Road, Basingstoke, RG24 8PR, UK
Tel: 01256 813000, Fax: 01256 479438.

The Association for Geographic Information
12 Great George Street
London SW1P 3AD
Telephone: 0171 334 3746
Fax: 0171 334 3791
Email: agi@geo.ed.ac.uk

the AGI - working for the geographic information community

Alastair Macdonald

AGI Chairman 1995

Since its foundation in 1989, the AGI has had a clearly focused mission to spread the benefits of geographic information and GIS to the wider community and to help all users and vendors of GIS. Over the past twelve months, the AGI has continued to make major contributions to many of the key issues of concern to the GIS community.

One key activity has been the high level forum between the Government and the AGI to discuss issues on the accessibility of public data. Four meetings of the AGI Government round table have been held: on general availability issues, marketing and pricing, legislative issues such as privacy and copyright, and standards. The findings of the round table will be taken up with Government: as the exercise was set up with the explicit blessing of the Prime Minister, we have confidence that there will be worthwhile and tangible results. Certainly there is a surprising degree of consensus between the public and private sectors, and one key point to emerge is that one of the greatest beneficiaries of any resolution of the issues will be the public sector itself.

The wider spread of GIS and the wider adoption of desktop platforms has meant that many users of GIS are newcomers to the technology. The AGI has had to adapt to reflect this. Our membership has traditionally encompassed a very broad spectrum from vendors of hardware, software and data, to users in central and local government and government agencies, the utilities and the wider general commercial market. Our aim has been to attract new groups of specialist interests and we have successfully established new special interest groups.

For example, the health special interest group (SIG) has conducted a major survey of GIS within NHS health trusts and organised a well attended seminar on the benefits of GIS. The group has also helped organise a one day conference during our annual event at AGI '95. The business SIG has also organised a session at AGI '95 and two regional meetings in Bristol and Leeds. Our newest group, the marine and coastal zone management SIG, has attracted an outstanding level of support and will be active in the future. These new activities are additional to the well established environmental SIG and the survey and mapping group. The Young Gis'ers (YOGIs) have been active too, and produced the post-AGI '94 and AGI '95 conference newsletters.

We made substantial changes to the AGI conference and exhibition to reflect market needs. The revamped conference has been particularly successful in attracting delegates, consolidating the AGI's position as the organiser of the definitive GIS event in the UK. The event has absorbed a considerable amount of management and membership time and it was decided by Council that it would be in our members' and the industry's best interests if we were to enter into a partnership with a professional organiser to ensure that the event got the investment resources needed to ensure its future success. This will have the added advantage of guaranteeing the AGI a significant amount of income for the next few years. AGI will continue to organise the conference element of the event and its continuing quality is thus assured.

In 1994, AGI joined EUROGI with some misgivings mainly concerned with the high cost and with the danger of duplication of activity. However, EUROGI's first year has proved our fears to be unfounded and Council formally confirmed its support for EUROGI in June. This decision was reinforced in October when the Department of the Environment, on behalf of the Government, undertook to pay half the AGI annual subscription with immediate effect. One development which has fuelled our respect was EUROGI's involvement in the consultation on the EU's strategy document *GI 2000 – Towards a European Geographic Information Infrastructure* which may eventually have important implications for the European geographic information community.

None of this would be possible without the heavy involvement of many of our members. As a small organisation, with limited resources, we do rely on our members to get things done. The AGI's achievement are *their* achievements. As Chairman, I see at first hand the effort being put in. I would like to thank formally all those members who have served on committees and in other ways. More important, I, and my successor as Chairman, look forward to your continued support and to encouraging more of the industry to become active members. The AGI has an ambitious agenda, and it is only by exploiting our collective muscle that we can continue to make progress in the exciting and dynamic world of geographic information.

geographic
information
1996

editorial

David R Green

Centre for Remote Sensing and Mapping Science,
Department of Geography, University of Aberdeen,
Elphinstone Road, Aberdeen AB9 2UF, Scotland, UK.
Tel: (01224) 272324
Fax: (01224) 272331
Email: d.r.green@aberdeen.ac.uk

and

David Rix

MVM Consultants plc., MVM House, Oakfield Road,
Clifton, Bristol, BS8 2AL, England, UK.
Tel: (0117) 974 4477
Fax: (0117) 970 6897
Email: davidrix@mvm.co.uk

introduction

Each year the Association for Geographic Information (AGI) Sourcebook for Geographic Information combines a selection of invited chapters on different aspects of GIS and the related technologies with comprehensive UK GIS directories. The reasoning behind the continuation of this successful format is twofold. Firstly, it provides the reader with an up-to-date overview of a number of current GIS topics and issues - in other words a 'snapshot' of the 'hot topics' of the year; secondly, it provides the directory user with a comprehesive list of GIS information (e.g. about applications software, vendors, products and services) and contacts (telephone and fax numbers, electronic mail, and postal addresses). Although largely centered on UK-based information, European information is also included where possible.

In many respects the directories component of the Sourcebook is very like a telephone directory, albeit one that focuses on a very specific theme, but one that is significantly easier to query. Like the telephone directory it has become a vital part of the professional office as an up-to-date source of information. In the 'GIS World', where technology is evolving very rapidly, access to such information is invaluable for those who are new to GIS and who wish to know more; those who wish to acquire details from a software vendor, and even those who need to contact a colleague quickly. Moreover, the Sourcebook is coordinated by representatives of the UK body responsible for GIS, the Association for Geographic Information (AGI), is updated annually, and is aimed specifically at the UK GIS community.

chapters

The rationale behind the content of the AGI Sourcebook chapters in the 1996 edition has been to take a single GIS theme and then to explore various different aspects and issues within this broad area.

With the rapid and continuing growth in commercial applications of GIS to the business, marketing and retailing sector, the Editors considered it appropriate to examine a number of different aspects and issues pertinent to this theme. Additional factors influencing their decision were the number of international conferences and exhibitions offered on GIS and Business over the past 12 months, and a decision by the AGI'95 Event committee to focus on this very theme for the AGI'95 Conference and Exhibition.

Three broadly defined sections are identified in the Sourcebook: Data, Applications, and Emerging Technologies.

In the first section of the book entitled Data there are three papers, each one of which examines a different aspect associated with data sources, data supply and the issues of data ownership. The first paper by Gina Fisher of the Ordnance Survey (OS) examines the direction the OS is now taking with respect to supplying data to the business community. Ralph Robbins (Automobile Association) and Mary Short (Kingswood Ltd.) raise some important questions about the misuse of digital data emphasising the importance of the Data Federation (or DataFed) initiative. Nick Walker of the Dolphin Consulting Group discusses the use of census and non-census data in demographic and lifestyle analysis for business GIS and examines the different types of delivery system used for market analysis applications, and the data sources now available to supply market analysis capability in GIS systems. In the final paper of this opening section Jonathan Shears (Erdas UK Ltd) and John Allan (Erdas UK Ltd) examine the potential of spatial data originating from new digital photogrammetric software for a range of GIS applications.

The second section of the Sourcebook, Applications, examines a cross-section of the wide ranging commercial applications of GIS in the world of business. Alan Bourke, of Scottish Homes, considers the growing popularity of Windows-based PCs for GIS using the case study of Scottish Homes. This paper presents a very useful 'down to earth' user-based perspective on the application of GIS software in the work environment, highlighting some of the requirements of the applications specialist, including reference to the user-interface. Laszlo Bardos, MapInfo Inc., also examines the potential of the desktop GIS. Visualisation and graphic data analysis are two of the unique possibilities for GIS applications in sales and marketing. This paper, in particular, stresses the importance of the visual map graphic in the workplace which can be used as the basis for planning and decision-making. An interesting point made by Bardos is the ease with which it is possible to use low-cost GIS technology to create a simple map directly relevant to the inquiry - one that is effectively free of the clutter often associated with more traditional paper maps not created for the purpose. The paper cites a number of examples to illustrate the sorts of applications possible. Peter Sleight, a partner in the Target Market Consultancy examines GIS applications in the retailing and marketing sector of the economy and some of the key issues. The paper specifically considers the close relationship between GIS and the geodemographics 'industry', in particular Lifestyle data. Peter Mingins, Intergraph, discusses a more specific application of GIS within the UK Regional Electricity Companies (RECs) for providing supply of service and customer care. The application of GIS is examined in the context of Business Process Re-Engineering. Viewing it as a corporate modelling and management tool, which is part of the whole business system, Mingins outlines a more integrated role for GIS in the business process. Following the success of the road accident analysis system in GIS, Tom Lithgow, Senior GIS Development Consultant with Lothian Regional Council's Transportation Department, outlines the the use of GIS for its road maintenance managment system; specifically examining issues of data volumes, operationality, and portable information handling for 'on-site' validation.

The final section, Emerging Technologies, addresses some of the recent areas of development in GIS and the related technologies already beginning to have an impact on GIS. Four papers in this section examine topics ranging from the growing importance of the Human Computer Interface (HCI) to applications software in the workplace, the emergence of orthophotography, to OpenGIS, and finally the much hyped promise of the Information Superhighway. David Green (University of Aberdeen) and David Rix (MVM Consultants plc) present a brief overview of the role of the Human Computer Interface (HCI) examining some of the reasons why this aspect of GIS is of growing importance to the GIS user. It is suggested that recent developments e.g. the World Wide Web (WWW), coupled with the requirement for access to information will pose a significant challenge to GIS vendors to develop user-interfaces in the future to meet demands. Richard Markham (MVM Consultants plc) considers the potential applications of digital orthophotography to GIS. In this paper the author suggests that, as a result of recent developments in hardware and software technology, the photographic image will find new roles in commercial applications for GIS where the more traditional cartographic images have proved to have interpretation difficulties for users unfamiliar with cartographic symbology. These images are not without their problems, however, and file size and sources of error can prove difficult.

John Glover, (Intergraph UK) presents an overview of OpenGIS, a topic that has found growing interest over the past 12 months. The final paper touches upon the links between an ever increasingly familiar topic, the Information Superhighway and GIS. Tackled in the last Sourcebook from an academic perspective, by Professor Michael Goodchild (NCGIA), Jim Crowder (Digital) now examines the subject from the commercial viewpoint. An inevitable connection waiting to be established, the Information Superhighway offers many possibilities for widespread access to geographical information resources via a user-friendly interface. Still at an early stage in its development, there are, however, some problem areas in need of solution e.g. the commercial value of data, and copyright, security, topics already touched upon earlier by Robbins and Short. Despite this, Crowder is positive about the futue of the Information Superhighway ending by stating – the cost is low, the technology is widely available, all we are lacking is the will.

Throughout the chapters, a number of common topics and themes seem to reappear. These include the recognition that the technology has arrived and the general desire for access to data and information in support of business requirements. The key challenge facing the industry, however, is the provision of access to both data and functionality, from the desktop, in a form that is appropriate to the non-technical user. If GIS is to develop from its traditional role, as a research tool, into a legitimate tool for the business user then a fundamental requirement will be ease of access to information and ease of use of the functionality. Research into HCI together with the emerging standards for OpenGIS may prove to be the watershed in bringing GIS to the desktop.

Users, any users, should be able to select software components which are appropriate for their business requirements, and from these components make use of information pertinent to the business. OpenGIS may at last provide users with access to software components which can be integrated to their busines requirements where in the past such users may have been locked into inappropriate technologies supplied by centralised IT departments whose selection criteria may have been biased more by fashion than by function.

Research into the HCI may remove the lingering barrier of technology to the OpenGIS initiative. Whatever the merits of OpenGIS, if the everyday user is restricted, by whatever means, in fully utilising the functionality provided by the software component, then we may reasonably consider that the utilisation of the component will have failed. Assuming the vendors and application developers commit to this concept, what will be the implications of 'plug n'play' technologies how will the industry rationalise this progress to the demands from some quarters for the certification of GIS 'professionals'?

This is likely to be an ongoing debate.

Ordnance Survey meets the needs of the commercial business user

Gina Fisher

System Suppliers Marketing Executive,
Ordnance Survey, Romsey Road,
Maybush, Southampton, SO16 4GU.

Tel. (01703)792042; Fax. (01703) 792208

abstract

Ordnance Survey (OS) is committed to improving its understanding and appreciation of the GIS market. This paper outlines some of the ways in which the OS have responded to the evolving GIS market.

The evolution of the GIS market is described in terms of three waves – each wave heralding the development of a new family of GIS applications. The two strategies of Product Development and the establishment of a Co-Marketing Channel are introduced in order to demonstrate the increasingly flexible, market-driven response from the Ordnance Survey.

KEYWORDS: ADDRESS POINT; OSCAR; MERIDIAN; PRODUCT DEVELOPMENT; GIS APPLICATIONS; COMMERCIAL; VALUE ADDED RESELLERS

introduction

Early in 1995 the OS conducted significant market research in order to investigate the spatial information requirements of a wide range of industries. This study represented the largest and most comprehensive evaluation of the GIS market undertaken to date.

The results are being used to ensure that OS can provide a responsive and flexible service, appropriate to the requirements of the evolving GIS market. One of the most striking findings of the research study was the changing composition and character of the market place. The GIS market of the mid 90s is significantly different to that of the early 90s – specific differences are expressed in terms of the size, type and number of organisations requiring access to spatial information. Furthermore, the research indicated that the GIS market is continuing to develop and evolve – market maturity has not yet been realised.

the evolution of GIS applications

The first wave – inventory applications

The seed-bed of GIS development within the UK was in the utilities, local government, and certain central government departments – organisations which remain the core markets for digital map data today. A market of traditional map users, these organisations justified their GIS investment on the basis that it would enable them to perform their mapping functions more economically than manual methods. This early wave of GIS development was characterised by data collection and inventory type applications.

GIS were employed to address simple data queries of a 'what is where?' and a 'where is what?' nature.

The second wave – analytical applications

The second wave of GIS applications introduced improvements in the analytical power of GIS. The majority of GIS users today have developed applications which typically involve the interrogation and analysis of two or more layers of information. A variety of spatial and statistical techniques are applied to the data sets in order to identify and explain the 'why of where ?'.

The third wave – business and commercial applications

Recent experience suggests that a third wave of GIS applications is beginning to gain momentum within the market – a wave of applications designed to support activities within the commercial business market. Increasingly, commercial organisations are beginning to evaluate the functionality of GIS, in terms of its ability to provide solutions to every day business problems and to fully support their business activities. Drawing upon market research findings, and recent experience, it appears that this new wave of GIS applications will drive some significant changes in the structure and composition of the GIS market. Two of the most tangible changes are outlined below:

Type of organisation

The GIS market is going to become increasingly heterogeneous in terms of its industrial composition. The utilities, central and local government have traditionally been the core industries which have fuelled the development of the GIS market. However, the new wave of GIS applications will sweep along a whole new raft of organisations interested in utilising the power of GIS.

Organisations working in the fields of insurance and banking, retail, transport, market research and advertising are all beginning to appreciate the value of viewing their information in a spatial context.

Type of GIS user

The new commercial business GIS user will typically be the director of insurance underwriting, the bank manager, or the marketing manager. These users do not have an established tradition in organising and managing their information through map use. They are not likely to be interested in different data structures, data transfer formats, or learning a new programming language. Their key requirement is for a desktop working solution, which can start providing market or business intelligence immediately.

The Ordnance Survey response

The OS have responded to the changing nature of the GIS market in a number of different ways. This section describes two areas in which market-driven responses have been introduced. These responses fall into the areas of: 1. Product Development, and 2. Distribution Channels.

product development

Precise geo-referencing tools

A basic information requirement of the commercial business GIS user is understanding the nature and composition of their market. The widespread adoption of bar coding, the trend of purchasing goods and services with credit cards, and the introduction of customer loyalty schemes creates a data rich environment. The desire to improve understanding of the locational characteristics of customers and competitors has led many organisations to evaluate GIS technology. The integration of customer information with other data sets, in a geographical context, allows businesses to evaluate:

What is going on?

Where?

and,

Why it is happening?

Through the provision of Address-Point™ OS have provided the commercial business GIS user with a powerful tool for market analysis. Address-Point™ provides a full national grid reference for each and every business and public postal address in England, Scotland and Wales. The product provides the definitive tool for the identification and use of precisely located addresses, upon which a wide range of commercial business applications can be built. For example, insurers have traditionally used the postcode district or sector as the basic spatial unit for risk assessment.

Hence the premium quoted for an individual property would have been based upon the 'average' risk applied over the entire postal district or sector of several thousand houses. Through developing a detailed understanding of the spatial distribution of hazards, such as flooding, subsidence, storm damage and crime, and using Address-Point™ to precisely locate each property, insurers can develop rating structures which accurately reflect the 'actual', rather than the 'average' risk to an individual property.

A family of products

Improved awareness of digital mapping has meant that an increasingly heterogeneous group of individuals and organisations are using GIS for an expanding range of applications. This inevitably has led to a multiplication in the specific data requirements of the market place. Data requirements vary widely, and must be tailored to fit the application and the realities of the software and hardware. For example, some applications will need only local geographic data, others may need regional coverage, whilst other users will require data at a national scale. The OS are committed to meeting the diversity of demand which is being generated from the commercial business market. The Oscar® family of products provides one example of the way in which the OS are developing flexible customer-driven responses. Oscar is the name given to a family of digital road centre-line products which have been derived from a single definitive source. The specification of each of these products was defined on the basis of detailed market research, producing a family of products, distinguished on the basis of their accuracy and information content.

Consider the requirements of the ambulance trust and the large retail store. Both require a digital representation of the road network. However, their applications, and hence specific requirements, are very different. The ambulance service wishes to incorporate the road network into an emergency response system. It is critical that every road and pedestrianised street are represented. Oscar Asset-Manager, the most detailed dataset, meets the needs of the application. The retail store requires a road network to build into their planning information system, in order to evaluate the accessibility of proposed retail sites. Levels of detail and resolution are not so critical to the success of their application, and Oscar Route-Manager satisfies their data requirements. The Oscar® family is the first 'family of products' launched in response to a highly diverse customer demand pattern. The success of this family of products is indicative of one way in which the GIS user community is driving the OS new product development programme.

hybrid products

In 1994, the OS identified changing patterns of demand in the GIS market place, and responded quickly through the development of a new vector data product called 'Meridian' introduced to the market in late 1995.

Meridian was developed to satisfy market demand for a 'mid-scale' vector product, falling between the large-scale Land-Line product and small-scale Strategi product. Meridian satisfies the requirement for a mapping product which can provide the commercial business GIS user with a broad overview of their operational area – a product which is neither too detailed, nor too general, for regional scale analysis.

Through the development of Meridian the OS have introduced a fresh approach to product development. The product is not derived from a single definitive source, rather it is composed of a range of features which have been sourced from a variety of different databases, structured to produce a new product for the GIS market.

Distribution channels

The introduction of the Co-Marketing Channel illustrates a second area in which the OS have developed a flexible response to the changing requirements of the GIS. The Co-marketing Channel was established at the end of 1993 to manage the marketing of digital data products to the commercial business user through Value Added Resellers, and to manage relations with System Suppliers and Consultants in the GIS market.

the OS initiative

The OS has been successful in developing strong business relationships with a variety of organisations who are taking up the OS Initiative and becoming registered Value Added Resellers (VARs) of Ordnance Survey data. The OS Initiative was set up in order to develop the enormous potential for digital map data in new markets. Studies have shown that around 80 per cent of business and commercial data is geographically related. The scope for map-based systems to fully exploit the information held by organisations is therefore enormous. The Co-Marketing team have already been successful in developing relationships with VARs, who have put together specialist packages, invariably involving innovative use of the data, 'tailored' to satisfy specific requirements for end users not previously attracted to direct data sales. The following examples illustrate a number of ways in which Ordnance Survey VARs are meeting the requirements of the business commercial market.

Data agents and data distributors

An increasing number of VARs are acting as agents and distributors for OS small-scale data. These agents may be organisations who source and market a wide range of different digital data sets, in a range of different formats, providing a 'one-stop-shop for data'.

Alternatively they may be System Suppliers acting as 'single source suppliers', who market the data in a software solution targeted at specific markets. These types of solution can already be identified in the desk-top mapping industry, whose general philosophy is one of the provision of flexible, low cost, geographic visualisation and analysis solutions, that are as easy to use as a spreadsheet or database. However, without maps this type of software is not very productive, hence the OS are encouraging suppliers of desk-top technology to supply OS small-scale mapping with their software. The advantages to the customer are clear – single source supply of both software and data in a proprietary format.

Value-added services

VARs are also active in utilising our data products to provide a bureaux or value added service to customers. One example is the PhoneLink plc Tel-Me service, an On-Line Windows-based information system providing essential business information to the commercial market. One component of this On-Line service is the Mapper Street Level 'how to get there' service which is based upon Oscar® and other Ordnance Survey context mapping. The advantage to the customer is that they do not have to hold the full national database on their system, and are able to enter into a 'pay-as-you-use' charging structure.

support to systems suppliers and GIS consultants

The Co-Marketing Channel also provides technical and marketing support to the network of system suppliers and GIS consultants. Increasingly the OS are working in collaboration with software suppliers, providing a technical and marketing support service. Sample and demonstration data are available to System Suppliers for testing system development, and for demonstrating their applications and solutions to the commercial business market.

the challenge to Ordnance Survey

The OS are committed to supporting the development of appropriate GIS applications within the commercial business market. The objective is to raise awareness of the business benefits of digital mapping, and to offer appropriate digital map data products and solutions designed to meet the diverse information requirements emerging within the GIS market.

data to drive GIS – a dynamic resource

Ralph Robbins

Head of Cartographic Research,
Automobile Association Developments Ltd.,
Fanum House, Basingstoke, Hampshire,
RG21 2EA, England, UK.
Tel. (01256) 492906
Fax. (01256) 494651

and

Mary Short

Managing Director, Kingswood Ltd.,
449, Chiswick High Road,
London, W4 4AU
Tel. (0181) 994 5404
Fax. (0181) 747 8047

abstract

This paper examines the pivotal role of digital map data in Geographic Information Systems, and highlights a recent tendency among some users to downplay the significance of the data. It shows that producing and maintaining digital maps is an expensive and ongoing task that needs funding from the GIS world in order to continue. The paper goes on to consider the copyright and intellectual property issues surrounding the use of digital map data, and examines the impact that recent legislative and technological developments could have. It concludes that a responsible attitude among data users is the key to long-term growth of GIS.

KEYWORDS: DIGITAL DATA; DATA THEFT; INTELLECTUAL PROPERTY RIGHTS; COPYRIGHT; DATAFED

introduction

Underlying the whole concept of Geographic Information Systems (GIS) is a paradox. Mapping has made possible the very existence of GIS, yet recently there has been a tendency in some quarters to marginalise the importance of the pure cartographic elements of GIS.

Digital map data in particular is starting to be taken for granted. Instead of seeing it as a distinctive and valuable resource, some GIS practitioners have started to regard it almost as a disposable commodity. There seems to be an assumption that map data will be more or less freely available; and this attitude is often accompanied by a corresponding disregard for the high cost and complexity of gathering the data, of keeping it up to date and accurate, and of processing it for commercial use.

The reality is that digital mapping underpins the whole of the GIS revolution. What you choose to analyse with GIS is infinitely variable, but map data is an essential component.

This paper considers the development of map data in GIS, the issues underlying its use, and the steps being taken by producers to ensure that the significance of map data continues to be recognised.

digital mapping: mainstream or marginal?

To some extent the apparent marginalisation of map data was perhaps inevitable. It reflects a natural tendency in the GIS world to give priority to the use of the map data, not its source. GIS is after all dynamic. It is not about mapping as such; it is about enhancing a whole range of analytical tasks by adding new depth and precision to the spatial aspects of the work.

To cartographic suppliers this trend is exciting. As far as they are concerned it has drawn mapping into the commercial mainstream. The potential increase in revenue presents enormous opportunities and challenges; it means there should be more funds available for even better research, development and refinement of the product. These benefits in turn should feed back to the GIS world in terms of improved and more finely-targeted mapping: a classic business circle.

However, if map data suppliers are denied due recompense for their work the circle is broken and benefits to the GIS world will not come. While this may not yet be a serious threat, lately there has been increasing evidence of map data misuse, piracy and downright theft. Perhaps more worrying than any individual incident is the fact that a climate should exist in which these activities are considered in any way acceptable. Custom and practice can quickly overturn inhibitions about such things if there is no check.

computer technology – opportunity and threat

It is of course computers that have given map data its new and potent force and have also raised the problems of data control. Until the advent of digital technology map data, in the modern sense of the term, did not really exist, and nor did data theft. Map publishers might occasionally plagiarise other printed material, but the results were not hard to identify and the output of the relatively small community of map producers was in any case fairly easy to monitor. The advent of digital map data and GIS has changed all that.

The use of map data has quickly spread into the realms of business, industry, government, public services. Unlike published maps, the end products of GIS are seldom on public view; and even when they are, the cartographic elements may be concealed by a customised presentation unique to the specific implementation. This means that even though GIS is steadily permeating every walk of life, the mapping elements that make up any given instance can be extremely difficult to monitor.

Within the province of computer-based mapping, there was an early perception that rasterised versions of printed maps were particularly desirable, since they were attractive to look at. But these were inconsistent in quality and not particularly useful for analysis – or ripe for pirating. It was the development of scaleable vector data that really opened the door to GIS. By storing individual topographical features separately, vector data allowed analytical software to reassemble the map selectively, feature by feature. Maps based on vector data could also be produced in a whole range of cartographic styles, from simple outline form to detailed full-colour productions resembling printed output.

However, this very flexibility has probably been one of the keys to misuse of the data. The problem is that the end result often has no standard appearance to identify its origin. That depends on the power of the GIS software and the user's ingenuity. In some cases, the temptation has been for users to focus their attention on the prime analytical tasks, and to dismiss the underlying map data as incidental.

cartographers rise to the commercial challenge

If the cartographic world invites criticism over any aspect of this issue, it is perhaps for being almost too open in its attitude. Historically, in the relatively small cartographic community, there has been an instinct to share knowledge while respecting the integrity of individual suppliers.

There was a similar openness when the first digitisation was being done in the 1970s and 1980s. The Automobile Association (AA), for instance, made no secret of the fact that its digitisation work was prompted initially by the desire to speed up and enhance the process of preparing its own printed atlases. Selling the data to outside users was originally low on the agenda.

However, the digitisation coincided with the emergence of GIS as a new discipline, and cartographers soon realised they held a key to its growth among outside users. Although some suppliers were quick enough to place a value on the map data they were now able to sell, they may not all have attached equal importance to the task of tailoring it to their markets or adding value to it. Perhaps understandably, some GIS customers formed the opinion that their interest was being exploited by cartographers who saw revenue from map data as no more than marginal income.

If once true, though, this attitude among the cartographers is now long gone. Early on, for instance, the AA appointed an external software specialist, Kingswood Ltd. to market its AutoMaps data. Raw digital data does not necessarily come in a form suitable for further use, and the AA recognised that it takes a focused effort of this kind to separate out its components, translate file formats and generally make it commercially accessible.

Map data sales increased, but it soon became clear that there was no quick commercial return to be made. The digitising of maps has proved an extremely lengthy, painstaking and expensive process for all concerned in the industry, and although computers are an essential element in the task, there is still a large degree of manual input at various stages. Geographical features have to be plotted with minute care on a digitising tablet, and then the resultant coordinate structure has to be cross-checked and verified to ensure accuracy.

It is now clear that digitising could not have continued at the rate it has done in recent years without the promise of income from beyond the immediate world of mapping. In other words, external sales of digital map data have become fundamental to its continued development.

perception and reality – conveying the true value of map data

The challenge for cartographers has been to convince external customers of the intrinsic value of map data. Compared with other forms of statistical material, maps suffer because of a general perception that they belong in the public domain. Much of what is shown on maps is observable by anyone; it is not secret or privileged information. Moreover, the general shape of the landscape, and even the broad pattern of the road system, tends to be relatively unchanging from year to year, and much of it is represented with reasonable accuracy on maps that are now out of copyright. It is perhaps not surprising that some users query the need to pay what can seem a fairly hefty price simply to ensure that this apparently 'free' information is brought up to date. However, there are two important cost factors to consider in the equation. One is the high investment cartographers have to make in their core business. Techniques such as satellite imaging, aerial photography and the basic research are not cheap, but the cost can be kept down if they are paid for by every user of the data, not just by customers for the printed atlases that are produced with the research.

This philosophy is exactly the same as is applied by any market research organisation. To take a simple parallel, you could determine the location of every supermarket in the country by driving around looking for them all; but to avoid that effort it would be perfectly normal practice to pay someone else for the information instead. You would certainly not expect to be given the material free, as a right. Buying digital mapping is a generic instance of the same concept.

The other cost factor associated with map data is that of the digitisation. In a sense this represents the packaging of the basic material, and is unique to the individual map data supplier. To take another parallel instance, Britain's telephone numbers are not secret or subject to any access fee, but the aggregate of phone numbers contained in BT's directories is protected by copyright. The value is not in the individual components of the dataset, but in the dataset as a whole, and in the work that goes into compiling and maintaining it.

the law

Where does the law stand on all this? Legislation on data in recent years has tended to focus on protecting the rights of access of individuals to data about themselves, and to restricting wholesale dissemination of sensitive information.

Protection for suppliers of digital data has been less evident, and in the main they have had to draw on established principles of copyright law. In particular, the suppliers have assumed intellectual property rights to their material, much as an engineer would over an original design which could not be patented.

Until recently that contention had not been widely put to the test, but this year a strong indication of the likely interpretation by the courts came in two cases instigated by the Ordnance Survey. Two map publishers were accused of using OS and other copyright map data in their products without paying the appropriate royalties or acknowledging the source. In one case extensive evidence was seized under an Anton Piller order (which allows unannounced access to premises).

One of the accused companies admitted liability and agreed to pay the missing royalties and costs. In the other case, the High Court found in favour of the OS, and an injunction was granted preventing further production or sale of any of the products concerned. The two cases have been greeted by the cartographic world as a landmark assertion of the rights surrounding the use of digital map data. These are, however, fairly extreme instances of blatant fraud. In most cases, misuse of digital map data takes a more low-key form. It is likely to emerge where a licensed user inadvertently channels the data into some internal application that falls outside the licence terms – particularly when the end result is some form of printed map. While clearly less damaging to the supplier than deliberate pirating of the data, such misuse is nevertheless in some ways equally insidious. It contributes to a general belief that map data is somehow 'up for grabs', and that once acquired it can be used for any purpose.

treading a delicate path

As map data suppliers, committed to expanding the use of GIS, we have a difficult path to tread in all this. Clearly we have no wish to make customers feel hidebound by over-restrictive licensing agreements and are keen to avoid appearing heavy-handed in our attitude to them. It is in no one's interest for suppliers to bear down on users over every imagined infringement of the terms – technical or real. At the same time, we believe it should be made plain to all map data users that they are dealing with a valuable commodity that demands respect and requires judicious use.

The key is a reasonable and fair attitude. In selling AA map data, for instance, we avoid annual licences involving constant renewal; we prefer a one-off arrangement, and in drawing up the licence we place the emphasis on defining exactly what the user is entitled to do with the data. If necessary the terms can always be amended later by mutual agreement. Pricing is another key to a responsible attitude to map data. If the cost of the data is too high, users may feel they are being treated unreasonably. Rightly or wrongly, the result can be a sense of injustice that actually encourages misuse and piracy. At the other end of the scale, if the price is too low it may tend to devalue the commodity, fostering the attitude that the product must have little value and can therefore be copied with impunity.

moves to promote awareness of the value of map data

The question is how the GIS world can encourage a responsible attitude to data. In the parallel field of computer software, strenuous efforts have been made in recent years to establish a climate in which such copying is recognised as both morally wrong and legally indefensible. In Britain, the Federation Against Software Theft (FAST) has been highly successful in bringing the issue to public attention, and, because it is an industry-wide initiative, it avoids the stigma of appearing to be merely a lobby group protecting individual suppliers' interests.

Digital data has slipped through that net, but this year saw a new initiative intended to fill the gap. The Data Federation, or DataFed, has been created to promote a wider understanding of the significance and value of data – and especially the kind of data used in GIS. It covers not just pure digital map data, but also other forms of information such as demographic data, lifestyle data and all forms of terrain data.

DataFed aims to promote a broad media debate on the issue, and to raise awareness of the value of digital data throughout the business community. Among its key objectives is to encourage businesses to apply self-imposed audits on their use of data – just as many of them have now done over the use of software. From its inception, DataFed has been supported by several leading data providers including the AA and BT.

the communications revolution: a new element in the equation

However, just as some kind of stability is emerging in the world of digital data, there are signs now that the situation could be thrown back into turmoil by the information technology revolution. The massive explosion of interest in the so-called information superhighway poses potentially far-reaching questions about the long-term viability of keeping a check on the use of any kind of data. Whatever national controls there may be on what can be disseminated electronically, they can be circumvented by the international nature of computer communications. There is always a risk that an irresponsible user somewhere in the world will place data on line illicitly for all to use.

Besides, communication technology is not the only threat to the security of map data. Some western countries are already less protective of data sources than others. Recent court rulings in France, for instance, suggest that the law in that country is unwilling to afford digital map data the same copyright protection as hardcopy mapping. Around the world, the whole issue is now coming under debate. On one side is the view that safeguarding investment in data-gathering helps to underwrite future development; on the other is the assumed sanctity of concepts such as freedom of access to information. Against this background, the prospect of preserving a viable income stream from map data sales becomes progressively slimmer.

However, the picture is probably not as bleak as it might seem. GIS seems likely to remain a specialised discipline within the broader business world. It should therefore continue to be possible to identify key groups of GIS users or potential users, and direct a clear message towards them about the value of map data they are using. But as the use of GIS spreads, it will be more and more difficult to identify individual users within the overall community. So the role of DataFed and similar initiatives could become even more important in getting the message across.

Lately the wheel has come full circle in the world of digital map data, with recent emphasis being placed on rasterised display map data. This is finding growing appeal in applications such as command and control systems, where the priority is presentational appearance rather than layer control or minute analysis. Unlike early scanned images, the latest raster data from the AA is derived directly from the underlying vector data, and is therefore consistent and precise. Yet the screen displays look just like printed maps.

In terms of possible copying and data piracy, the prospects for raster data seem mixed. To the extent that the data is distinctive and relatively fixed in appearance, illicit use should be easy enough to recognise. But because it is attractive and easy to display, raster data should widen the appeal of GIS among users who were not interested in the complexities of handling vector data; and with that new interest could come new temptations to copy and misuse. It could therefore become even more important to convey the message that the data has a finite value, and cannot be kept up to date without fair recompense from users.

conclusion

So what should the GIS world make of this issue? Clearly it will do little good to the cause of GIS if users are perpetually badgered for payment by the map data suppliers, or feel constantly under suspicion over presumed misuse of the data. What is needed is a responsible attitude all round.

Perhaps it will be enough in these relatively early days of GIS if users simply keep in mind that every piece of data has a source – and that someone, somewhere, had to pay to assemble it.

For every new GIS project initiated, it is reasonable to enquire about the source of the data involved, and query any that is unacknowledged. GIS is nothing if not dynamic, and it needs ongoing research investment to ensure that the data which underpins it continues to meet that need.

census and lifestyle data in market analysis GIS

Nick Walker

Dolphin Consulting Group,
DCG, 10 Collingwood House,
Dolphin Square, London SW1V 3ND.
Tel. (0171) 798-8465;
Fax. (0171) 798-8692

abstract

Data sources available to supply market analysis capability in GIS systems are no longer just the tried and trusty Census. This chapter commences with an overview of 'locality' marketing, and the demographic and behavioural data that make it so effective. It then outlines the nature of the Census and other relevant data, particularly that deriving from Lifestyle databases – and looks at their strengths and weaknesses. After examining the way that 'classification' systems are developed from the data, the chapter reviews the types of delivery system used for market analysis applications. The chapter concludes with a brief look at likely developments in the immediate future.

KEYWORDS: MARKET ANALYSIS; CENSUS; LIFESTYLE; GEODEMOGRAPHICS; TARGETING

introduction

A developing trend during the 1980s was a shift in the approach of consumer marketing from mass marketing to 'niche' (or micro) marketing. This has been characterised by a focus on individual customers and their characteristics, in terms of where they live – or 'locality' marketing. Various factors have driven this trend, particularly:

♦ ever extending ranges of products (and services)

♦ increased fragmentation of lifestyles, and association of lifestyle with consumer behaviour

♦ media fragmentation: new technologies have increased the ability to reach consumers selectively by type and locality

♦ changes in computing technologies have meant reduced cost, increased performance, and more refined and focused data collection.

Marketing maturity varies by industry sector, but a common driving force behind the growth of market analysis has been the need for businesses (and, indeed, public services) to improve the effectiveness of their marketing and communication programmes. Return on marketing investment strategies can, and must be, optimised by precise targeting and monitoring of results to ensure continual re-focusing.

The growth of 'locality' market analysis (and its many applications such as target marketing, direct mail, retail location, sales prediction etc.) has been associated with the development of Geodemographics and, more recently, 'Geolifestyles'. We are attempting to examine and predict consumer behaviour – we are seeking to answer questions such as:

♦ What kind of people buy my product, and where do I find more of them?

♦ How do I evaluate relative performance of the stores in my branch network?

♦ Where should I site my next outlet?

♦ How do I estimate the market size for my product in this area?

♦ Where do I advertise to best approach my target market?

♦ How do I maximise efficiency of allocation of sales territories?

These problems have been expressed in terms of business objectives, but techniques to solve them are, of course, equally applicable to many public services (such as planning of emergency services). In fact, applicability is to any situation where the characteristics of people are involved – their demographics, lifestage, lifestyle and attitudes – and where existing or potential demand is to be examined.

This paper examines the primary data sources available to address these issues and the computer systems which will use them. These systems are perhaps best termed, generically, 'Geographic Decision Support Systems'. 'GIS' tools used in this context cover a broad spectrum; from full GIS functionality to basic mapping, all normally using some form of the data described in this paper which supply measurements of demographic, socio-economic, behavioural and even psychographic characteristics. From a history of use of only the decennial Census, the mix of usable information has been enriched by a variety of non-census data. Of particular significance has been the possibilities arising from the 'Lifestyle Databases', as they have reached 'critical mass'–lifestyle data are discussed separately, in acknowledgment of their growing importance.

the census data

Collection and products

The Census of Population for Great Britain is carried out by the Office of Population Censuses and Surveys (OPCS) in England and Wales, and the General Register Office in Scotland (GRO(S)). The Census is paid for from Government funds, although the marginal cost of preparation of outputs is passed on to users of the statistics. The most recent Census was held in April 1991 and was the nineteenth full Census in a series commencing in 1801. The Census is by means of a self-completion questionnaire (completion is legally compulsory), involving 19 questions on each person, and five questions on the housing itself. More than 55 million people were recorded from 23 million households – a massive task. To help ensure satisfactory completion rates, there are strict confidentiality rules, including thresholds on the size of output areas supplied.

The Census provides a unique source of data on the national population. It is a snapshot for a single day, consistent and comprehensive, which can be examined at geographical levels down to individual neighbourhoods – termed Enumeration District (ED) for England and Wales and Output Area (OA) for Scotland. There are around 110,000 ED's each typically of 150 to 200 households and 35,000 OA's with even less households. At this level, over 9000 counts are available in tabular form. This dataset is termed the Small Area Statistics (SAS), amounting to nearly 1.5 billion data items. The data is also available for wards, postcode sectors and many other geographical areas.

The Census data cover a range of basic demographic and socio-economic characteristics, for example:

Total population by age, sex, and marital status; household composition, economic activity, socio-economic group, housing type, amenities and tenure, number of rooms and cars, educational qualifications, ethnic group, workplace and occupation.

Most of this information is supplied on a 100% basis, but some (such as socio-economic group) are processed on a 10% sample basis, due to the need to 'code' the answers from written responses.

Other relevant Census products include the Special Workplace Statistics, Special Migration Statistics and the Sample of Anonymised Records. These are powerful datasets and possibly underutilised in relation to their potential.

relevance and availability

It is the SAS information which is normally of most relevance to market analysis. This is produced by OPCS on magnetic media, but, in the private sector, will most often be supplied by Census agencies, re-processed to operate with appropriate analysis software.

The most commonly used geography for marketing applications is the postcode sector (of which there are around 9,000 in UK). The reasons for this are partly because this is the 'traditional' marketing geography, and partly because of the decreasing cost of the data. As the cost issue becomes less relevant, the benefits of more precise analysis become realisable. A typical marketing system will include, say, 50 to 200 variables, perfectly adequate for most purposes.

geodemographics

Census data is historically synonymous with the concept of geodemographics (the analysis of people by where they live – the mechanism of locality marketing) and most analysis involves some use of Census or Census-derived data. It is an enormously rich data source, and, used directly, can act as a powerful tool, particularly where selection criteria are well defined or where area analysis by census variables is appropriate.

Census counts are, by nature, non-categorical variables and for many applications of targeting and assessment of potential from market research data, it is far more convenient (and effective) to use the categorical approach supplied by Neighbourhood Classification systems. There are also, of course, 'visualisation' advantages and possibly better regional identification. With these systems a geographical unit (normally the ED or postcode) is 'labelled', typically with a selection of 40 to 60 labels. Classification systems originated as completely Census-based, but have increasingly incorporated other data in their construction. They are discussed in more detail later.

census strengths and weaknesses

Strengths of Census data in terms of effectiveness for geographically based market analysis include:

♦ The complete coverage provides consistency for both households and in geographic extent

♦ The data is inherently geographical (each ED is spatially referenced and linked to other geographic units

♦ The level of aggregation is sufficiently small to be appropriate to a large number of applications

♦ In context, the data are modestly priced

Weaknesses and inherent limitations of the data include:

♦ Many specific questions considered relevant to market analysis are not included (such as income). This can cause use of possibly flawed surrogates

♦ The accuracy and comprehensive nature is compromised in some areas by the 10% sample questions and by possible biased under-coverage

♦ The information in any case relates to demographic and social issues, not to the behavioural and attitudinal characteristics considered important in many marketing applications

♦ It is 'neighbourhood' based, and certainly in direct marketing this can be seen as a disadvantage compared with alternate information more directly focused on individuals

♦ The Census is itself only updated every ten years, creating a (perceived) ageing problem.

However, person and household count updates are produced; in any case there is considerable evidence that changes are small – areas change less than individuals, and people who move are replaced by people similar to themselves.

non-census data

A range of data relating to consumer demographic and behavioural characteristics has increasingly been used to enhance the traditional census-based approach. This applies both to the construction of geodemographic classification systems and, in many cases, to other forms of targeting and segmentation products.

The reasoning for use of these data in classification systems is considered in the later discussion of classification systems. Arguments are presented to support the incorporation of such data in general classifications, and for focusing on a particular theme, such as financial behaviour.

In considering data sources, they are generally evaluated against criteria such as:

♦ Breadth of coverage (how regional is it?)

♦ Statistical projectability (how uniform is the distribution?)

♦ Viability at a small geographic level (how feasible is aggregation without compromising use at the lowest level?)

♦ Contribution to predictiveness (how inter-related are the variables, do they make an independent contribution?)

♦ Database compatibility (how disparate are the data sources – for example Census and lifestyle data?)

♦ Adherence to data protection principles (is this a problem?)

♦ Updateability (how dynamic is the data – can it be updated regularly?)

♦ Reliability (will the data continue to be collected?)

♦ Relevance (is it correlated to consumer behaviour, and will it work for market analysis?)

A last point is, naturally, also very important – what is available (and can its use be negotiated)? Certain data elements (such as income) have obvious intuitive significance, but have not been collected on the Census (although wealth surrogates are, of course, used). Of recent time, income measures at postcode level, derived from lifestyle data, have become available.

These issues, relating to the viability and usefulness of data, have been raised to illustrate the considerations—we make no attempt to judge particular data sources on the criteria.

The list of typical data sources below is not at all exhaustive.

CREDIT ACTIVITY: Very large amounts of transactional information have been built up by several credit-referencing bureau, based on PAF or the Electoral roll (see below). The data can be aggregated to postcode level to supply a 'credit activity indicator'.

COUNTY COURT JUDGEMENTS (CCJ'S): This information summed for households at postcode level supplies a further measure of credit worthiness, in terms of credit risk.

THE ELECTORAL ROLL (ER): The register of electors has been used for mailing purposes for many years and holds around 44 million names. Whilst the ER itself is segmented geodemographically (via the postcode), organisations holding the ER have derived 'variables', largely due to its dynamic nature. Examples are: household composition, young voters, recent movers, length of residence, mobility. Analysis of names is used, not just for household composition, but for lifestage and lifestyle prediction. Again, this information can be summed to postcode level.

POSTCODE ADDRESS FILE (PAF): Continually updated by Royal Mail, PAF contains some 24 million addresses, residential and non-residential. In addition to counts of households by postcodes, it can be used to help identify the presence of farms and flats and the extent of business premises within postcodes. By examination of changes, PAF is used to measure 'migration'. In addition to the PAF address information, the 'POSTZON' file supplies information at postcode level for 1.6 million postcodes—the National Grid Reference is of particular significance for marketing systems.

UNEMPLOYMENT DATA: From the Department of Employment, this information has successfully been used to enhance classifications, although it is produced only at postcode sector level. It is very frequently updated.

POPULATION UPDATE ESTIMATES: From Local Authorities and OPCS, the data can be used to help calibrate population and household updates. For classifications that are mainly Census-based, stable demography will, of course, be assumed.

MARKET RESEARCH SURVEYS: Cross-tabulation to market research data has always been very important to geodemographics. The ability to link via classification systems (made possible by 'coding' respondent postcodes with the classification) supplies a powerful method of assessing 'penetration' and 'potential'. The Target Group Index (TGI) is perhaps the best known of the of this type of survey, and, in addition to a vast amount of information on product usage (consumer goods, financial services, holidays, etc.), it includes media usage of all kinds. There are, of course, many other useful research sources cross-tabulated by classifications, for example; Superpanel and Homescan (grocery purchases), TMS (clothing and footwear), FRS and MFS (financial services), NRS and BARB (media research). Market research data is important in several areas. The cross-tabulations are often a key part of market analysis systems, supplying area based market estimates and targeting, media planning and many other capabilities. The data itself can also act as a powerful tool in development of classifications (particularly 'market specific' variations).

The previous section has described some of the more common 'extra' data sources. There are many others (such as company director records, retail accessibility, insurance ratings) that have successfully been used to assist in identifying important themes for neighbourhood classifications. One final important source of information is also often the most powerful–real customer data. The most focused and effective targeting and segmentation tools will quite likely result from development of predictive models based upon analysis of the characteristics of known customers. However, appropriate data is not always available, and it may be a relatively expensive approach.

lifestyle data

There are many sources of information that could be considered as describing 'lifestyle'; and even more in terms of surrogates. For the purposes here, however, we are referring to what are known as 'lifestyle databases', and the information which can be derived from them.

Lifestyle databases

Lifestyle databases are populated from self-completion 'lifestyle questionnaires', which supply records of individuals associated with an extensive range of demographic and behavioural information. The questionnaires are distributed using a variety of methods, such as product registration, mailings, magazine inserts and door-to-door distribution. In most cases, some form of incentive is used, such as prize draws, money-off coupons and special offers. These are considered to improve response rates.

Questionnaires typically include questions to supply:

♦ A range of demographics (age, sex, occupation, family composition, etc. –and importantly, income)

♦ Information on hobbies and interests

♦ Extensive motoring details (including insurance)

♦ Information on product consumption, including in some cases for 'sponsored' questions (for which the results may not be available).

The range and extent of questions used (and the resulting database variables) is impressive, although the detail varies considerably, depending largely on the origins of the various operators. As an example, the National Shoppers Survey questionnaire used by CMT (see below) contains around 110 question topics, 1300 'tick-boxes', and generates more than 300 'variables' per individual. Below are a few points about the more established lifestyle database operators (having the larger databases). Some degree of care must be taken in considering volumes: ageing, 'opt-out', validation and de-duplication all affect actionable volumes, depending on the application. In practice, considerable effort is invested in validation (and de-duplication) against the electoral roll.

NDL INTERNATIONAL: NDL commenced operations in the UK in 1985 and say that they have received a total of over 19 million responses, at a rate of over 3.5 million per year.

Even after rationalisation NDL claim around half of UK households (meaning more than 10 million), including information on over 80% of postcodes. The NDL data originates mainly from product registrations and results in a database called 'The Lifestyle Selector'.

CMT: Computerised Marketing Technologies has been operating in the UK since 1987. Its core questionnaire is called the National Shoppers Survey (NSS), which uses magazine inserts and other means, with incentives such as prize draws and reward packs. 'Behaviourbank', the resulting database, is claimed to hold around 8 million individuals and 5 million households.

ICD: international communications and data has operated in the UK since 1988. Its core survey (called 'facts of living') is distributed in similar manner to the NSS and a sister survey called IS mailed from the electoral roll. The ICD database probably holds around 3 million households.

These are the more established companies with the largest databases (NDL and CMT in fact have the same owner). There are an increasing number of other organisations collecting lifestyle data through similar surveys, including CCN (Chorus), Dudley Jenkins (Consumer Surveys Ltd) and many more.

lifestyle products

The main activity traditionally associated with these lifestyle databases has been targeted mailings ('list rental'). The sheer volume of the databases and the highly relevant information ('hard facts') has ensured their success in this area. The targeting is assisted by 'profiling'; the matching of customer records to the database and examination of the characteristics of the matched records. The mechanisms for this vary; scoring systems (even interaction detection) are used–but the idea is to establish criteria for selections off the lifestyle database.

The potential for development of these data has extended the opportunities both for direct mail and geographic market analysis.

database 'extension'

Using the Electoral Roll as a base, the major lifestyle companies have attempted to extend the predictivity of the data to a much larger universe. ICD were the first with the 'National Consumer Database' using modelling techniques on variables such as 'house value' and income. NDL developed 'The Lifestyle Network', their approach including use of the geodemographic classification 'Define'. CMT have joined in with 'Consumerbank'.

products for market analysis

To extend the use (and value) of their databases, the lifestyle companies have increasingly explored the use of their data in various forms of market analysis and information systems–sometimes in collaboration with vendors such as geodemographic agencies.

CMT have in fact spawned MIC (Marketing Information Consultancy) in order to develop this type of business with CMT (and NDL) data.

An early example was 'Checkout', a grocery product developed by CCN and CMT to use the NSS data in CCN's market analysis system, followed by similar products for DIY, financial and motoring applications. CMT also offers the 'SCAN' family of PC based products which supply analysis for stores, financial products etc., in addition to demographic and lifestyle variables.

More recently, the NDL and CMT data have been merged to produce the 'Lifestyle Census', which holds the common information from these sources. They say that this database has over 100 pieces of information on more than 13 million households. A product derived from this is 'FIND', a directory of income details (based on 'real', not surrogate, data), available nationally at postcode level. NDL also produce a postcode sector based product (Lifestyle Market Indicator) supplying sector profiles for about 70 variables from the Lifestyle Selector. This dataset has in fact been used in several 'GIS' systems.

geolifestyles

Clearly, as the lifestyle databases have grown (and reached 'critical mass'), there is the opportunity for development of 'classifications' for general segmentation and targeting–a term which has been used (in addition to geolifestyles) is 'Behaviourgraphics'.

A recent development was the joint venture between Infolink (now Equifax Europe) and NDL, resulting in 'Portrait', constructed using a number of datasets, but significantly, no data from the 1991 Census. Data incorporated in Portrait includes: NDL lifestyle data (such as income, hobbies, age, car ownership, occupation, and many others), unemployment statistics, credit searches and CCJ's, company directors and electoral roll derivations.

Portrait can be used in the same way as the 'normal' geodemographic systems (described later)–it is assigned to postcodes nationally. The classification prefixes clusters with income and age, resulting in 175 'types'. A product such as Portrait becomes possible because of the size of a database such as NDL's, covering over 80% of all unit postcodes and enabling a sufficiently robust solution. Portrait is now only marketed by MIC for the NDL/CMT group.

lifestyle strengths and weaknesses

The strengths of lifestyle data are quite convincing, particularly in a marketing context.

♦ The data is current, and continually updated– meaning that products are updateable

♦ It is relevant, based on 'hard facts'

♦ Lifestyle variables can be used simultaneously and directly connected. This means that, in principle, some of the Census 'neighbourhood' limitations are avoided (the ecological fallacy)

♦ Key variables such as income and behavioural measures are present (not available on the Census)
Possible weaknesses of the data include:

♦ It is not anonymous (a weakness depending on the application)

♦ There are doubts expressed as to the representativeness; both in terms of bias of the respondents versus the total universe, and inherent bias in the questionnaires

♦ In some areas, there is likely to be patchy (or very little) coverage

♦ It may not be suitable for some non-marketing applications

To some extent, of course, it is up to the lifestyle companies to demonstrate that the techniques used to overcome bias and coverage are robust, particularly in the context of geolifestyles as a 'general' consumer classification.

Lifestyle data have attracted their share of criticism. However, in terms of marketing ability (for example, predictiveness for targeting) there is no doubt that these are powerful tools.

The availability of derived products for market analysis (particularly in 'GIS' systems) is relatively recent, but we can expect considerable growth as it becomes established; in market analysis systems, in integrated marketing information systems, and in enhancement of customer data.

classification systems

History

Geodemographic classification systems have been available in the UK for over a decade. The first commercial product was 'Acorn' (from CACI), followed by PIN (and FINPIN) from Pinpoint Analysis, 'Mosaic' from CCN, SuperProfiles from CDMS, and later DEFINE from Infolink; and several others. All use Census data; some solely (such as Acorn), whilst others (such as Mosaic and Define) use a mixture for (perceived) reasons discussed below. The newest product (PORTRAIT) uses no Census data at all. There has been no proliferation of classification systems, however, there being no increase after the 1991 Census, in spite of an increase in the number of Census agencies.

Why classifications?

All these systems are, in fact, intended as 'neighbourhood' classification systems, even when at postcode level. They could, in fact, be considered as 'data reduction' methods to aid ease-of-use, or to supply a compromise for those unable or unwilling to use more extensive data. However, there are other quite rational reasons for producing classifications. Firstly, consider the ways of describing people and neighbourhoods; below are some of the issues.

Neighbourhood description: The different types of neighbourhood identifiable reflect common experiences that have resulted in distinguishable cultures. They are not a statistical abstraction caused by combinations of scores on measurements (such as Census variables); the measurements in fact reflect the nature of the neighbourhood.

Examples to illustrate this are many (R. Webber, CCN uses 'military bases': the 'character' of the area (and its consumer behaviour) are more than the sum of the demographic characteristics).

Surrogates: The use of surrogates (commonly for wealth) can be misleading. For example, 'percentage of detached houses' could be used as a measure of affluence, but those areas with the highest levels are very often rural with low disposable income.

Explaining behaviour: many differences in behaviour (and attitudes) cannot be explained adequately by demographic variables. For example, the same basic demographics could apply to a young professional living in a cosmopolitan high density inner suburb as to a similar person preferring to commute from provincial areas. There will be many differences in attitudes, values and behaviour.

Influence: In any case, values (of people with similar demographics) will be influenced and re-enforced by the local community.

There are, of course, flaws in the above arguments; for example:

♦ At a pragmatic marketing level, does it always matter? This is possibly supported by the relatively good performance of simple variables such as lifestage.

♦ Many of the issues above are related to attitudinal and motivational considerations. Will a geodemographic (or even geolifestyle) classification always discriminate adequately on this basis?

More straightforward justifications for classification systems are:

Ease of understanding: most of the interpretation has been done, and is embodied in the description. If necessary, the 'label' can be described in terms of the component variables and other data sources.

Ease of use: A simple (normally) univariate categorical classification enables ease of profiling (particularly for customer profiling). It also greatly simplifies the 'coding' of data–an important example is for coding of market research data in order to supply the linkage for geographic analysis.

Methods

The brief description here relates to a Census-based classification, although the same principles (and problems) apply to the 'mixed' solutions using non-census variables. The ingredients consist of a selection of demographic variables. These tend to be similar between systems (although the emphasis may change for a 'market specific' classification), and normally include age, social class, housing composition and tenure, occupation, car ownership – typically 60 plus variables will be used. The most common processing method employed is cluster analysis–using these variables for the output areas. This will result in the required number of groups of areas such that each group is as far as possible dissimilar from other groups. A variation is the use of 'Principal Component Analysis' (PCA), the resulting 'theme' variables then being used for clustering. The reason for use of PCA is usually to avoid 'noise'–although some argue that it is not required with the correct clustering approach. The classification types can then be described in terms of the variables used to create them; assisted by overlaying of other data (such as market research data on media and products), to give a range of behavioural and psychographic attributes. Physical examination may improve insight into what neighbourhoods have been identified. Descriptions vary from housing orientated to more imaginative labels ('Bohemian Melting Pot' and 'Chattering Classes' are courtesy of Mosaic systems).

Whilst cluster analysis is the common choice, genetic algorithms, neural networks, and other methods could be used. In fact, a genetic algorithm was used in 'calibration' of the Mosaic solution against key market measurements.

Why use non-census data?

We have described previously possible sources of data suitable for enhancement of census data. The advantages of the use of multiple data sources are generally considered as:

♦ The census, although comprehensive, does not cover some key topics relevant to consumer marketing. Non-census data can avoid misleading surrogates

♦ Most non-census data can provide information at a finer level than ED's; normally to postcode level (albeit, with less than perfect coverage in some cases). This lower level discrimination means that the classification can be implemented at that level

♦ Non-census sources are usually updated frequently (often dynamically, and at least annually). This allows new areas to be classified accurately and demographic changes to be identified during the inter-censual period

♦ In terms of relevancy, information such as retail accessibility may be important for goods and service provision

♦ For market specific classifications, the appropriate data supplies the required focus

Products

The table below illustrates the parameters for some of the 'general purpose' classification products currently on offer, indicating the data sources used. Systems usually have several levels of cluster groupings–these are indicated.

Table 1
Parameters for
some of the
'general
purpose'
classification
products
currently on offer
indicating the
data sources
used

Product	Supplier	Input Vars	Clusters	Non-Census Data
ACORN	CACI	79	6/17/54	None
MOSAIC	CCN	87	11/52	Credit data, CCJ's, ER, PAF, Retail access, Directors, et al.
Superprofiles	CDMS	120(+130)	10/40/160	TGI, ER, Credit data, CCJ's
DEFINE	Equifax	146	10/50/1050	Credit data, ER, Unemployment,
Neighbours + Prospects	Eurodirect	48	9/44	None
PORTRAIT	MIC	69	10/40/175	NDL Lifestyles, Credit data, CCJ's, Dirs, Unemployment, et al

In the case of many of the suppliers, regional and market specific variations exist (for example, there are financial and Scottish versions of Acorn and Mosaic).

Classification summary

CENSUS DATA ONLY: The strengths rest largely with the full and consistent national coverage–a purer solution; the weaknesses are perceived as the lack of updateability, and the size of the neighbourhood.

CENSUS WITH OTHER DATA: This has many of the advantages of the above; but usually with additional discrimination at postcode level. It may be more focused on consumer behaviour and updateable frequently (at least, in part). Weaknesses include possible inconsistency and incomplete coverage for the non-census components; and updates are still subject to the static census component at ED level.

NON-CENSUS DATA ONLY: There are advantages in recency, updateability, with directly relevant lifestyle measurements used as input–meaning less surrogates and imputation.

Implementation is at postcode (or even household) level. As a downside, there may be doubts as to representativeness (both for the population and geographically), and it may be less suitable for non-marketing applications.

There are, of course, negative arguments in each case. In tests on targeting and segmentation performance, each product has shown its strong and less strong cases.

Overall there is not a great variation in performance. In any case, there is the school of thought that says that the best method of all (where usable) is likely to be 'raw' census variables (and, presumably, raw lifestyle variables). Lifestage has been seen to be a particularly effective census variable; and one would expect good performance in some cases from lifestyle variables such as income. Non-census data facilitate the development of market (or product) specific classifications – there is a strong argument for the merits of this 'segmented classification' approach; and performance gains will usually result – particularly with the involvement of customer data.

applying the data

Linkage

Not all of the data we have discussed is automatically suitably geographically referenced (postcode based data is, of course, by its National Grid Reference). Thus, the census is referenced to an ED, but since this is not a geography commonly used in market analysis, then it is very often associated with postcode sectors and with unit postcodes. To do this, an ED to postcode directory is required, and from 1991 this has been available as a direct link from the census form postcode (in previous times, this was carried out by proximity analysis).

In actuality, the user of a market analysis system is normally not concerned with such issues. Census data is most often supplied in units of postcode sectors (most applications can operate at this level). For postcode level analysis (such as customer profiling), a postcode 'directory' is used to relate to classifications This linkage will have been pre-prepared in the case of a census only system. Postcode level systems will always operate at unit postcode and sector aggregation levels. A link to ED's will only be needed if Census data (or a classification) are held at that level.

How data are used in geographic market analysis

RAW DATA: Census data is frequently used in its raw form, at varying geographical levels (postcode sector is probably the most common). Much of the other data discussed would be unsuitable in its raw form. 'Raw' counts (or penetrations) from lifestyle data need to be nationally extrapolated and smoothed; but, once done, can supply direct, effective measurements at postcode or sector level.

CROSS-TABULATIONS: We have already mentioned the use of research data such as from TGI; and the information on lifestyle databases can also be accessed like this (that is, by coding the actual records with a geodemographic classification. The cross-tabulations for the products or behaviour required for an application will normally be stored on the system and used dynamically to calculate potential and rank areas by attraction.

GEODEMOGRAPHIC AND GEOLIFESTYLE CLASSIFICATIONS: As we have discussed, the range and power of classifications have rapidly increased; and with the inherent advantages of ease of use and understanding, we can expect development to continue. Classifications are normally stored as a linkage to postcodes (and perhaps ED's) and in aggregated form by postcode sector.

Geography level

We have mentioned the preference for use of postal geography in market analysis 'GIS' systems. Whilst an obvious reason for use of postcode sectors, at an early stage, was the economics of storing and processing at a lower level, technology has overtaken this as an argument in itself. Ideally data is available for processing at the most accurate level possible. However, there are still reasons for use of postcode sectors (a sum of unit postcodes), for example:

MAPPING: Thematic mapping of postcode sectors (with aggregated attributes by sector) is for many the most sensible portrayal; smaller units can be messy and difficult to interpret.

MARKETING 'CURRENCY': The sector is used as a convenient unit in so many areas of marketing (often for logistic or cost reasons). Use of the sector is required for compatibility.

Applications

The broad objective of market analysis systems (or geographic decision support systems) is to allow more and better decisions; and geodemographic and geolifestyle analysis provide powerful tools for achieving this across a wide range of organisations. The assessment of existing and potential demand to improve targeting of services is as applicable to most public services as to consumer products and services in the private sector.

The functions and use of market analysis in a multitude of applications, such as targeting, branch planning, site location, and media planning, are discussed in Sleight (1993).

delivery systems

The purpose of a 'GIS' computer system is, at its simplest, to act as a 'delivery system' for the reports and maps which will be output from market analysis. The coverage below does not attempt to be exhaustive, but to supply a flavour of this marketplace.

History

From 1988, the larger market analysis agencies (such as CACI and CCN) have supplied systems, offering analysis, reporting and mapping functions. These systems tended to be in-house developed and strong on specialised analysis and simple modelling (to a large extent reflecting a subset of the bureau services). GIS facilities, as such, were limited mainly to add-on mapping. More recently, some agencies (who tend to supply these kind of systems, at least, for the middle to lower end of the market) have taken different routes; some using a base GIS orientated platform. For example, Infolink (now Equifax) used the SIA product Datamap as a vehicle, meaning that quite powerful GIS facilities are available, if required. Other systems use products such as MapInfo, which for many applications can supply (with add-on functionality) perfectly adequate features and performance. At the top end of the scale, large retailers have tended to favour more 'serious' GIS systems from the traditional suppliers, usually with a significant investment in customisation and generally including comprehensive site location modelling capabilities.

The 1990s has seen a steady growth in the use of in-house systems, aided not just by advances in computer technologies, but by a wider understanding of the methodologies employed and the benefits achievable.

Low to mid-range systems

Segmenting the marketplace on price is difficult because a large component of price will be for the data and functionality installed; varying greatly according to the application. We will thus consider two main groups.

ESTABLISHED MARKET ANALYSIS VENDORS: CACI produce their Insite product and CCN their Mosaic Systems. These products are developed in-house; and are broadly similar in purpose, supplying a range of analysis, data manipulation, reporting and mapping functions. They both have a repertoire of add-on modules, the CCN system probably being more modular. Both tend to be configured specifically for each customer, but this is to a certain extent true for all the systems because of varying tasks and data required. As a newer offering, Decisionmap (from Equifax) is based on SIA's Datamap. It has a range of market analysis functions and potentially higher GIS functionality (such as support for raster maps). Eurodirect has its Demograf system which achieves modest functionality at a modest price and CDMS uses the Prospex system for its wares. All of the systems mentioned here are committed to their proprietary classification systems (respectively, Acorn, Mosaic, Define, Superprofiles and Neighbours and Prospects) and census data is usually also offered.

OTHER VENDORS: Other systems do not necessarily fall into the same mould. The Data Consultancy offers their Illumine product using a MapInfo platform. Demographics are normally supplied by census data and a choice of classification system (including Superprofiles). This type of system offers, at a good price, analysis, a wide range of data, MapInfo functionality, and good interfaces (CACI now offer a similar product). There are many other low price mapping/GIS systems (with relatively low GIS functionality) available that can act as platforms for low/mid range market analysis systems–for example Mapbase and Spansmap. At a slightly higher base price (with more built-in analysis capability) are systems such as Tactician and Atlas GIS. Cenario (from Capscan) stands out as an exception. It is a new completely home-grown product and currently only uses 'raw' census data (off CD-ROM).

Larger systems

Implementations using software systems with extensive GIS functionality have tended to be installed with large retailers by suppliers such as ESRI (Arc/Info), Laserscan, Smallworld and others. These systems have usually involved substantial investment in customised systems and data. The demographic data will probably include substantial raw census data, with a relatively large selection of variables; although classification systems may also be used.

Other system data

An effective geographically-based system will of course require other data; topographical, definitions of 'geographies', and attributes. A vast range of such data is available; the range, quality and accessibility steadily improves. Typical data that may be required include:

♦ Coastline, waterways, railways, towns, road network–in vector form

♦ Boundaries depicting postal, administrative, media, and other geographies

♦ Point locations (postcodes, store locations)

♦ Raster backdrop data and many other geographically referenced data are readily available

The impact of the dramatic improvements in price/performance of desk-top computing and storage is, of course, affecting this marketplace. Low price reasonably functional software can easily be used as a platform for function and data rich systems. With 'Windows' style technology, effective interfaces between systems can be built; for example with interactive postcoding software. A limitation in terms of data availability, of course, is the lack of mobility of the classification systems.

However, we should put these systems in context. The requirement for true GIS functions is relatively small in the mass of low to mid-range implementations; it is only for the more demanding 'high-end' applications that there is normally such a need.

looking ahead

There has often been a tendency for market analysis systems to be isolated within organisations from other marketing and management functions. Not only may this be logistically inefficient, but the full value of corporate cross-information may not be realised.

There have perhaps been reasons for this – the 'different' nature of the applications and data and the lack of easy interfaces. With more open systems and a wider understanding of the objectives involved, we can expect greater integration with other 'marketing' data and systems (such as marketing databases and marketing information systems) and, indeed, with the corporate information systems strategy. To assist with this we should expect improved aids for integration and ad-hoc requirements.

In terms of technical features, we can expect multi-media aids and useful, easy to use GIS features becoming more commonly available on low-priced systems. The systems will become more user-friendly and provide improved user interaction (eventually, expert systems?).

We can expect (and can already see) the introduction of Pan-European and International capability. The ability to action simultaneous marketing strategies across a diversity of socio-economic scenarios is a powerful tool.

We can expect increased exploitation of the best of current established data sources. The various data that we have discussed will complement and re-enforce rather than compete.

Solutions will use data appropriate to the problem, and more refined methods will allow greater input of business issues. Predictive analysis methods will become more flexible with the availability of more and improved information.

In this paper we have discussed data that are primarily demographic and behavioural in nature. For many, measures of attitudes, beliefs and motivations are also of key concern; particularly for the presentation (creative) side of marketing. This type of measure has been termed 'psychographics' and, whilst it is not a new approach, it is only recently that a segmentation and targeting product (called 'Psyche'), has been launched, by CCN. The Psyche classification has been extrapolated nationally at postcode level; and 'geopsychographics' is with us. Similar data to that used in Mosaic was used to assist the model for national dispersion from the research data. Psyche classifies seven Social Value Groups to describe people as 'Survivors', 'Explorers' etc.; and, like lifestyle data, it relates to individuals. This product illustrates the continual developments in the quest to understand consumers and markets. Again, geopsychographics will perhaps take its place alongside geodemographics and geolifestyles to re-enforce and enhance.

Lastly, we should mention the next Census (in 2001). It is to be hoped that there will be additions of questions of direct relevance to market analysis (such as income); but, in any case, the Census will no doubt successfully deliver another round of very high quality data.

conclusion

There are many important issues that have only been mentioned briefly in this paper; for example, the question of 'updates' to all these data. Even with updateable data, it can be a surprisingly troublesome task–just at headcount level.

We have also only outlined the world of marketing GIS systems. In fact, the growth of their use has probably not reached the expectations of several years ago. There are many reasons for this; probably partly the nature of the marketing world, partly budget restrictions. There is also the suggestion that many vendors overly focus on the technology, not the business benefits.

A reasonable question to ask is: 'Census or Lifestyles'? We have outlined the strengths and weaknesses, but the conclusion is one of cohabitation. Lifestyle data have reached the 'critical mass' level of coverage and are becoming a powerful aid to consumer analysis.

They must become an integral part of the marketing mix, providing ever increasing insight into consumer behaviour.

references

Leventhal, B., Moy, C. and Griffin, J., 1993, *An Introductory Guide to the 1991 Census*, NTC Publications Ltd.

Raper, J., Rhind, D. and Shephard, J., 1992, *Postcodes: The New Geography*, Longman Scientific and Technical.

Sleight, P., 1993, Targeting Customers: *How to use Geodemographic and Lifestyle data in your Business*, NTC Publications Ltd.

Dugmore, K., 1992, 1991 Census: Outputs and Opportunities, Geographic Information 1992/3 (*The Yearbook of the AGI*), London, Taylor and Francis.

Sleight, P., 1995, Explaining Geodemographics, *Admap* Issue 347.

Fischer, C., 1994, Lifestyle Databases, MRS CIG annual Geodemographic Seminar.

Webber, R., 1993, Micromarketing with Geographic Systems, ABMRC Seminar.

Winters, P., 1994, Choosing Data for Geodemographic Analysis, MRS CIG annual Geodemographic Seminar.

Concise descriptions of many of the terms used in this discussion can be found in:

Berry Consulting, 1992, *A Glossary of Direct Marketing Terms*, London, Direct Marketing Association (UK) Ltd.

softcopy photogrammetry and its uses in GIS

J C Shears and J W Allan

ERDAS International, Telford House, Fulbourn, Cambs CB1 5HB, U.K.
Tel: +44 1223 880802
Fax: +44 1223 880160

abstract

Two years ago, the term 'softcopy photogrammetry' was almost unheard of in GIS circles. Today, the availability of low cost softcopy photogrammetry systems, such as IMAGINE OrthoMAX, has opened up a vast range of data provision and updating options to GIS users. The two primary datasets created by softcopy photogrammetry are terrain data, in the form of a Digital Terrain Model (DTM), and an orthorectified image (Orthoimage), which is a georeferenced image, free from any sensor or relief distortion. This paper discusses why Softcopy Photogrammetry is needed and briefly covers the digital processes involved in the production of such data, together with a comparison of these processes with traditional manual methods. It closes with a description of the type of GIS projects currently integrating softcopy photogrammetry.

KEYWORDS: DIGITAL TERRAIN MODEL; ORTHOIMAGE; DIGITAL TERRAIN ELEVATION DATABASE; 3D GIS

the importance of terrain data

There are a large number of military and commercial GIS applications that rely entirely on the ready availability of digital terrain databases. Their success or failure is dependent on the timely production and ultimate accuracy of the terrain models that are fed into them. The military applications range from simulation, mission planning and mission rehearsal to terrain referenced navigation and weapons guidance systems. Commercial applications include land use monitoring and assessment, such as the EC MARS program and base mapping for oil and gas exploration activities. At a time when budgets are under scrutiny, there is now a greater need for digital terrain data to be generated more cost effectively, whilst still being made available in a timely manner but also without sacrificing or compromising the accuracy of the data.

With this in mind, new softcopy photogrammetric techniques have evolved which go some way to providing a solution for this. To understand how this has been achieved, it is necessary to look at conventional techniques for building terrain databases. There are three common methods, the most popular of which is using traditional analytical photogrammetry. An alternative is digitising contours or spot heights from hardcopy maps and creating a surface from them. The derivation of surfaces from digitised height features, will involve a degree of interpolation, hence the surface will be inherently more generalised than a photogrammetric compilation. Given that maps and charts are themselves derived from aerial photogrammetric surveys, then any errors in the photogrammetric compilation will also be propagated once they are digitised, hence true photogrammetric compilation will always provide a more accurate terrain database than map digitising.

A third source of terrain data is to use existing products. Terrain databases do already exist in a digital form and are available as standard products like the US Defence Mapping Agency's Digital Terrain Elevation Database (DTED). Similar products are also available from other mapping organisations, such as the USGS and the UK Ordnance Survey and a number of other National Mapping Agencies throughout the world. DTED has been the most widely available dataset for military applications and is widely accepted, but its predominance masks some inherent problems associated with it and also with other digital elevation model (DEM) products.

With DTED, the resolution of the height data is fixed and is generally available at either 100m spacing (Level 1) or at a nominal 50m spacing (Level 2). However, due to the higher resolution of Level 2, it is only available to authorised users on a restricted basis. Whilst 100m may be suitable for broad area applications, higher resolution DEM data is needed to provide greater levels of detail to fulfill the potential of terrain based applications. The Level 1 product represents a very generalised view of the terrain even at 100m resolution and some of the finer terrain detail is lost. This is sometimes done deliberately to protect compilation sources and provide collateral against national sources or because there was simply insufficiently accurate source material available at the time of compilation. Either way, the relatively poor resolution constrains the potential capabilities of the applications.

This highlights the second major drawback as the user has no control over the accuracy and quality of the DEM. DTED is compiled from whatever sources were available for a given area using either photogrammetric extraction techniques or digitised mapping. RMS figures are provided for both horizontal and vertical accuracies but in some instances the application may demand higher levels of accuracy and detail. Mission planning projects for example, require higher quality at terminal locations whereas a lower quality may suffice for en route positions. The user needs to be able to both specify the resolution required and vary it as required for the application as well as the ability to edit the DEM to increase the accuracy if needed. The same criteria apply to commercial applications, where accuracy has an impact upon commercial decisions, rather than human lives.

the photogrammetric process

Photogrammetry has established itself as the main technique for obtaining precise three dimensional measurements. It involves the use of overlapping images to recreate the original stereo geometry of each adjacent pair of images, from which precise three dimensional measurements can be derived. Conventional photogrammetry involves the use of specialist and expensive plotting equipment to mimic the stereo geometry at the moment of image exposure using optical trains. The operator first has to measure calibrated points on the film, either fiducials or reseau marks, to establish a relationship between film space coordinates and model space coordinates. The machines are set up or 'oriented' using a pair of original hardcopy diapositives in left and right stage plates. Each stage plate can be positioned with respect to each other and oriented in x, y and z using threaded spindles to emulate the precise attitude and position of each diapositive with respect to each other. In this way any roll, pitch or yaw in the taking camera or satellite can be recreated to replicate the attitude and position of each image at the moment of its exposure. At this point the images are said to be in relative orientation. Absolute orientation, based on real world coordinates, requires the operator to observe and measure known ground control points in the model space as well.

Once oriented, all residual y-parallax will have been eliminated, allowing the operator to view the model in stereo, a projective geometry termed epipolar. When viewed in stereo, conjugate image points appear in different positions in each of the images. This 'apparent' movement of the imaged point is due to the movement of the observer (in this case the aircraft or the satellite platform) and is known a parallax. Its measurement forms the basis of determining height. The only remaining parallax will be in the x direction, the amount of x-parallax being a function of height. Using a half mark etched in the optics of each lens, the operator can 'float' the point and move it in a vertical direction. By placing the point 'on the ground' , individual features in the model can be heighted.

This process has some fundamental drawbacks when compared to digital techniques. Firstly, it is all based on very specialised hardware. It is largely mechanical (analogue) although some plotters can be upgraded to include linear encoders powered by servo-motors (analytical) which will drive the operator to pre-defined points for measuring. Both analytical and analogue machines however are designed to carry out these single specific tasks and cannot be used for other applications. Secondly, the process is a highly skilled one which requires many hours of training and hence increased staffing costs. Most of the operations are also very labour intensive, particularly the collection of height data as each point has to be visited and measured individually. Experienced operators can measure anywhere between six and ten points a minute and like all manual work, it could only be maintained at the desired accuracy for a specified period of time, certainly no more than eight hours maximum. This will also contribute significantly to overall production costs.

automated DEM generation

With the advent of sophisticated photogrammetric software and ever increasing and inexpensive computer power, softcopy photogrammetric workstations to a large extent replace the human operator and automatically create the DTM by means of digital image processing. With production speeds in excess of 150 points per second, the DTM production time is significantly reduced. The history of digital photogrammetry can be traced back to the late 1950s, since which time photogrammetry has undergone a tremendous change and softcopy photogrammetry now offers the potential to generate terrain databases with greater speed, at lower cost and with less training and photogrammetric skill than ever before.

The major difference between digital and conventional photogrammetric systems is that images used in digital systems are in digital format and hence suitable for processing by computers. If conventional aerial photographs are used, then they will need to be scanned prior to input into the system. The systems can also make use of image data collected digitally, such as satellite imagery. In this context, the SPOT satellite is the most commonly used as it currently provides the highest resolution stereo overlap coverage. However other digital CCD cameras could also be used.

As with conventional analytical instruments, digital photogrammetric workstations carry out the same orientation process in order to model the original stereo geometry. The principles used are exactly the same, but the implementation is faster and offers greater ease of use through intuitive software interfaces. There are a variety of automated tools based on cross correlation of image patches to locate and measure fiducials in the image, tie points, pass points and ground control points. The correlator can be trained to recognise and measure fiducials for various camera types and, with the exception of observing a minimal amount of ground control, the entire orientation process is automated, requiring very little attendance and operator time.

The area where most research has been concentrated is that of automated DEM collection. Sophisticated algorithms have been developed to replace manual collection and whilst there are differences between various collection algorithms, the problem of automating the process of DEM capture has generally been solved.

The methods that are mostly used are either area-based or feature-based matching techniques using correlation of small image templates between image pairs. Once oriented, the software computes the coefficients of a set of rational polynomials which summarises the stereo geometry. These are used by the DEM correlator to emulate the projective geometry of the cameras. The normalised cross correlation approach discussed here is an area based algorithm that digitally correlates points based on tonal variations present in each image. Areas that have high tonal and textural variation will be correlated very quickly as the correlator uses the high frequency components to correlate on. Image content is the single most contributing factor to correlation success. Low frequency areas will generally be correlated slower, although the templates will automatically increase in size until sufficient texture exists in the template to allow correlation. The correlator will visit and attempt to height every point in the DEM and if a correlation cannot be accurately computed, then a height is interpolated. The interpolation is performed by a weighted technique based on radial distances of points in a neighbourhood.

Evidence has shown that collecting hierarchically from a very coarse resolution through successively finer levels, reduces the possibility of generating a false fix. By reducing the scale of the imagery, changes in elevation are less pronounced in image space, thus the effect of height variations are minimised and searches over broad ranges of elevation are quicker. Also correlations at reduced scales tend to reduce the confusion between similar appearing objects by locking into gross areas which include the objects. The heights derived from each level are used as an estimate for the next higher resolution collection level.

editing tools and accuracy

Central to generating an accurate DTM is the ability to edit the computer generated model. This is required either where man made features (with sharp edges) need to be highlighted, cliff edges need to be added as 'breaklines' or simply where the computer has failed to find suitable matching points from which to generate a height. Correlated points will be designated a 'quality figure' based on a user-defined set of signal-to-noise ranges, as either 'good', 'fair' or 'poor'. Failed attempts at correlation are labeled as 'interpolated' and are all rankings are made available to the operator to give assistance for the editing stage.

Whilst the automated collection will generate heights based on statistical correlations, it is important for the sake of ensuring accuracy that the height values can be validated by the operator. The software displays all the points at their correlated positions in a stereo view, colour coded to allow a rapid visual inspection of the whole model.

The editing tools provide an interactive method of modifying the height of points deemed to be in error by the operator. In this way, the resulting DEM has both a statistical statement of accuracy and one that has also been verified by an operator using their skill and judgment. As the accuracy of the correlation is dependent on the accuracy with which the stereo geometry was computed, the software provides a full summary of all mathematical calculations including standard deviations for all final computations of camera positions and attitude.

derived products

As well as being used in a range of military and commercial spatial analysis applications, the terrain database can be used to generate additional products, such as orthoimages. These are images that have been corrected for displacements due to relief variation and sensor imperfection. In any imaging system, each imaged point will have a particular perspective geometry and in order to view each pixel in an orthogonal projection (i.e. from a nadir view, as if each pixel were being viewed from directly above) the effects of terrain have to be removed. The DTM is used to model the relief variation present in the image and each pixel in the raw image is resampled into an orthogonal projection which the user can define.

The orthoimage is vital, especially if perspective views are to be rendered for mission planning or visualisation as this will ensure that all features are in their true position with respect to the underlying elevation model and curious effects such as rivers flowing uphill can be avoided! It is also vital for use an highly accurate, up to date base map for commercial applications, including database updating.

uses within GIS applications

Uses with GIS applications can be divided into two primary types; those requiring height information (the DTM) or a derivative (slope, aspect) and those requiring high precision base maps (the orthoimage), either for backdrops or as a source of vector data.

Many spatial modelling applications, such as site location and route planning require height derived 'layers' as part of the process. For example, where new housing development projects are being planned, using soil type combined with slope can show areas where land slippage may occur. In route planning in military applications, slope again is important as certain vehicles may only be able to negotiate low angle slopes. One area where aspect (ie south facing, north facing etc) is important is in vineyard location – it is important that vines are planted at the optimum location to produce the best quality grapes!

DTMs can also be used in visualisation, specifically in environmental and military applications. The siting of new facilities can first of all be generated using the spatial analysis described above. The proposed site could then be viewed in 3D, with the facility 'added' to the DTM, either by adding a polygon of the appropriate value or 'height' in mono view or by accurately adding the height of the facility in stereo. This enables the user to check on its visibility from surrounding areas. Viewshed analysis can also be used in the opposite sense to show if your own location can be seen from other areas for the purposes of concealment.

Orthoimages, as described above, provide the most accurate (and up to date!) base maps of all. Many natural resource management applications now simply use a symbolised base map instead of a complex vector based map. The old adage 'a picture is worth a thousand words' could easily be changed to 'a pixel is worth a thousand vectors' in this instance! However, the largest demand for orthoimages lies in the data provision aspect of GIS.

An orthoimage can be used for generating vector map and other measurement information directly from the computer screen. In the past this has to be done by the photogrammetrist on an analytical stereo plotter using stereo imagery, as this was the only way to get accurate x, y & z map information, unaffected by relief distortion. Today, the operator can simply use the mouse to digitise vector map information (and attributes) directly from the orthoimage on screen, using the computer as a 'monoplotter'. This de-skills the entire vector generation process and hence reduces the cost of creating the database. As this is normally the major cost component of a GIS system , softcopy photogrammetry is a way of reducing that cost. With the standard ERDAS IMAGINE software able to digitise, clean and build true ARC/INFO coverages, IMAGINE OrthoMAX is a invaluable asset to any ARC/INFO GIS.

3D GIS

Finally, a glimpse into the future. The computer world as we know it is becoming a 3D world. No longer are simple planimetric views enough, with users demanding perspective views and real time flythroughs. In Autumn 95, ERDAS will be releasing its own real time 3D Viewer, which will allow DTMs generated in IMAGINE OrthoMAX (or from anywhere else) to be flown around in real time. Vectors (such as ARC/INFO coverages), symbols and annotation can also be draped and flown around. One unique feature of the software will be the 3D GIS capability, which allows the 3D image to be queried in real time. Essentially, it will provide all the functionality of a 2D GIS but in 3D! This is the first step in a new direction in GIS where the real world can be modelled, analysed and queried in 3D on the desktop.

conclusion

It is easy to appreciate the advantages that are brought to the production of terrain databases and derived products such as orthoimages with the introduction of well developed software algorithms, combined with the increasing availability of powerful desk top workstations. It is however fair to say that despite these tremendous advances, there is still a significant caution in the user community and there are many published technical evaluations that bear witness to this. Without doubt however, digital systems are here to stay and are being constantly improved. They have already proved that they offer significant improvements in production throughput and their ability to operate with minimal operator intervention will certainly mean that reduced production costs can be easily achieved. It is anticipated in the future that fully automated systems will be available, requiring not only less operator time, but less skilled operators. As systems become easier to use, so the technology will be more accessible to a wider range of end users, who traditionally were excluded from undertaking photogrammetric projects by virtue of the technical complexity and the level of operator training.

From the application engineer's perspective, it is now possible to generate digital terrain models and orthoimages as and when required on standard commercially available hardware. More importantly, the information can be generated on demand to the exact density, area and quality required by the particular project, and as all GIS users know, having the correct data in place on day one is the first major step towards a successful project.

using GIS for housing planning in scotland

Alan Bourke

GIS Coordinator Scottish Homes,
Thistle House, 91 Haymarket Terrace,
Edinburgh, EH12 5HE.
Tel. (0131) 479 5303
Fax. (0131) 479 5252
email: bourkea@scot-homes.gov.uk

abstract

Scottish Homes GIS provides the means by which Scotland is analysed, using a wide range of data sets and spatial analytical techniques. This is done as part of a carefully defined 'corporate planning framework' and helps decide housing investment priorities throughout Scotland. Recent technological advances have helped further disseminate this technology in Scottish Homes.

KEYWORDS: HOUSING SYSTEMS ANALYSIS; CORPORATE PLANNING FRAMEWORK; WINDOWS NT; CENSUS OUTPUT AREAS (COA); REGISTER OF SASINES

GIS in scottish homes

Back in 1989 when Scottish Homes was formed, two new members of staff with GIS backgrounds got together and managed to get GIS specifically incorporated into the then developing corporate Information Technology Strategy. A report was written (Bourke, 1989) which highlighted the need for GIS, and funds were found to undertake a pilot study. Its aim was to help decide which data sets and analytical techniques were required to analyse Scotland. These in turn would provide operational managers with a firm analytical basis on which to base key investment decisions (Bourke and Robertson, 1992).

Scottish Homes is the Government's housing agency in Scotland and invests over £300 million per annum in social housing. Clearly there is a need to identify areas and groups of people who require access to social housing. Almost all of the questions asked by managers have a clear spatial element and therefore require GIS to answer them. From an early stage in the pilot it was clear that while hardware and software considerations were important, data and analytical requirements were more critical. Our prime goal then, was to determine key questions being asked by the organisation; a GIS could then be chosen to meet these requirements. Data collection, system specification and software/hardware selection, therefore flowed directly from an assessment of operational requirements. True, there were hardware considerations; one thing the pilot showed was that a PC system could not serve as a corporate GIS – an important requirement in Scottish Homes. By seeking to make the GIS corporate, PC-based systems had to be excluded as they were simply not up to the task either in terms of hardware or software. This effectively put GIS beyond the experience of most staff in the organisation – after all, most people in Scottish Homes are housing experts not computer experts. The 'corporate equals UNIX' decision therefore made GIS the preserve of so called 'experts' until very recently.

Clearly the advent of Windows NT and the huge growth in power of PCs has changed all that. The effects of these developments on GIS in Scottish Homes will be examined later on in the chapter.

what are the key questions in scottish homes?

Since GIS system specification flowed from a clear understanding of key operational requirements it was necessary to determine the nature of these requirements. Almost the entire range of analytical techniques and data requirements are represented in the following sample questions for one sample area – Shortlees Estate, Kilmarnock:

♦ 'What is the total population of the Shortlees area in Kilmarnock?'

♦ 'Where are there significant clusters of lone parents above the Scottish average, in Shortlees?'

♦ 'What are the house prices in that area?'

♦ 'Is Shortlees a self contained housing market area?'

It was necessary to examine each question carefully and break them down into a number of clearly defined steps. Each step would yield data and analytical requirements which in turn formed a detailed system specification. From this Scottish Homes chose Smallworld GIS which is now used operationally in the organisation and can answer these, and many more complex, questions. An examination of each question will serve to highlight how Scottish Homes actually uses its GIS operationally.

identifying a study area

The primary requirement for any GIS is to provide a map skeleton upon which analysis can take place.

♦ 'What is the total population of the Shortlees area in Kilmarnock?'

In order to answer this question the issue of map coverage has to be addressed.

Map coverage

Often, areas to be analysed do not conform to standard administrative boundaries – this is particularly the case when studying towns, villages or housing estates. In this example, Shortlees is a housing estate on the outskirts of a town, and has no formal administrative boundaries. In order to define the extent of the estate it is necessary to define it using a map base held within GIS. Being a 'national' agency, analysis is required across Scotland, so pan-Scottish map coverage is required at a range of scales. Such maps do not necessarily have to be bang up-to-date since maps are generally used as a 'backcloth' upon which other data is draped. Obviously, to purchase all Ordnance Survey maps for Scotland would be prohibitively expensive, so the decision was made to purchase raster maps at 1:50,000 scale for all of Scotland and at 1:10,000 scale for most urban areas. These are becoming increasingly out of date; the Ordnance Survey have, however, recently signed a service level agreement (SLA) with government departments and agencies in Scotland allowing them access to key OS products for much more reasonable fees. The effect of this on GIS usage on Scottish government departments and agencies remains to be seen, however it is likely to be significant. Certainly it means that in Scottish Homes, obliged by financial considerations to make-do with a static raster map base, they can now look forward to using regularly updated, digital map data.

An important issue relating to mapping is developing the ability to output quality maps from GIS. For a long time Scottish Homes could only produce map output in black and white on A4 paper.

Despite all the technology, the perception of GIS throughout an organisation depends, to a large extent, on quality mapping output. This was severely limited until an A0 Colour inkjet plotter was purchased. The proposition that 'coffee stains on an airlines seats indicates poorly maintained engines'(Scottish Homes, 1993) is clearly relevant here in that 'poor quality mapping indicates poor quality GIS and poor quality analysis'. Certainly, in Scottish Homes the production of quality mapping raised the perception of GIS.

One problem related to covering the whole of Scotland, is finding the location of areas for analysis. To help with this an Ordnance Survey gazetteer based on 1:25,000 Pathfinder maps has been developed.

Defining study areas

Having obtained a map base it is necessary to define the Shortlees Estate boundary. This largely coincides with the location of houses owned by Scottish Homes. This however, may not coincide with other organisations' view of what constitutes the Shortlees Estate. The boundary can be considered 'informal' in terms of the wider world therefore census data is not available for that specific area. In order to determine the population of Shortlees it has to be constructed using smaller census areas.

In Scotland, the smallest spatial unit for which 1991 Census data is available is the Census Output Area (COA) (38,000 COAs in Scotland). Small areas 1991 Census data for Scotland together with COA boundaries are used in the system as is the analytical capability to amalgamate these 'building blocks' into larger areas. COAs may not match informal boundaries. One way around this is to develop the ability to 'factor' data for COAs however the veracity of this technique at small area level is doubtful (Dale, 1991).

Having defined Shortlees from census 'building blocks' it is then necessary to be able to display the relevant census table containing the data required. In this case Census Table 1 holds population counts and the population of Shortlees, as defined, is 1923 persons.

filtering census analysis to fulfil specific criteria

Often, managers ask questions which depend upon selecting specific areas fulfilling certain criteria. Questions such as:

♦ 'Where are there significant clusters of lone parents above the Scottish average, in Shortlees?'

To answer this, all the analysis detailed above is required as is extra data and functionality. Firstly, levels of lone parents in Scotland as a whole must be calculated – 1991 Census data at the Scotland level (Local Based Statistics) are required for this. Also required is the capability to create calculations based on individual census cells, as well as the ability to store and recall them.

Creating calculations

In this case, Table 31 of the Census contains the data required. The calculation is created and run against the Shortlees study area (mean percentage lone parents is 13%). It is necessary to filter these results so that only COAs with greater than 13% lone parents are selected. Enhancements to this technique now allow the filtering of multiple variables and can incorporate non-census data.

Clumping areas

A further requirement is to determine areas with clusters or 'clumps' of particular characteristics. Clearly, because Shortlees is a small area, all areas are 'clustered'. If the analysis covered a much larger area, then users must be able to filter out areas not adjoining each other in 'clumps' of census output areas or other polygons. It must also be possible to select how many areas constitute a 'clump'. Scottish Homes GIS allows the user to specify the size of such 'clumps' and filters the data accordingly. Clumping is clearly a purely 'spatial' function which augments census manipulation techniques. Using it, managers can, and do, identify significant groupings of particular households types or population characteristics.

Manipulating point data

In order to help determine whether these lone parents in Shortlees can afford housing a manager may well ask:

♦ 'What are the house prices in that area?'

To answer this, data from the Register of Sasines is used. The Register of Sasines is unique to Scotland where all land and property transactions have been continuously recorded since 1617. Details of ownership, price mortgage and conditions of sale are given (Williams and Twine, 1991). Key to its importance is the fact that:

♦ the register is comprehensive

♦ it extends back almost continuously back to 1617

♦ they are public registers.

These paper records are available in a computerised form (Appendix 1) and have been purchased and loaded into GIS.

Spatially referencing point data

In order to spatially analyse this point information, it is necessary to provide each transaction with a grid reference. The Postal Address File (PAF) allows the batch matching of address data with grid references though the overall 'hit' rate is just 70%. In other words just 70% of addressed held in the Sasines database are found in the PAF. This clearly has implications for data quality and accuracy.

For background on this see Endnote 1. (Bourke, 1995; Burroughs, 1994).

Displaying point data

Smallworld GIS allows this data to be loaded and displayed on a map base. Individual transactions can be displayed as a symbol where the darker the colour the higher the price – each point symbol also displays data from the 'price' field in the sasines database. A number of data representation techniques can be selected for the sasines relating to symbol colour, size and shape, all to aid with visualisation.

Generalising point data

Because of the large number of property transactions in Scotland (10,000 per month) the GIS can generalise point information to form average prices for areas. In this case, point in polygon functionality is used to select points that fall within each COA, and find the average.

Point in polygon analysis

Having analysed house prices in Shortlees a manager may wish to test the area to see if it can be termed a Housing Market Area:

♦ 'Is Shortlees a self-contained housing market area?'

In order to carry out this task, the GIS performs point in polygon analysis to extract only those property transactions within the defined area. It then applies various calculations to the results and returns an answer, either yes or no, together with statistics (the area is self-contained). For infomation relating to this analysis refer to (Scottish Homes, 1993). This technique automates what would otherwise be an extremely tedious task, using conventional database or spreadsheet techniques. The results help managers to identify areas where there is a localisation of housing demands (or needs) stemming from the necessity of choosing housing within accessible distances of workplaces.

getting the basics right

Such questions are representative of the kind of operational issues typically raised throughout Scottish homes. The GIS itself has additional data sets and functionality but the ability to perform functions detailed form the foundations upon which the entire system is built. These functions and data sets are currently operational within Scottish Homes and form an important part of an overall planning framework. The nature of this relationship with strategic planning is examined next.

Table 1 summarises the key data sets and analytical techniques outlined above.

Table 1
Data sets and
procedures
used in
application

Data Sets
Census data (1981 & 1991)
Register of Sasines
OS Raster Maps 1:50,000 and 1:10,000 scales
OS Gazetteer
Census boundaries; Administrative boundaries; Scottish Homes boundaries
Census Analytical Techniques
Define areas using census building blocks 'COAs, Pseudo Postcode Sectors, Local Authorities
View census tables for defined areas
Create calculations using census data, save and store these
View key statistics for area
Map the results of census analysis
'Filter' data to exclude COAs with less than average of variable
Create new area from 'filtered' data
Apply cluster analysis to find 'significant' clusters
Create new 'area set' with results of clusters
Sasines Analytical Techniques
Display individual property transactions on map base using visualisation tools (colour, size, shape etc.)
Generalise point data to areas
Perform 'point in polygon' analysis to select only property transactions falling within areas of interest
Run 'containment' analysis for area to find out if it is a self contained 'housing' market area

strategic/corporate use of GIS for planning

Smallworld GIS was chosen as the one which would best allow us to perform those functions detailed above. Crucially, GIS is a key part of an organisational 'Corporate Planning Framework'.

While GIS development was underway, an organisational 'Corporate Planning Framework' was being constructed, and GIS was adopted as a key mechanism in its implementation. Scottish Homes overall direction is outlined in an annual 'Strategic Plan'. Each administrative district must then produce a 'District Plan' showing how they will implement the strategic objectives locally. In order to produce these plans, districts must analyse their areas using a technique called Local Housing Systems Analysis (LHSA). Such analysis takes place within formal guidelines described in a 'Best Practice Guide for these plans, districts must analyse their areas using a technique called Local Housing Systems Analysis (LHSA). Such analysis takes place within formal guidelines described in a 'Best Practice Guide for Local Housing Systems Analysis'. See Endnote 2. (Scottish Homes, 1993). This publication gives comprehensive guidance on data sources, updates, alternatives, analytical possibilities and interpretative options to use when analysing the housing system in an area. GIS provides the data repository for, and the means to carry out, many analytical techniques specified in the Best Practice Guide.

GIS is therefore seen as part of the formally agreed process of planning within the organisation. Most GIS developments are done in the context of, and for, this process. GIS is also used for ad-hoc mapping requests but unlike in many local authorities, this is not its prime purpose, nor is the map base of sufficient quality to be useful to produce accurate, up to date mapping for detailed areas.

the impact of new technology

Clearly, staff in Scottish Homes are housing experts, not computer experts. The aim of all GIS developments has been to make its use as 'push button' as possible. In fact, my inspiration comes from those town maps you find at the seaside, where you can press a button and a light comes on on a map background showing where the swimming pool or the local park is. Indeed, the 1991 'corporate equals UNIX' decision, made GIS into a 'specialist tool' for use only by designated, specially trained and possibly specially interested people – it was never viewed as a tool for occasional user. It was certainly never viewed as just another icon among others tools, on a desktop PC, or as easy to use as a seaside town map.

It is important to remember that when the organisation was formed in 1989 there were few PCs around and most applications were mainframe based. The corporate IT strategy changed all that putting a PC on almost every desk – the result was that staff have become more computer literate, and increasingly happy to make use of such technology in their everyday work – on a PC. Using GIS in the UNIX environment proved an insurmountable barrier to some staff. Recent software and hardware developments have changed all that.

The introduction of Microsoft Windows NT coupled with the availability of massively powerful PCs now means that 'corporate' doesn't necessarily 'equal UNIX'; now corporate can equal PCs running Windows NT – we can now have our 'seaside' map. SmallworldWide Systems Ltd. recently ported their Smallworld GIS to run in the Windows NT environment, and Scottish Homes have recently implemented this. PC-based systems have been installed in several new sites around the organisation. The difference in ease and levels of use between these new 'Windows' sites, and existing 'UNIX' sites has been remarkable.

GIS is no longer the preserve of 'experts' but sits on a PC for general use. Staff, who in the past required GIS analysis but would never attempt it themselves, have begun to 'experiment' with GIS, based on their own knowledge of MS Windows. One important effect of this is that decision makers now view GIS differently – it is now no longer an icon – 'the GIS' – it is simply viewed as a tool alongside other IT tools to be used when required.

Because of this it is used more frequently and more efficiently by more people – including decision makers, some of whom are willing to 'have a go'.

Until now, smaller GIS and map management tools have been PC based. Now that full-blown, serious corporate GIS are being placed on PCs, via Windows NT, perhaps the GIS community is dependent upon Bill Gates and his Microsoft company for the true integration of GIS onto the desktop in organisations. (See Endnote 3). It may be so, but in Scottish Homes it means that the original vision set out in the 1989 report recommending GIS has now been achieved – a 'serious' corporate GIS, on a PC, linked to the network, alongside other useful software in a familiar environment on the desktop!

conclusions

The identification and analysis of areas whose residents are likely to be considered Scottish Homes 'customers' is crucial for the business. Major data sets and analytical tools currently used in GIS were chosen directly from an analysis of typical questions likely to be asked by managers.

Thus, system choice was based on organisational needs, not software or hardware driven. Analytical and interpretative techniques used in GIS have been formalised in a 'best practice' guide which ensures that GIS delivers nationally agreed analysis within a corporate planning framework. Clearly any organisation wishing to make full use of GIS must place their system within such a formally agreed framework. Key to the success of GIS, has been its ability to deliver analysis unavailable elsewhere. It is true however, that the 'corporate equals UNIX' ethos of the early nineties held back full acceptance of GIS, erecting a 'technological barrier' which to some staff proved insurmountable. The advent of Windows NT (and Windows 95) on ever more powerful PCs has made 'serious' GIS power available to staff in a familiar desktop PC environment. This is helping to break down important barriers to its acceptance within the organisation. GIS in Scottish Homes, like in the wider GIS community, is therefore becoming less focused on 'applications and developments in technology – which are increasingly being taken for granted... rather, issues of data diffusion in the workplace...' are more important (Masser, 1995). Thus, in Scottish Homes we have come full circle: data dictated system choice, which in turn was constrained by technology. Now, technology is no longer a constraint, data and its dissemination are the key to GIS success.

references

Bourke, A., 1989. A GIS Strategy for Scottish Homes. *Scottish Homes*. September 1989.

Bourke, A., and Robertson, K., 1992. GIS in Housing: A Pilot Assessment. *Scottish Homes Occasional Paper*. March 1992.

Bourke, A., 1992. Data and GIS in Scottish Homes. *Scottish Homes*. June 1992.

Bourke, A., 1995. Data Accuracy and Quality in Scottish Homes GIS. *Scottish Homes Smallworld User Manual*. January 1995. pp. F1-F29.

Burrough, P.,A., 1994. Accuracy Issues for Future GIS. *Paper* presented to AGI'94 Conference, Birmingham. November 1994.

Dale, P., 1991. *Issues in the 1991 Census*. Social Statistics Unit, City University Northampton.

Leslie, S., 1995. *Mapping Awareness*. March, 1995. p. 12.

Masser, I., 1995. The Diffusion of GIS – The Chorley Report Eight Years After. *Mapping Awareness*. May 1995. pp. 20-21.

Peters, T., 1995. In Search of Excellence.

Scottish Homes, 1993. Local Market Analysis and Planning in Scottish Homes – *A Best Practice Guide*. pp. 24-30.

Williams, N., and Twine, F., 1991. A Research Guide to the Register of Sasines and the Land Register in Scotland: A Report to Scottish Homes. *Scottish Homes Technical Information Paper*. No. 3. 1991.

endnotes

1. PAF-matching Sasines data has an overall 'hit rate' of around 70% though this does vary; in urban areas 80% hits can be expected while in rural, 50% is the norm. For more detail see, Bourke (1995) and Burrough (1995).

2. For a detailed description of how Scottish Homes defined housing market areas and self containment see Scottish Homes (1993).

3. For the influence of Microsoft on GIS you only have to reflect on Bill Gates statement at the International COMDEX exhibition in the US in November 1994. He said: ' I see the growth of spatial data and its analysis as being one of the key areas for development in the near future'. Shaun Leslie (Leslie, 1995) said: ' GIS is entering the PC age. It is surely significant that Microsoft will soon be including a simple... mapping package as part of its basic PC software suite, Microsoft Office. Many more people in future will have ready access to GI analysis tools to help them run their businesses.'

appendix

17 fields of information relating to each property transaction in Scotland are available in Sasines data

Class of Sale:	Residential
Property Details:	
Characteristics:	
Street Number:	94
Street:	Hillfoot Road
Flat Position:	
Town:	Ayr
Postcode:	KA7 3JZ
Registration Number:	9401488
Month of Sale:	October
Year of Sale:	1988
Sasine Number:	25414
Sale Price:	£42,000
Sale Code:	Private Sale
Buyers Name:	Mr & Mrs V Sweeney
Town of Origin of Buyer:	Ayr
Distance Moved:	0
Sellers Name:	David & Raymod Shirra

mapping for success: the impact of desktop mapping on sales and marketing programmes

Laszlo Bardos

European Marketing Director,
MapInfo Corporation, Centennial Court,
Easthampstead Road, Bracknell,
Berkshire RG12 1JA, England.
Tel. (01344) 482888
Fax. (01344) 482777

abstract

Desktop mapping – a tool widely used in the US for visualising and analysing data – is emerging in the UK and Europe as a valuable tool for sales and marketing specialists in all business areas. At its simplest, it enables companies to plot information on a map to gain a quick visual overview of their position. At its most complex, it helps them to analyse unexpected relationships between locations and events that can provide a real insight into operational performance and future business prospects. This chapter explains how desktop mapping works and demonstrates several business situations where it has made an important contribution to sales and marketing success.

KEYWORDS: DESKTOP MAPPING; VISUAL PERSPECTIVE; DECISION MAKER; SALES AND MARKETING ANALYSIS; TERRITORY PLANNING

introduction

If every picture speaks a thousand words, why bother with the figures? Given the option, most people would prefer to look at a graphical chart or map, than a ream of dry spreadsheet printouts or column reports. Viewing data on a map brings the data to life. It becomes easy to understand and – because it allows you to see data in relation and proximity to other data – it makes comparisons easy – highlighting trends, patterns and relationships that may otherwise be missed.

By integrating and displaying information held in a variety of databases and spreadsheets, desktop mapping allows people to visually compare and analyse different sets of data. Companies can combine information about their facilities, sales territories, customers and prospects and leading competitors on a single map, as well as demographic and industry statistics and details of the areas most likely to generate future income.

With such a visual perspective, decision-makers can develop and implement sales and marketing strategies far more effectively and then allocate resources to areas of greatest potential. Sales analysis, territory planning, sales quota setting, market and site analysis, direct response campaigns, direct sales, advertising and recruitment: are all essential functions in which desktop mapping is making a vital contribution to performance, customer service and overall profitability.

early days

The database management systems (DBMSs) introduced in the 1980s certainly helped improve most organisations' understanding of their business, which had a corresponding effect on their profitability. As a result, over the last decade, sales and marketing professionals worldwide have been able to access and extract information about practically any customer, prospect or marketplace. However, they have lacked the graphical dimension to display and understand this information in a quick visual format.

Until recently, sales analysts, marketing managers and planners routinely spent days at a time analysing complex statistics from row and column reports and spreadsheets. Critical geographical relationships between companies and their customers were either left unexplored or plotted manually on unwieldy paper maps, graphs or charts. Such maps were difficult to handle, hard to interpret, hopelessly inflexible and almost impossible to share among users.

For some companies this meant that marketing resources were committed to customers who were unlikely to respond, sales territories were drawn up without enough knowledge of the main customer and prospect locations, and new sites were planned in areas that were too far away from growth centres to be of practical use – maybe with the lure of a development grant to sway rational judgement further.

As the power and functionality of the PC has advanced over the last decade, affordable mapping systems have emerged on the desktop. These are providing sales and marketing managers with the graphical tools they need to help them plot relationships between various sources of information on digital maps, charts and graphs.

how does desktop mapping work?

Most desktop mapping systems use information stored in spreadsheets and databases as well as a wide range of statistical data and maps. These include detailed street maps, boundary maps (including county, district, borough and ward information and postcode details), demographic and census data, economic and population statistics and retail sales statistics.

Users can attach data directly to maps via graphical symbols such as points, lines and polygon shapes. By simply pointing to a symbol or point, the database information it represents can be called up. Information can include customer address details; monthly and annual sales figures; details of outstanding debt; and quantities of data about business locations and customer sites – all of which can then be analysed easily.

Sales and marketing managers can create thematic maps using a range of colour codes and fill them with patterns that represent specific information, according to the set of values they choose.

For example, they may want to show customers with high purchase volumes in red and territories with low sales volumes in yellow. All they need do is select what they want to view and how they want to view it; the desktop mapping system will automatically find all the information and plot the data on a computerised map.

sales analysis

Desktop mapping is playing a particularly useful role in corporate sales analysis. Sales reports were previously compiled from a range of databases in tabular rows and columns, meaning that they were often too unwieldy to be analysed deeply, leaving critical information unexamined and ignored. Companies without a visual source of analysis were obliged to rely on instinct to determine their sales territories, and as a result many companies maintained unprofitable territories for too long, while others failed to approach potentially profitable business areas with the necessary levels of support and personnel.

With desktop mapping, companies can actually see the productivity level of each sales territory, salesperson and distributor. In addition, they can chart and rank their customers for every product and service and gain a clearer picture of their position in the marketplace.

Brann Direct Marketing, one of Britain's top direct marketing consultancies, helps clients define a target customer base, segmentation strategies and even communications strategies that will help create loyal customers and build business. Brann has developed a suite of software tools based on desktop mapping technology to allow its customers to manipulate and 'drill-down' into their own customer data. Brann's clients use the software to build profiles of their customers and cross tabulate the data against a whole list of variables such as income, geography, social class and newspaper readership.

territory planning

Territory balancing is one of the most popular and valuable of the desktop mapping applications used today by sales and marketing managers. By overlaying sales territories, sales representatives, customers and even potential customers, sales and marketing managers can use desktop mapping to consider the merits of existing and potential sales areas.

For example, this type of information could reveal that a company has too few sales people to service the potential within an existing territory. The same technology allows new territories to be created to correct the imbalance and optimise the company's ability to reach its customers.

Brann has developed a five-step approach which analyses the existing customer base; uses a cluster of geodemographic techniques to assess demand for products within geographic areas (including postcode sectors, towns and counties); compares the client's current sales territory definitions to the customer geoprofiles using desktop mapping to discover how good they are; uses Brann's own model to produce a realigned set of territories; and modifies these idealised territories with desktop mapping.

sales quotas

As mentioned earlier, sales professionals who typically design territories can use desktop mapping to establish their company's sales quotas. Since these quotas were traditionally arrived at through an exhaustive analysis of often inaccurate sales figures and probable market potential, they have often been inadequate yardsticks for measuring success. As companies were unable to view their revenue sources clearly, they seldom gained a complete picture of their market and usually based sales goals on past performance rather than realistic territory potential.

Many are now using desktop mapping to turn this process around. One recent example is a major US manufacturer which used a desktop map to develop regional, district and territory quotas. After charting existing customers, distributors and mail order houses (as well as all the Fortune 500 companies), the company's marketing specialist overlaid its established regions, districts and territories and then plotted direct customers, distributors' customers and mail order customers. Each of these was assigned a colour code to represent annual business purchases. Additional symbols were used to signify each distributor and mailing house, together with details of their annual sales.

With all this data in view, the sales manager was able to use the desktop mapping system's RDBMS to analyse detailed statistics from individual regions, districts and territories. Using the 'what if?' function, the manager then graphed the percentage of sales from the largest region to the smallest territory and assessed the market potential in each area. Using a map-based representation of the results the company was able to formulate realistic sales quotas that senior managers had faith in, and the sales team believed were achievable. The result was increased productivity and performance.

market and site analysis

Other companies use desktop mapping to achieve a better understanding of their chosen markets. They view sales, marketing and demographic data on digital maps, graphs and charts and can produce more precise marketing programmes as a result, which are more finely targeted to customer segments to maximise the return on the marketing budget.

Customer demand analysis is also taking a giant step forward. Once performed using large, complex GIS or paper maps, this analysis can now be easily done at the desktop. It is time-efficient and enables critical investment decisions to be made quickly and knowledgeably.

There are many examples of businesses which have used desktop mapping in this way. One, a leading US-based sand and gravel supplier, owns several rock quarries and coal mines and routinely acquires new ones to maintain supplies for customers in certain geographical areas. In the past, finding new sand and gravel sources was a laborious job. The company used enlarged road maps, manually plotting all quarry, coal and gravel reserves, existing customers and competitors. This map was always muddled and planners often missed significant buying opportunities and lost revenue to rivals. Since switching to desktop mapping, the company has been able to take advantage of every buying opportunity. By simply juxtaposing its own product sources with customer, competitor and available reserve locations, it now has a clear view of every investment option.

direct response campaigns

Desktop mapping is ideal for targeting and tracking suitable prospects for direct response campaigns. As well as displaying the best potential sources for a direct mailing, a loose insert or a telemarketing programme, desktop mapping systems can also chart customer responses.

direct sales

Now that companies can capture relevant visual market data, they can distribute maps displaying major prospects in each territory to their sales team.

One credit card company produces maps of customers for all its salespeople. Sales managers use the mapping system to draw a fifteen mile radius around each territory, plotting all major retail outlets within the circle on a digital street map. The marketing analyst who developed this application reports an average 25 per cent increase in daily sales calls and a seven per cent sales increase. Not only this, but the company has saved money on unprofitable customer contacts, dramatically reduced unproductive driving time, and maintained a constant level of travel expenses despite the higher sales levels.

The same company distributes sales leads from its various promotional activities by plotting them on a map of sales territories and sending these to the appropriate sales managers and reps. Fewer are lost or misdirected, all are acted on more quickly and the company is gaining a stronger position with new clients.

advertising

Some organisations use desktop mapping to support advertising decisions. Combining a visual display of key markets with a view of media outlets in the same areas, helps them use their advertising budgets more effectively.

One newspaper with several 'area' editions routinely presents visual demographic data to its advertisers. The marketing manager prepares a map displaying data for each area, based on customer criteria. The sales rep then delivers the map to the customer, who decides in which editions to place an advert. This process has particularly helped advertisers serving highly specific market segments. Using it, the newspaper has often discovered useful demographic information that has won it some loyal clients. One of these, a sports shoe manufacturer, was surprised to discover that an area thought to be populated mainly by senior citizens on limited incomes also had a growing population of highly salaried 30 to 45 year-olds with children. As a result, the company decided to advertise in the newspaper edition covering the area.

conclusion

Tighter budgets and an increased reliance on micro marketing disciplines have awoken organisations in every industrial sphere to desktop mapping. Banks, manufacturers, retailers and service organisations across the world are beginning to realise the many business benefits that this technology offers. Desktop mapping is highly functional and readily affordable. Placed in the context of today's business environment and the dramatic advances that characterise the computer age, it is a tool that sales and marketing managers looking to the next century may well consider themselves unable to manage without.

GIS applications in the retail and marketing sector

Peter Sleight

Partner, Target Market Consultancy,
Woodlands, Woodlands Close,
Holmer Green, Bucks. HP15 6QG, U.K.
Tel. (01494) 712371
Fax. (01494) 714203

abstract

GIS has come relatively recently to the retailing and marketing sector of the economy but has made rapid progress in the last few years. This paper reviews some of the issues involved, and the uses to which GIS is being put. In this sector, GIS operates in close relationship with the geodemographics 'industry'; recently, Lifestyle data has joined census data as a raw material for use in geodemographics. The way these data may be used in GIS applications will be explained. Finally, the concentration is on consumer marketing and retailing. The principles are similar for business-to-business, but the data sources are different.

KEYWORDS: GEODEMOGRAPHICS; LIFESTYLE DATA; TARGETING; GRAVITY MODELLING.

introduction

The retail sector and the manufacturing/marketing sector have a number of things in common; the most important of which is their customers – consumers. Knowing the geographical distribution of both actual and potential customers can bring great benefits. GIS is an essential aid to this understanding.

For retailers, the concept of the catchment area of a store is crucial – the geographical area from which the bulk of the store's custom is derived. A frequent convention defines this as the area from which 80% of the store's customers travel to that store. GIS can help retailers to visualise their catchment areas, and relate them to catchment areas of competitive stores; thus displaying a representation of supply and demand across geography. The store locations are the supply points, the residential areas represent potential demand. Converting potential demand into actual demand, and satisfying that demand, is what makes both retailers and manufacturers successful.

Manufacturers may not be particularly interested in retail catchment areas (although those manufacturers selling into retailers might be, in the sense of helping their retail customers 'sell through' to consumers), but they should be similarly interested in the demographic characteristics of residential areas. The concept of geodemographics (a fusion of geography with demographics) has brought home to marketers the geographical pattern of actual and potential consumer demand in a way that was previously unusual.

Manufacturers can use that greater geographical knowledge in two ways; to try to influence their actual and potential customers directly (through targeted 'local' media), and/or in negotiation with retailers, by attempting to influence stocking policy and via 'tailor-made' promotions with their retail customers.

definitions and explanations

Some definitions are helpful at this point:

Geodemographics is the classification of people by where they live. Census EDs are enumeration districts, averaging some 150 households, the smallest units of area for which UK census data are published. Lifestyle databases are large databases of individuals which have been sourced from 'lifestyle questionnaires'.

Geodemographics suppliers, with their main classification systems, include:

♦ CACI   (Acorn, Financial Acorn)

♦ CCN   (Mosaic, Financial Mosaic)

♦ CDMS   (Superprofiles)*

♦ Equifax   (DEFINE, PORTRAIT)

♦ Eurodirect   (Neighbours & Prospects)

   * Superprofiles also supplied under licence by The Data Consultancy, and by Business Geographics

Lifestyle database companies include ICD, CMT and NDL International (the latter two having now come together within the Calyx Group, and having been joined by MIC, a sister company which markets 'The Lifestyle Census'). Recent market entrants are CCN (with 'Chorus') and Consumer Surveys Ltd. (with 'Lifestyle Focus').

It is beyond the scope of this chapter to go into technical detail, but a brief outline places these data sources into context. The neighbourhood classification systems (Acorn, Mosaic, etc.) are the product of cluster analysis having been applied to demographic data (predominantly sourced from the census of population) with the objective of identifying and locating distinctly different neighbourhood types. All residential postcodes can then be coded with the classification in question, enabling (for example) customer files to be profiled, and market research surveys to be analysed, by the classification.

Historically, the lifestyle databases have derived much of their income from the rental of mailing lists, sourced from the wealth of data they hold about individuals. However, the largest of these databases have now reached sufficient 'critical mass' to be usable as a data source for geodemographics. For example, the 'Lifestyle Census' now marketed by MIC contains information on roughly one–half of U.K. households. Demographics are one key component of the information held – this source can either complement or substitute for the demographics from the census of population. Each source has its own strengths and weaknesses; it is likely that we will see a combination of these data sources in future geodemographic products.

how are these data applied using GIS?

There are two distinct types of data: individual data, and aggregated data. Individual (or household) data are the product of the lifestyle databases, whereas geodemographics inevitably use aggregated data. In practice, it is very rare for lifestyle data to be located at individual level on a GIS. One reason is the high cost of such an approach, which would make such an application non-viable for retail or marketing applications. The other, very straightforward reason for not 'spatially referencing' data about individuals is that the applications in question do not need this precision. Individual records will be aggregated to a higher level, either a unit of postal geography (such as the unit postcode, averaging some fifteen addresses, or the postcode sector, averaging approximately 2,500 addresses) or to a census enumeration district, or ED, averaging about 150 households.

These data are generally used in GIS at some level of area aggregation. The most frequently occurring unit of geography used in marketing and retail applications is the postcode sector. For many purposes these are convenient geographical areas, and are frequently used as units for aggregation into retail catchment areas, sales territories or distribution areas. Geodemographic classifications are constructed either at a census ED level, or at a unit postcode level, depending on the type of data used; even the ones that are constructed at unit postcode level may still be generally used at ED (or 'pseudo' ED) level.

Whether units of area of postcode sectors, or of EDs, are used depends on the level of accuracy required. Clearly, targeting at ED level (c. 150 households) is more accurate than targeting at postcode sector level (c. 2,500 households); and it is generally true that small areas tend to be much more homogeneous (in terms of their demographics) than large areas. Set against this increased accuracy is the cost element (it generally costs considerably more to hold data at ED level, or similar) and the data storage implications. Holding demographic data summarised as profiles at postcode sector level take much less storage than equivalent data held at ED level. So, 'horses for courses'.

Thus GIS can hold the geodemographic data for the areas in question, can manipulate those data and produce maps and reports. Typically, geodemographic classifications are held as 'postcode directories'; files of all GB (or UK) postcodes, some 1.5 million in total, each with a neighbourhood classification appended. Profiling software will enable a postcoded customer file to be matched against the directory, and a customer profile produced. This will show the penetration of customers by each neighbourhood type, compared with a base area (which might be GB, or a defined area, region etc.).

For area profiling, the classification will be held at one of the levels discussed above (unit postcode, census ED or 'pseudo ED', or postcode sector). Then the system will allow an area to be defined, usually in a wide variety of ways (e.g. drivetime, customer penetration, gravity model, or more simplistically, be describing a polygon or radius) and will then sum the units of area contained within this new boundary, and produce an area profile. In the case of postcode sectors, where the data are held at that level, the data will be in profile form. Unit postcodes and EDs are usually held as centroids in the system and will be picked up by 'point in polygon' search within a boundary.

Profiling can perhaps be illustrated with a simple (indeed, simplified) example. Using PiN, at the twelve-fold 'group' level, and for illustration, only using four of the 12 PiN types:

A   Rural
C   Upwardly Mobile Young Families
D   Affluent Households
F   Suburban Middle-aged or Older

The table below is fairly typical of the layout of profiles provided by Geodemographics agencies. It is a customer profile for a fictional product:

Table 1
Customer profile –
product base X:
Base – GB

PiN Type	Profile Count	Profile %	Base %	Index
A	150	7.5	5.1	147
C	1,040	52.0	14.1	369
D	750	37.5	10.0	375
F	60	3.0	17.0	18
Totals	*2000*	*100*	*46.2*	

(Note – this is slightly unrealistic, in that *all* the customers are in four of a total of twelve PiN types; nevertheless, it shows the principle of profiling clearly.)
To explain the columns:
(a) 'Profile Count' is a count of the customers in each of the PiN types. This will have been derived automatically by the profiling system, by analysing customer postcodes.
(b) 'Profile %' is the percentage represented by each PiN type of total customers; these numbers always add down to 100.
(c) 'Base %' is the percentage represented by each PiN type of the base area, in this case Great Britain. Most profiling systems allow the base to be specified.
(d) 'Index' is simply an expression of 'Profile%' divided by 'Base %', multiplied by 100. If the index = 100, this would indicate that the percentage of *customers* in the PiN type was exactly equal to the percentage of *population* in that PiN type, in GB – i.e., on the national average.

Table 1.

Earlier it was mentioned that there is a choice between using geodemographics in a neighbourhood classification form, and as 'raw' variables within areas (variables such as age, sex, household tenure, family composition, etc.). Area profiling using demographic variables is very similar in approach to that using classification systems. Again, the variables may be held at different area levels, similar to classifications (EDs or postcode sectors are most common), and having defined a study area, the system will aggregate all the units of area contained within that boundary, and express the total as a profile.

area potential modelling

One of the main uses of geodemographics is for modelling small area potential, for a variety of products and services. The concept is simple; given that geodemographics provides the demographic make-up of units of geography, if we can quantify the demand potential in demographic terms, then we can provide a measurement of that potential within these units of area. A market research source is generally used to generate the information (which should ideally contain value or volume data, if it is to be utilised in a potential model). Depending on the market concerned, the information might be, for example, value of holdings (for a financial services product); the differential rate for each neighbourhood type is then calculated, if a classification is being used.

Thus, consumer-based profile information is generated for the product or service in question. The other part of the exercise consists of producing an area profile (using the same neighbourhood classification) of the study area in question. Then the system will 'multiply through' the value per household for each neighbourhood type for the product in question, taking account of the penetration of the product by neighbourhood type, and produce an estimate of potential for the study area. A similar method involves using customer file information instead of market research information; producing a geodemographic profile from the customer file (not forgetting the 'value' component), and then applying this profile to an area to derive a measure of potential. Finally, lifestyle data provides yet another alternative source for the consumption data, on which the area potential model can be based.

applications

Broadly speaking, applications can be split into two types:

♦ Targeting

♦ Area potential estimation

Both applications are centred on profiling. The methodology inherent in area potential modelling was outlined above. Targeting using geodemographics starts from a very similar standpoint; by establishing the profile of existing customers, one can identify their characteristics, and then apply this knowledge in locating people with similar characteristics. So, produce a geodemographic profile of your customers for the product in question (if possible, refined to your 'best' customers, however defined), and then search to establish the locations of potential customers, i.e. people with a similar profile. Again, targeting using lifestyle data works on the same basis. Profile existing customers, and find 'lookalikes'.

Targeting using a GIS, unsurprisingly, can work in two ways; either by finding locations, or by finding areas, that meet the criteria specified. In this context, locations will generally be the centroids of census EDs, areas will often be postcode sectors, or other units of postal geography. So for example, within a particular study area, one may wish to locate:

♦ EDs having a particular neighbourhood classification, which we have established as being a key target, or

♦ EDs having (say) over 50% of households containing children, or

♦ postcode sectors containing 80%+ of owner–occupied housing, or

♦ postcode sectors falling within the catchment areas of Tesco stores, which contain above a defined proportion of specified neighbourhood types

A commonly used technique is 'postcode sector ranking', where all the sectors within a defined area (e.g. a geographical region or a store catchment area) are sorted into rank order on the targeting criterion in question, and the 'best' (above a cut – off point) are selected. This technique has been used virtually as long as geodemographics have been around, in the targeting of door-to-door leaflets, coupons or samples. The cut-off point is the total number of items to be distributed in the defined area; thus wastage of items can be minimised.

Media targeting

National media are generally targeted using national market research (for example, BARB for commercial television, National Readership Survey (NRS) for print media). However, the more local the media, the more relevant is GIS. Normally, the media owners themselves, or intermediate agencies, will populate GIS with their data, in order to make a selling case to their clients. Some examples:

Posters

The poster medium lends itself particularly to a GIS approach. Proximities of poster sites to, for example, named retail outlets, car dealerships or other sales or service points, is clearly relevant, and can be calculated and displayed on a GIS. Similarly, the road network is crucially important, and information about traffic flows can be incorporated. Finally, geodemographics can be used to model potential demand for the products or services in question, and although this is not always straightforward (for example, trying to relate modelled demand to traffic flows) this demand element may be added to the model. It is also possible to look at actual customer penetration, from customer files or lifestyle databases; so the targeting of posters can be considerably enhanced.

Local/regional newspapers

Free newspapers are particularly susceptible to geographical analysis, as they are generally distributed by 'postal geography', and will be delivered to virtually all households within the postcode sectors which form their delivery areas. Thus the technique of postcode sector ranking, referred to earlier, is often employed to target inserts into those sectors that best match the target market.

Local and regional 'paid for' newspapers have a less well-defined circulation area than 'frees'; but it is still possible to digitise the circulation area of each title, from the information held by newspaper management. The reason this may be useful is that retailers, who form a large share of the advertising revenue for this medium, may wish to compare the 'footprints' of distribution areas ('frees') or circulation areas ('paid fors') with the catchment areas of their own stores, to ensure a good fit between their actual or potential customers, and the geographical distribution of their advertising messages.

Door-to-door distribution

The distribution 'sectors' for door-to-door distribution are generally based on postcode sectors and, as mentioned earlier, the use of the postcode sector ranking technique is widespread in this medium. A geodemographic profile of the product of service in question is derived (from analysis of customer data, or from cross-tabulation of a market research source, such as BMRB's Target Group Index). This profile is used as the basis for ranking the sectors in the area concerned, with the 'best' sectors (in terms of most desirable demographics) at the top. Thus, the most cost-effective distribution plan can be calculated.

Where a retailer is commissioning the distribution, or where a manufacturer wants to take account of the location of stockists of his product, the ranking is conducted within the catchment area of each store concerned. Unless, of course, a 'blanket' distribution within catchment areas is required.

Other media

Direct mail is, of course, the most 'targetable' of media, capable of targeting to individuals at their home locations. However, GIS is rather 'overkill' for this degree of precision! Given an address and a stamp, the Royal Mail will deliver. Certainly, geodemographic targeting, and particularly selections from lifestyle databases, take their place alongside conventional list rental as sources for 'cold mailings'.

The boundaries of local commercial radio transmission areas can be defined and digitised, thus incorporated into GIS. Of course, a precise boundary is rather a notional thing but, nevertheless, it may be valuable to see how this form of 'media geography' fits against other areas, for example, marketing areas or retail catchments.

Finally, although commercial television can be a national medium, in practice it is delivered via a series of twelve regions. It may be useful to define the boundaries of TV regions, which are available either from BARB as overlap areas, or from ISBA as non-overlap areas. Again, this geography can be compared on the GIS with other geographies of interest to the marketer or retailer.

area potential estimation

This application occurs across a diverse range of end-users; all types of retail (including financial services retail), 'network' operations such a utilities and cable T.V., and many consumer marketing companies, whether practising direct marketing, or selling through retail outlets.

The use of geodemographic and lifestyle data for market segmentation, and for learning about the characteristics of customers, is outside the scope of this paper. However, the geographical aspects, which are clearly more relevant to GIS, are dealt with below. First, marketing applications.

Sales territory planning

GIS can hold and organise the various elements which contribute towards the planning of sales territories. Again, we are assuming a consumer product or service (although, as stated at the outset, business-to-business follows the same principles, just different base data). The demographic characteristics of consumers should form the basis for territory planning (giving the theoretical potential for demand). Often, postcode sectors are taken as convenient units of area for manipulation. Current sales calls are 'geocoded' and located; potential calls are also shown. Using suitable software, the territories can be optimised, taking account of travel distances and times, and assumed workloads.

Managing retailer relationships

These days, in grocery retailing the contact between manufacturers and retailers tends to take place at head office level, between buyers and key account negotiators. The old days, of salesmen calling upon stores, have long gone. So the manufacturer needs to have good, store-based information with which to discuss product potential and stocking policy. The retailer will have EPOS-based sales data, so can have an 'information advantage'; but if the manufacturer has sound data about catchment potential, he can have a dialogue about market share, joint promotional strategy, and ultimately, the targeting of local promotional activity. Without such data he is left relatively naked! The same principle can apply in other markets, where manufacturers rely on retailers to sell their products through. GIS can hold the relevant data, the models can be run, and the results displayed - in map form, as well as tabular output. We all know how powerful maps can be at making a sales case.

Advertising and promotions

This chapter has previously considered the way in which GIS can be used to target local advertising and promotion. The best prospects for the product or service are located and targeted, thus minimising wastage.

This exercise can be conducted on a larger scale, for example, at the level of television areas, if it is intended to advertise in some regions, not in others. The potential return can be calculated from each T.V. region, set against the costs, and this information can inform the decision.

For companies with their own customer databases, analysis of the geographical pattern of customer penetration can be very useful when deciding on the best/most cost efficient method of stimulating more sales. Direct mail is by no means the only route! Other media may prove more cost-effective, but it is necessary to see the patterns before this becomes clear.

Targeting direct sales

Companies selling direct to the public can find immediate benefits from adopting a GIS/ geodemographic approach. By identifying their best prospects in geodemographic terms, they can locate areas of best potential, and concentrate their sales efforts accordingly. They can also produce a base measure of latent potential, 'other things being equal', against which they can set sales targets and reward achievement.

Perhaps an obvious example of this is home improvement products such as double glazing, where the type of housing (and indeed the demographics) in an area make a great difference to the potential for business. But the cable T.V. operators have also found that an analysis of the geodemographics within their franchise area have given solid pointers to those areas of greatest potential for profitable business; not so much for initial 'sign-ups', but more importantly, for customers who stay with them – and pay their bills!

retail applications

The importance of location to retail outlets is such as not to need elaboration. So GIS has a key role to play in assessing store location strategy. Some of the major retailers have developed very sophisticated site location research units (indeed, many years before GIS as such were used within them), but the arrival of GIS has served to automate many processes, and to hold and display the data; essentially, to integrate the system.

Retail location analysis

The locational issues vary by type of retailer. For example, large 'destination' retailers such as grocery superstores or DIY 'sheds' will clearly require sites that are easily accessible by car, that provide sufficient space for sales area, storage area and large car park, and that have a sufficient catchment to generate 'enough' of the right kinds of customers to make the required profit. Depending on operator, they might look for areas currently undersupplied (if there are any of those left!) and try to avoid major competition. Or they may welcome current competition, believing in their ability to win the inevitable battle!

'Comparison' retailers, or branches of banks and building societies, by contrast, will want to be on High Streets or in shopping malls in close proximity to competitors. The whole nature of the shopping experience (in the case of comparison retailers, not banks!) requires a repertoire of similar outlets nearly. The trading area of the shopping centre will need to generate sufficient of the 'right' sort of customer for them to succeed.

To meet these, and other types of retail requirements, a GIS set up with geodemographic data and locations of present stores/shopping centres will then need either data on current shopping patterns, or software to generate such. This may be based on gravity modelling principles (see later), on drivetime software, or simply on some form of proprietary 'shopping geography'. The more sophisticated retailers will have developed forecasting models, which they will have refined over time. For some purposes, a fairly basic 'site screening' model will be adequate, capable of giving a top-line forecast for candidate sites, sufficient to decide whether to reject them or subject them to more detailed analysis.

A more sophisticated model than the site-screening model will estimate market share, within a trading area, for the site in question. It requires similar input data, but in addition will require an estimate of squarefootage of selling area of the relevant merchandise category in the trading area, and information on sales/sq.ft. for competitive retailers.

Store performance analysis

Using similar input data to that used in retail location analysis, the performance of stores or branches operated by a retail organisation can be assessed. Very often, the data collection phase of store performance analysis is by far the most time-consuming part of the job. Use of a GIS to store the additional data collected (and to generate the 'demographics' for each catchment area) will assist the process.

There are two main methods in use; regression analysis, and gravity modelling. The objective in the regression modelling approach is to build a model which explains the performance of the stores/branches in the network. The input data comes into three categories:

(i) within the defined catchment for each outlet:

- total population

- population in core target market

- demographic

- geodemographic

- workplace population (if relevant)

- potential expenditure on products/services in question

(ii) data about the store/branch in question, e.g.

- sales area – product range (if this varies)

- size of frontage – decor/fascia (if this varies)

- other store-specific features

- branch turnover

- car parking (adjacency)

(iii) 'locational' variables, such as:

- competitive presence, proximity/size

- type of location (high street, mall, edge-of-town, etc.)

- 'pitch' of site (primary, secondary)

- accessibility to pedestrian flows, or vehicular traffic flows, as appropriate

- adjacency to 'anchor store' (if relevant).

Once these data have been collected, a multiple regression model is built, testing variables until an optimal solution is found. The analyst will want to explain store performance (in this case, turnover is the 'dependent' variable that we want to predict) by a particular combination of variables in the model. Normally, the model will be built on one sample of stores in the network, and then tested on a matched sample.

Once it is 'robust', it can be run on the complete network, and under–and over-performing stores identified. The model effectively predicts an 'other things being equal' turnover; it should be able to identify where there is room for improvement in the network, and which branches are already getting as much turnover out of their trading area as can be expected.

Gravity modelling

Gravity modelling (otherwise known as spatial interaction modelling, or impact analysis) is a dynamic method of retail modelling, whose aim is to build a model of the interaction of supply and demand within a retail marketplace. In essence, it models the 'gravitational pull' of each store or shopping centre against its competitors. The basic principle is that 'other things being equal', people will tend to visit the nearest store, with the time and cost of travel being a factor militating against visiting stores further away. Of course, things are not quite this simple; the attractiveness of an individual store or centre may persuade people to travel further. So a model is built for a region, and the supply-points within this region are located, the road network is superimposed, the residential areas (and their demographics) are shown. Some form of flow data is needed to calibrate the model (e.g. credit card or store-card transactions, market research data, or retail transaction data). Once set up, a series of 'what if' analyses can be run, to look at the effect of opening new stores in particular locations, or closing existing stores, or perhaps adding selling area to existing stores. The important point is that the effect of particular actions (their 'impact') can be modelled in advance of taking those actions, and the trading network can be optimised. A crucial issue may be the degree to which a new store will 'cannibalise' existing stores in the same network, as against its effect on competitive stores. Gravity modelling should be able to predict the outcome.

As gravity modelling will predict turnovers for existing and new stores, it can also be used for store performance analysis and market share analysis. Its very nature means that it is best operated within a GIS environment. It certainly takes full advantage of GIS functionality.

conclusions

GIS has been adopted by both retailers and manufacturer/marketers. The most sophisticated systems tend to occur at the 'top end' of the retail market, where the high cost can be justified in the context of the opportunity cost of getting store location decisions wrong. However, hundreds of systems are in use across a broad range of retail and marketing organisations, some being quite small capacity single PC systems. These are perfectly adequate to the task in situations where the size of the data manipulation task is not too onerous (for example, where demographic data is held at postcode sector level). Given the importance of geography to both retailers and consumer marketers, the role of GIS is likely to increase.

references

Sleight P., 1993, *'Targeting Customers'*, How to use Geodemographics and Lifestyle Data in Your Business, NTC Publications Ltd.

re-engineering of business processes in regional electricity companies

Peter Mingins

Senior Manager, Intergraph UK,
Delta Business Park, Great Western Way,
Swindon, Wiltshire, SN5 7XP.
Tel. (01793) 619999
Fax. (01793) 618508

abstract

This paper considers the application of GIS within the changing UK Regional Electricity Companies (RECs) and examines Business Process Re-engineering and argues a case for the adoption of GIS as one of the tools that can be supportive of critical business processes. The paper argues that technology and understanding in the field of GIS is allowing the visions of the past to become the business advantages of today.

KEYWORDS: REGIONAL ELECTRICITY COMPANIES; BUSINESS PROCESS ENGINEERING; BUSINESS MODEL; PROCESS OBJECTIVES

introduction

Contracting margins, competition and tighter regulatory control are forcing UK Regional Electricity Companies (RECs) to examine ways of providing greater efficiencies in the supply of service and care to the customer. The cultural change from a producer to customer led market, have made demands on providing greater efficiencies of service. These service efficiencies are demanded against a requirement to reduce costs. Shareholders have grown to enjoy and expect sustained income from investments. In some utilities, including some RECs, exceptional growth has occurred. In order to maintain the share values and defend the large cash reserves and hence guard against acquisition, many utilities have attempted to diversify into non-regulated activities. Many of these activities are outside the core regulated business. This diversification has included investment in hotel chains and particularly in the context of the RECs, high street retailing.

RECs have substantial investment tied up in asset infrastructure. These distribution networks are spread over very large geographic areas. The inherent infrastructure investment, coupled to the costs of maintenance and planning, is enormous. It is perceived that this substantial area of cost will provide the largest benefits in terms of savings.

islands of automation

Many utilities' Geographic Information Systems (GIS) are confined to the records drawing office. They were installed as point solutions and specifically designed to automate an isolated function. Not surprisingly, these islands of automation have never matured into integrated corporate systems and have typically remained the domain of the engineer or technician. As a result, the business benefits of capitalising on the earlier investment in data modelling, collection and maintenance, have never been fully exploited. Individual functions that require the sharing of information and the co-operation of the differing personalities involved is frequently a source of trouble. An opportunity exists in many RECs to provide corporate-wide distributed information systems that are built around business processes.

what are business processes and why re-engineer ?

So what is Business Process Re-engineering (BPR)? What are business processes and why are Geographical Information Systems supportive of them in the Electricity Supply Industry?

The vast majority of companies owe at least part of their structural origins to the Industrial Revolution and the regimented hierarchical structures that existed to manage labour intensive human activity systems. The requirements of today's RECs are to provide a flexibility in the organisational structure that accommodates change and maximises profit. Customers, competition and change are creating a new business world, and it is becoming apparent that enterprises designed to operate in one environment cannot be fixed to work well in another. RECs have moved from a culture of accountable public servants to organisations where customers, competition and change, demand flexibility and fast response.

'Re-engineering is the fundamental rethinking and radical redesign of business processes to achieve dramatic improvements in critical contemporary measures of performance, such as cost, quality, service and speed.'

Business process re-engineering does not mean tinkering with what already exists or applying incremental changes that respect the basic structures. It involves determining the business drivers that are essential by providing processes that are supportive of them. It involves not selective change but wholesale changes. BPR requires from employees a different 'mind set' and a conscious reshaping of an organisation behind a new corporate vision which encompasses the market and the customer.

'A business process is a collection of activities that takes one or more kinds of input and creates an output that is of value to the customer.' (Gower).

Some utilities have attempted to improve profitability through diversification. They have bought businesses outside the regulated core activities. This kind of thinking distracts companies from making fundamental changes in the work they actually have to do. The current trend in RECs is to step back from diversification and to re-focus on how efficiencies can be gained from the core business, that of supplying and distributing electricity.

A typical methodology for BPR would involve the development of a business model and with it a vision of process objectives. In doing this the key business processes will be identified. It is these processes in a REC that are vital. It is these processes that by their very nature will cross departmental boundaries. They will demand supportive, enterprise wide automation to allow supportive information flow. In the previous 20 years, investment in automation in the electricity supply industry has been high. Automation has not been a complete answer to the problems of organisational structure. True, computers can accelerate productivity of some tasks, but fundamentally if the same jobs are being done then no fundamental improvements in the organisation's overall performance are achieved. Task oriented jobs in today's world of change are becoming obsolete.

reduced operational and engineering costs through increased efficiency

Geographic Information Systems frequently provide the modelling and management tools to allow RECs to model the necessary aspects of the network that allow for power analysis and leaner engineering. Providing an engineer with the facility to conduct 'what if' analyses on a portion of planned network represented as a computer model, allows him the luxury of iterative design until the optimum design requirements are met. Real savings can be achieved by applying an optimum solution rather than over engineering. Examples of RECs not having to build planned sub-stations exist and as a result cost benefits can be clearly highlighted for maintaining a geographic model of the asset base.

Higher return on infrastructure investment

It is essential to any organisation to know what assets it possesses, what their value is and, in the case of such a wide geographic spread, knowing where they are. It would seem fundamental that a business, any business, needs accurate information about its costs in order to manage its profitability. This type of information in a REC is likely to be inaccurate, expensive to collate or at best out of date. An accurate value for the infrastructure will allow a tighter management of capital expenditure.

Increased control on capital expenditure

The ability to manage capital expenditure will allow the opportunity to defer expenditure, as such profitability and share value can be better controlled. This allows for a tighter managed control on overall profitability.

Improved quality of customer services

Fundamentally, as all RECs supply electricity to customers, the differentiator in a competitive free market will be the quality of the service provided and the efficiency and cost with which it is achieved. The Electricity Supply Industry (ESI) is changing from being producer focused to very much a customer focused industry. This stimulus to improve quality of service will increase as a function of competition and regulatory pressure.

why now?

The climate is right. A pressure to change and a pressure to compete is being created. The Holy Grail of a totally integrated GIS, supportive of many functional areas centred around the valuable asset register, is moving towards reality. This reality is happening as the old functional silos within RECs are destroyed and the business focus has become more process-orientated. The re-engineering has reduced the departmental and functional politics allowing the sharing of information across departments and even across organisations. This trend will accelerate as competition increases.

changing attitude within the electricity supply industry

The maturing attitude and the acceptance for the need to change has provided an opportunity for GIS to be acknowledged as a cost effective and viable technology. The UK RECs are littered with GIS pilots that are point solutions. They have, for political and technical reasons, never flourished. As the RECs come out of re-engineering and GIS is being successfully implemented then examples of hard GIS derived benefits are becoming more frequent.

Technological innovation

Technical computer graphics systems, including GIS, have for years developed on a parallel track with commercial IT. Computer graphics has now joined the main IT stream. Graphics are no longer the specialist application. They are an accepted and demanded requirement throughout mainstream IT. Technological advances have also increased the corporate viability of GIS. The growing acceptance by IT groups that GIS is in the mainstream of information technology is adding to the momentum. Wide acceptance of Graphic User Interfaces and the hardware technology to support both office automation and technical applications on the same seat are fundamental. This facility at a low cost is essential if valuable data is to be shared and become supportive of valuable business processes.

Integration of geographic information is necessary on two levels. Firstly, it is necessary that information is integrated at a corporate level. Corporate data structures are based around technologies that support maximum usage and exploitation. It is costly capturing electricity asset data. The case for investment must be based on making it available to support cross-company processes rather than single functional areas. The acceptance of RDBMS technology is providing this opportunity. Applications are ephemeral, as business requirements change applications to support them will change. What is fundamental is that sufficient investment is made in the data model and data integrity that allows for maximum exploitation and flexibility.

Issues surrounding data management and quality are becoming better understood. The need to provide quality data on which to make better business decisions is crucial. These decisions within a business process have to be made by individuals who have access to salient information. Until recently, within many office environments a requirement for two platforms was needed: one to do technical computing CAD, GIS and TIM (Technical Information Management) etc. and one to support office automation tools like word processing and spread sheets.

The improvement and user acceptance of the Graphic User Interfaces allied to the exponential performance in Intel based hardware has allowed these different functions to coexist in the same hardware/software environment. This trend in one-per-desk technology, is beginning to allow the sharing of data at the desktop level by utilising the underlying supported structures of the operating systems i.e. DDE (dynamic data exchange), OLE (Object Linking and Embedding) in the Microsoft Windows environment. This technology is now available at a price point that allows staff in RECs to take maximum advantage of valuable information in a cross-departmental and functional context.

business processes likely to have a GIS component

The following business processes will benefit from the adoption of a distributed information system which incorporates GIS as a component not a separate adjunct:

♦ Asset Management

♦ Customer Information Systems

♦ Forecasting and Planning

♦ Optimal Network Design and Analysis

♦ Work Management

♦ Loss of Supply Management

♦ Materials Management

♦ Facility Model Management

♦ Network Management and SCADA

conclusions

Regional Electricity Companies must protect their investment by investing in open technologies. It will provide them with the flexibility to procure the latest and best hardware and to run on standard open commercial software. The value of the Distribution Information System is the data and the accessibility of that data. Operating systems like Windows95 and Windows NT support 32-bit technical applications and allow them to coexist and share data with many other Windows supported applications.

Literally thousands of applications run on these operating systems providing the most flexibility in choosing software. In these environments the business/office automation tools and technical applications run on the same computer and have the same intuitive interface. RECs are rapidly changing. Management are striving for efficiency. Techniques like BPR have allowed a re-focusing on processes that support organisational objectives rather than a fragmented departmental approach. GIS technology is supportive of business processes. The BPR philosophies can be realised by using GIS to integrate existing data, and to support information flow which is supportive of business objectives. With technical advances in processor speeds, operating systems and software combined with the mind set change in RECs due to BPR activities, an opportunity now exists. This opportunity will be realised by providing geographic data as part of an overall information system and not just as a proprietary encapsulated island of automation.

GIS hits the street

Tom Lithgow

Senior Consultant (GIS Development),
Lothian Regional Council,
Transportation Department,
18-19 Market Street, Edinburgh,
EH1 1BL, SCOTLAND.
Tel. (0131) 4693762
Fax. (0131) 4693773

abstract

The success of its road accident analysis system in GIS prompted Lothian's Transportation Department to use GIS for its road maintenance management system. The accident system has a 'ready made' dataset of accident records from the police and is used to develop strategies for accident prevention. In contrast, the maintenance system is operational in nature using data which is consequently very volatile. In addition, the volumes of data involved in the maintenance system are an order of magnitude greater than the accident system. To ensure that the information used is as accurate as possible Lothian has used the latest advances in portable geographic information handling which allow validation 'on-site'.

KEYWORDS: HIGHWAY MAINTENANCE; MANAGEMENT SYSTEM; ROAD ACCIDENT ANALYSIS SYSTEM; CHAINAGE; INVENTORY COLLECTION

introduction

Lothian Regional Council's Transportation Department, as a roads authority, has responsibility for the publicly maintained roads and footways which fall within the Region's boundary. With around 700,000 residents, some 210,000 of whom enter and leave the city of Edinburgh each day, the pressure on the road network is extreme. The public road network in Lothian extends to around 4200 kilometres. On this there are in the order of 1-1.5 million individual inventory items for which the Department is responsible, e.g. 107,000 street lamps, 20,000 drainage gullies, and 2,000 bridges. It is estimated that the network has a replacement value of £2.6 billion. Given an annual maintenance budget which currently stands at less than 0.75% of that value, it is important that those limited funds are directed to maximise their benefit.

The popular perception of a roads authority is that its sole function is filling potholes in the roads and maintaining the fabric of the road network. Whilst road maintenance is one of the important roles of the Transportation Department, and indeed accounts for a large part of its annual budget, its primary aims are to promote road safety, the economic development of the Region and improvements in the environment, and hence the quality of life, through its transportation policies.

As a part of its work, the Department is responsible for policing others' activities on the road network. It receives 20,000 notices of opening from the utilities every year and issues around 4000 permits for operations on the street covering skips, cranes, portacabins, hoardings, scaffolding, etc. Events such as The Edinburgh Festival, the Tall Ships Race, Hogmanay Celebrations, and other parades and demonstrations, have to be managed and co-ordinated along with all the other activities on the streets.

Although individual systems exist for the management of much of the above, there is currently no automated method of correlating all of the various activities and resources involved in management of the streets.

history

In the late 1980s the need was recognised for a comprehensive computerised system to handle the information required to manage the road network. However, because of the wide diversity of activities undertaken by the Department and the limitations of the technology then available, no effective progress was made towards fulfilling this need. It was finally recognised that the most appropriate solution was the use of GIS. However, at that time GIS was still in its relative infancy and as an option was considered a high risk venture.

Around the same time, activity in GIS was increasing dramatically elsewhere. The interest of the utilities had accelerated the Ordnance Survey's (OS) digitising programme to the extent that map data was becoming available in useful quantities. Prior to this involvement, the OS's digitising programme had stretched well into the next century. The Chorley report had been published. The local authority associations had formed the national Geographic Information Steering Group (GISG) under the auspices of LAMSAC. In addition, there was a growing awareness that GIS was becoming an area of activity of increasing potential and importance. In the midst of this, Lothian Region's GIS Steering Group was set up with representation, initially, from each of the Council's technical departments. The chief aims of the Group were to investigate the use of GIS, raise general awareness of GIS in the Council's departments and to ensure that a common approach was taken throughout the Regional Council to the development of GIS and the handling of geographic information.

The Group's first initiative was to investigate the use of OS digital mapping as an alternative to paper maps by setting up a pilot digital map library. A conscious decision was taken at that stage to specifically exclude any investigation of GIS to avoid complications. The pilot concluded successfully and the digital map library, in an expanded form, now provides for the Regional Council's large scale mapping requirements. Following on from this, attention moved to the use of GIS itself.

Experiences of computer developments in general and the observation of other organisations' experience, specifically with the implementation of GIS, indicated strongly that the adoption of GIS as a corporate resource would carry with it a high risk of failure. It was also seen that where data capture for a development was involved, this typically accounted for 60-90% of overall project costs. In such cases, this initial overall implementation cost had a tendency to mask the specific benefits which GIS would bring to the project in the longer term. Bearing the above in mind, two projects were selected for piloting which were relatively small scale, had well defined scopes and which had existing datasets available. In this way, the actual benefit which development in a GIS environment would bring could be much more easily established. The projects selected were Census analysis using the 1991 Census data and a road accident analysis system which uses the STATS19 data from the police. Lothian already had well established vetting procedures for the accident data which attached accurate grid references to each accident. Both of the pilots were highly successful and have been developed into full working systems which are in continuing use and development. It is reputed that the accident analysis system is the most advanced in existence.

Having seen the success of the GIS pilots, it was decided that the time was ripe to proceed with the development of the road management system. The original concept of the system was the creation of a large central database. This concept had been formulated in the mid to late eighties, before the micro-computer had been able to have a marked effect on local authority operations, and had been overtaken by the events of the intervening years. Computerised systems had been developed on a piecemeal basis throughout the Department. As a result a more appropriate approach to the implementation of a road management system was seen to be through the use of GIS as an integrating mechanism. Although most of these systems had been developed in a non-spatial manner because of the lack of available technology to do otherwise, it was considered a relatively simple task to add georeferences to the information handled by these systems.

inventory collection

The initial task in setting up a system for road management is the collection of a complete inventory of the items for which the Department has maintenance responsibility. Traditional road management systems use the road network as the basis for recording inventory, and items are referenced by chainage along the relevant road section and their position on the road cross-section. Although it is possible to relate items referenced in this way geographically, experience of systems where this has been attempted show that it is not possible to achieve a satisfactory cartographic representation without an unacceptable level of manual intervention.

The philosophy mirrored in setting up contracts for the inventory collection was, therefore, to provide a dataset which accurately represents the road inventory in a descriptive sense and appears cartographically correct when displayed against a digital map background. Initially it was thought that, where features appear on the map, measurements on site could be dispensed with in favour of measurements taken from the map. However, early field trials showed that such dimensions could not be relied upon. Either the overall cartographic accuracy of the maps was inadequate or, in some instances, changes on the ground were not reflected on the map. This has meant that the inventory collection has had to be dual in function, firstly collecting measurements and other attribute information about the inventory items and secondly digitising their position and geometry relative to the Ordnance Survey features.

strategic versus operational

Both of the pilot applications implemented were essentially strategic in nature. They use historic information as the basis for analysis on which future strategies are based. This is apparently a common element in most of the 'first generation' GIS developments. The accuracy of the information handled in these systems is relatively easy to manage because it is relatively static or has only a small volume change occurring. This is the case with both the census and the accident data.

A system for managing the road network differs from the pilot applications in that it has both strategic and operational functions to fulfil. In accordance with the 'garbage in, garbage out' principle, the quality of decisions made, based on the analyses of such systems, can only be as good as the quality of the information used for those analyses. It is, therefore, important that mechanisms are in place to ensure that the quality of information gathered in the field is high. One of the difficulties with the road management system is that the volumes of data involved are an order of magnitude or more higher than, for example, the accident system, the data itself is much more dynamic in nature with a higher proportional rate of change. This poses a significant problem in ensuring an acceptable level of accuracy of the data in the system, as the volumes involved preclude, for practical purposes, any large scale secondary vetting of the information.

Some of the strategic operation of a road management system is related to the road inventory quantities, e.g. grass verge cutting, gully cleaning, and road marking renewal. The value of these analyses will decline from the moment of inventory collection as changes are made to the inventory. Specific action must, therefore, be taken to ensure that the inventory remains up to date. Other, and more complex, strategic issues relate to changing conditions on the road network, e.g. deterioration of the road structure. These issues require accurate information on road condition to be gathered regularly over time.

Historically, highway information systems have tended to be text based, both in the office and in the field. The staff responsible for collecting the data used in these system used text-based portable computers or even paper based collection methods which were later transcribed. These give little or no assistance in verifying the accuracy of the data collected, and in many cases, the information flowed in only one direction, so that the staff collecting the information saw little or no benefit from the operation. In such cases there was a tendency to view the task of information gathering as peripheral to the 'real' job and as such the effort put into that task reduces. There followed a consequential reduction in the quality of the information collected.

It is clear that in order to ensure that the information held in the system is of high quality, and that the decisions made on the basis of that information are of a high quality, the staff involved in the collection of that information must have an interest in ensuring that this is the case. In essence, the viability, in the medium to long term, of the strategic aspects of the operation of the system, is dependent upon the successful operational use of the system 'at the coal face'.

road inspection

The inspection staff are essentially the 'eyes and ears' of the Department. They are multifunctional in their jobs carrying out routine inspections of the condition of the roads, investigating complaints and accident claims from members of the public, utility operations, and new road construction under the road construction consents (RCC) procedures. As mentioned earlier in this chapter, a number of systems have evolved over a period of time and many of these feed information to the inspection staff. Although the systems in the office may well be computerised, the information which the inspectors receive is invariably in a paper form.

Having made the decision to develop the road management system in a GIS environment with a matching GIS-based road inventory, the further logical progression was that the system to be devised for the inspection process should similarly be GIS-based. The emergence of pen-based portable technology and the amalgamation of this with digital mapping software made this possible.

system development

The office based system has been developed around the Oracle/Exor Highway Maintenance Manager System and Smallworld GIS, Lothian Region's IT Department (Lothian Information and Technology Services (LITS)) having 'project managed' the development of the system. The experience which they have built up through the pilot projects and subsequent developments has meant that they could handle the GIS front-end of the system. However, for the integration of the Maintenance Manager System and Smallworld they appointed Informed Solutions Ltd. who are experts in the integration of systems, particularly in the GIS environment.

The initial development of the Inspection System covers Inventory inspection and defect recording. The aim of the system is to reduce the amount of paperwork required to be processed by the inspectors to a minimum, freeing them to concentrate on the more productive aspects of their work. A feature of many existing computerised inspection systems is that they are based around the recording of the existence of road defects. They tend not to allow the recording of details necessary to describe the repair required. This has the effect of requiring either a second visit to the site or taking notes, on paper, during the first visit, to collect the details required. The system being developed in Lothian is intended to overcome this shortfall by giving facilities to collect all of the information required to effect repairs to defects. This falls into three categories; a description of the defect being considered, a detailed description of the work required to repair that defect and a location plan.

future developments

The initial development of the system covers routine inspection of the road network and recording of the defects found. In order to make the system as acceptable as possible to the staff required to use it, it is important that further functions are included in the system to assist them in their work. The following have been identified as having a high priority for inclusion.

Utility operations

As stated in the introduction around 20,000 notices of street opening are registered through the SUSIEPHONE system each year. This represents five openings per kilometre of road, but the scale of these openings can vary from a single small excavation to a track running the length of a street. It is recognised that such openings are major source of road defects. At present, the ability of the inspector to identify the utility responsible for a particular reinstated excavation is severely limited, as SUSIEPHONE is a purely text-based system. As most notifications have effectively a postal address, it is considered practical to use that in conjunction with the Ordnance Survey's ADDRESS-POINT and OSCAR products to attach meaningful locations which can be shown against a map background. In this way, the 'owner' of any particular defective reinstatement can more easily be found.

Clarence

CLARENCE (Customer Lighting And Road ENquiry Centre) is a free phone system which takes calls from members of the public on road issues. It currently takes in the region of 54,000 reports of defects per annum. Each one of these generates a paper report which is passed to the relevant inspector for investigation and report back. This amounts to the handling of 54,000 pieces of paper per annum by only 32 inspectors. It is intended that each defect report will be 'tagged' with a geographic reference which will then allow it to be allocated automatically to an inspector's beat area and also to be displayed graphically against the map base. This will have the added advantage of allowing easy identification of both multiple reports of the same fault and recurring faults.

Other information sets

The above are two of the more urgent needs, but there are many more systems which are currently isolated and/or do not have the benefit of usable geographic referencing. These include; the adopted streets register, which is currently stored in a card index, containing verbal descriptions of the areas of street; a data base of permits for operations and occupancy of the streets; Road Construction Consents procedures, which follow the progress of new road construction by private developers to ensure that the standard of construction used meets the requirements of a street to be adopted for maintenance by the roads authority; road project area definitions; traffic flows; condition surveys; Traffic Regulation Orders (TROs); Temporary Traffic Regulation Orders (TTROs); and planning applications.

conclusion

When looking at the path the development of GIS has taken, there are parallels that can be drawn with the development of computing as a whole. The computer systems emerging in the post war years were complex pieces of extremely high-cost equipment. They lived in dust filtered, temperature regulated, humidity reduced, security monitored computer rooms and required a brigade of white coated experts to both program and to operate them. Naturally, access to such machines was heavily guarded, and, for many years, computing was available only for very specialist and controlled activities. In local authorities and other large organisations, computing typically was a finance department function. Mainframes were (and still are) ideally suited to the batch processing typical of such routine activities as payroll, invoice payment, and rates collection.

When other departments started to take an interest in computing, it is little wonder that their approaches were treated with suspicion. They were viewed, perhaps justifiably, as an unpredictable threat to finance's well-ordered computing. As a result non-financial computing, when it was allowed at all, was often relegated to off-peak use. Restrictive run-time parameters would often be applied to avoid any unpleasant effects on the 'real' business of the mainframe.

In response to these restrictions and to an increasing demand for computing from a wide range of areas, lower cost computing started to emerge. This, first of all, took the form of timeshare facilities on bureau mainframes, and later of in-house mini-computers (mainframes where the budget would stretch). This was the start of the devolution and democratising of computing away from centralised mainframe systems and restrictive controls placed on the user. Computer users were starting to have control of their own destiny. The emergence of the micro-computer and its subsequent proliferation into all areas of endeavour was the final death warrant for the dominance of mainframe computing for all but the bigger data processing operations.

The development of GIS has followed a similar path. Early Geographical Information Systems were complex, expensive and required a significant investment in equipment and expertise. They were developed and used by specialists for predominantly strategic applications. Systems have and are gradually downsized until it is now quite feasible to operate a reasonably competent GIS on a portable computer away from the office.

It has not been practical on cost and functional grounds to develop operational GIS until such time as this level of downsizing was available. With the recent GIS advances in PC and more particularly pen-based portable computing, access to geographic information and spatial analysis is much more readily available. Geographic information can be handled as such in the field, giving the benefits of the ease of access to information this allows along with immediate and much more valuable verification of information held or gathered at the point of collection. This raises the value of the work done in the field, reduces the validation overheads, which are so often a major part of major office based GIS. The result of all of this is that analyses of much higher quality can be achieved on which a much higher reliance can be placed.

access to information: the human computer interface and GIS

David R. Green

Centre for Remote Sensing and Mapping
Science, Department of Geography,
University of Aberdeen, Elphinstone Road,
Aberdeen, AB9 2UF, Scotland.
Tel. (01224) 272324
Fax. (01224) 272331
Email. d.r.green@aberdeen.ac.uk

and

David Rix

MVM Consultants plc., MVM House,
Oakfield Road, Clifton, Bristol,
BS8 2AL, England.
Tel. (0117) 974 4477
Fax. (0117) 970 6897;
Email. davidrix@mvm.co.uk

abstract

The Human Computer Interface (HCI) is an important consideration in the modern work environment and will play an increasing role in the successful implementation of GIS technology. The importance of the user interface extends beyond simple consideration of the Windows environment to include aesthetics and the design of the screen layout, and dialogue terminology. As GIS becomes more widely applied there will be a growing demand for a range of far simpler interfaces to GIS to allow the user to concentrate on the task(s) at hand rather than having to contend with the complexity of the underlying technology. This paper considers the importance of the HCI, through the growing use of IT in the workplace, and the evolution of the interface. It looks at the objectives of interface design and describes some of the basic principles of the interface. The paper concludes with an assessment of the relevance of interface design to GIS, the issues to be considered for the future development of interfaces for GIS, and considers the role of the user, rather than the computer scientist, in the successful development of such task-based interfaces.

KEYWORDS: HUMAN COMPUTER INTERFACE; GIS; WORK ENVIRONMENT; END USER; TASK-BASED INTERFACE

introduction

Over the years a considerable body of research has been devoted to developing the Human Computer Interface (HCI) of microcomputer and workstation software (e.g. Baecker and Buxton, 1987; Bodker, 1990; Ellis et al., 1993; Ellis, 1993; Patton, 1993; Shneiderman, 1993; Davies and Medyckyj-Scott, 1994). The most notable developments have been the Apple Macintosh and subsequently the MS-Windows interface, NT Windows, and OpenWindows, the most recent newcomer being Microsoft's Windows95 launched in August 1995.

To a large extent the development of the 'windowing' environment has helped to overcome some of the 'usability' problems encountered with previous software operating systems. MS-Windows has, for example, helped to standardise many aspects of, e.g. screen layouts, menus, and the commonality of menu terminology between software, making certain tasks, e.g. file handling, printing, and display, far more straightforward. The point and click approach to menu and option selection has been a tremendous boon to the less computer literate users facilitating ease of use. The end result has effectively been a 'screen template' which has successfully struck a balance between the need to provide a standard, familiar, and intuitive work environment, but also one that can be customised to suit individual requirements.

But, despite all the years of research, and the widespread adoption of a windows-based environment, many of the current applications software interfaces continue to meet with very varying degrees of success. While the 'end-user's lot' has clearly improved dramatically in recent years, there are still many problems in need of attention as far as using software applications are concerned, e.g. to assist in the work flow; to help avoid the incorrect use of software menu options; to overcome the apparent failure of software to do things (that is, where the user is not able to 'see the wood for the trees'); poor error messaging; and so on.

These particular sorts of problems appear to be the result of a combination of factors: the continuing evolution of software interfaces (many of which are seemingly becoming more and more complex, not just in functionality but also in appearance, and in some cases appear to lack adherence to the basic original MS-Windows format (see also comment by Hewson (1995, p. 18)); a growing number of variations on a theme, e.g. Windows95, Windows NT, OS/2, all similar in concept, but with their own particular quirks; a growing user-base, which has changed in terms of background and experience, but many of whom are still not really computer-literate in the truest sense; an apparent temptation by software designers to continue to load the interface with every functionality possible, using a combination of menus, buttons, dialogue boxes etc...; and despite the development of simpler alternatives, e.g. Multimedia and Hypertext interfaces, a continuing dominance by, e.g. MS-Windows, in the marketplace.

Nowhere are these problems currently more apparent and more acute than in the case of GIS software. As far as design (and the problems associated with producing a satisfactory interface design) is concerned, GIS interfaces probably suffer to a greater extent than many other more familiar desktop software (e.g. word processing). There are a number of reasons for this: the functionality and sophistication of the technology in the operational/functional sense is probably greater and more complicated than that associated with more common desktop applications. GIS software is finding an increasingly important role in the workplace which makes it commonplace and yet the operational functionality is still not familiar to many, or really likely to be in the long term; and GIS is increasingly being pushed towards an 'on the desktop' selectable icon, e.g. Autoroute etc., on the MS-Windows toolbar.

As GIS technology becomes more and more widely used and applied in the workplace, and increasingly by a more diverse group of end-users, with widely differing backgrounds, knowledge, understanding and experience, successful adoption, implementation, and use of GIS will, in part, depend upon the interface(s) (in the broadest sense) provided as the intermediary or 'go-between' the software and the user. Furthermore, the importance of the HCI will take on new dimensions as the computer hardware and software and IT technology continues to evolve; as the technology becomes more commonplace; as many different types of GIS (ranging from the fully functional to the browsing tool) are developed; and as the GIS tool becomes ever more sophisticated in its analytical, presentational, functional, and visualisation capabilities. Additional pressure will also be placed on GIS software interfaces as other applications software evolves, and as other technologies, e.g. the World Wide Web (WWW), develop, and growing use is made of, e.g. Multimedia and Hypertext interfaces, with multiple links and image windows.

why is the HCI important?

Software interfaces are undeniably very important to all computer users in a number of ways. Not only do they influence our perception of the software (how easy the software is to use and to work with), and also the purchase of certain 'brands' of software (one word processor versus another), but ultimately they affect the operational use of the software in the workplace as the environment in which a user may spend a greater or lesser amount of time undertaking a wide variety of tasks ranging from the simple to the complex, e.g. browsing, retrieving, and analysing. In this respect they affect the user learning curve, performance speed, error rates, and overall satisfaction with the computer software. Extensive research has revealed the following (e.g. Foley et al., 1984; Booth, 1989):

- Poorly designed interfaces can lead to degraded user productivity, user frustration, increased training costs, and the need to redesign and reimplement

- When a person uses, e.g. an interactive graphics system to do work, they want the system to virtually disappear from consciousness so that only the work and its ramifications have a claim on their energy

- Ergonomic quality to the interface is vital to avoid boredom, panic, frustration, confusion and discomfort

- Interfaces affect the time taken to undertake a task, the accuracy with which it is performed, and the pleasure derived from it

- Learning time, recall time, short-term memory load, long-term memory load, error susceptibility, fatigue susceptibility, naturalness, boundedness are all affected by the interface

- In the workplace, the effects of the interface on stress, lower work rates, decreased job satisfaction, absenteeism are of social importance

- In extreme cases, interfaces can lead to dissatisfaction and conflict within organisations

In a broader context there are many other reasons which make the user-interface an important consideration.

growing use of IT in the workplace

The spread of IT use in the workplace can be attributed to:

- the availability of the right technology at the right time, e.g. the colour monitor, the graphics card, the software, processor speed, RAM capacity, and so on. These have all coincided with changing demands for information in the workplace

- the new technological developments, more compact desktop, portable and pen-based computers, lower hardware and software costs, individual affluence, marketing (e.g. advertising (Microsoft, IBM, Intel)

As people have made increasing use of computers to undertake a wide range of tasks, from the elementary (e.g. word processing), to the more sophisticated (e.g. calculations and statistics with spreadsheets, data storage with databases, graphics and DTP, to desktop mapping with software packages for cartography (simple mapping software and clipart) and desktop Mapping/GIS (e.g. Autoroute, MapInfo, Lighthouse)), it is inevitable that more emphasis has been placed on the HCI in order to accommodate end-users and to improve the ease with which it is possible to use and work with computers.

the role of information technology in the workplace

Focus on the importance of the human computer interface and the development of a standard interface which links into other software from the desktop has also been influenced by the growing realisation by, e.g. academia, industry and government, of the importance of information technology in the workplace (Booth, 1989). It is also recognised that the technology is both sophisticated and complex, and as more people, who are not computer specialists, come 'on-stream' to use computers for a wide range of applications there is considerable potential for people to 'get lost', which is potentially detrimental to the longevity and use of computer software. There is an underlying need therefore to ensure that the potential of the technology is realised.

the specialist versus the casual user

As IT has become increasingly popular and affordable, and the number and variety of commercially available software packages have grown, the importance of interface provision for the user has grown considerably. Growing use of IT in the workplace (at home and/or in the office), computer games for leisure, and more recently the Internet (e.g. via CompuServe), has meant that a wider cross-section of people (from many different backgrounds) are buying and applying software, many of whom are not IT or computer specialists. Computers have now become the realm of the applications specialist, the layman, and people who use the computer only occasionally (the casual user), and perhaps solely for simple tasks such as browsing, viewing, playing – what has been termed by Booth (1989, p.3) as the 'discretionary user'. This has in turn placed greater pressure on the software suppliers to provide software that is easy to use and which can be applied with little effort in the workplace.

education and training

More and more people are also becoming better informed about IT, and more computer literate via education, training, and re-training. Consider, for example, the large number of PC magazines now on the newsagents' shelves, and the advertising of computers and software on TV and in magazines, e.g. the Sunday Times magazine, for the workplace and home. The people now being targeted are literally the 'user generation'.

increased competition in the marketplace

Although the choice of software packages was once limited to only a few, the marketplace has become increasingly diverse and competitive and there are now far more examples to choose from. Besides choosing a software package on the basis of its functionality (suitability for a task), uptake of a particular software application by users is highly dependent upon the ease with which the software can be used to undertake a specific task and other related tasks in the workplace. This is largely determined by the software interface. In this respect a final purchase may be influenced by software reviews, word of mouth, a visit to a software exhibition, a vendor demonstration, and so on, all of which will reveal how easy an interface is to use, how it looks and feels.

increased expectations

As awareness of the technology and its role in the work place as a facilitator of tasks has grown, so increased expectations have resulted. People want to be able to use (all) the software, (a) available to them within the same environment, (b) interactively, i.e. being able to move from one software package to another, e.g. cutting and pasting tables, graphics etc. (c) easily without the need to refer to the manuals or help files, (d) with the minimum learning curve, (e) without the need to recap, and (f) without knowing much about the internal workings or logic of the software.

information, the information age, and access to information

Whether or not we are in the Information Age (see e.g. Liebenau and Backhouse, 1990) there is no doubt that we are now not only able to gather far more information than ever before, but also with the aid of computers we are able to gain access to more information in one place (under one roof) than ever before; to select from this information, view it, and ultimately to use it in a variety of different ways, e.g. text documents, slide shows, simulations etc. We now have a vast information resource at our finger tips.

Access to information is important and becoming increasingly so. In any walk of life we require information, and multiple different types of information are necessary for both decision-making and planning. Many new developments in computer technology, e.g. communications and multimedia, now offer infinitely greater possibilities for delivering information in a variety of different formats to a user on a desktop. Perhaps the most recent, and obvious, illustration of this is the potential opened up by the Internet and the various accompanying WWW browsers, e.g. Mosaic and Netscape, the ease of access to information that we now have on our desktop, in the form of text, and graphics, sound, video clips, taken from multiple different sources, and the ability to cut and paste this information wherever we want. As can be seen from Figure 1, this type of interface provides all the tools necessary to utilise the information, e.g. to load, print, move backwards and forwards etc. Working within an MS-Windows environment also permits integration of this information into other software. A number of examples can be cited: the retrieval of a text document from a remote source which can then be cut and pasted into a report; the retrieval of an image which can be processed using digital image processing software, and subsequently cut and pasted into a report or document; the retrieval of raw data which can then placed into a spreadsheet and used for subsequent analysis. One particular advantage of the WWW (Kleiner, 1994) is that access to information is hypertextual, which provides an easy means of exchanging information, and it also enables movement of large chunks of information from remote areas to local ones.

Figure 1
The Netscape
Interface

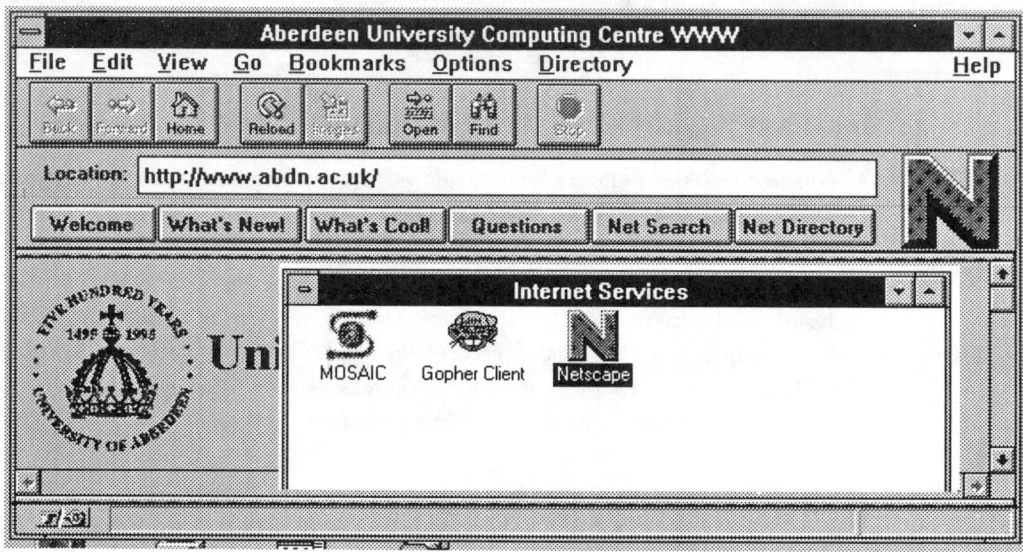

interface design

A direct outcome of the various influences and developments outlined above has been the increasing pressure on the quality and design of the software user-interface, the so-called Human Computer Interface (HCI). Therefore, one of the biggest research problems now facing software specialists is to develop the interface communication with the user.

Interface design is not trivial, however, and involves a multidisciplinary approach, drawing upon understanding, concepts, and inspiration from computer science, audiovisual media, industrial design, computational linguistics, artificial intelligence, cognitive science, cognitive psychology, social psychology, software engineering, human factors, ergonomic research, audiovisual design, organisational psychology, mathematics, and even the graphic and editorial design of conventional paper publications (Lynch, 1995; Booth, 1989; Foley et al., 1984). Interface design is also a highly creative process that integrates intuition, experience, and a careful consideration of many technical issues.

Design of the 'ideal' HCI, however, is difficult to say the least, and it is almost impossible to design an interface to suit everyone. There is no such thing as the 'average user'. Instead, software interfaces have to cater for many individuals, users with different requirements, characteristics, expectations, and even personal preferences.

HCI evolution

In the last decade the Human Computer Interface has evolved very rapidly (Ellis, 1993). The interface types usually considered are:

♦ command line interface (CLI)

♦ menu

♦ image (GUI)

♦ other

Command Line Interface: This is the archaic solution, such as is found in MS-DOS and Unix and was utilised by many of the early GIS innovators. The interface is easy to develop but it is now generally accepted that it is inadequate. The problem is that the user needs to develop expertise in the vocabulary and syntax of the language used in the interface. The language used was often cryptic in the extreme. While this may have proved acceptable in a tightly controlled environment, where users would have access to a small number of common commands, in the context of GIS on the desktop such an interface is now of historical interest only.

Menu: the command line interface has gradually been superseded through the use of on-screen menus. This solution simply uses screen menus to explicitly declare the options available to the user. There are many different examples of menu-interfaces. These include hierarchical and non-hierarchical menu systems, single, multiple, binary, extended, nested, pop-up and pull down, multiple choice, linear sequence, tree structured, and embedded menus. Whichever system has been implemented the operational context is that the user is presented with a series of options and makes a choice. However, while this interface design solves the problems of vocabulary and syntax, it presents its own problems. To present the menu option in full would risk filling the screen with few options. To avoid this interfaces may include reduced (or cryptic) instructions which may reduce the effectiveness of the interface to the user. The use of menus for multiple options also has shortcomings. The greater the extent of the menu the more cumbersome is the use of the menu. Such an all encompassing menu structure may also include options which are not pertinent to the user. The alternative, which has been typically adopted by GIS vendors, is the 'nested' menu. This interface design provides the user with primary menus which may call up secondary or tertiary menus which may be a function of the user's application requirements. This solution creates its own problem and that is menu 'navigation'. Unless experienced, the user can rapidly become lost in traversing the menu tree.

Image (GUI): a GUI offers users a Windows, Icons, Mouse and Pointer (WIMP) environment, with resizeable windows, scroll bars, icons, pop-up and pull-down menus, dialogue boxes, and interactive buttons. Input and selection is via a combination of keystrokes and the use of a mouse or other pointing device. Interaction is usually through the direct manipulation of graphical objects on a screen. A GUI includes interaction metaphors, images and concepts used to convey function and meaning on the computer screen, the detailed visual characteristics of every component of the graphic interface, and functional sequences of interactions. It determines the 'look and feel' of a system, the shape of windows, buttons, and scroll bars, how you resize things, and how you edit files etc. (Quin, 1995).

The value of the GUI comes into its own based upon the fact that human communication is largely visio-manual. It is user-centred, task oriented, establishes the common ground between the system and the user, and depends on the direct manipulation of objects, making the assumption that the user is familiar with objects (Michels, 1995). It allows the illustration and use of real world objects and the idea of real-world metaphors. It also provides an opportunity to have an interface in line with the human's real-world e.g. the user of an Apple Macintosh carries out tasks in the manner they are familiar with, and they think in terms of the application domain rather than those of the medium of computing.

The graphic icon has an important role to play in this type of interface. Research by organisations such as Apple and Microsoft has focused on the use of images (icons), typically supporting nested menus, as a way of minimising the problems inherent in other types of interfaces. Typically images will be designed to pictorially represent a given process, for example a printer image for the print process. A selection of images representing the most commonly used processes will be provided to the user via a default screen display. When the user wishes to run a process the image is selected by the user (via the pointing device) and the process is then activated. This design of user interface represents the current state of the art and is typified by Apple and Microsoft Windows-based applications. However, while it offers many advantages over previous interfaces, there are still limitations. In many respects we have come full circle in providing a few 'cryptic' commands for commonly called processes – in this case the encryption being in the form of an image (icon). Providing for a wider range of functionality through a greater selection of icons may only serve to confuse as it may be difficult for users to learn and remember the differences between icon types if the number exceeds twenty (Galitz, 1989). Even though the usability of such an interface for normal business use would now be questioned, for the most part the GIS 'world' has yet to achieve even this level of usability.

Other

Multimedia/Hypermedia: Multimedia interfaces are a derivative of the GUI. Users of, e.g. multimedia computers, interact with information in a novel way which has no precedent in paper document design (Lynch, 1995). The potential of this type of interface lies with the advantages of flexibility offered for the novice and infrequent user. Increasingly the multimedia interface is finding growing popularity in the home, whilst multimedia touchscreens are finding acceptance for Public Information Systems (PIS).

Hypertext or Hypermedia systems use a 'visual interface model' (Kacmar, 1993). These interfaces use navigational methods of interaction. As noted (p. 85): 'The user must use visual, mental, and physical skills to recognise those objects which are or are not associated with links and to activate links in order to navigate hypertexts.' Hence, the interface component and interface-application relationship play a more critical role in attempting to simplify human-computer interaction in hypermedia environments. The power of hypermedia lies in the relationships (links) which exist among the objects and it is in the use of these relationships that users truly benefit.

Virtual Reality: The question of virtual reality (VR) interfaces is increasingly being mentioned in the literature, although there are few real concrete examples to offer as yet for users of e.g. GIS, which is perhaps due to an over conservative approach by the GIS community (Jacobson, 1994). The concept of the virtual interface, or the immersion interface, is to allow the computer to become 'human-like' rather than vice versa. According to Jacobson (1994, p.9):

'The Virtual Interface is about to free GIS from today's technological, representational strictures, giving it new power and importance in the realm of decision making.'

Mention is made by Jacobson of Worldesign's WorldSpace™, which it is believed will 'heighten user-understanding', allow comparison of alternative scenarios, and the integrated study of whole systems.

Adaptive Interfaces and Games Interfaces: The final types of interfaces briefly considered here are the so-called adaptive interface and the more familiar games interface. In the former, users enter their own commands that are recognised by the computer rather than using a specific syntax (Booth, 1989). Interfaces for games software are somewhat unique in that interaction with them is relatively unpredictable compared to the more predictable tasks associated with other types of software (Shneiderman, 1993).

what does the term HCI include?

Design considerations for user interfaces usually encompass the whole computer work environment. At the most generalised level this includes the:

♦ Hardware – Input, Display, Output

♦ Software – Dialogue Language

hardware and software

In an operational sense this includes a combination of the physical (active) and the visual (seeing or viewing) work environment, for example:

♦ menus

♦ windows

♦ keyboard

♦ mouse

♦ sound

♦ information channels

The hardware and software components of the interface environment complement each other in terms of navigation, operation, and communication. The software component enables the navigation and use of the functionality, determines the visual appearance (number and position of windows, shape, type of menu, customisation, complexity, overlapping windows, use of colour), the layout (aesthetic, functional and uncluttered, providing all the tools (and only the ones) necessary, and analysis and communication via visual display). Human perception is complex and affected by many factors e.g. age, response time, the hue brightness and saturation of colours, colour contrast, low or bright light. The degree of logic and ordering in command sequence design, the extent to which user intuition has been considered in the location of 'on-line' help, the understanding or perception and memory when designing icon shape, even the degree of on-screen clutter, will all affect user performance. However, the hardware component is also important, because it dictates the user impressions of software in terms of e.g. response time, and display rates.

At a more detailed level, a number of aspects should be considered.

Usability

One often talks about the usability of the system. According to Booth (1989) there are many definitions of what this includes. Michels (1995) defines the following usability attributes:

♦ Learnability

♦ Effectiveness

♦ Efficiency

♦ Memorability

♦ Errors

♦ Satisfaction

♦ Flexibility

♦ Attitude

Other factors include the hardware, and ergonomic factors, and the work environment (Michels, 1995).

Operation

Operation covers the use of the software for applications use – the day to day working with the system. As software becomes more task specific, operation of the software will increasingly be governed by the particular task, that is the functionality will be customised for a job. It will be increasingly necessary to customise the interface to suit the tasks, the users and the applications. Already this is happening, e.g. information systems with all the functionality required by the task; using the browsing interfaces; the viewing type of interface, and so on. In such cases operations are cut to the bare minimum to suit the user task and only the user task.

Navigation

One of the most important aspects of using a computer system relates to the ease with which a computer, and subsequently its software, can be opened, loaded and run, accessed, and used. A major part of this involves navigation, not just in the exploratory sense of the word, but also relating to the ease with which it is possible to link and trace the appropriate logical steps involved in undertaking a task – the 'workflow'. Consideration should also be given to the installation interface for the software.

The problem associated with navigational aspects lies with being able to define all the tasks that a user may wish to undertake, and in addition the many different pathways.

Error messaging

As far as possible a system should be robust and 'immune' to user-mistakes which can cause subsequent difficulties. However, where necessary, any error messages responding to a user mistake, e.g. through pressing the wrong key or selecting the incorrect step in an analysis, must be easy to interpret (the rule to follow is 'assume nothing') and easy to rectify (the user should be able to get out of a situation easily at the very least, or should be pointed in the right direction, either automatically or with the aid of prompts, preferably without a system crash and loss of work and a requirement to 're-boot' the computer system). In some cases where recovery is possible care must also be taken with the options offered to ensure that the user chooses the correct one.

Analysis

The interface must also be able to accommodate a wide range of users who wish to apply the technology for a wide range of different tasks (see e.g. Medyckyj-Scott, 1994), sometimes very complex tasks which may be difficult to comprehend without the appropriate background or subject knowledge. There is a need to ensure that provision is also made for guidance with regard to the application, e.g. perhaps through the use of a simple example, the same sort of assistance that might be provided for a word processor or statistical package.

Problem solving

The design of the interface is also important in the context of problem solving. A carefully constructed interface can be used to aid the user undertaking a task with software. For example, the use of software for a specific application may necessitate that stages be followed in a certain order so as to achieve the objective; the on-screen location of menus, availability of options, and ordering of menus options can all help to guide the user through a task.

Visualisation, displaying data and communicating information

Accessibility to data/information must be manageable, and comprehensible (Medyckyj-Scott, 1994). However, many users have great difficulty in extracting a lot of information from vast volumes of data. Therefore there is a need to use the techniques of visualisation to 'manage' such large volumes of data. (Shneiderman, 1993). In recent years sophisticated graphics and visualisation tools have become part of many software packages. Users now have very powerful tools to hand to enable them to view data in many new and exciting ways. User-controlled animation techniques can further increase comprehension. However, the whole approach to the problem of displaying data (transforming data into information) is one that needs considerable care (see e.g. Dickinson, 1973). There are certain ways to optimise the display of data and information that are well established and better than others. This requires user-guidance and care, e.g. to map data. A good example is the choice of map type, e.g. thematic, symbol and so on. Help, in a variety of forms, is therefore required to ensure that the user works with the appropriate tools for the application at hand, e.g. the visual display of spatial concepts.

Once again, the interface design can aid the system user. Careful structuring and design of the interface can provide the guidance necessary to visualise data in an appropriate manner; right from the stage of choosing the display type through to a series of different visualisations in separate windows, which may include an examination of the raw data and the graphic, e.g. in Microsoft Works for Windows both the spreadsheet and the graph can be tiled or cascaded allowing users to see both representations. In the case of more sophisticated visualisations, the problem is made more difficult by the variable interactions likely dependent upon the tasks.

objectives of interface design

The primary objective of the interface is the need to promote usability of a system. Specifically this refers to the ease with which users can access the application or system functionality provided by the software. Various criteria for measuring such usability have been defined (see e.g. Markham and Rix, 1995) Such criteria can only be judged against the context of the existing workflow patterns and the education, skills and motivations of the intended users. For any interface to be effective it is important that time is devoted to identifying the characteristics of each class of potential user and that the systems analysis process is sufficient to identify these requirements. Adequate documentation of the business requirement must be a primary requirement of any interface design.

documenting the business need

For any design, the business requirements, i.e. the tasks to be performed – the business objectives to be achieved, must have priority over the interface. A task can be defined as an activity that, from the user's perspective, has clear beginning and end points. Tasks are undertaken to achieve a pre-defined objective. Evolving a systems design based on tasks is an important first step in translating a user's mental map into a system design. Such an approach can provide designers with a clearer understanding of what the system is to be used for and, more importantly, how it is to be used. It can also make the outputs of the design process more accessible to the user. Task analysis is the formal name for the methodology of studying what users need to do and the ways in which they do their tasks (see e.g. Rheingold, 1990). Building GIS systems around tasks is the first step towards evolving a more appropriate GUI design.

Table 1a
Advantages
and
disadvantages
of the various
different types
of interface

CLI:

Advantages

- ♦ potential for quick interaction
- ♦ little screen space required

Disadvantages

- ♦ user unfriendly (long commands, special syntax, unfamiliar terminology)
- ♦ memorisation needed
- ♦ command line interfaces are unforgiving with regard to user mistakes (Michels, 1995)
- ♦ confusion of syntax
- ♦ inconsistencies
- ♦ arbitrary syntax
- ♦ errors entering commands
- ♦ mismatch between task and concepts syntax (Shneiderman, 1993, p.40)

MENU:

Advantages

- ♦ shorter learning curves
- ♦ structures decision-making
- ♦ semantic knowledge remains in long-term memory compared to syntactic knowledge which is short-term/volatile
- ♦ full-screen interfaces can make use of function keys since there are relatively few and they are easy to remember – interaction accelerators (Michels, 1995)
- ♦ don't need so much knowledge, reduce memorisation of commands, reduce training, and structure the user's decision-making. (Shneiderman, 1993, p.67)
- ♦ pie menus have advantages – reduce target seek time, lower error rates (Shneiderman, 1993, p. 79)
- ♦ forms can mirror the logic of the system for structuring input
- ♦ number of options on menus – aim to communicate user's intent to the computer as quickly as possible

Disadvantages

- ♦ slows down the frequent user
- ♦ consumes available screen space (Shneiderman, 1993, p. 90)
- ♦ rapid menu display rate

Table 1b Advantages and disadvantages of the various different types of interface	**IMAGE/GUI:** **Advantages** ♦ intuitive ♦ reduce the learning curve ♦ graphically and aesthetically appealing ♦ icons desirable ♦ recognition memory rather than recall memory ♦ continuous feedback is given – this can give the user a sense of control ♦ direct manipulation interfaces are object oriented in nature ♦ help to concentrate on task (Shneiderman, 1993, p. 17.) ♦ mastery, competence, ease of learning, confidence, enjoyment, eagerness, explore (Shneiderman, 1993, p. 18) ♦ learn functionality quickly ♦ carry out a wide range of tasks ♦ little syntax to be remembered ♦ reduced error rates because of less syntax ♦ rapid feedback ♦ exploration is encouraged because of ease of system use ♦ initiation of actions provides confidence
	Disadvantages ♦ difficult to find enough icon symbols ♦ screen resolution may affect recognition and interpretability ♦ icon interpretation depends on user experience ♦ system resources required ♦ visual representations not always an advantage over text ♦ consume screen space ♦ knowledge of the real world required ♦ guided only by what we know may mean that exciting potential to interact/explore differently will be lost

basic interface design principles

In the development of graphical user interfaces certain default principles have evolved:

Direct manipulation: of screen objects. Under this principle screen objects remain visible during user operations with the effect of operations on objects being immediately visible.

Point and click (or see and point): provides user interaction with the screen through selecting objects and performing activities through the use of a pointing device, typically a mouse. Such an approach is based on the general form of user action, noun then verb, i.e. select an object of interest (the noun) and then choose the action(s) to be performed on the object (the verb). Both of these operations assume that users can see on the screen what they are doing and that they can point at what they see.

Consistency: in the interface allows users to move between applications (or tasks). Consistency should be applied to both the appearance of the interface and to the behaviour of the interface.

Wysiwyg (What You See Is What You Get): requires that the functions (or processes) needed to perform the selected task are available to the users without the need for the user to search for the function or to create the function as a product of other functions.

User control: of operations establishing a balance between providing users with the capabilities required to perform tasks with security of functionality and data – best achieved through the use of 'alert' boxes.

Feedback: is required to provide users with information regarding operations initiated by the user. Feedback from any user action should be immediate with meaningful information concerning the initiated process and the current status and anticipated duration of the process. Feedback should be direct and simple and in a form that will be understood by the user.

Forgiveness: through all processes within the application being reversible where feasible – users should feel comfortable and not threatened when using the system to perform the task. (For more details on these concepts see Apple Computer, Inc. 1992, pp 4-12)

the relevance to GIS

What is the relevance of all this to GIS? To begin with GIS is an application of IT, albeit with application software that has been designed for a very specific task. In many respects GIS integrates a range of more familiar software e.g. spreadsheets, databases, and CAD. As GIS technology continues to grow in popularity (as the value of the technology is realised) and it continues to migrate towards the non-specialist user, so the importance of the HCI will increase. In many respects the HCI will be even more important to GIS than for other applications software. Different types of GIS, with different objectives, and aimed at different users, will require different interfaces. The likely result will be a hierarchy of interface levels designed to accommodate a hierarchy of users. This will be a fundamental requirement if GIS is to achieve the industry goal of simply becoming another icon on the integrated desktop.

the current situation

The apparent failure to date in the GIS revolution may be attributed to a number of reasons. However, a leading contender is that even now GIS are perceived as being difficult to use. Traditionally GIS procurement has been driven by an internal organisational 'Champion'. Typically such a 'Champion' would represent a 'technical' discipline and most likely would occupy a position fairly low down on the organisation 'totem pole'. Viewed from this perspective the Champion sees GIS as less of a business solution and more of a career solution. Objectives for the system tend to be very narrow and, again, typically many implementations of GIS have been limited to one or two users, generally used to display and print a few OS maps, and generally never tend to progress beyond the pilot stage. While the limited vision and objectives of the 'Champion' may prove to be a contributory factor much of the reason why GIS fails to develop is that they have been difficult to use, demanding a high level of technical competence from the user. What are the reasons for this?

Research has demonstrated that much of this apparent difficulty can be attributed to a fundamental failing in the 'developer – user relationship' (see e.g. Markham and Rix, 1995). Software analysts and developers are unlikely to have an understanding of the business processes to be accommodated by the users and rarely will they understand the 'mental maps' users have of their systems or even acknowledge their levels of skills and motivations. From the developer's viewpoint users themselves have tended to be vague and unsystematic in articulating their requirements to developers, cocooned in the belief that rapid prototyping and 'waterfall' development methodologies would provide an opportunity to 'tune' the requirements. As a consequence developers have tended to concentrate on processes rather than tasks, on what users do and not how they do it – the end result being functionally rich, complex, process-oriented systems and not workflow-based task-oriented solutions.

This failure to incorporate the user as part of the design solution often leaves, at the GUI level, a system meeting a narrow set of developer-focused design parameters, i.e. a working system yet not fully effective at the operational level. This was less of an issue in the recent past when GIS was considered the domain of the specialist user and any working GIS was guaranteed to promote the user to conference speaker and committee delegate. The nineties ushered in the need for business pragmatism and it is no longer an acceptable business goal to invest large amounts of money in providing a couple of users with the facility to print off OS map extracts. To gain greater business acceptance and wider distribution in the transaction process-based application areas the design of GIS systems must quickly evolve to take greater cognisance of the usability issues. Quite simply, if GIS is to evolve it must become part of the integrated desktop. To achieve this it must evolve from being a technical discipline for specialists to become a general tool for all users. This will only be achieved through the provision of 'task-based' interfaces designed to support business processes.

the issues

The issue of systems interfaces is not one that must be addressed at the end of the application development life-cycle, but a problem of definition which should be integral to the initial problem analysis stage. Rarely at this stage is a description of the intended users included, with their skills and capabilities and even motivations documented. Without this baseline requirement GUI design is rarely produced within a framework which is directly traceable to the intended user. Typically, as a result, the user 'gets what he/she gets', functionally capable, operationally wanting. Although prototyping is often seen as a mechanism for getting the overall look of the system right it is often presented as a *fait accompli*, albeit allowing room for some minor iteration; a situation that owes more to the limited knowledge base of the programmer, and perhaps the time already invested, than a necessary limitation of the system.

simplicity and complexity: future interfaces for GIS

As GIS becomes more popular and the technology begins to find more applications, growing attention will need to be paid to the different types of 'GIS' that are gradually evolving and the requirements of the different end-users who need spatial data and information in the workplace. Different types of users with different requirements, will need different types of interfaces. Essentially these interfaces will be developed to support pre-defined tasks equivalent to business processes. A refinement on the task interface will be the need to support users migrating between tasks and having access to a hierarchy of task interfaces.

At present, although many GIS now operate within one or more of the industry standard windows interfaces, most have their own terminology, menus, screen layout, appearance, and so on. This is inevitable. Accepting that there are different types of GIS, which may range from the simple desktop mapping package to the multi-functional corporate system, to a great extent most of these systems are very similar, being based around a number of basic principles and functionality. Beyond this they all have their own little quirks, the result of the programming team, their perception of the market, the degree to which they have considered the interface, and so on.

However, as GIS has developed there is evidence that different types of interfaces are being developed to cater for specific applications in the marketplace. Whilst GIS has been and for that matter still is, used by a select few, the user base has grown considerably over the last few years. An increasing number of users wish to use spatial data in the form of information for decision making and planning (see for example Green and Kemp (1993)). These can be sophisticated systems which may require a complex interface, or simple systems, where the interface requirements may be basic. An example of this latter requirement are public information systems which must cater for a wide range of potential users who require information presented in a clear and unambiguous fashion. The time available to find and retrieve this information is usually short, and the environment in which it is retrieved is often 'disturbed' and 'open', which means that the interface must be simple to use and the instructions and pointers easy to assimilate rapidly.

A growing number of such systems are leaning towards the hypercard/hypertext type of interface with its simple point and click operation. With a minimum of instructions it is possible to provide the user with access to information without having to learn the system. This is of particular importance for systems where access is infrequent. Examples of the application of this concept in everyday use are those of EFT (cash) machines widely used by Banks and Building Societies and 'Pay Before You Exit' car parks. Meeting even a most basic requirement, such as paying for car-parking, can assume Herculean proportions when the system designers fail to design the interface from the perspective of the inexperienced or casual user.

As GIS develops, e.g. for public information systems, the design of interfaces will assume more importance. Although potentially less complex, public information access systems are equally as demanding in terms of the interface design aspects as for other systems required by other users because, as Shneiderman and993, p.219) notes, they have:

♦ little idea what to expect

♦ their information needs are poorly formed

♦ they feel inadequate if not able to use the system

The first or opening screen for any system is critical and will often act as the 'make or break' point. It needs to be both an 'attention getter' as well as being aesthetically pleasing, plus there is a need to provide instructions for getting started (Shneiderman, 1993). Potential system users also need to be able to find recognisable landmarks (reference points that are easily remembered/recalled in terms of the context) to allow them to move back to where they were if they make a mistake. In effect these act as a 'safety net'. For example, use of simple terminology and push-buttons e.g. RESTART, NEXT, BACK, FORWARD. This adds a certain amount of confidence to its use, a feeling of security and safety for the user, e.g. knowing that the user can't break the machine, whilst consistent design aids speed of comprehension. In other words the environment is established at the outset.

Figure 2
The Smallview
Interface

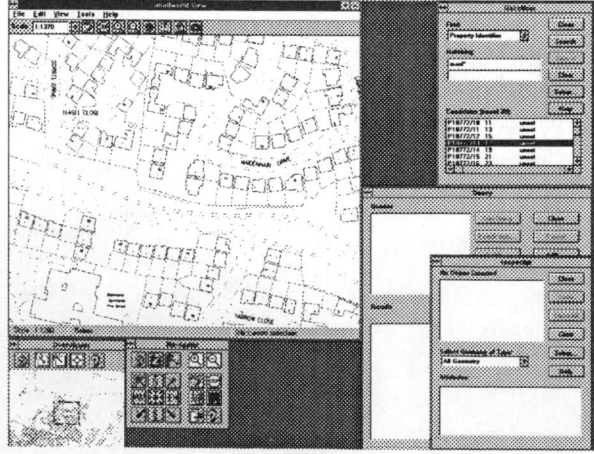

Figure 3
The ArcView 1
Interface

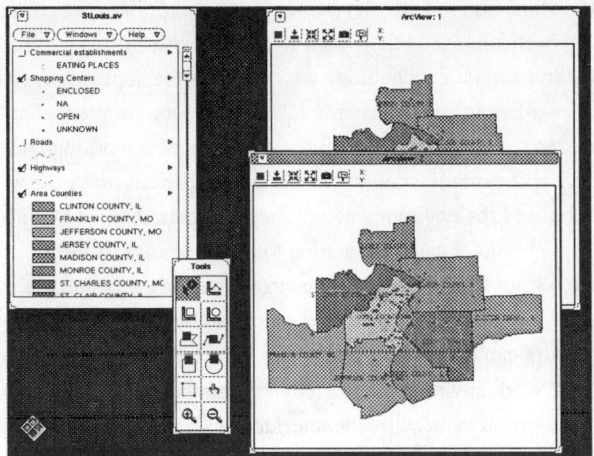

Figure 4
The COMPAS
Interface

Already in the case of GIS this type of user is being accommodated in the form of software package interfaces, e.g. Smallview (Figure 2) and ArcView (Figure 3). Others, e.g. Raal et al. (1995), have explored similar interfaces for specific applications such as marine and coastal information systems. Similarly the Hypercard-based COMPAS system (COMPAS,1992) (Figure 4), and a Hypercard-based system developed for the Norfolk and Suffolk Broads Authority by Gardener and Paul (1993) are other examples. On a larger scale the Louisiana Geological Survey has developed networked access to coastal data and information (LCGISN, 1994).

Access to information is a growing trend. There are many potential new roles for GIS, whereby spatial data is prepared and distributed in an easy to access format. Although this is not the only type of GIS, the need for interfaces that aid the user in a specific application area will grow in the next few years.

summary and conclusion

HCI are already well developed in the theoretical sense and to some extent in the practical or applied sense. But most researchers agree – whether from the IT side or GIS side – that a great deal more research is still needed to accommodate the rapidly evolving technology and the end-user. One approach to a solution is the need for improved communication links between the potential user and the designer which will lead to, for example, an interface that will adjust itself to the required task through the appropriateness of the interaction techniques with the task (Booth, 1989). Such application-based interfaces are now provided by some third party application developers and are becoming increasingly common. However, it is unlikely that the vendors will commit to this level of functionality while GIS is still viewed as an application in its own right.

Whilst the adoption of GIS technology is currently being driven by applications, the ultimate control of continued GIS uptake and subsequent use in the varied workplace will still be controlled by the availability of the hardware and software technology which acts as the application enabler. Although much potential may be seen well in advance it usually takes some time for the technology to develop in that direction, and for someone to see the potential. Interfaces provide a working framework for applications. Whilst different software may run within a similar basic environment with similar tools to permit the modification and use of the environment and specific tasks associated with the individual software, task functionality will require careful planning to facilitate ease of use for a wide range of users. As is already the case within MS-Windows the interface can be minimised or maximised, preferences can be enabled or disabled as required, Windows can be placed where they are considered to be convenient, screen colours can be selected, and so on – all of which enable the system to become a user-friendly and accessible work environment, thereby stripping away all the complexity of the task application. In many ways the problem faced by the interface designer is very similar to that of an abstract artist – the screen design/work of art must get to the essence of the user requirement/scene – and convey only the necessary information.

references

Apple Computer Inc, 1992. *Macintosh Human Interface Design Guidelines*. Addison-Wesley. New York

Baecker, R.M., and Buxton, W.A.S., 1987. *Human-Computer Interaction: A Multidisciplinary Approach*.

Bodker, S., 1990. *Through the Interface: A Human Activity Approach to User Interface Design*. Lawrence Erlbaum Associates: New Jersey. 169p. + Appendix.

Booth, P., 1989. *An Introduction to Human-Computer Interaction*. Lawrence Erlbaum Associates:London. 268p.

COMPAS, 1992. NOAA's Coastal Ocean Management, Planning and Assessment System. *User's Guide for the Texas Product*. U.S. Department of Commerce, NOAA. 6-3 Pages.

Davies, C., and Medyckyj-Scott, D., 1994. Introduction: The Importance of Human Factors. Chapter 20, in Hearnshaw, H.M., and Unwin, D.J., 1994 (eds.) *Visualisation in Geographical Information Systems*. John Wiley and Sons: London. 243p. pp. 189-192.

Dickinson, G.C., 1973. *Statistical Mapping and the Presentation of Statistics*. 2nd Edition. Edward Arnold, London. 196p.

Ellis,S., Kaiser, M., and Grunwald, A., 1993. *Pictorial Communication in Virtual and Real Environments*. Taylor and Francis, London. 615p.

Ellis, S., 1993. Prologue. In, Ellis,S., Kaiser, M., and Grunwald, A., 1993. *Pictorial Communication in Virtual and Real Environments*. Taylor and Francis, London. 615p. pp. 3-11.

Foley, J.D., Wallace, V.L., and Chan, P., 1984. *The Human Factors of Graphics Interaction Techniques*. IEEE CG & A. November 1984. pp. 13-48.

Galitz, W., 1989. *Handbook of Screen Format Design*. QED Information Sciences Inc. Wellesley, Mass. pp 173-177.

Gardener, L.A., and Paul, R.J., 1993. Developing a Hypertext Geographic Information System for the Norfolk and Suffolk Broads Authority. *Hypermedia*. Vol. 5(2):119-143.

Green, D. R., and Kemp, A., 1993. A GIS May be the Answer to Every Pipeline Manager's Dream. *Proceedings*. AGI93: 1.26.1-1.26.10.

Hewson, D., 1995. Better Handling, Better Help – Applications. *Windows 95 Supplement*. The Sunday Times/PC World. 27th August 1995. p. 18.

Jacobson, R., 1994. When Understanding Matters: The Virtual Interface and GIS. *Proceedings* of the Eighth Annual Symposium on GIS – GIS'94: So – Now What? – Decision Making with GIS – The Fourth Dimension. Vancouver, British Columbia, Canada. February 21-24 1994. Volume 1. pp. 9-10

Kacmar, C.J., 1993. Supporting Hypermedia Services in the User Interface. *Hypermedia*. Volume 5(2):85-101.

Kleiner, K., 1994. What a Tangled Web they Wove. *New Scientist*. 30th July 1994. No. 1936. pp. 35-39.

LCGISN, 1994. Louisiana Coastal GIS Network *Newsletter*. December 1994. Vol. 4(2).

Liebenau, J., and Backhouse, J., 1990. *Understanding Information: An Introduction*. Macmillan: London. 125p.

Lynch, P.J., 1995. *Interface Design in WWW Systems* – Yale C/AIM WWW Style Manual.

Markham, R., and Rix, D., 1995. GIS Ergonomics and the Problems of Interface Design, *Proceedings* Joint European Conference on Geographical Information, March 26-31, 1995. The Hague, Netherlands. pp 539-546.

Medyckyj-Scott, D., 1994. Visualisation and Human-Computer Interaction in GIS. Chapter 22 in Hearnshaw, H.M., and Unwin, D.J., 1994 (eds.) *Visualisation in Geographical Information Systems*. John Wiley and Sons:London. 243p. pp. 200-211.

Michels, S., 1995. Human Computer Interaction. *Thesis*. The Netherlands.

Patton, P., 1993, Making Metaphors: *User Interface Design*. ID 40 (2):62D66.

Quin, L., 1995. *Open Look GUI* FAQ 01/94: General.

Raal, P., Burns, M.E.R., and Davids, H., 1995. Beyond GIS: Decision Support for Coastal Development – A South African Example. *Proceedings of COASTGIS'95*. First International Symposium on GIS and Computer Mapping for Coastal Zone Management. University College, Cork, Ireland. February 3rd-5th 1995. pp. 273-282.

Rheingold, H., 1990. An Interview with Don Norman in Laurel, B (Ed) *The Art of Human Computer Interface Design*. Addison-Wesley. New York. pp. 5-10.

Shneiderman, B., 1993. (ed.) *Sparks of Innovation in Human-Computer Interaction*. Ablex Publishing Co. New Jersey.

the role of digital-orthophotographs in GIS

Richard Markham

Director, MVM Consultants plc.,
MVM Consultants plc, MVM House,
Oakfield Road, Clifton,
Bristol, BS8 2AL, England
Tel. (0117) 974 4477
Fax. (0117) 970 6897

abstract

With recent advances in computer hardware and software, digital orthophotographs now offer users of digital mapping and GIS a far more realistic view or visualisation of the landscape. This paper briefly defines a digital orthophotograph and outlines their history.

Various different applications are outlined, including map revision, thematic mapping, and Digital Terrain Models (DTMS). The paper concludes by examining a number of problems and issues that still need to be addressed in the context of digital orthophotography.

KEYWORDS: DIGITAL ORTHOPHOTOGRAPH; MAP REVISION; THEMATIC MAPPING; DIGITAL TERRAIN MODELS

introduction

For many users of digital mapping and GIS, systems remain difficult to use, not least because the underlying vector or raster map bases used by GIS systems are difficult for non-specialists to understand, and they require a great deal of interpretation. There is also the added complication that base mapping, by its very nature, will be out of date and is often subject to generalisation by the cartographer. Help is on the horizon in the form of digital orthophotographs, digital photographs that provide true colour imagery of the landscape.

the digital orthophotograph

An 'orthophotograph' is a photographic image that has been differentially rectified (i.e. 'adjusted' to several thousand control points produced from a DTM) to remove any distortion due to recording geometry (position and tilt). and relief displacement. Digital ortho-images can easily be manipulated and their accuracy, in particular relative accuracy, can be very high. The high degree of correction applied to an ortho-image enables it to be directly superimposed onto a map to accurately delineate the boundaries of observable topographic features. An orthophoto can therefore be used as the backdrop to a vector map digitised from conventional analogue maps or can be used as a 'map' itself against which application data can be overlaid.

Digital orthophotographs, as geo-coded images, provide significant advantages over their analogue image and line mapping counterparts, as they present to the user a more accessible and more accurate representation of the real world than previously available. In particular the greater similarity of the screen image with the landscape improves readability for the user by improving orientation (ie the user can quickly recognise obvious features, landmarks etc.) and improves understanding by reducing the level of interpreted content. For the more specialist user, ortho-images provide a whole range of improvements on their analogue counterparts, for example they provide multiscale, temporal and spectral properties.

Whilst the potential of digital ortho-imagery has been promoted since the mid-1970s to date it has failed to have a significant impact in the GIS or digital mapping marketplace. In part this was due to the lack of powerful computers with sufficient resources to hold, manage and manipulate the images; in part the high cost of acquisition has meant a only a select client base could afford it; until recently the availability of ortho-maps was limited to monochrome images which were not seen as having significant advantages over raster copies of standard analogue mapsheets; and finally, because of the lack of flexible applications software necessary to take the ortho-image data into real-world applications systems, e.g. GIS systems were not capable of integrating the images together with other vector based application data.

In recent years, however, hardware and software technologies have developed to a point where the management of large libraries of ortho-images have become a practical proposition. However, it has to be acknowledged that potential problems still exist to the extent that the cost of output devices necessary to match screen resolutions is high, and orthophotos create very large data files which have implications for processing capacity, storage and archiving. From the applications software perspective the arrival of 'hybrid functionalities' in GIS for handling both vector and raster data together, and in volume, means that the opportunity for integrating orthophotos into GIS has finally become a reality. In addition, the availability of image scanners, high performance computers and peripheral devices (plotters) have led to higher production speeds, improved image quality, high accuracy and stability. In commercial terms this translates into better and broader accessibility at lower costs. The result is an increasing interest in replacing both analogue ortho-images and digital line mapping with digital ortho-photography.

orthophotos - the background map for visualisation

Whilst many people see the benefits to thematic mapping as being the greatest impact orthophotos will make to GIS, for most users (and in particular the army of future users) the number one priority will be the ability to relate what is seen on the computer screen with the mental map they have of the real world. Traditional map-based systems, depending as they do on products derived from basic line mapping, are heavily dependent on cartographic symbolisation and generalisation to convey the identity of feature type, patterns and texture. This requires the user to have a knowledge and appreciation of the symbol conventions used and an ability to orientate the requirement against these conventions. Digital orthophotomaps make the need to employ complex cartographic symbolisation as an aid to interpretation redundant It is in using the orthophoto map as background for visualisation purposes that users of GIS systems will derive the greatest benefit.

Recent experience has demonstrated that the primary requirement of the average GIS end user is not for complex spatial analysis functionality, but for simple locational referencing. i.e. the requirement is to establish the location of objects and features and the locational relationship between objects and features e.g. locate a house in relation to a utility pipe, a factory in relation to a highway and so on.

Where previous base mapping is not already available digital ortho-images provide an efficient and economical mechanism for establishing the all-important reference base. However, to be useful to the end-user in its digital form, processing systems utilising the ortho-image data must provide for quick display, efficient and simple data handling, geocoded linkages to appropriate database references, an ability to merge with standard raster or vector mapping, to access descriptive attributes and simple image enhancement functions.

data capture and map updating (revision)

One of the perennial problems for users of GIS and digital mapping systems is to ensure that the base mapping data is available, current and complete. Unfortunately in many countries (including the UK) base mapping data - particularly that in the rural areas- is either missing all together or of questionable quality. Organisations reliant on this data are either therefore heavily dependent on having their own external data sources to supplement the shortfall, or are forced to undertake, often in a piecemeal fashion, field surveys or traditional aerial photography based analogue mapping. Because of their geometric properties, digital orthophotography is a cheap and efficient way of undertaking data capturing and updating (revision) of vector mapping, In addition it can be used as a direct substitute where raster backdrops are the norm.

For data capture, geometric information can be measured while non-geometric information, eg number of highway lanes can be added directly to a database. Orthophotos can be used for deriving original databases of natural features (water courses, woodland etc.) and cultural features (manhole covers, lamp stands etc.). Where data exists, orthophotos can also be used for quality control and verification of other data capture methods, as a control for completeness and for the control of changes.

thematic mapping

Digital ortho-photomaps present the world as it is with its lack of continuity, uniformity and texture. In the world of thematic mapping the previous digital processes involved accessing and merging data from different sources, and overlaying it on traditional vector or raster based line mapping. The process is often made unnecessarily complicated by the inaccuracy inherent in the way classes of observed phenomena are defined. For example, forests, soils and even land parcels always seems to adjoin each other with an unnatural uniformity that derives from the need to digitise sharp boundaries. Such activity, even with user induced checks and balances, still leave the user's ability to model the real world severely compromised.

Ortho-photos inject realism into the thematic mapping activity by improving accuracy of content and context, thereby both extending and enhancing the analytic possibilities. When added to appropriate image processing technology it can be used to support faster and more accurate semi-automated digitisation of features.

variable scale mapping

Producing digital ortho-photos for mapping at large scales (1:500-1:2500) also provides the opportunity to use the same digital information to generate a wide range of mapping products at smaller scales. It is not untypical for an organisation to require mapping at 1:5000 and 1:10,000 scale for strategic planning purposes whilst at the same time needing 1:500 for detail site planning. The ability to derive multi-scale mapping from a single source of photography is not only cost effective, but also ensures commonality and referential integrity between any derived datasets, all of which is of high value if GIS is to be applied corporately in potential user organisations. Adopting the digital approach also helps to reduce data redundancy.

digital terrain models (dtms)

The DTM is produced as part of the process of correcting the image distortions caused by relief displacement. Ortho-images do not provide full 3D information (due to the lack of pixel information on vertical surfaces). Therefore 3D objects (buildings, trees etc.) that exist on the terrain are distorted but can they can provide the user with valuable context within GIS applications by providing a wider range of perspectives on the topographic surface than can be achieved simply from viewing a 2D map. The potential for draping the ortho-image over the DTM thereby creating textured 3D views of the landscape can be particularly valuable in applications requiring animation and simulation; for example, flood risk assessment, highway planning and urban development schemes. As an added development stereo-orthophotos can be used as the basis of fully fledged 3D applications.

other possibilities

The ability to integrate digital ortho-images into Geographic Information Systems (GIS) offers significant potential beyond being used as a presentational backdrop for data or digitisation. The list of potential applications to which digital ortho-images can be put is virtually limitless. Any application in which detail of the topography is an issue then digital ortho-images have a role to play.

Application opportunities identified include improved Data Integration where, in addition to providing a constant and consistent validation of spatial data displayed against it, ortho-images can provide a stable base for registering other image data, e.g. SPOT and Landsat TM images.

As coloured digital orthophotos are multi-scale, -spectral and -temporal they can be treated like other remote sensing data. A second area is in change detection applications. Since DTMs remain relatively stable, creation of new ortho-images and comparison with older ones can be easily acheived.

Areas of change can be readily identified both from the images and in combination with other GIS information, leading to potential enhancements in 'What if' applications and simulations. Digital ortho-imagery also has the potential for an important role in the following:

♦ Compilation of Area, Regional and National Land Bases

♦ Update and revision and quality control of Area, Regional or National Land bases

♦ Inputting as the presentational backdrop for other data

♦ Data Capture of thematic and cultural data

♦ Creation of photomontages employing ortho-imagery and third party visualisation tools

integrating digital orthophotos into GIS – issues still to be addressed

Although much of the technology exists today to enable end users to take Digital Ortho-images into GIS systems, there are still a number of issues outstanding that need resolution before digital ortho-images become generally accepted as fundamental components of the GIS solutions.

Standards

The GIS industry is belatedly moving to accepting adherence to standards as an integral element in achieving sustainable solutions. At the international level the OPEN GIS foundation in the USA is bringing together leading suppliers to agree on transfer formats and integration tools that will allow free transfer of data between systems and the co-existence of multiple system architectures, whilst in the UK and in Europe great emphasis is being placed on data exchange standards at the national level. In the area of ortho-imaging a range of different raster formatting standards exist including TIFF, GT., LRD, SUNRAS, RLC and JPEG. It is important when selecting a GIS to establish that the utilities for importing the required formats, or for converting to the required formats, are available.

GIS functionality

Until recently much of the software used for managing, displaying and manipulating digital-ortho-images was to be found in dedicated software modules that existed outside of standard GIS products. This, however, is becoming less of an issue as the functionality within hybrid (raster- and vector-based) GIS systems becomes available. It is also the case that many of the specialist suppliers have built translators and exchange formats to allow their software to co-exist with the mainstream GIS products. This allows for specialist tasks such as raster editing and feature extraction etc. to be completed in the appropriate dedicated environment whilst the outputs are freely available for distribution within the GIS.

To take full advantage of ortho-imagery GIS systems must provide functionality to handle both vector and raster data. Within this context 'handling' refers to the capture, management, display and analysis of both data types concurrently. Minimum functionality required includes quick display and refresh rates, appropriate data structures for efficient data handling and simple image enhancement. For the more sophisticated applications higher level image processing tools are required including raster editing, neighbourhood operations, map compose, semi-automatic image analysis and feature extraction.

Hardware issues

The underlying hardware configurations currently used to promote digital ortho-images tend to be based around high powered CPUs, large memory requirements, dedicated graphics accelerators and with high resolution monitors. The cost of these devices (£20-30k) has to date limited the potential growth in the market for digital orthophotography to those organisations with very specific and specialist requirements and those who are able to get a positive result to the cost-benefit equation. But what of the future?

The wider acceptance of digital orthophotomaps as a data source within GIS systems will remain, for some time, highly dependent on the availability and cost of processing units. The market for GIS is still predominantly workstation based although the clear trend is toward the desktop PCs, reflecting the user community's desire to maximise the value of GIS investments by spreading access to all aspects of an organisation's activity. Falling prices and the availability of new generations of processing power (DEC Alpha, Intel Pentium and IBM/Apple Power PC), suggest distributed access to digital orthophoto databases from the desktop is now a realistic expectation, therefore bringing it within reach of a wider user community.

Managing the data

Digital orthophotos are generated from high precision scanning of analogue images and are usually supplied to a user via computer disk or tape. The scale of the original photography will determine the level of detail in the image and the resolution of the scanning employed will determine the overall size of the output files that need to be imported to and managed by the computer system. To give some orders of magnitude 1: 6000 scale photography generating a 1:1250 mapping and scanned at 22.5 microns (c1200 dpi) will generate an uncompressed digital map file of around 50Mb, at 7.5 microns (3600 dpi) the map file grows to around 450Mb. Files of this magnitude, even when compressed at ratios of 10:1 are still difficult to manage on-line in computer systems. Handling such large volumes requires not only appropriately designed data structures but also an ability to use the latest image compression technologies.

The recent acceptance of the JPEG standard and the availability of commercial compression/decompression boards will in time reduce the extent of this problem.

The problem of erroneous data

One of the potential difficulties with using digital ortho-photomaps is the presence of erroneous or inconsistent information in the photography. Such imperfections can be characterised by:

1 The presence of people, vehicles, events or other foreign objects captured by the photography. Whilst it is not necessarily a serious problem there are occasions where the presence of such foreign bodies either detracts from the overall context of the scene or in fact obscures the important information e.g. a row of cars can obscure the location of manhole covers, roadside gullies etc. Such problems can only effectively be resolved by a combination of site survey editing of the raster image which is a time consuming and potentially expensive solution.

2 Colour discontinuities that occur when the photography has taken place under different weather conditions, vegetation phenology or lighting conditions. Resolving these issues have become less of an issue with the availability of radiometric calibration algorithms. Such post processing may, however, add significantly to the cost of data acquisition.

conclusions

Digital ortho-photography has a significant potential to improve the overall quality and accessibility of data held within GIS, particularly now that colour imagery can be made available easily at price levels acceptable to a broad market, and now that image data can be integrated into a variety of GIS systems for display and manipulation. There is a clear, if not well articulated, demand for digital ortho-image maps to be used both as an adjunct to existing traditional line mapping and as a direct replacement for traditional topographic maps. This demand will, however, (despite the marketing efforts of the data suppliers and the availability of shrink wrapped datasets such as the 'Cities Revealed' series of aerial photo databases) only translate into firm sales when the user community can be convinced that the demonstration capabilities can be delivered into real world situations, (i.e. that systems can hold and manage digital ortho-imagery covering large areas) and be integrated into real world business solutions. To satisfy demand in this area, there is a requirement to develop application based software to incorporate the imagery into GIS systems in general and into individual business applications. In practice this means system designers establishing the most effective data structures for storing the image data, and amending graphic user interfaces used within applications to access and exploit the data.

Once this has been achieved the full potential of digital orthophotography as a primary source of information for the GIS User will finally be realised.

references

Balsavias, E, (1993). Integration of ortho-images in GIS. In: Fritsch, D., Hobbie, D. (eds): *Photogrammetric Week '93*: pp. 261-272; Stuttgart, Wichmann: Karlsruhe.

Cadoux-Hudson, J (1993). Imagery in GIS , in Editors Shand P & Ireland, P.J., *The 1994 European GIS Yearbook*, pp. 143-146, NCC Blackwell and Hastings Hilton Publishers Ltd.

Grenzdorffer G., and Bill, R. (1994). Digital Orthophotos for Mapping and Interpretation in Hybrid GIS Environments EGIS/Mari '94, *Proceedings of the Fifth European Conference on GIS*: pp. 1845- 1856, EGIS Foundation, Utrecht, The Netherlands.

Griffin W and Sedgwick, F. (1995). Digital Orthophotos: The basis for cost sharing. *Earth Observation Magazine*. March 1995, pp. 40-43 Littleton Colorado.

the integration and interoperability challenge – the need for OpenGIS

John Glover

Intergraph UK, Delta Business Park,
Great Western Way, Swindon, Wiltshire,
SN5 7XP.
Tel. (01793) 619999
Fax. (01793) 618508

abstract

This paper discusses some of the issues now challenging the GIS community. With the increasing need to share maturing data sets in a dynamic business environment, it is important to understand the difference between data transfer and interoperability. The paper shows that with the emergence of new object-based technologies, GIS databases and software can be fully integrated into traditional business applications. By introducing the efforts of the OpenGIS Consortium and their initiative to develop an Open Geodata Interoperability Specification, it is hoped that discussion in the United Kingdom and Ireland will be stimulated around the need for OpenGIS.

KEYWORDS: OPENGIS; DATA TRANSFER; INTEROPERABILITY; OBJECT LINKING AND EMBEDDING

introduction – the enterprise today

In a time when organisations are having to concentrate on their core business objectives, the acceptance of GIS into the IT framework should greatly assist in the integration of heterogeneous workflows and databases. However, due to the dichotomy of needs, it is often found that the user's demands for focused applications can frequently be in conflict with the organisations demand for integration and compatibility. To safeguard the proposed investment, much time and effort is spent developing site specific IT strategies and defining the standards that need to be adopted. As technologies advance this process has to be revisited and updated.

GIS is recognised by most to be a powerful tool that can integrate diverse data themes through its primary function of providing a spatial index to which these are referenced. This in itself can adversely affect the implementation of a GIS as many political and technical boundaries have to be crossed. Occasionally users will rename a project, e.g. 'Digital Mapping', to avoid issues and conflicts with the 'GIS Working Team', who may be bogged down investigating the quagmire of vendor offerings. This is often done for the very innocent reason of achieving a specific project goal, whose time frames are tight and focused by commercial objectives.

From this several negative results may occur:

♦ The project is stopped due to ongoing investigations

♦ A niche product is purchased and the enterprise investment in data is jeopardised by the possibility of 'locked' proprietary formats and data structures

♦ The project is expanded beyond the initial scope to provide a safety net for future projects therefore providing a complex solution to what could be a simple problem

To solve a particular business need several hardware and software modules will be required to interface and exchange data. There is a tendency for each application not to rely on other third party modules and to include in its environment proprietary functions such as windowing, database interfaces, plotting and programming tools. This produces monolithic applications, that duplicate functionality without offering interoperability, and causes major restrictions to efficient product maintenance and enhancement. In the event that agreement on interoperability with a third party application is resolved, individual product enhancement and version compatibility become an expensive headache for both vendor and customer.

Due to the technical nature of GIS applications they become isolated from the mainstream IT and business functions and appear to be 'Islands of Automation'. With emerging technologies, such as Microsoft's Windows operating systems, it can now be seen that technical and business applications exist within a single environment, in which the possibility of integration and interoperability via standards such as OLE (Object Linking and Embedding) and COM (Common Object Model), have become a reality.

the vision

Ideally, heterogeneous workflows should be constructed component-by-component to accomplish an objective, without committing to oversized expensive systems. In order to do this, the interfaces to these components would have to be published so that a user could purchase them as 'Plug-and-Play' modules. This would allow total freedom to acquire the most suitable software, instead of becoming locked into a specific vendor computing environment. In a similar way, the user would be able to swap-out components as better, more cost effective, solutions became available. Commercially, vendors will have to provide quality components at very competitive prices.

There are two fundamental types of interoperability mechanisms. The first set is based on open data (formats, translation standards, data exchange and data models). The second set is based on open process (service requests, query mechanisms such as SQL, and data model mapping). If these are in place then OpenGIS applications would be able to transparently share data. This would provide for the integration of distributed heterogeneous databases from organisations such as the Ordnance Survey, the Land Registries, Environment Agencies and Local Government, without the need for awkward and slow translation. In accessing this data it would be essential to maintain transparency of location, data structure and source software system.

For example, a user should be able to locate a parcel boundary on their GIS application and then, via the network, reference required themes, such as base mapping, surface geology, environmental constraints, flood hazards, land ownership, planning zones and transportation routes. In order that this can be achieved, it will be necessary to provide the user with an intelligent context sensitive search facility to browse related available data sets both geographic and Multimedia. A standardised 'look-and-feel' interface, such as MS Windows, would be essential to minimise the skill level required.

With the identification of 'ownership' and improved communications networks on the 'Information Super Highway' it would be possible to remove costly data duplication. In order that the decision process can be assisted it will be important for the data to be accompanied by quality and integrity assurances. By implementing this level of integration and interoperability it will be possible to fuse GIS thoroughly into business applications and widen the use and potential of geographic information. This evolution of GIS applications will only be possible if the investment in legacy data is protected and the data providers and software suppliers fully commit themselves to the vision of OpenGIS.

so what is the problem?

The GIS industry has now matured to such a degree that comprehensive datasets exist. For example, the completion of the Ordnance Survey of Great Britain's large scale digital database and the many utility companies that have become members of the well deserved '100% club' (data capture completed). The largest investment these organisations have, is not hardware or software, but the data itself.

For most organisations, departments and individuals the initial scope of their project was not to share data but to manage the assets they owned. However, most now realise the need to exchange GIS data with others as a service, whether it be at commercial rates or as part of inter-organisation co-operation. The problem is, there is no such thing as 'GIS data', it is Intergraph data, ARC/INFO data or some other vendor specific binary format and often users feel they have created 'islands of information'. This presents the users with the problem of importing and exporting data (if allowed). During this process, careful attention needs to be paid to the different data structures and schemas of the host and client systems in order that this data is correctly represented in its' destination environment.

It is difficult to see from this how dynamic data exchange is possible whilst the barrier created by proprietary data formats exist. Opening up vendor formats is the consensus process leading to the development of a common data model. Intergraph, along with some other GIS vendors such as Genasys, have recognised the critical need to do this. Others need to follow suit and make their formats public domain so that OpenGIS can be more than a dream and GIS data sharing a reality.

There are many other details such as data accuracy, resolution, quality, precision and age that need to be transferred. Also required are explanatory notes regarding the meaning of the attributes and how the data should be interpreted. Usability and reusability can often be restricted by copyright issues that, unless relaxed or resolved, will hamper a wider use of geographic data.

Figure 1
Islands of
Automation

the need for standards?

When considering the issue of open data we need to investigate data formats, translations, exchange and transfer standards. GIS data formats, within a particular vendor's system, may be open to processing by the various modules that the vendor supplies, but due to the proprietary nature of the format no other process is allowed direct access. Processes that enable file-to-file translation allow for data to be openly moved between a maze of formats. To preserve information as it progresses in a particular workflow, it is necessary to ensure that either a common data model or file-to-file mapping is used. In order that GIS applications can dynamically exchange data, a vast number of translators are required (both directions for each pair of systems i.e. ARC/MGE, MGE/ARC). These can only exist if the vendors share information on internal binary structures.

In order to simplify this, the user community is striving to achieve 'openness' through sets of official standards rather than using the de facto formats of particular vendors. Standards such as NTF (National Transfer Format) in the U.K. and SDTS (Spatial Data Transfer Standard) in the U.S. allows a single agreed format to be used. To avoid data loss, these formats must be able to hold all the types of information from each of the individual GIS formats. Using these standards it is often required to translate in batch due to the granularity of data unit being transferred. This does not allow for dynamic data exchange and requires a degree of version control. The fundamental differences between data transfer and interoperability are summarised in Table 1.

	Transfer	**Interoperability**
Scope	data, no process	data and process
Data Unit	data set	object (data set or lower)
Communication	blind (1-way)	negotiated (2-way)
Integration	in target system	in server or during communication

Transfer Standards are important, but the needs of integration and interoperability require a greater degree of openness. Robin G.Fegeas of the U.S. Geological Survey states (Fegeas, 1995):

'If we can build on the past and meet the new challenges – with patience, for there is considerable work ahead – we can realise our goal of a spatial data infrastructure based on geodata interoperability and transfer standards.'

emerging technologies and open systems

To allow for open interoperability and integration of both data and processes, new industry standards are emerging for object technologies: CORBA (Common Object Request Broker Architecture) as specified by the OMG (Object Management Group) and OLE (Object Linking and Embedding) developed by Microsoft; OLE, familiar to most Windows users, is the technique of working within a compound document centric environment, where the data being manipulated may be held in another server document. In this way updates are dynamically processed through to the client document and the required changes are made. (Figure 2).

Figure 2
Compound document

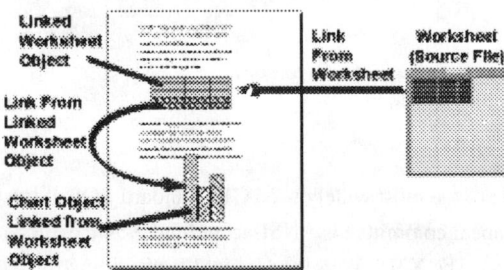

Recently announced by Microsoft was the provision of 3D OLE extensions for Design and Modelling Applications that will allow 'Plug-and-Play' of technical applications such as CAD and GIS, to work within Microsoft's Windows operating system. If applications are OLE enabled, the building of business applications integrating GIS technologies becomes a reality.

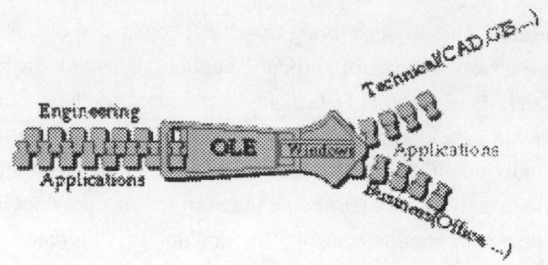

For example, an engineer wants to compile a report of new road scheme. First he generates the body of the text, some of which is filled in from an attribute database. The compensation figures are shown in a financial spreadsheet table that is linked to a chart showing land use percentages. With the new extensions, the engineer can now integrate a plan of the route from the GIS showing the thematic display of land to be acquired. If the map is linked, the engineer would be able to click on it and zoom in and out and enquire on the database using the GIS toolbar that is dynamically loaded. If GIS vendors open up their data structures by providing OLE data servers, this engineer would have the means to reference other geographic themes, such as surface geology, from data suppliers on the 'Information Super Highway'.

OLE also includes a protocol based programming language called COM (Common Object Model). The use of COM components allows the building of systems with 'Best-of-Breed' modules. This is the same approach as buying a Hi-Fi system. You would investigate the best unit for each function, usually from different vendors, with the understanding that if a better unit becomes available you can append or exchange.

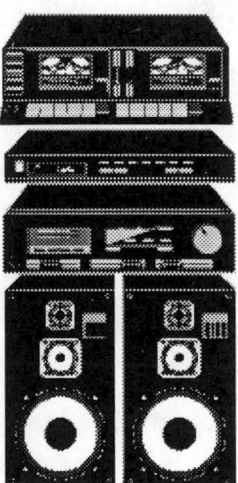

Improvements are also being made to the most widely used GIS standard SQL (Structured Query Language). The industry GIS technical committee is ANSI-accredited X3L1, which is a subcommittee under X3 (Information Technology). The X3L1 Working Group 2 is providing the technical support to the development of GIS extensions to SQL with Multi-Media and known as SQL3/MM. This is important to the GIS community, as finally the building of a unified distributed database that can handle multidimensional and spatiotemporal data sets will allow seamless data integration. This standard is due for publication in 1996, and database vendors are already starting to develop and test the server systems, which are themselves, object enabled.

With improved communications networks, object applications, object operating systems, object data servers and faster and cheaper hardware technologies the vision discussed is closer than you may think! It is now essential that users, vendors, data suppliers and system integrators come together to break down the barriers to OpenGIS.

the OpenGIS consortium (OGC)

The OGC is a recent initiative, designed to organise resources and GIS community support for the development of OGIS (Open Geodata Interoperability Specification). OGC fully supports the concept of OpenGIS that can be defined as:

'the full and free integration of spatial data and geoprocessing resources in the distributed environment of the national and world-wide geoprocessing-geodata community.'

The following statements can be found via the OGC Home Page on the World Wide Web (WWW) (OpenGIS Consortium, 1995):

'OGC was initially founded to create interoperability specifications in response to wide-spread recognition of the following problematical conditions in the geoprocessing and geographic information community:

♦ The multiplicity of geodata formats and data structures, often proprietary, that prevent interoperability and thus limit commercial opportunity and government effectiveness

♦ The need to co-ordinate activities of public and private sectors in producing standardised approaches to specifying geoprocessing requirements for public sector procurements

♦ The need to create greater public access to public geospatial data sources

♦ The need to preserve the value of legacy GIS systems and legacy geodata

♦ The need to incorporate geoprocessing and geographic information resources in the framework of national information infrastructure initiatives

♦ The need to synchronise geoprocessing technology with emerging Information Technology standards based on open system and distributed processing concepts

♦ The need to involve international corporations in the development and communication of geoprocessing standards activity, particularly in the areas of infrastructure architecture and interoperability, in order to promote the integration of resources in the context of global information infrastructure initiatives'

In order that this can be achieved:

'OGC has enlisted the active involvement of official representatives from more than thirty organisations, including: GIS vendors, such as Intergraph, Genasys II, Graphic Data Systems (GDS), MapInfo, PCI, and ESRI; computer vendors such as Sun Microsystems and Hewlett Packard; and federal agencies, such as ARPA, the Defence Mapping Agency, the COE, USDA Natural Resources Conservation Service, U.S. Geological Survey, NASA, and NOAA. Universities, including the University of California at Berkeley, UCLA, Rutgers, and the University of Arkansas are supporting the project with considerable research activity. The Object Management Group (OMG), the pivotal consortium managing development of object technology standards, is also a member of OGC.'

'The Executive Director of the OGIS Project is appointed by the OpenGIS Consortium board to manage the OGIS development process. Additional staff resources include a Technical Director, and Directors for Technical Programs, Research, and Communications. Within this framework the OGIS Project is organised as a Management Committee, a Technical Committee and the OGIS Testbed. The membership fee structure distinguishes between the four following classes of members: Principal Members belonging to the Management Committee; Technical Committee Members; Testbed Members; and Associate Members, non-voting participants, both commercial and academic, who may attend meetings and subscribe to the project's information sources.'

To make sure the results from the OGIS Project are not purely academic, there are a series of Testbed projects designed to refine the specification through prototype implementations. Each Testbed will have specific objectives controlled by specific mechanisms to make sure the OGIS Project is productive.

In order to assist the OpenGIS Vision the OGIS Architecture will provide:

- A single 'universal' spatio-temporal data and process model that will cover all existing and potential spatio-temporal applications

- A specification for each of the major database languages to implement the OGIS data model

- A specification for each of the major distributed computing environments to implement the OGIS process model

By providing the above interoperability standards, the OGIS will provide the means for creating:

- An interoperable application environment consisting of a configurable user workbench supplying the specific tools and data necessary to solve a problem

- A shared data space and a generic data model supporting a variety of analytical and cartographic applications

- An interoperable resource browser to explore and access information and analytical resources available on a network'

The final OGIS specification is expected to be passed on to the ANSI and ISO bodies by September 1996.

OGC corporate contact information

Membership Inquiries: dschell@ogis.org

Technical Questions: kurt@ogis.org

Information Requests/Publications: lmckee@ogis.org

will OGIS be the future of GIS?

It was intended that this paper should help stimulate discussion in the United Kingdom as to the future of OpenGIS. It is hoped that via the AGI (Association for Geographic Information) the GIS community will show enthusiasm for OpenGIS by establishing links with the OGC. Much interest has already been generated in the United States and in Europe via EUROGI, championed by their President Michael Brand.

The Editor of Geo Info Systems in January 1995 wrote (Editor, 1995):

'We believe that OGC is doing important work in fostering the development of the Open Geodata Interoperability Specification (OGIS) which in the most simplistic terms will allow GIS users to easily access a variety of geoprocessing tools and data sources – and raising the level of the discussion about the concept of open geographic information systems.'

It is recognised there are some critics that feel the level of hype is exceeding the results achieved and do not share in this vision of openness. They are concerned about the impact of OGIS as stated by the Director, Fred Limp, of the Centre for Advanced Spatial Technologies and National Centre for Resource Innovations (Limp, 1995):

'Because OGIS will eliminate competitive advantage based on proprietary data formats, GIS vendors will need to become more competitive in providing processing capabilities and ease-of-use, which means that users will spend less money for more functionality.'

However, by unlocking the current and future reserves of geographic data the GIS market place will grow exponentially as more users access geodata. By using components we will be able to integrate GIS into specialist business applications. In this way the Integration and Interoperability challenge will be satisfied and we will all share the benefits of an OpenGIS.

references

Editor, 1995. OpenGIS. *Geo Info Systems*. January 1995.

Fegeass, R.G., 1995. OGIS - Building on SDTS. *Geo Info Systems*. January 1995.

Herring, J.R. What is an OpenGIS Application? *Intergraph Open GIS Pack*.

Limp. F., 1995. State and Local Governments - Key Beneficiaries of Open GI. *Geo Info Systems*. January 1995.

OpenGIS Consortium. *World Wide Web Page*. http://ogc.igis.org

mapping the information superhighway

Jim Crowder

Digital Equipment Corporation,
Digital Park, Imperial Way,
Worton Grange, Reading, RG2 0TE,
England.
Tel. (01734) 203470
Fax. (01734) 868711

abstract

The availability of computer-based information over networks in recent years has led to two significant problems. The first is the question of freedom of information and open government. The second is the one of charging for data. With the rapidly increasing use of the Internet and WWW browsers by the general public to access many different types and sources of information such questions are becoming increasingly important. In the context of GIS this technology is also providing a unique opportunity to widen the appeal of GIS and to show the benefits to a broader community. The cost is low and the technology is already widely available. This chapter describes some of the possibilities of using GIS on public computer networks, such as the Internet and JANET, in order to share information and allow access to public resources.

KEYWORDS: GIS; COMPUTER NETWORK; INTERNET; WORLD WIDE WEB; COPYRIGHT; DATA TRANSFER STANDARDS; SHARING INFORMATION

introduction

GIS and the data it uses are inextricably linked to computer networks. It is a rare GIS that is built on a single database containing all the data required for the system to be useful and it is a rarer one that only allows a single user on a single computer. We are all familiar with the use of GIS on computer networks to access and share data albeit within a single company or organisation. This chapter describes some of the possibilities of using GIS on public computer networks, such as the Internet and JANET, in order to share information and allow access to public resources.

internet history

In a few years time, we will be able to look back on the late 1990s as the time when huge amounts of information became available to everyone with access to a computer. With the advent of the World Wide Web (WWW) as an available technology, anyone with a PC and a modem is able to 'surf the web' and browse the vast amounts of information available there. This has led to a re-defining of both country borders and ideas on freedom of information, since information not published in one country, for example data on the safety of certain medicines, may be available in another, and can be accessed by the WWW. In our arena, this might also include maps that are classified in the country of origin, but can be copied from a WWW site in another country perfectly legally. In fact with computer networks, there is no need even to know where the host computer is.

The Internet was originally started in the 1960s, as a computer to link US military and research establishments, to allow effective collaborative research without spending large amounts of money on travel. Ten years ago, it is estimated that there were 5,000 Internet users world-wide; there are now believed to be about 30 million. In the absence of established networking standards in the early 1970s, the Internet Protocol (IP) was defined from previous military standards for use on the Internet. This became the TCP/IP that we have grown to know and love with its domains and naming conventions.

With the growth of UNIX systems, and since TCP/IP is the standard networking solution for UNIX, the use of the Internet spread across the world, and more people became exposed to its hostility towards users. The WWW was originally created at the CERN research laboratory in Geneva as a way of organising and retrieving data, and as a part of this, more user-friendly tools were designed to allow people to access the data that had been made publicly available on various computers on the Internet.

Around this time, the name changed from the Internet, with its connotations of being difficult to use, to the World Wide Web, which sounds much more accessible. These user-friendly tools are now available for most types of computer, from PCs and workstations to text terminals such as VT220s and IBM terminals at minimal cost. Also, the well-known software companies such as Microsoft and Adobe provide authoring tools for people to make information, including graphics, available on the WWW (the tools themselves are also published on the WWW).

access to information

This availability of information leads to two significant problems. The first is the one mentioned above, the question of freedom of information and open government. The second is the one of charging for data since not only is information available for browsing freely, there is also an opportunity to provide subscription and credit services for more commercially valuable datasets that might have taken considerable cost and resources to build. Current technology does not yet support charging for transactions on the Internet, although it is currently a major issue being investigated by a number of commercial companies, the most well known of which is Barclaycard.

Barclaycard have recently launched a home shopping service on the WWW, with support from other well-known high street stores. They are investigating how people might pay for purchases they make using their credit card number without compromising security, as there are few agreed standards on data security on such a public network. Their current conclusions are that, in general, since people are used to giving out their credit card number on the telephone, then the Internet should offer a similar level of security or better, and will be acceptable. Only time will tell if this is considered acceptable, although security methods are rapidly becoming more effective. This same approach can be extended to map data, although generally users of such data will be known to the data provider and then standard credit arrangements could be applied.

One of the major problems when buying commercially available data such as maps or satellite images are its suitability. Maps can be out of date very quickly, and while the Ordnance Survey is making every effort to keep their available data current, there are times when this might not be sufficient. Also, the exact land coverage of a satellite image, or the extent to which it is obscured by cloud cover, is not easily described in a catalogue. For a potential purchaser of such data there are significant benefits in viewing the data from some form of catalogue in low resolution, before buying. The Internet can provide this facility.

A few years ago, Digital Equipment took part in a pilot project in the US to allow unbundled Landsat Thematic Mapper (TM) data to be previewed at low resolution by PCs using standard telephone lines to connect to a host system. These images could then be ordered on-line or, if your credit was good and the data was required immediately, could be down-line loaded across the network. Users found this a useful service, and the data suppliers gained significant benefits from the reduction in the number of images that were returned as being unsuitable. This service has now been extended to the Internet through the WWW, and users can search and preview the images using standard web browsing software, and then order the images they want, again either to be copied across the Internet, or posted in the traditional way.

In the USA recently, a number of federal and state agencies have set up clearing houses on the Internet to supply geospatial data and metadata. Those existing at the time of writing include the National Wetlands Inventory, The State of Montana and the US Geological Survey (USGS). The latter also provides metadata describing geology, water and mapping datasets and can be browsed using a standard web browser. They also offer a spatial searching mechanism for specific datasets based on the Z39.50 Internet protocol.

The reasons for the US being more advanced than the UK providing information on the Internet are primarily twofold. The first is that a local telephone call there is usually provided free of charge, so the connection costs to the network for private users and small companies are minimal, and the second is based in the principle of freedom of information, which is that government information should be publicly available at minimal or no charge unless there are good reasons for not doing so. This drives the information holders such as government agencies to find ever cheaper ways of disseminating that information. It is estimated that a map costs about $9 to distribute by post, whereas data that is stored on a computer anyway can simply be linked to the Internet, so the incremental cost of providing that data on the Internet is low.

For freely available data, this approach is fine. However, when the data has some commercial value, then the copyright must be protected. When the only means of copying data is the traditional method of copying computer tapes, then the likelihood of copyright infringement is probably low. When all an unscrupulous user needs to do is to provide network access to the data, then the problem is different. This is very similar to the problem of software licensing, but is more complex to resolve since a computer program can easily be set to expire after a certain date, or number of uses, and request that a new licence be entered or even require a 'dongle' to be used. Unfortunately, data cannot be treated in the same way, and it is not easily possible to restrict the data to licensed users. The Ordnance Survey (OS) now provide map data on CD-ROM and are preparing themselves for such problems. There is a commercial map-maker in America which adds spurious roads to its maps so it can identify copies, and although the OS do not reveal their approach to security, it is at least on a similar level.

internet opportunities

It is possible that utility companies could share their network and maintenance data both with each other and with local government. This would allow more effective planning and use of street works, one of the ideas behind the Computerised Street Works Register. An additional benefit would be the enhanced safety of the utility companies knowing where each other's plant was when performing maintenance on their own plant. However, this would depend on some agreed standards for data transfer between systems, which are currently being worked on by both vendors and users, and a higher network bandwidth than is generally currently available. Also a new binary data transfer format would have to be defined, as a text based data format such as NTF would not be appropriate for this sort of operation because of the large volumes of data associated with sharing plant and records information, and the performance advantages to using a binary file format. This is under consideration by both EUROGI and its US equivalent, since it is one of the factors inhibiting the spread of GIS.

Before this is to happen in the UK, the security aspects of allowing this type of data to be available on a public network would have to be overcome, and, since many utility companies are now involved in supplying more than one utility, the competitive aspects of allowing say two gas suppliers to share information would need to be examined. It would be unreasonable, for example to require British Gas to share its network plans with other, potentially competitive, energy suppliers, or British Telecom to share its network data with the cable television industry.

Another potential use of public networks by GIS applications is the growing number of municipal GIS being built. There are a growing number of systems that are being designed to provide information to the public, that could easily be developed to supply that same information on the WWW. There is no reason why the whole concept of the WWW should not be extended to provide geographic information. In fact many current GIS, especially those which are simple information providers, are little more than images of maps with 'hot spots' allowing additional detail or information to be displayed, which is basically the same as is provided by WWW viewers such as Mosaic and Netscape. Alternatively, if that is considered too restrictive, then why not make data available in formats that can be read by publicly available PC viewers such as Arcview, which ESRI are currently making available on the WWW at no cost to the user? An alternative would be to include such technology as Arcview into the Web browsing software.

Local authorities who are concerned about public access to their internal data networks should look at the example of Hampshire County Council, who allow School Governors who work for IBM access to HANTSNET, their own internal network, so that they can work more closely with their schools. If this is politically possible, and data security can be managed much more effectively now than a few years ago, why can't more authorities allow their council tax payers access to their information across the Internet?

conclusions

I have deliberately ignored most of the security issues of computer networks, as they are general problems encountered when access is allowed to data, either on computers or not. At a library we accept the need for books to be stamped, and only to be issued to members of that library, why shouldn't some of the same criteria apply for information held on computers? Similarly, if you left your keys in the front door when you went out, you might not be surprised if some of your possessions had been stolen, or at least tampered with. Nevertheless, we tend to be surprised when the same happens to computer information.

The opportunities are huge; with the rapidly increasing use of the Internet and the WWW by the general public, it would be a great pity if the GIS community chose not to take advantage of these developments and was left behind. The chance to widen the appeal of GIS and to show the benefits to a broader community are staring us in the face and should not be ignored. The cost is low, the technology is already widely available, all we are lacking is the will.

trade
directories
1996

What precisely does AGI mean?

In short, AGI means business.

It means business because our success in promoting GIS and their benefits to the business community at large gives AGI (literally, the Association for Geographic Information) member companies greater opportunities to succeed.

Each member, large and small, enjoys the many benefits of the events, information, lobbying and discounts all available through membership of the Association for Geographic Information.

We mean business in our approach too. Every opportunity to champion the GIS cause is exploited - single-mindedly and relentlessly.

Join the AGI today and reap the benefits. Call Shaun Leslie on 0171 334 3746. It could mean a lot to you and your organisation.

ASSOCIATION FOR GEOGRAPHIC INFORMATION

guide to the use of the directories

The information contained in the directories 1 to 40 that follow has been collated from responses to the AGI survey which was carried out during April to August 1995. Apart from directory one, the *full trade directory*, which contains extended entries, no charge has been made for inclusion in the directories. The descriptive detail contained in the *full trade directory* has been provided by the organisation concerned.

In creating the directories, the AGI has taken due care in assembling and transforming the information provided by the organisations that participated in the survey. The AGI has not edited the information but has endeavoured to ensure its validity. The AGI can take no responsibility for the accuracy of the details included. The reader of the directories is responsible for independently determining the exact position of the products and services detailed within the directories.

The sort order of an organisation's name is based on the first word of the name as provided by the organisation concerned. As a result a name that begins with *The* will be in one of two positions e.g. *Geographical Association, The* appears under G whereas the *The LGMB* appears under T.

Directory 1 the *full trade directory* also provides an indication of an organisations involvement with the AGI. The AGI logo appears along side the organisations entry if the organisation is either a Sponsor Member or a Corporate Member of the AGI. The extent to which an organisation is involved in the AGI can be determined from consulting the AGI directories contained in the miscellaneous reference material section of the book.

Some organisations may appear more than once in each directory. There may be two reasons for this:

♦ the organisation has offices in a number of different countries;

♦ the organisation provides different products and services from different locations. These products may have different markets and attributes.

The directories have been constructed to enable ease of use to locate a product or service from a number of starting positions. For example:

♦ if only the trade name of a product or service is known, directory 2, *product trade names* can be used to locate organisations that provide the product and service of that name. Alternatively directory 2 can be used to locate all those organisations that provide the same product;

♦ if only the name of the organisation is known directory 40 *supplier contact* directory can be used to establish the contact name within the organisation and their electronic mail address, telephone and fax numbers;

♦ if the requirement is to locate an organisation within a geographic area, then directory 39 *location* directory can be used to locate organisations within a postal area;

- if the requirement is to locate a service provider that is not a provider of a product then directory 15 *independent providers of services* should be used instead of directory 14 *providers of services* which contains all organisations that provide services irrespective of whether they also provide products;

- if the requirement is to locate an organisation that provides an off-the-shelf application directories 29 to 33 can be used;

- if the requirement is to locate a product or supplier User Group directory 38 *user group information* can be used;

- if the requirement is to locate an organisation that is BS5750 accredited directory 28 *organisations quality* accredited can be used;

- if the requirement is locate a GI recruitment organisations directory 24 *providers of staff recruitment* services can be used;

- if the requirement is to locate an independent project manager that conforms to the PRINCE methodology then cross correlate directory 15 *independent providers of services* with directory 26 *conformance to project management methodologies*;

Once a list of potential organisations have been obtained from an initial scan of the directories further information about the organisations maybe obtained from Directory 1 in the organisations extended entries, the World Wide Web by obtaining the WWW address from directory 9A *publications & supplier details on the World Wide Web*, or from the organisations' advertisements which may appear within the source book as detailed in the index to advertisers which will be found at the back of the book.

The meaning of the majority of terms used within the directories can be ascertained by consulting the GIS dictionary contained in the miscellaneous reference material section.

In directories 29 to 33 an *off-the-shelf application* is meant to be one which has a data model, graphic symbology, entities and attributes pertinent to the application already set up within the product. It is possible that some of the responses received from the survey do not conform with this definition of an off-the shelf application. Some organisations may have used a wider definition. i.e. their product can be used for a particular application rather than being already configured for that application.

The various directories can be cross correlated to improve the accuracy of the information.

The information about data products is contained within four directories, directories 4 to 7. The data sets have been loosely grouped to fit the following broad definitions :

data set	group
land feature data	group 1 in directory 4
map data	group 2 in directory 5
address related data	group 3 in directory 6
demographic data	group 4 in directory 7

Information about software products is contained within three directories, directories 10 to 12. The software products have been loosely grouped to fit the following:

software products	directory
generic or complete system software suites	directory 10
supporting software libraries for data import,	
manipulation and processing	directory 11
system software	directory 12

Information on the markets supported by the individual organisations is broad and is contained within directories 34 to 37. Organisations that have indicated that they support all markets have been included in directory 33 under all markets. The markets supported have been loosely grouped to fit the following:

market supported	group
applicable to all markets	group 1 in directory 34
government	group 1 in directory 34
retail, finance, marketing	group 2 in directory 35
land & coastal oriented	group 3 in directory 36
miscellaneous	group 4 in directory 37

Feedback on the terms, content of directories, accuracy of information is welcomed. Please return your comments to the AGI, 12 Great George Street, London SW1P 3AD.

Worked examples:

Example 1:   Locating a Project Manager

Requirement:

To locate an independent GIS Project Manager that can Project Manage the implementation of a GIS. The Project Manager is to be within the Southeast of England, that specialises in the Utility sector, conforms to the PRINCE methodology and is a member of the AGI.

Solution:

Step 1 -   Consulting directory 15 *Independent Providers of Services* and considering the column *Implementation Services* provides a list of 49 potential organisations out of 91. **Note :** If directory 14 *Providers of Services* had been used then the list would have contained 211 entries without testing for independence.

Step 2 -   Consulting directory 21 *Providers of Implementation Services* and considering the column *Project Management* the list of 49 organisations derived in step 1 is reduced to 45.

Step 3 -   Consulting directory *26 Project Management Methodologies* and considering the column *PRINCE* the list of 45 organisations derived in step 2 is reduced to 16. i.e. the organisation's name has to appear both on the list derived from step 2 and on the list of organisations listed under PRINCE.

Step 4 -   Consulting directory 36 *Suppliers Markets group 3* and considering the column *Utility* the list of 16 organisations derived in step 3 is reduced to 13.

Step 5 -   Consulting directory 1 *Full Trade directory* and checking for the presence of the *AGI logo* against each of the 13 organisations derived in step 4 reduces the list to 8.

Step 6 -   Continue to use Directory 1 and checking the address of each of the 8 organisations derived from step 5 reduces the list to 7.

        e.g.   Cambridge Computer Consultants
              Coopers & Lybrand
              Corbins Consultancy
              GEOBASE Consultants Ltd
              Kingswood Consultants Ltd
              PA Consulting Group
              Smith System Engineering Ltd

Step 7-   Using directory 40 *supplier contact directory* each of the organisations listed under step 6 above can be contacted for further information.

The process detailed above provides a manageable list for a tender or quotation process with a balanced set of organisations. Further information could be gained about the organisations by consulting the AGI Papers directory to establish whether the organisations contact point has presented papers at an AGI conference which may provide an insight into the named individual and the organisation.

Example 2: Locating a GIS Supplier
Requirement:
To locate a GIS vendor which has an established base within UK Local Authorities, has a range of off-the-shelf Local Authority applications which are compliant with standards BS7567 and BS7666 and which can provide a turnkey solution.

Solution:

Step 1 - A quick look at directory 3 *product suppliers* shows 168 organisations supply complete systems. Consulting directory 13 *suppliers of complete systems* and consulting the *GIS* column provides 119 organisations. These two lists are too large to commence the correlation (directory join). An alternative approach is to consult directory 27 *conformance to standards* first. This provides a list of 8 organisations which meet the standard requirement.

Step 2 - Consulting directory 29 *off-the-shelf Local Government Applications* and correlating the 8 organisations obtained in step 1 reduces the list to 5. A further refinement of the list would be to include only those organisations with two or more off-the-shelf applications. This reduces the list to 4.

  e.g. ESRI (UK) Ltd
      Land Aspects Consultancy Ltd
      Sysdeco (UK) Ltd
      System Options Ltd

Step 3 - In order to check the validity of the above list, further information could be gained by consulting the directory *AGI papers by subject* under the keyword *Local Government applications* and looking for recent surveys that may have been presented at the AGI conferences. Consulting the directory provides several surveys the most recent of which is *The take up of GIS in Local Government: The LGMB/University of Sheffield Project, Paper 14.2 presented at AGI94 by Masser. I., & Cambell. H.* Obtaining and consulting this paper shows that the list does represent over 35% of the installations within Local Government organisations in 1993.

Example 3: Locating a Data Supplier
Requirement:
To locate a supplier of data which is quality accredited to BS5750 or ISO9000 and which supplies gazetteer data.

Solution:

Step 1 - Consulting directory 6 *suppliers of data sets group 3* and considering the column *gazetteers* provides a list of 40 organisations.

Step 2 - Consulting directory *28 organisations quality accredited* and considering the columns *BS5750* and *ISO9000* provides a list of 97 organisations which if correlated with the 40 organisations obtained from step 1 provides a list of 6 names.

Step 3 - Consulting directory 3 *product suppliers* and considering the data products column and ensuring that the organisation only offers data then the list obtained from step 2 reduces the list to 1. e.g. Land Aspects Consultancy Ltd (Parkman Group)

directory 1:
full trade directory

A.L.Downloading Services

Voysey House
Barley Mow Passage
London
W4 4PT

telephone: 0181 994 5471
fax: 0181 994 4959
email: sales@aldown.algroup.com

contact: John Farrant
product: Hardware, Complete
Systems
services: Data

A.Rutherford Ltd

62 Barton Road
Haslingfield
Cambridge
CB3 7LL

telephone: 01223 872646
fax: 01223 872646

contact: Allan R Rutherford
services: Audit, Data,
Implementation, Management
Services

ACDS Graphic System Inc

80 Jean Proulx
Hull
Quebec
Canada
J8Z 1W2
telephone: +1819 770 9631
fax: +1819 770 9267

contact: Jean-Guy Laplante
product: Software, Complete
Systems
services: Data,
Implementation, Media,
Management Services

Action Information Management Ltd

 Ashton Road
Hilperston
near Trowbridge
Wiltshire
BA14 7SZ

telephone: 01225 777288
fax: 01225 751616

contact: Tony Hay/John Page
product: Data, Hardware,
Publications, Software,
Complete Systems
services: Data,
Implementation, Management
Services

AI(M) Ltd a leading developer of
fast, easy to use, PC based desktop
mapping software, utilising
integrated raster and vector map data
for DOS, Windows and NT.

Active Software Ltd

 Datum House
Roentgen Road
Basingstoke
Hampshire
RG24 8NG

telephone: 01256 56629
fax: 01256 56708
email: 100140.472@compuserve.com

contact: Paul Smith
product: Data, Software,
Complete Systems
services: Data,
Implementation, Management
Services, Procurement,
Training

Adept Scientific Micro Systems Ltd

6 Business Centre West
Letchworth
Hertfordshire
SG6 2HB

telephone: 01462 480055
fax: 01486 480213
email: atlasgis@adeptscience.co.uk

contact: Stephen Hawkins
product: Data, Publications,
Software, Complete Systems
services: Data,
Implementation, Training

Advent Imaging Ltd

Rotten Row
Hambleden
Henley-on-Thames
Oxfordshire
RG9 6NB

telephone: 01491 411566
fax: 01491 411577
email: sales@advent.co.uk

contact: Clare Bamforth
product: Software

AGFA UK

27 Great West Road
Brentford
Middlesex
TW8 9AX

telephone: 0181 231 4141
fax: 0181 231 4957

contact: Paresh M Patel
product: Hardware, Software,
Complete Systems
services: Implementation,
Media, Procurement, Training

AiC Analysts

Sheraton House
Castle Park
Cambridge
CB3 0AX

telephone: 01223 300044
fax: 01223 302005
email: 100067,1364@compuserve

contact: Trevor Jarvis
services: Data, Management
Services, Procurement,
Recruitment, Training

ALLM Systems & Marketing

21 Beechcroft Road
Bushey
Watford
Hertfordshire WD2 2JU
telephone: 01923 230150
fax: 01923 211148
email: apritchard@cix.compulink.co.uk

contact: Alan Pritchard
product: Data, Publications
services: Data

The GLOBAL GAZETTEER is a
database of 750,000 records of
worldwide places including
information on latitude/longitude,
postcodes, population,
administrative areas, telephone
codes, area, height, etc.

ALTEK Corporation

12210 Plum Orchard Drive
Silver Spring
Maryland 20904-7802
USA
telephone: +1301 572 2555
fax: +1301 572 2510

contact: Shawn Richards
product: Hardware

AM/FM International – European Division

P.O. Box 6
CH-4005 Basel
Switzerland
telephone: +41 61 6915111
fax: +41 61 6918189

contact: Ing.Hans J.Festen
product: Publications
services: Training

AND Mapping B.V.

Schiedamsedijk 44
3011 ED Rotterdam
The Netherlands
telephone: +31 10 433 3440
fax: +31 10 414 0606
email: @andmap.nl

contact: John Heofnagels
product: Data
services: Data

Aneberie CAD

Elmfield
Layer Breton
Colchester
Essex
CO2 0PR

telephone: 01206 331215
fax: 01206 330313

contact: Ralph Massie
product: Software

Anglian Engineering & International Consultancy

Survey House
Old Market Street
Thetford
Norfolk
IP24 2EQ

telephone: 01842 750329

contact: Reuben Hickin
services: Management
Services, Procurement
Resident Engineering, Project
Management Land Survey,
Feasibility studies for Civil
Engineering sitework

AP³ Imaging Services Limited

Unit 3B
Acton
Sudbury
Suffolk
CO10 0BD

telephone: 01787 378242
fax: 01787 374017

contact: Peter Wigmore
product: Data, Hardware,
Publications, Software,
Complete Systems
services: Data,
Implementation, Media,
Procurement

APIC Systemes

'Le Baudran'
25 Rue de Stalingrad
94742 Arcueil
France

telephone: +331 496 99090
fax: +331 396 99293
email: info@apic.fr

contact: Jean-Pierra Rogala

APIC SA

 The Stables
Bishop's Stortford
Hertfordshire
CM23 2BN

telephone: 01279 466966
fax: 01279 466788
email: info@apic.fr

contact: Nick Chisnall
product: Software, Complete
Systems
services: Management
Services, Procurement,
Training

Apic Systems is the designer and the publisher of tools for exploiting geographic information.
The tools are based on the SPACE development language. As a high-level language, SPACE manages the storage and access to spatial objects, geographic, topological and graphical analytical processing.
APIC is a geographic information system generator. It can use multiple databases containing several gigabytes of vector and/or raster data. Applications developed using APIC are totally portable across the different platforms supported.
SPACE/Motif introduces a new approach to the development of spatial applications. SPACE/Motif is used to develop applications that use a distributed architecture. By separating the various parts of an application, SPACE/Motif enables developers to incorporate the management and processing of spatial data into any information system running under Motif. SPACE/Motif also introduces inter operability to spatial applications (access SPACE to C++ and, symmetrically, access C++ from SPACE) as well as access to relational database.
SPACE/Windows brings all the power of the spatial data processing available in Apic Systems UNIX-based products to a Microsoft Windows environment. Linked to object modelling and management of the spatial continuum, the SPACE language breaks down the barriers between UNIX platform performance and PC usability. With SPACE/Windows, large-scale databases containing hundreds of thousands of objects can be stored and processed locally.

ARC Systems Pty Ltd

Level 8
55 Sussex Street
Sydney
New South Wales
Australia
2066

telephone: +612 290 2400
fax: +612 261 3472

contact: Harry Clarsen
product: Software
services: Implementation,
Training

Ashtech Europe Limited

 Blenheim Office Park
Long Hanborough
Oxfordshire
OX8 8LN

telephone: 01993 883533
fax: 01993 883977

contact: Barrie Hogarth
product: Hardware, Software,
Complete Systems
services: Implementation,
Training

Assist Applications Limited

3 School Lane
Yardley Gobion
Towcester
Northamptonshire
NN12 7UL

telephone: 01908 543323
fax: 01908 543324
email: axis@assistap.demon.co.uk

contact: Steve Kurle
product: Hardware, Software, Complete Systems
services: Data, Implementation, Training

Audifilm Girona S.L.

Mas Homs
17181 Aiguaviva
Spain

telephone: +34 72 242611
fax: +34 72 242311

contact: Mr Joan Font
product: Software, Complete Systems
services: Data, Implementation, Management Services, Recruitment, Training

Autodesk Ltd

Cross Lanes
Guildford
Surrey
GU1 1UJ

telephone: 01483 300077
fax: 01483 304556

contact: Sales Department
product: Software

Babtie Shaw & Morton Limited

 Shire Hall
Reading
Berkshire
RG2 9XG

telephone: 01734 234780
fax: 01734 310268

contact: Chris Gower
services: Data, Management Services, Procurement

Bartholomew

 77 – 85 Fulham Palace Road
Hammersmith
London
W6 8JB

telephone:
0181 307 4065
fax: 0181 307 4813
email: barts_twr@geovax.ed.uk

contact: Dave Benson
product: Data, Publications, Complete Systems
services: Data

Bartholomew, a high quality cartographic publisher for more than 150 years, is now a world leader in digital mapping. Its databases are available in both vector and raster

formats for use in GIS and other software packages and include:
World 1:20M, 1:10M and 1:5M;
Europe 1:1M;
GB 1:250K and Greater London 1:5K.

Baymont Technologies Inc

14100 N. 58th St.N.
Clearwater
FL 34620
USA

telephone: +1813 539 1661
fax: +1813 539 1749

contact: William Reid
services: Data

Beacon Dodsworth Ltd

 90 The Mount
York
YO2 2AR

telephone: 01904 638997
fax: 01904 638999

contact: Simon Perry
product: Data, Software, Complete Systems
services: Data

ProSpex for Windows is a powerful, flexible market analysis GIS with a macro language, drive time analysis module and Super Profiles geodemographic data.

Bentley Systems UK Ltd

L'avenir
Opladen Way
Bracknell
Berkshire
RG12 0PF

telephone: 01344 412233
fax: 01344 412386
email: kevin.twigger@bentley.nl

contact: Kevin Twigger
product: Software
services: Implementation

Binnie Black & Veatch

 Grosvenor House
69 London Road
Redhill
Surrey
RH1 1LQ

telephone: 01737 774155
fax: 01737 772767
email: igbush@binnie.demon.co.uk

contact: Ian G Bush
services: Data, Management Services

Outstanding Features
Hydrodynamic modelling for river

basin and coastal zone management studies, contingency planning for hydraulic hazards, environmental impact assessment studies and visualisation, asset surveys and assessment.

Services supplied
Binnie Black & Veatch are international consulting engineers and contractors with a force of 2,400 professionals in the environmental field. We offer GIS, 3D surface modelling and visualisation services to a wide range of environmental and civil engineering projects using GIS tools linked to a broad spectrum of environmental software packages.

Birkbeck College London

 7 – 15 Gresse Street
London
W1P 1PA

telephone: 0171 631 6485
fax: 0171 631 6498
email: d.unwin@uk.ac.bbk.geog

contact: Professor David Unwin
product: Data, Publications
services: Data, Implementation, Management Services, Training

For details of our Master's degree course, short courses, year-long principles of GIS course, and our range of customised courses contact Merle Abbott, 0171 631 6471

BKS Surveys Ltd

 47 Ballycairn Road
Coleraine
County LondonDerry
BT51 3HZ

telephone: 01265 52311
fax: 01265 57637

contact: Jon McNally
product: Data
services: Data, Management Services

BKS specialise in creating customised datasets. Our commercial focus is on contracting with customers for the delivery of key information resources, specifying the technical content to fit in with the customer's specific goals and being flexible in our approach in order to adhere to the constraints which apply (technical, commercial or economic).

Our skilled and professional work force provide services in :
* Data Conversion
* Data Scrubbing
* Digital Mapping and Terrain Modelling

* Photographic services
* Digital Imaging
An ISO 9001 approved company,
and approved contractor to Ordnance
Survey, BKS has earned a reputation
for providing quality data for
virtually any hardware and software
configuration.

Bradly Associates Ltd

Manhattan House
140 High Strret
Crowthorne
Berkshire
RG11 7AT

telephone: 01344 779381
fax: 01344 773168

contact: Peter Kelly
product: Software

British Geological Survey

 Kingsley Dunham Centre
Keyworth
Nottinghamshire
NG12 5GG

telephone: 0115 936 3100
fax: 0115 936 3200
email: k_ald@uk.ac.nkw.va

contact: Dr.Alan Dobinson
product: Data, Publications,
Software, Complete Systems
services: Data,
Implementation, Media,
Management Services,
Training

BS International Consultants

12 Beauly Crescent
Kilmacolm
PA13 4LR

telephone: 01505 873563
fax: 01505 873563

contact: Bob Stirling
services: Audit, Data,
Implementation, Management
Services, Procurement

Bull Information Systems Limited

3700 Parkside
Solihull Parkway
Birmingham
B37 7YT

telephone: 0121 717 0777
fax: 0121 626 1550
email: nsheath@uk22p.bull.co.uk

contact: Nigel Sheath
services: Audit, Data,
Implementation, Management

Services, Procurement,
Training

Business Information Management

14 Kings Avenue
Denton
Newhaven
East Sussex
BN9 0NA

telephone: 01273 515018
fax: 01273 515557
email: 100411.3323@compuserve.com

contact: Rob Mahoney
services: Audit, Data,
Implementation, Management
Services, Procurement,
Recruitment, Training

Business Information Management
Business Information Management
specialises in the provision of
independent GIS Consultancy
services and takes pride in a proven
track record of delivering genuinely
impartial advice.
Business Information Management
provides a wide range of services to
organisations throughout Europe.
We undertake projects that cover all
aspects of the GIS life cycle, from
the earliest conceptual design to post
implementation analysis.
Our aim is to ensure that our clients
receive the best professional advice
available and our philosophy is
always to work closely with them to
ensure that the most appropriate
solution is delivered to meet their
specific business needs.

Byers Engineering Company

6285 Barfield Road
Atlanta
Georgia 30328
USA

telephone: +1404 843 1000
fax: +1404 843 2000
email: 74677.1174@compuserve.com

contact: Bonnie Owen
product: Software, Complete
Systems
services: Audit, Data,
Implementation, Management
Services, Procurement,
Recruitment

C.A.Design Services Ltd

 The Design Centre
Hewett Road
Gapton Hall
Great Yarmouth
Norfolk
NR31 0NN

telephone: 01493 440444
fax: 01493 442480

contact: Matt J Tuohy
product: Hardware, Software,
Complete Systems
services: Audit, Data,
Implementation, Management
Services, Procurement,
Recruitment, Training
Founded in 1984 we have grown into
one of the largest data conversion
bureaus in the UK and offer
customised total hardware/software/
data solutions.

CACI Limited

 CACI House
Kensington Village
Avonmore Road
London
W14 8TS

telephone: 0171 602 6000
fax: 0171 603 5862
email: jrae@cacipo.caci.co.uk

contact: John Rae
product: Data, Publications,
Software, Complete Systems
services: Data,
Implementation, Management
Services, Procurement,
Training
Understanding your market and
customers is the key to success. You
need to know your markets' size,
what types of people may be your
customers, and where to reach them.
InSite puts it all at your fingertips.
InSite is CACI's powerful desktop
market analysis and mapping
system, designed specifically for
analysing markets and customers. It
comes with a range of datasets and
easy to use analysis capabilities,
widely used in retail, finance,
utilities, media and public services,
it is the UK's leading PC market
analysis system.
CACI is an international high
technology services corporation.
Founded in 1962, the company
employs over 3,000 people in 65
offices worldwide.
CACI has operated in the UK since
1975. Our business has grown
consistently year on year and CACI
is now a market leader in its three
primary areas of operation: market
analysis, information systems and
direct marketing.

CACI's corporate objective is the effective interpretation of information to help our clients understand and increase the efficiency of their business.

CAD – Capture Limited

Whitebirk Estate
Blackburn
Lancashire
BB1 5UD

telephone: 01254 583534
fax: 01254 665528
email: info@cadcap.co.uk

contact: Sarah Pickering
product: Hardware, Publications, Software, Complete Systems
services: Data, Implementation, Media, Management Services, Procurement

CAD-Capture provides scanning services for raster and vector GIS or CAD systems. Specialising in small scale & contour mapping. Clients include County Councils and Public Utilities.

CAD R&D Centre Limited

69A Shiptchenski prohod str.
P.O. Box 112
Sofia 1113
Bulgaria

telephone: +359 2 705257
fax: +359 2 703556

contact: Plamen Mateev
product: Data, Software, Complete Systems
services: Audit, Data, Implementation, Media, Management Services, Procurement, Recruitment, Training

CADAC Ltd

 14 Saxon Business Centre
Windsor Avenue
Merton
SW19 2RR

telephone: 0181 543 3411
fax: 0181 543 6844

contact: Paul W Bennett
product: Data, Hardware, Software, Complete Systems
services: Data, Media, Recruitment, Training

CADAC are providers of GIDE (Graphical Interface Data Environment). A low-cost GIS for data collection and asset viewing based on Bentley Systems Microstation.

CalComp Limited

3 – 5 Ruscombe Park
Ruscombe
Reading
Berkshire
RG10 9NU

telephone: 01734 320032
fax: 01734 341215

contact: Sales
product: Hardware

CAM – Centre for Analysis & Modelling Limited

200 Alaska Building
61 Grange Road
London
SE1 3BH

telephone: 0171 232 1111
fax: 0171 237 4247

contact: Gurmukh Singh
product: Data, Software, Complete Systems
services: Audit, Data, Implementation, Media, Management Services, Procurement, Training

Cambashi Ltd

52 Mawson Road
Cambridge
CB1 2HY

telephone: 01223 460439
fax: 01223 461055
email: 100431.3342@compuserve.com

contact: Mrs Jenny.R.Jacobsberg
services: Management Services, Recruitment

Cambridge Computer Consultants (UK) Ltd

 18 Oaklands
Fenstanton
Huntingdon
Cambridgeshire
PE18 9LS

telephone: 01480 469577
fax: 01480 466784
email: ccc_uk_ltd@online.rednet.co.uk

contact: Colin Hookham
services: Data, Implementation, Management Services, Procurement

Cambridge Computer Consultants is an experienced independent GIS Consultancy specialising in information management issues and the provision of application development services.

Cambridge Market Intelligence

London House
Parkgate Road
London
SW11 4NQ

telephone: 0171 924 7117
fax: 0171 403 6729

contact: Peter Bomer
product: Publications

Carl Bro Group

Newton House
Newton Road
Leeds
West Yorkshire LS7 4DN
telephone: 0113 2620000
fax: 0113 2620737

contact: Dr Chris McDermott
product: Data, Software, Complete Systems
services: Data, Media, Training

Carl Zeiss Limited

P.O. Box 78
Woodfield Road
Welwyn Garden City
Hertfordshire AL7 1LU
telephone: 01707 331144
fax: 01707 373210

contact: Mr.E.H.Wickens
product: Hardware, Software, Complete Systems
services: Implementation, Training

CARTograph Ltd

The Eden Centre
47 City Road
Cambridge CB1 1DP
telephone: 01223 67818
fax: 01223 464142

contact: Nigel Payne
product: Data, Software
services: Data, Implementation

Cartographical Services (Southampton) Limited

Landford Manor
Stock Lane
Landford, Salisbury
Wiltshire SP5 2EW
telephone: 01794 390321
fax: 01794 390867

contact: John B Waterman
product: Hardware, Complete Systems
services: Data, Implementation

Cartwright Associates

Woodpeckers
Canterton Green
Lyndhurst
Hampshire SO43 7HF
telephone: 01703 812472
fax: 01703 812472
email: jac@caas.demon.co.uk

contact: Jac Cartwright
product: Data, Software,
Complete Systems
services: Audit, Data,
Implementation, Media,
Management Services,
Procurement, Training

An Information Systems
Consultancy offering supporting
services for GIS on all platforms,
including strategic studies,
acquisition and implementation,
project management, education, and
data acquisition/conversion.
We specialise in spatial analysis and
modelling in the fields of
environment, agriculture,
meteorology, energy resources,
water management and civil
engineering, demography, health and
marketing.

CATALIST

10 Manor Park
Redland
Bristol
BS6 7HH

telephone: 0117 923 7113
fax: 0117 923 7166

contact: Nigel Lang
product: Data, Software,
Complete Systems
services: Data, Management
Services, Training

GIS applications for business
* Retail or Branch network
optimisation;
* Sales Territory Planning
* Marketing, Customer Profiling;
plus the definitive Petrol Station
database.

CCN Marketing

Talbot House
Talbot Street
Nottingham
NG1 5HF

telephone: 0115 941 0888
fax: 0115 934 4903

contact: Elaine Peters
product: Data, Publications,
Software, Complete Systems
services: Data, Management
Services, Training

CCTA – The Government Centre for Information Systems

Rosebery Court
St Andrews Business Park
Norwich
NR7 0HS
telephone: 01603 704844
fax: 01603 704817
email: pmiddleton@ccta.gov.uk

contact: Pat Middleton
product: Publications

CCTA is responsible for stimulating
and promoting the effective use of
Information Systems in support of
the efficient delivery of business
objectives and improved quality of
services by the public sector.

CDD Ltd

Marlborough
Spurlands End Road
Great Kingshill
High Wycombe
Buckinghamshire
HP15 6HY

telephone: 01494 713769
fax: 01494 713769

contact: Derek Pavely
services: Data,
Implementation,
Procurement, Recruitment

CDR Group

Birchfield Hall
Aston Lane
Hope
Sheffield S30 2RA
telephone: 01433 621282
fax: 01433 621292
email: 100436.1330@compuserve.com

contact: Martin Waters
product: Software, Complete
Systems
services: Data,
Implementation, Management
Services, Training

Chiltern Digitising Services

Whitehouse Lodge
Brightwell Baldwin
Watlington
Oxfordshire
OX9 5NT

telephone: 01491 612581
fax: 01491 612930
email: 75337.2741@compuserve.com

contact: Dr.M.J.Lowing
product: Complete Systems
services: Data

Chroson Ltd

6 Dove Close
Andover
Hampshire
SP10 5PB

telephone: 01264 336339

contact: Mr.N.C.Adnitt
services: Implementation,
Procurement, Training

Citywise

17 Rathbone Street
London
W1P 1AF
telephone: 0171 636 5448
fax: 0171 636 5451

contact: Tim J.Craine
product: Data
services: Data,
Implementation, Training

Citywise provide commercial
occupier and building based
information to the property industry.
The Cityview application allows
desktop interrogation of data sets at
individual building level.

Cliffe House Associates

34 Meadowbank Avenue
Nether Edge
Sheffield
South Yorkshire
S7 1PB

telephone: 0114 285 0663
fax: 0114 285 0663

contact: Peter Clegg
services: Audit, Data,
Implementation, Management
Services, Procurement

CMG Computer Management Group (UK) Ltd

Telford House
Tothill Street
London
SW1H 9NB
telephone: 0171 233 0288
fax: 0171 799 2017

contact: Andy Downie
services: Data,
Implementation, Management
Services, Procurement

CMG is a leading European IT
services and products supplier.
CMG's GIS focus is Consultancy
and application development (Small
business partner) for the utilities
market.

Cobham Digital Services Limited

6 Brook Farm Road
Cobham
Surrey
KT11 3AX

telephone: 01932 868133
fax: 01932 867024

contact: Roger K Dollimore
product: Data, Hardware, Software, Complete Systems
services: Data, Implementation, Management Services, Procurement, Training

CODEC Facilities Limited

Cambridge House
3 Newbold Street
Royal Leamington Spa
Warwickshire
CV32 4HN

telephone: 01926 330112
fax: 01926 316728

contact: Dennis J.Caldwell
product: Software, Complete Systems
services: Implementation

Colorgraph (UK) Ltd

Unit 2, Mars House
Calleva Park
Aldermaston
Berkshire
RG7 4QW

telephone: 01734 819435
fax: 01734 815197

contact: Simon Johnson
product: Hardware, Software

ColourMap Scanning Ltd

93 – 99 Upper Richmond Road
London
SW15 2TG

telephone: 0181 789 0737
fax: 0181 780 2663

contact: D.J.Brooker
product: Data
services: Data, Media

Computer Aided Development (CADCORP) Ltd

Sterling Court
Norton Road
Stevenage
Hertfordshire SG1 2JY
telephone: 01438 747996
fax: 01438 747997
email: 100113.2367@compuserve.com

contact: Nicola Radford
product: Software, Complete Systems
services: Implementation, Training
CADCORP SIS (Spatial Information System) is a GIS environment for Windows. Products range from map viewing and printing to full GIS. Software developers toolkit available.

Computer Graphic Suppliers Association

8 Canalside
Worcester
Worcestershire
WR1 2RR

telephone: 01905 613236
fax: 01905 29138
email: 100013,427@compuserve.com

contact: R.Crumpton
product: Publications

COMSULT

67 Goldington Road
Bedford
Bedfordshire
MK40 3NB

telephone: 01234 342401
fax: 01234 328609

contact: P.N.Careless
product: Software, Complete Systems
services: Implementation, Management Services, Procurement

Concurrent Appointments International

27 Field Close
Harpenden
Hertfordshire
AL5 1EP

telephone: 01582 712976
fax: 01582 764858

contact: Alan Carnell
services: Recruitment
Concurrent Appointments is a unique specialist provider of recruitment services to the entire GIS industry. Our database of candidates is the largest in Europe.

Conic Systems

8 Mayfield Terrace
Edinburgh
EH9 1SA

telephone: 0131 667 2728
fax: 0131 667 2728
email: klatchman@conic.com

contact: Mrs Kailash Watchman
product: Software, Complete Systems
LOCATORGIS is a program for displaying updating maps, and attribute information. It supports a wide variety of surveying techniques and instruments such as a camera.

Consensus Information Technology Ltd

The Paddock
Handforth
Cheshire
SK9 3HQ

telephone: 01625 537777
fax: 01625 539621

contact: Janet Sheard
product: Software, Complete Systems
services: Implementation, Management Services

Construction Industry Computing Association

1 Trust Court
Histon
Cambridge CB4 4PW
telephone: 01223 236336
fax: 01223 236337
email: robhoward@constcom.demon.co.uk

contact: Erik G Winterkorn
product: Publications
services: Audit, Data, Management Services, Procurement, Training

Coopers & Lybrand

1 Embankment Place
London WC2N 6NN
telephone: 0171 213 2841
fax: 0171 213 2850
email: helen.mounsey@coopers.colybrand.gold400.gb

contact: Dr Helen Mounsey
services: Implementation, Management Services, Procurement

Corbins Consultancy

agi 50 Stanford Road
Brighton
East Sussex BN1 5PR
telephone: 01273 553110
fax: 01273 389497

contact: Chris Corbin
services: Audit, Data, Implementation, Management Services, Procurement

Corena A/S

Askerveien 61
1370 Asker
Norway
telephone: +47 66 794500
fax: +47 66 794590

contact: Stuart Hodgson
product: Software
services: Implementation

Council of European Professional Informatics Societies

7 Mansfield Mews
London
W1M 9FJ
telephone: 0171 637 5607
fax: 0171 637 5607

contact: Mrs Peta Walmisley
product: Publications

Cray Systems

agi 110 Fleet Road
Fleet
Hampshire GU13 8BE
telephone: 01252 816816
fax: 01252 812163

contact: Guy Pullen
product: Data, Hardware, Software, Complete Systems
services: Audit, Data, Implementation, Management Services, Procurement, Recruitment, Training
Providing systems and Consultancy by applying GIS, remote sensing and other key technologies with expertise in many areas including government, transport, travel, and the environment.

Cromwell House Technical Services

Cromwell House
78 Manor Road
Wallington
Surrey SM6 8RZ
telephone: 0181 647 1686
fax: 0181 773 3110

contact: Paul L.Asquith
services: Data

CSI

7 Meadowfield Park South
Stockfield
Northumberland NE43 7QA
telephone: 01661 842741
fax: 01661 842288

contact: Gilbert H Scott
product: Hardware, Software, Complete Systems

CZ Scientific Instruments Ltd

PO Box 43
1 Elstree Way
Borehamwood
Hertfordshire
WD6 1NH
telephone: 0181 953 1688
fax: 0181 953 9456

contact: Nigel Harding
product: Hardware, Software, Complete Systems

DAT/EM Systems International

1935 Merrill Field Drive
Anchorage
Alaska 99501
USA
telephone: +1907 274 3681
fax: +1907 272 6413
email: jrogers@datem.com

contact: Jim Cucurull
product: Hardware, Software, Complete Systems

Data Base Builders

1 Lawrence Road
Ramsey
Ramsey
Huntingdon
Cambridgeshire
PE17 1UY
telephone: 01487 813745
fax: 01487 813745

contact: Douglas Cross
services: Audit, Implementation, Management Services, Procurement

Data Collection Ltd

Sunshine Corner
Ancton Lane
Middleton-on-Sea
West Sussex
PO22 6NN
telephone: 01243 587390
fax: 01243 587390

contact: Steve Batchelor
services: Data

Data Dictionary Systems Limited

16 Tekels Avenue
Camberley
Surrey GU15 2LB
telephone: 01276 23519
fax: 01276 676670
email: awpl@applelink.apple.com

contact: David J.L.Gradwell
services: Data, Management Services
The key Data Dictionary Systems Limited GIS project in 1995 was the development of a geospatial data model for the Military Survey.

Dataflow Information Systems

The Rackham
Bristol BS1 4HJ
telephone: 0117 927 2466
fax: 0117 929 0768

contact: Clare Dorey
services: Data, Implementation, Training

Dataman Computer Solutions UK Ltd

The Old School
Hunsinyobe
Wetherby
Yorkshire LS22 5HY
telephone: 01423 358226
fax: 01423 358262
email: potts@dataman.co.uk

contact: Paul Potts
product: Hardware
Selling a wide range of UNIX hardware equipment for Sun SPARC, HP9000, SGI, IBM RS6000 and DEC workstations including memory, disc storage and networking equipment

Dataquest Europe Ltd

Holmers Farm Way
High Wycombe
Buckinghamshire HP12 4XH
telephone: 01494 422722
fax: 01494 422742
email: pgartzen@dqeurope.com

contact: Ms Petra Gartzen
product: Publications
services: Management Services

Datatechnology Datech Ltd

Sidcup Technology Centre
Maidstone Road
Sidcup
Kent DA14 5HU
telephone: 0181 308 1800
fax: 0181 308 0802
email: alistair.brook@datech.co.uk

contact: Alistair Brook
product: Hardware, Software, Complete Systems
services: Implementation, Management Services, Training

Dataview Solutions Ltd

40 – 42 Parker Street
London WC2B 5PQ
telephone: 0171 404 0640
fax: 0171 404 0664
email: geoff@dataview.demon.co.uk

contact: Geoff Kendall
product: Data, Hardware,
Software, Complete Systems
services: Data,
Implementation, Management
Services, Training
Named as MapInfo 'Worldwide
Partner of the Year 1994', Dataview
provides support, training, data
sourcing, Consultancy and
development services, focussing on
development of innovative
applications.

DCL Consulting

14 Bailey Mews
Auckland Road
Cambridge CB5 8DR
telephone: 01223 314888

contact: David Litton
services: Data, Training

Derek Hunter & Partners Ltd

Cromford Mill Business
Centre
Matlock
Derbyshire DE4 3RQ
telephone: 01629 822100
fax: 01629 822030
email: 100345,3001@compuserve.com

contact: Derek Hunter
product: Software, Complete
Systems
services: Implementation,
Management Services,
Recruitment

Design Computer Aids Limited (DeCAL)

16/2 Timberbush
Leith
Edinburgh
EH6 6QH

telephone: 0131 553 3159
fax: 0131 553 5121

contact: Lorraine Sinclair
product: Data, Hardware,
Publications, Software,
Complete Systems
services: Management
Services

Digital Equipment Corporation

Digital Park
Reading
Berkshire RG2 0TE
telephone: 01734 203546
fax: 01734 204757

contact: Malcolm Wicks
product: Hardware,
Publications
services: Implementation,
Procurement

DM Management Consultants Ltd (DMMC)

19 Clarges Street
London W1Y 7PG
telephone: 0171 499 8030
fax: 0181 948 6306

contact: Dr Peter
M.Thompkins
services: Audit, Data,
Implementation, Management
Services, Recruitment,
Training

DMAP Ltd

26-28 High Street
Fenstanton
Cambridgeshire PE18 9JZ
telephone: 01480 497673
fax: 01480 492281

contact: Paul Holroyd
product: Data, Software
services: Data,
Implementation, Media

DMV Consultants BV

P.O. Box 1399
3800 BJ Amersfoort
The Netherlands

telephone: +31 33 682300
fax: +31 33 682601

contact: R.Beck
product: Data, Software,
Complete Systems
services: Data,
Implementation, Media,
Procurement, Recruitment,
Training

Dolphin Consulting Group

10 Collingwood House
Dolphin Square
London
SW1V 3ND

telephone: 0171 798 8465
fax: 0171 798 8692

contact: Nic Walker
product: Data

services: Data,
Implementation, Media,
Management Services,
Procurement

Dotted Eyes

1c Plymouth Road
Barnt Green
Birmingham B45 8JE
telephone: 0121 445 6150
fax: 0121 445 6150

contact: Jamie M Justham
product: Data, Hardware,
Publications, Software,
Complete Systems
services: Audit, Data,
Implementation, Media,
Management Services,
Procurement, Training
Independent consultants. Developers
of Bartholomew maps on CD-ROM
and Ordnance Survey data sets. Add-
in programs for MapInfo include
LANDLINE, TRUESCALE,
HATCHING and BACKDROP.

Dowling Associates Limited

Lingside House
Thornthwaite
Harrogate
North Yorkshire HG3 2QX
telephone: 01943 880332
fax: 01943 880634
email: bdowling@cix.compulink.co.uk

contact: W.J.Dowling
product: Software
services: Implementation,
Training

Dr Stanley Port

44 Busbridge Lane
Godalming
Surrey GU7 1QD
telephone: 01483 421970
fax: 01483 861023

contact: Stanley Port
product: Publications
services: Management
Services, Procurement,
Training

EA Technology

Capenhurst
Chester CH1 6ES
telephone: 0151 347 2451
fax: 0151 347 2135
email: ggm@eatl.co.uk

contact: Gary Marsden
product: Data, Publications,
Software
services: Data,
Implementation

Earth Observation Sciences Ltd

Broadmede
Farnham Business Park
Farnham
Surrey
GU9 8QL

telephone: 01252 721444
fax: 01252 712552
email: matthews@eos.co.uk

contact: Matthew Stuttard
product: Hardware,
Publications, Software,
Complete Systems
services: Data,
Implementation, Management
Services, Procurement

Earth Resource Mapping

Blenheim House
Crabtree Office Village
Eversley Way
Egham
Surrey
TW20 8RY

telephone: 01784 430691
fax: 01784 430692
email: brian@ermapper.co.uk

contact: Brian Talbot
product: Software
services: Training

ECM Selection Limited

The Maltings
Burwell
Cambridge
CB5 0HB

telephone: 01638 742244
fax: 01638 743066
email: postmaster@ecmsel.co.uk

contact: Michael Gernat
services: Recruitment

Effective Solutions (Data Products)

Robert House
Station Approach
Romsey
Hampshire
SO51 8DU

telephone: 01794 514233
fax: 01794 514244

contact: Graham Collins
product: Data, Hardware,
Publications, Software,
Complete Systems
services: Data,
Implementation

Elstree Computing Ltd

133-139 Page Street
Mill Hill
London NW7 2ER
telephone: 0181 906 5656
fax: 0181 906 5666

contact: Steve Pittard
product: Hardware, Software,
Complete Systems
services: Data,
Implementation, Media,
Procurement, Training

Empress Software UK

Godalming Business Centre
Woolsack Way
Godalming
Surrey
GU7 1XW

telephone: 01483 861990
fax: 01483 860064

contact: Dennis Flavell
product: Software

Enghouse (UK) Limited

18 The Business Village
Tollgate
Eastleigh
Hampshire
SO53 3TG

telephone: 01703 615228
fax: 01703 615253

contact: Simon Crowley
product: Data, Software,
Complete Systems
services: Data,
Implementation,
Procurement, Training

Environment & Planning Library

9 New Hall Avenue
Broughton Park
Salford
M7 4JY

telephone: 0161 708 9799

contact: Harry Z A Orenstein
product: Publications,
Software
services: Data

EOSAT

4300 Forbes Boulevard
Lanham
Maryland 20706
USA

telephone: +1301 552 0525
fax: +1301 794 4243

contact: Annamarie De Carlo
product: Data, Publications

EPS – Essential Planning Systems Limited

Suite 200
6772 Oldfield Road
Victoria
B.C.
Canada
V8M 2AZ

telephone: +1604 652 8895
fax: +1604 652 8896
email: marketing@eps.bc.ca

contact: Alison Malis
product: Software, Complete
Systems
services: Implementation,
Training

ERA Technology Ltd

Cleeve Road
Leatherhead
Surrey
KT22 7SA

telephone: 01372 367028
fax: 01372 367099

contact: Peter Tucker
product: Complete Systems
services: Implementation,
Training

ERA-Maptec Ltd

5 South Leinster Street
Dublin 2
Ireland

telephone: +3531 676 6266
fax: +3531 661 9785

contact: Paul Kidney
product: Data, Publications
services: Data, Media,
Training

ERDAS (UK) Ltd

 Telford House
Fulbourn
Cambridge
CB1 5HB

telephone: 01223 880802
fax: 01223 880160
email: jshears@erdas-uk.demon.co.uk

contact: Jonathan Shears
product: Software, Complete
Systems
services: Implementation,
Training

ERDAS are world leaders in
imaging GIS technology. ERDAS
IMAGINE™ is the preferred choice
for any GIS application involving
imagery or raster data, however
large or small.

ERTEC

11 Kingswood Crescent
Kingswells
Aberdeen AB1 8TE
telephone: 01224 740324
fax: 01224 740324

contact: David R Green
services: Data, Training
ERTEC specialises in environmental
applications of Remote Sensing, GIS
and Cartography.
Expertise includes small-scale aerial
photography, file formats, map
design, colour specification,
education/training.

ESR Cartographers Ltd

Emerald House
30/38 High Road
Byfleet
Surrey
KT14 7QG

telephone: 01932 348981
fax: 01932 344882

contact: Alan Smith
product: Data, Publications,
Complete Systems
services: Media

ESRI (UK) Ltd

 23 Woodford Road
Watford
Hertfordshire
WD1 1PB

telephone: 01923 210450
fax: 01923 210739
email: csmith@esriuk.com

contact: Carole Smith
product: Data, Hardware,
Software, Complete Systems
services: Data,
Implementation, Management
Services, Procurement,
Training

ARC/INFO® The World's GIS™

A sophisticated suite of spatial
analysis and modelling tools with
* easy-to-use data automation tools
* powerful data management
capabilities
* Superior cartographic display and
output
* Compliance with industry standard
for data exchange formats
* An extensive set of included
utilities for all aspects of GIS

PC ARC/INFO® The World's Leading Desktop GIS™

A full-featured GIS for DOS or
Windows-based PCs. PC ARC/INFO
is used by organisations to create,
edit, analyse, and display geographic
data on microcomputers.

ArcView® GIS for Everyone™

Allows users to access and examine
geographic information through easy-
to-use, point-and-click tools.
ArcView provides desktop mapping
functionality, tabular data
management, and support for
multimedia data types. ArcView
software's object-oriented application
development environment,
Avenue™, enables users to customise
the ArcView interface and to create
custom applications.

ArcCAD® GIS for AutoCAD™

Links ARC/INFO, the world's
leading GIS software, to AutoCAD,
the world's leading CAD software.
ArcCAD provides an integrated
environment for traditional users of
GIS and CAD in applications such as
civil engineering, transportation
planning, and facility management.
ARC/INFO Extensions
ARC NETWORK™ modelling tools
route optimisation
ARC TIN™ 3-D display for
volumetric analysis
ARC GRID™ raster geoprocessing
tools
ARC COGO™ coordinate geometry
for survey data management
ArcExpress™ accelerate graphic
display
ArcSTORM™ comprehensive
database manager
ArcScan™ raster data preprocessing
and vectorisation
ArcPress™ graphics rasterisation
and plotting

EuroDirect Database Marketing Ltd

Onward House
2 Baptist Place
Bradford
West Yorkshire BD1 2PS
telephone: 01274 737144
fax: 01274 741126

contact: John K Dobson
product: Data, Software,
Complete Systems
Eurodirect is a major provider of
GIS marketing analysis software,
based on a Windows platform, its'
family of products include
DemoGraf*, Neighbours &
PROSPECTS and the US Residents
Database.

European Business Mapping

25 West Drive
Brighton
East Sussex BN2 2GE
telephone: 01273 702957
fax: 01273 673979

contact: Bruce Mackay
product: Data, Software
services: Management
Services

European Geographic Technologies BV

P.O. Box 99
de Waal 15
5680 AB Best
The Netherlands
telephone: +31 4998 93385
fax: +31 4998 92078

contact: Yiannis Moissidis
product: Data, Hardware,
Publications, Software
services: Data,
Implementation

Eurosense Technologies N.V.

Nervierslaan 54
B-1780 Wemmel
Belgium
telephone: +32 2 460 7000
fax: +32 2 460 4958

contact: Nancy Schryvers
product: Data, Hardware,
Software, Complete Systems
services: Data,
Implementation, Media,
Management Services,
Training

Evox Facilities Ltd

12 Hillfield Road
Redhill
Surrey RH1 4AP
telephone: 01737 764137
fax: 01737 764137

contact: Phil Assender
product: Data
services: Data, Management
Services, Procurement

Fairbairn Services Limited

 Fairbairn House
Ashton Lane
Sale
Manchester M33 1WP
telephone: 0161 976 3536
fax: 0161 969 5131
email: ks@fairbairn.co.uk

contact: Mr Kym Soni
services: Audit, Data, Implementation, Media, Management Services, Procurement, Recruitment, Training

FastCAD GIS Ltd

10 Cotham Road South
Cotham
Bristol
BS6 5TZ

telephone: 0117 942 8195
fax: 0117 942 8196
email: fcadgis@dircon.co.uk

contact: Louise Solomon
product: Data, Hardware, Software, Complete Systems
services: Data, Implementation, Management Services

FastCAD GIS Ltd

26 Greenhill Crescent
Watford Business Park
Watford
Hertfordshire
WD1 8XG

telephone: 01923 240216
fax: 01923 228796

contact: Miss Victoria Gregory
product: Hardware, Software, Complete Systems
services: Data, Implementation, Training

FileNet Ltd

The White House
57 – 63 Church Road
Wimbledon Village
London
SW19 5DQ

telephone: 0181 944 5111
fax: 0181 944 5146

contact: Penny Cooper
product: Software, Complete Systems
services: Data, Implementation, Management Services, Procurement, Training

Flynn & Rothwell

Thomas Tredgold House
231 London Road
Bishop's Stortford
Hertfordshire
CM23 3LA

telephone: 01279 507346
fax: 01279 758219
email: 100566.3365@compuserve.com

contact: Linda Elkins
services: Data, Implementation, Management Services

Foto Res

Centro Comercial 'El Descubrimiento'
10.001 Caceres
Spain

telephone: +34 27 216455
fax: +34 27 216455

contact: Jorge Fabricant
product: Data, Hardware, Publications, Software, Complete Systems
services: Data, Recruitment, Training

G.L. Consulting Ltd

Braeside
8 Agates Lane
Ashstead
Surrey
KT21 2NF

telephone: 01372 272937
fax: 01372 279362

contact: Dr Les W Thorpe
services: Audit, Data, Implementation, Procurement

Gamma Ltd

20 Westward Square
Dublin 2
Ireland

telephone: +3531 6713066
fax: +3531 6713593
email: gamma@iol.ie

contact: Fearoal O'Neill
product: Data, Software
services: Data, Implementation, Management Services, Training

Gardline Infotech

 Burlingham House
Hewett Road
Gapton Hall Industrial Estate
Great Yarmouth
Norfolk
NR31 0NN

telephone: 01493 442544
fax: 01493 441200

contact: David Pettit
product: Software, Complete Systems
services: Audit, Data, Implementation, Media, Management Services, Recruitment, Training

GARDLINE INFOTECH's Services
* Map digitising
* Records conversion
* Data Processing and plotting
* Scanning
* Consultancy and Training
* Contract Staff and Equipment
* BS/EN/ISO 9002 Accredited Services

GEC Marconi Research Centre

Avionics Laboratory
West Hanning Field Road
Great Baddow
Chelmsford
Essex
CM2 8HN

telephone: 01245 473331
fax: 01245 475244
email: tony.rye@gmrc.gecm.com

contact: Mr.A.J.Rye
product: Data, Software, Complete Systems
services: Data, Implementation

Genasys II Limited

 1st Floor, Parkway One
Parkway Business Centre
300 Princess Road
Manchester
M14 7LU

telephone: 0161 232 9444
fax: 0161 232 9453
email: johnt@genasys.co.uk

contact: John Tarleton
product: Software, Complete Systems
services: Implementation, Management Services, Training

The Company
Fast gaining a reputation as the worlds most innovative OpenGIS supplier, **Genasys** have been providing real business solution to real business problems since the mid 1970s.

Specialising in completely understanding the needs of its users, Genasys have established a worldwide customer base. Customers in government, military, telecommunications and petrotechnical organisations around the world, have discovered the commercial benefits of the fully OpenGIS approach which the company pioneered.

Genasys provide a full range of

support services including Consultancy, installation, software support and training.
Proactive about quality, the company are committed to Total Quality Management and are accredited to ISO 9000 and TickIT quality standards.
With over 4000 users and a network of offices and partners throughout the world, Genasys lead the way in innovative, integrated spatial solutions.

Genasys product family
The latest collective software release of the Genasys family of products continues the company's innovative approach, providing a set of products which enables the integration of text, vector and raster data, image management, data capture, document management and other applications under a single multi-purpose presentation environment.

GenaMap
GenaMap is a comprehensive, topological, vector/raster Geographic Information System (GIS) designed for today's business needs. GenaMap addresses all requirements, from digitising and analytical operations, to producing quality hardcopy maps. It provides the ability to directly link with external relational databases.
* Open Systems design
* Flexible Client-Server Architecture for Integration within a wider IT environment.
* Easy-to-use
* Full 2D/3D analytical raster and vector processing catering for a wide range of data types.
* Unique concept of Spatial Views provides ability to produce rapid 'virtual maps', saving valuable disk space.
* Full support for external RDBMS links including Oracle and Ingres.
* Multi-level Application Development Access – easy for a beginner, but powerful enough to satisfy expert users.
* In-built Graphical User Interface (GUI) – **Genius II.**

GenOSmap
GenOSmap provides a user friendly interface to the location and identification of UK Ordnance Survey (OS) map sheets. Based on a user defined point, area or cursor defined rectangle, GenOSmap will provide details of all OS maps within the user specified zones for all scales. Input can also be generated using OS map sheet names, wild cards, prefixes or gazetteer entry using suitable

databases.
* Full support for all OS map scales.
* Mapsheet based search of wild card selection.
* Support for other data sheets such as town plans, activity maps, etc
* Compatible with OS gazetteers and digital datasets
* Supports NTF 1.1/2, DXF, TIFF and other data formats
* Automated map sheet generation
* Networkable and compatible with corporate databases

Genius II
Genius II is a fully customisable Graphical User Interface (GUI) front end providing consistency across all Genasys products. An interactive, windows based applications building and editing tool, Genius II enables users to build application menu interfaces quickly and easily by pointing and clicking – no need for intricate, time consuming programming and compiling. Genius II also provides a front end to third party products as well as Genasys software applications.

General Register Office for Scotland

 Ladywell House
Ladywell Road
Edinburgh
EH12 7TF

telephone: 0131 314 4254
fax: 0131 314 4344
email: 100431.2507@compuserve.com

contact: Peter Jamieson
product: Data, Publications
For the 1991 Census, GRO(S) digitised boundaries of Scottish postcodes, 1995 versions are now available. In 1996, quality will be further improved by:
* digitising directly against Ordnance Survey Land-Line, and
* directly linking boundaries and attribute data.

GEO-Marketing Systems Ltd (GMSL)

Landfall
Cold Ash
Thatcham
Berkshire RG18 9HY
telephone: 01635 872382
fax: 01635 871302

contact: Alan Odham/Phil Durbin
product: Data, Hardware, Software, Complete Systems
services: Data, Implementation, Management Services

Geo-Perfect TWI B.V.

P.O. Box 204
2740 AE Waddinxveen
The Netherlands

telephone: +31 1828 30477
fax: +31 1828 31280
email: hkersten@knoware.nl

contact: Geworge J.W.Lavigne
product: Data, Hardware, Software, Complete Systems
services: Data, Implementation

GEO-UK Ltd

Pantiles House
22 London Road
Bagshot
Surrey GU19 5HN
telephone: 01276 473579
fax: 01276 473603

contact: Roy Wood
services: Management Services

Geo/SQL (UK)

Fordbrook Business Centre
Pewsey
Wiltshire
SN9 5NU

telephone: 01672 562012
fax: 01672 63001

contact: Phil Williams
product: Hardware, Software, Complete Systems
services: Audit, Data, Implementation, Media, Management Services, Procurement, Training

Geo2 Consulting

Bosrand 19
3121 XA Schiedam
The Netherlands

telephone: +31 10 4712372

contact: Chris W.Nelis
services: Audit, Implementation, Management Services, Procurement, Training

GEOBASE Consultants Ltd

 28 Church Road
Epsom
Surrey
KT17 4DX

telephone:
01372 811225
fax: 01372 811226
email: geobase@cix.compulink.co.uk

contact: John R Rowley
services: Audit, Data, Implementation, Management Services, Procurement, Recruitment, Training
GEOBASE is a leading independent Consultancy providing services associated with Geographic Information and having specialist capabilities in several areas such as GIS Strategy, Property, Utilities and government applications. We deliver this broad range of services economically and flexibly by operating as a network of associated consultants and firms.

GeoData Institute

University of Southampton
Henfield
Southampton
Hampshire SO17 1BJ
telephone: 01703 592719
fax: 01703 592849
email: geodata@soton.ac.uk
contact: Chris Hill
product: Publications
services: Data, Implementation, Management Services, Procurement, Training
GeoData is an interdisciplinary research and Consultancy specialising in data capture, handling and analysis through GIS.
* Capture, Conversion, Translation
* Database and Applications Development
* SPANS Training.

Geodelta

Oude Delft 175
2611 HB Delft
The Netherlands
telephone: +31 1515 8188
fax: +31 1515 8154
contact: Jr. R.J.G.A.Kroon
product: Software, Complete Systems
services: Data, Procurement, Training

Geografix Limited

Hurricane Way
Norwich
Norfolk
NR6 6EW
telephone: 01603 788940
fax: 01603 788964
contact: Laurence Taylor
product: Data, Hardware, Software, Complete Systems
services: Data

Geographic Management Solutions Ltd

12 Turnpike Gate
Wickwar
Wotton-under-Edge
Glocestershire GL12 8ND
telephone: 01454 281802
fax: 01454 419417
contact: John Standerline
product: Software, Complete Systems
services: Procurement, Training
GMS provides easy to use data collection and map display software; full feature GIS; application development using spacewindows; software solutions Consultancy and training.

The Geographical Association

343 Fulwood Road
Sheffield S10 3BP
telephone: 0114 267 0666
fax: 0114 267 0688
contact: Graham Ranger
Subject Teaching Association aiming to further the study and teaching of geography, including the use of GIS, particularly in the school curriculum.

GeoInformation International

307 Cambridge Science Park
Cambridge CB4 4ZD
telephone: 01223 423020
fax: 01223 425787
contact: Elizabeth Wijnmaalen
product: Data, Publications
services: Data, Management Services, Training
Europe's leading provider of geographical information products and services to the professional and educational markets. GI's product range include date products, GIS magazines, GISTutor2, technical and non-technical books.

GeoMEM Software

1 High Street
Blairgowrie
Perthshire PH10 6ET
telephone: 01250 872284
fax: 01250 873290
email: sales@geomem.win-uk.net
contact: Mr.Marlon.P.Binner
product: Data, Hardware, Software, Complete Systems
services: Audit, Data, Training

Geometria GIS Systems House Ltd

Felso Zoldmali ut 128-130
H-1025 Budapest
Hungary H-1025
telephone: +361 250 0989
fax: +361 250 1231
email: 73501.173@compuserve.com
contact: Mr Tibor Tenke
product: Data, Software
services: Data, Implementation, Management Services, Training

Geoplan (UK) Ltd

14/15 Regent Parade
Harrogate
North Yorkshire
HG1 5AW
telephone: 01423 569538
fax: 01423 525545
contact: Richard Ives
product: Data, Publications, Software, Complete Systems
services: Data, Implementation, Media, Management Services

Geops BV

Agro Business Park 36
6708 PW Wageninggen
The Netherlands
telephone: +31 8370 79636
fax: +31 8370 79704
contact: Ir.G.J.M.Kreuwel
product: Software
services: Data, Implementation, Training

GEOSOFT Ltd

3M Springfield House
Hyde Terrace
Leeds
West Yorkshire LS2 9LN
telephone: 0113 234 4000
fax: 0113 246 5071
email: sales@geosoft.co.uk
contact: Chris Inie
product: Software
services: Data, Management Services, Procurement

Geosystems

PO Box 40
Didcot
Oxfordshire OX11 9BX
telephone: 01235 813913
fax: 01235 813913
contact: Dr Roger F Templeman
product: Data, Publications
services: Audit, Data

Geotronics Limited

Mensura House
Blackstone Road
Huntingdon
Cambridgeshire
PE18 6EF

telephone: 01480 433555
fax: 01480 432480

contact: Alan Sharp
product: Hardware, Software,
Complete Systems

Geoview Systems Kft

Radnoti Miklos.u.2. V. em.
Budapest
Hungary
H-1137

telephone: +361 269 2099
fax: +361 112 6861
email: nikkel@bp.geoview.hu

contact: Istvan Nikl
product: Software, Complete
Systems
services: Implementation
Our Company was founded 1990 to
create a leading GIS application
development system (ADS). Since
then we have won approx. 30% of
the Hungarian GIS market with our
GreenLine object-oriented ADS.
GreenLine includes an object-
oriented database, offers a unique
interface, is user-friendly and 100%
portable.

GGP Systems Limited

12 Vincent Road
Croydon
Surrey
CR0 6ED

telephone: 0181 656 8562
fax: 0181 656 8562

contact: Mrs.A.Maxwell
product: Software

GID Ltd

1 Captains Gorse
Upper Basildon
Reading
Berkshire
RG8 8SZ

telephone: 01491 671964
fax: 01491 671964
email: andy@gid.co.uk

contact: Andrew Greener
product: Software
services: Data, Procurement

GIMMS (GIS) Ltd

30 Keir Street
Edinburgh EH3 9EU
telephone: 0131 668 3046
fax: 0131 668 2104

contact: M.A.Ferenth
product: Software, Complete
Systems
services: Management
Services, Procurement

GIS Services Ltd

 3 Ullenhall Road
Knowle
Solihull
West Midlands
B93 9JD

telephone: 01564 779656
fax: 01564 779656

contact: Stewart McAusland
product: Hardware
services: Audit, Data,
Management Services,
Procurement, Recruitment,
Training
Full range of Consultancy services
from awareness seminars to
implementation project planning and
management; supply of contract staff
and equipment for rental.

GISDATA

University of Sheffield
Western Bank
Sheffield
S10 2TN

telephone: 0114 272 0185
fax: 0114 272 2199
email: gisdata@sheffield.ac.uk

contact: Max Craglia
product: Publications
services: Implementation,
Management Services,
Training

GISL Limited

 1st Floor, Blenheim House
Crabtree Office Village
Eversley Way
Egham
Surrey
TW20 8RY

telephone: 01956 285077
fax: 01428 707132

contact: Justin Saunders/
James Cutler
product: Data, Software
services: Data,
Implementation, Media,
Management Services,
Procurement, Recruitment,
Training
Established Consulting company
providing GIS/Remote sensing

services and solutions to UK and
worldwide government and
commercial clients in natural
resources, environmental, coastal
and geoscience applications.

Glen Computing Ltd

 309 High Street
Orpington
Kent BR6 0NN
telephone: 01689 875577
fax: 01689 828735

contact: Derek.G.Prior
product: Data, Hardware,
Software, Complete Systems
services: Audit, Data,
Implementation, Media,
Management Services,
Procurement, Training
Glen provides a range of GIS
products, including MAPINFO.
Services include bureau digitising,
data conversion, bespoke application
development, system integration,
training and continuing support.

Global Surveys Ltd

2 Ridgacre Lane
Quinton
Birmingham B32 1ES
telephone: 0121 421 1414
fax: 0121 423 1480

contact: Alan F.Wright
product: Complete Systems
services: Data,
Implementation

GMAP Ltd

GMAP House
Cromer Terrace
Leeds
West Yorkshire LS2 9JU
telephone: 0113 244 6164
fax: 0113 234 3173

contact: Paul Kelley
product: Data, Software,
Complete Systems
services: Data,
Implementation, Management
Services, Procurement

Graphic Data Systems Corporation (GDS)

 Unit 8, Woking Eight
Forsyth Road
Sheerwater
Woking
Surrey GU21 5SB
telephone: 01483 725225
fax: 01483 725221

contact: Johanna Afors
product: Software
services: Implementation,
Management Services,
Training

The GDS family of object-based spatial data management software was designed to support all stages of the project life-cycle from planning to design, construction, maintenance and operation.

Graphical Data Capture Ltd (GDC)

 262 Regents Park Road
London
N3 3HN

telephone: 0181 349 2151
fax: 0181 349 4095

contact: Peter.M.Klein
product: Data, Hardware, Software, Complete Systems
services: Data, Implementation, Media, Management Services, Training

GDC has specialised in digital cartography since 1974. Combining the strengths of MapInfo, the world's no 1 desktop mapping solution with GDC's extensive range of local and international datasets ensures rapid and cost-effective implementation of Windows-based GIS solutions. Digitising, Training, Consultancy and Application Building complete the range of services available.

Graphics Online Limited

The Cooper Buildings
Sheffield Science Park
Sheffield
S1 2NS

telephone: 0114 279 7972
fax: 0114 275 3708

contact: Edwin Guiton
product: Data, Software, Complete Systems

Graphite Management Services Ltd

Sherwood Business Centre
7 Gregory Boulevard
Nottingham
NG7 6LD

telephone: 0115 969 1114
fax: 0115 969 1115

contact: G.Adrian & K.Hardy
product: Hardware, Software, Complete Systems
services: Data, Implementation

Graphtec (UK) Ltd

Environ House
Welshmans Lane
Nantwich
Cheshire
CW5 6AB

telephone: 01270 611234
fax: 01270 626733

contact: Peter Mitchell
product: Hardware

Greig Fester Limited

 Devon House
58 – 60 St Katharine's Way
London
E1 9LB

telephone: 0171 488 2828
fax: 0171 265 1234

contact: Andrew Mitchell
services: Data, Management Services, Procurement

Grove Projects Ltd

Grove House
27 Hammersmith Grove
London
W6 0NE

telephone: 0181 846 2459
fax: 0181 846 3388

contact: M.C.H.Sumner
services: Audit, Data, Implementation, Media, Management Services, Procurement

GTCO Corporation

7125 Riverwood Drive
Columbia
MD 21046
USA

telephone: +1410 381 6688
fax: +1410 290 9065

contact: Sales Dept.
product: Hardware

GTX Europe Ltd

GTX House
Cedarwood
Chineham Business Park
Basingstoke
Hampshire
RG24 0WD

telephone: 01256 843555
fax: 01256 246634

contact: Robert Brown
product: Complete Systems

Guild of Incorporated Surveyors

 1 Alexandra Street
Queen Road
Oldham
Lancashire
OL8 2AU

telephone: 0161 627 2389
fax: 0161 627 3336

contact: Brian Birchenall

H R Wallingford Ltd

Howbery Park
Wallingford
Oxfordshire
OX10 8BA

telephone: 01491 835381
fax: 01491 826352
email: pab%hydres.uucp@uknet.ac.uk

contact: Dr Peter A Bradbury
product: Software
services: Data, Training

Hall & Watts Systems Limited

Acorn House
Shab Hill
Birdlip
Gloucestershire
GL4 8JX

telephone: 01452 864244
fax: 01452 864194

contact: S.McCarthy
product: Data, Hardware, Software, Complete Systems

Hansa Luftbild

Elbestr 5
D-48145 Munster
Germany

telephone: +49 251 23300
fax: +49 251 2330112

contact: Hans-Dieter Arnold
product: Data, Software
services: Data, Implementation, Procurement, Training

Hitachi Home Electronics (Europe) Limited

Hitachi House
Station Road
Hayes
Middlesex
UB3 4DR

telephone: 0181 849 2092
fax: 0181 569 2763

contact: Mark Wilkin
product: Hardware

HJM Imaging Systems

3 High Street
Kislingbury
Northampton
NN7 4AG

telephone: 01604 39792
fax: 01604 30919

contact: Les Bootes/John
Hale
product: Hardware, Complete
Systems

HMSO Books

St Crispins
Duke Street
Norwich
NR3 1PD

telephone: 01603 695911
fax: 01603 696784

contact: Lisa Hallett
product: Publications
GIS publications published by
HMSO for CCTA, the Government
Centre for Information Systems:
Making GIS work for your Business
ISBN 0 11 330673 3 £40
Register of GIS in Government
ISBN 0 11 330632 6 £95
GI and GIS Standards
ISBN 0 11 330628 8 £35
Survey of GIS in Government
ISBN 0 11 330635 0 £40
Introduction to GIS
ISBN 0 11 330612 1 £30
GIS in Government: Realizing the
Opportunities
ISBN 0 11 330607 5 £25
GIS: A Buyer's Guide
ISBN 0 11 330606 7 £25
Available from HMSO Bookshops,
Agents (see Yellow Pages:
Booksellers) and through all good
booksellers.
Credit card orders
Tel: 0171 873 9090
fax: 0171 873 8200

HollyBush Software Limited

Innovation House
Dr William Price Business
Centre
Pontypridd
Mid Glamorgan
CF37 1TJ

telephone: 01443 482785
fax: 01443 482788

contact: Peter Jolly
product: Software

Hunting Aerofilms Limited

Gate Studios
Station Road
Borehamwood
Hertfordshire
WD6 1EJ

telephone: 0181 207 0666
fax: 0181 207 5433

contact: R.C.A.Cox
product: Data, Publications,
Complete Systems

Hunting Engineering Ltd

 Reddings Wood
Ampthill
Bedfordshire
MK45 2HD

telephone: 01525 841000
fax: 01525 405861

contact: Robert Brevitt
product: Software, Complete
Systems
services: Audit, Data,
Implementation, Procurement
Providers of turnkey solutions for
information management systems.
With worldwide experience of
system integration in a range of
specialist fields including military
planning and environmental
monitoring.

Hunting Technical Services Ltd

Thamesfield House
Boundary Way
Hemel Hempstead
Hertfordshire
HP2 7SR

telephone: 01442 231800
fax: 01442 219886
email: remote-
sensing.hts@cityscape.co.uk

contact: Graham Deane/Mike
Whitelegge
services: Data,
Implementation, Training
GIS Consultancy services including:
- Data Capture, Conversion;
- Integration of remotely sensed
data;
- Data analysis, training courses,
design and implementation of GUI
Using Proprietary packages.

Husky Computing Limited

Walsgrave Triangle Business
Park
Eden Road
Coventry
West Midlands
CV2 2TB

telephone: 01203 604040
fax: 01203 603060

contact: Christine Smith
product: Hardware, Complete
Systems
services: Training

I.S. Ltd

Maggs House
78 Queens Road
Clifton
Bristol
BS8 1QX

telephone: 0117 925 0553
fax: 0117 925 0663
email: 100044.745@compuserve.com

contact: Neil Quarmby/
Richard Selby
product: Software, Complete
Systems
services: Data,
Implementation, Management
Services

IBM UK Ltd

PO Box 31
Warwick
Warwickshire
CV34 5JL

telephone: 01926 464336
fax: 01926 311345
email: gbibm9t5@ibmmail

contact: GIS Manager
product: Hardware,
Publications, Software,
Complete Systems
services: Audit,
Implementation, Procurement

ICL Ltd

 127 Hagley Road
Edgbaston
Birmingham
B16 8LD

telephone: 0121 456 1111
fax: 0121 455 0358

contact: Alan Roden
product: Hardware,
Publications, Software,
Complete Systems
services: Data,
Implementation, Management
Services, Procurement,
Recruitment, Training

IMASS Limited

 Eldon House
Regent Centre
Gosforth
Newcastle-upon-Tyne
NE3 3PW

telephone: 0191 213 5555
fax: 0191 213 0526
email: Ursula.Cooke@IMASS.co.uk

contact: Ursula Cooke
product: Data, Hardware,
Software, Complete Systems
services: Data,
Implementation, Management
Services, Training

IMASS is an applications developer
and systems integrator specialising
in GIS for utilities, local
government, emergency services and
environmental sectors.

IME (UK) Ltd

Alloa Business Park
Alloa
FK10 3SA

telephone: 01259 210210
fax: 01259 217303

contact: Lisa Currid
product: Hardware, Software,
Complete Systems
services: Implementation,
Management Services,
Procurement

Infolink Decision Services Limited

Coombe Cross
Croydon
Surrey
CR0 1DL

telephone: 0181 686 7777
fax: 0181 680 8295

contact: Ian Liddicoat
product: Data, Software
services: Data

Informed Solutions Limited

11th Floor,
Regent House
Heaton Lane
Stockport
Cheshire
SK4 1BS

telephone: 0161 476 6716
fax: 0161 476 6710
email: markk@informed.co.uk

contact: Dr Mark Ketteman
services: Audit, Data,
Implementation, Management

Services, Procurement,
Training

Informed Solutions is an independent
Consultancy organisation providing a
comprehensive range of professional
IT consulting services from high level
strategic, business and operational
modelling to bespoke information
systems development and
implementation. We have extensive
experience in strategic Consultancy,
programme and project management,
systems integration and can provide
complete life cycle support.
Our focus is on effectively using
leading edge technologies such as
GIS, relational databases,
geodemographics and field data
systems to deliver maximum business
benefit to our clients, assisted as
appropriate by the pragmatic use of
structured design and project
management methods.
We work extensively within the Local
and Central Government, Utilities,
Telecommunications, Cable and
Retail sectors.

Ingecon B.V.

P.O. Box 164
6720 AD Bennekom
The Netherlands

telephone: +31 8389 16681
fax: +31 8389 18648

contact: Maarten van Heest
services: Data,
Implementation, Management
Services, Procurement,
Training

Institute of Hydrology

 McLean Building
Crownmarsh Gifford
Wallingford
Oxfordshire OX10 8BB
telephone: 01491 838800
fax: 01491 832256

contact: Mr.R.V.Moore
product: Data, Publications,
Software, Complete Systems
services: Data,
Implementation, Management
Services

Institute of Terrestrial Ecology

ITE Monks Wood
Huntingdon
Cambridgeshire PE17 2LS
telephone: 014873 381
fax: 014873 467
email: t.moffat@uk.ac.ite

contact: Timothy Moffat
product: Data, Publications
services: Data

Intera Information Technologies

200, 2 Gurdwara Road
Nepean
Ontario
Canada K2E 1A2
telephone: +1613 226 5442
fax: +1613 226 5529

contact: Robert Dams
product: Data, Software,
Complete Systems
services: Data,
Implementation, Media,
Management Services,
Procurement, Training

Intergraph (UK) Ltd

 Delta Business Park
Great Western Way
Swindon
Wiltshire SN5 7XP
telephone: 01793 619999
fax: 01793 618508
email:
pmingins@swindon.swindon.ingr.com

contact: Peter Mingins
product: Hardware,
Publications, Software,
Complete Systems
services: Audit, Data,
Implementation,
Procurement, Training

Intergraph offer the user a fully
scalable GIS solution on all-
Windows operating systems. Thus
allowing the full integration of
technical and business applications.

International Map Trade Association

5 Spinacre
Becton Lane
Barton on Sea
Hampshire BH25 7DF
telephone: 01425 620532
fax: 01425 620532

contact: Mike Cranidge
product: Publications

International Products

14 King Henry Mews
Enfield Lock
Enfield
Middlesex EN3 6JS
telephone: 01992 651695
fax: 01992 651695
email: 100342.12@compuserve

contact: John Lessing
product: Data, Hardware,
Software, Complete Systems
services: Data, Media,
Training

IT Southern Ltd

 Southern House
Lewes Road
Brighton
East Sussex
BN1 9PY

telephone: 01273 600444
fax: 01273 675299
email: ah78@solo.pipex.com

contact: Charles Gray
product: Hardware, Software,
Complete Systems
services: Audit, Data,
Implementation, Media,
Procurement

ITC

350 Boulevard 1945
P.O. Box 6
7500 AA Enschede
The Netherlands

telephone: +31 53 874444
fax: +31 53 874400
email: ilwis@itc.nl

contact: Hans Melos
product: Publications,
Software, Complete Systems
services: Implementation,
Training

ILWIS Software description
Vector data
Raster data
Point data
Data Base Management System
ILWIS is a PC-based package which
combines the benefits of image
processing capabilities, digitizing,
conversion between raster and vector
data, rasterbased spatial modelling
and tabular databases in a compact
system. One of its most powerful
features is the fast interactive
overlaying of multiple raster maps
and remotely-sensed images
integrated with attribute data analysis.
ILWIS offers data capture, import and
export routines with widely-used
vector-raster-atrribute data formats,
data management (querying, data
structure conversion, table management), visualisation, analysis (spatial
modelling and remote sensing) and
output/presentation options. ILWIS
offers the choice for a menu-driven or
command-line user interface.

Hardware
IBM compatible DOS PC's with
mathematical co-processor
Monochrome screen for user/system
interaction. Colour screen for
graphics display
In the fall of 1985, an ILWIS version
under Windows is expected to be
launched. It will be accompanied by
an extended set of training materials,
demos and a 'Getting Acquainted' to
help the user to familiarize him or
herself with the software.

ITS: Intertrade Scientific Ltd

Linford Wood Business
Centre
Sunrise Parkway
Milton Keynes
Buckinghamshire
MK14 6LP

telephone: 01908 676633
fax: 01908 666595

contact: Matt Costin
product: Hardware, Software
services: Implementation

J.C.White Chartered Land Surveyors

Lemanis House
Stone Street
Lympne
Kent
CT21 4JN

telephone: 01303 261212
fax: 01303 264040

contact: M P Heiman
services: Data, Media,
Recruitment

John D Leatherdale FRICS

10 Bartrams Lane
Hadley Wood
Barnet
Hertfordshire
EN4 0EH

telephone: 0181 449 0123
fax: 0181 449 0123

contact: John Leatherdale
services: Audit, Data,
Implementation, Management
Services, Procurement,
Recruitment, Training
Independent Consultancy since 1987
in GIS, land information
management and commissioning
surveys and mapping for
international development,
government, energy and construction
projects, and private sector.

Kalidor Europe (ALPS Electric (Ireland) Ltd)

Clara Road
Millstreet Town
County Cork
Ireland
telephone: +353 29 21212
fax: +353 29 21213
email: jcostigan@alps.ie

contact: Judy Costigan
product: Hardware
Withstanding severe environmental
conditions the 486-based Kalidor

K2100 pen computer is water and
dust resistant, has been shock and
drop tested and can operate in
temperatures up to 50 degrees
centigrade.
Weighing 1.5 Kgs it comes with a
170 MB independently ruggedised
Hard Drive, up to 16 MB RAM and
uses a NiMH battery.

Kamyco International

43 West Farzan Avenue
Vali-Aser Avenue
Tehran 19688
Iran
email: kamyco5-E

contact: Dr.Ahmadi
services: Data,
Implementation,
Procurement, Recruitment

Keele University

 Keele
Staffordshire
ST5 5BG

telephone: 01782 583078
fax: 01782 713082
email: michael@cs.keele.ac.uk

contact: Dr Michael F
Worboys
services: Training
Keele University has an
international reputation for training,
research and Consultancy in GIS.
The MSc in GIS recruits graduates
from all disciplines. Academic
partner Smallworld.

Kingswood Consulting Limited

 449 Chiswick High Road
London
W4 4AU

telephone: 0181 995 2050
fax: 0181 747 8047
email: 100322.205@compuserve.com

contact: Dan Rickman
services: Audit, Data,
Implementation, Management
Services, Procurement,
Training
We provide independent assistance
as a 'one stop shop' or specific help
for initial concept, strategy,
feasibility, financial analysis,
specification, system selection,
implementation, system integration.

Kirstol Ltd

Cheethams Park Estate
Stalybridge
Cheshire
SK15 2BT

telephone: 0161 338 7512
fax: 0161 338 8097

contact: Peter Dickinson
product: Hardware, Software,
Complete Systems
services: Data, Media

KJB Consulting

6 The Glebeland
Egerton
Ashford
Kent
TN27 9DH

telephone: 01233 756461

contact: Keith Burleton
services: Implementation,
Management Services,
Procurement, Training

Know Edge Ltd

33 Lockharton Avenue
Edinburgh
EH14 1AY

telephone: 0131 443 1872
fax: 0131 443 1872
email: 100042.3122@compuserve.com

contact: Robin A McLaren
services: Audit, Management
Services, Procurement,
Recruitment, Training

Know Edge Ltd is the oldest,
specialist LIS/GIS consulting
company in the UK providing truly
independent advice from strategy to
implementation. Worldwide
experience in formulating and
implementing corporate LIS/GIS
strategies within the private and
public sectors. Clients have
included: EU Statistics Office; EU
PHARE Programme; United Nations
FAO; Ordnance Survey GB;
European Utilities; Banks; and
Central & Local Government
organisations.
Application areas have included:
power & water utilities; financial
services sector; environmental
agencies; property management;
mapping services; cadastre and land
registration agencies.
Specialist services include: the
formulation of business lead
corporate strategies; feasibility
studies; support of tendering; system
benchmarking; implementation
planning including institutional and
organisational aspects; business
process re-engineering; project

management; and audits of mature
installations.
Nine years of GIS activities have
provided Know Edge with expertise
in the management, organisational
and technical aspects of GIS
projects. The multi-disciplinary
Know Edge team provides clients
with the essential ingredient of
practical knowledge of
implementation issues. The team is
augmented by a network of
internationally recognised experts in
creating consulting teams tailored to
the specific requirements of clients.
Over 90% of clients have invested in
LIS/GIS following feasibility studies
conducted by Know Edge Ltd.

KPMG

 8 Salisbury Square
Blackfriars
London
EC4Y 8BB

telephone: 0171 311 1000
fax: 0171 311 3311

contact: Richard Goodwin
services: Audit, Data,
Implementation, Management
Services, Recruitment,
Training

KPMG is one of the UK's largest
providers of professional services in
both the public and private sectors.
Our multi-disciplinary network of
GIS consultants covers all business
sectors, and offers an experienced
and effective team able to meet a
wide range of projects, including:
* IS and IT strategies
* Cost justification of GIS
* Data ownership and management
issues
* Human resources and management
of change
* Risk management
* Benefits realisation
* Programme management
* Executive awareness
* Feasibility studies
* Project Management
* System selection
* Procurement advice
* Implementation management
* Post-implementation reviews and
system audits

L.E.S. (Computer Services) Ltd

72 Croydon Road
Reigate
Surrey RH2 0NH
telephone: 01737 223899
fax: 01737 223911

contact: W.D.Rees

product: Hardware, Software,
Complete Systems
services: Audit, Data, Media,
Training

Lancaster University

 Lancaster
LA1 4YB

telephone: 01524 593762
fax: 01524 847099
email: ha001@lancaster.ac.uk

contact: Dr A.C.Gatrell
product: Publications,
Software
services: Data, Training

Land and Satellite Surveys

Unit 7, Hambridge Farm
Newbury
Berkshire RG13 2QG
telephone: 01635 49512
fax: 01635 523809

contact: Richard Andrews
services: Data,
Implementation, Management
Services

Land Aspects Consultancy Ltd (Parkman Group)

 Parkman House
Lloyd Drive
Ellesmere Port
South Wirral L65 9HQ
telephone: 0151 356 1666
fax: 0151 356 1119
email: jbridge@parkman.co.uk

contact: Janet Bridge
product: Data
services: Data, Media,
Management Services,
Procurement

Lantmatenet GIS-centrum

Vasagaton 11
S-111 20 Stockholm
Sweden

telephone: +46 8402 1700
fax: +46 8791 7151

contact: Ann Grengmark
product: Data, Software,
Complete Systems
services: Data,
Implementation, Media,
Management Services,
Recruitment, Training

LASCO Ltd

Station Approach
Arundel
West Sussex
BN18 9JL
telephone: 01903 882466
fax: 01903 882599

contact: D.H.Rollason
services: Data, Recruitment

Laser Technology International Ltd

27 Dudley Road
Tunbridge Wells
Kent TN1 1LE
telephone: 01892 863351
fax: 01892 863431

contact: Peter Fasey
product: Hardware

Laser-Scan Ltd

 Cambridge Science Park
Cambridge
Cambridgeshire CB4 4FY
telephone: 01223 420414
fax: 01223 420044
email: annef@lsl.co.uk

contact: Jenny Nash
product: Hardware, Software, Complete Systems
services: Data, Implementation, Management Services, Training

Over 25 years experience in digital mapping and GIS technology, providing a stable, yet versatile platform along with solutions to meet your needs now and in the future.

Leica UK Ltd

 Davy Avenue
Knowlhill
Milton Keynes
Buckinghamshire MK5 8LB
telephone: 01908 666663
fax: 01908 609992

contact: Deborah Saunders
product: Hardware, Software, Complete Systems

LiveChart

21 Premier Way
Abbey Park
Romsey
Hampshire SO51 9AQ
telephone: 01794 518085
fax: 01794 518086

contact: Garry Symes
product: Data, Hardware, Software, Complete Systems
services: Data, Management Services

Logica UK Limited

 Medina House
Business Park 4
Randalls Way
Leatherhead
Surrey
KT22 7TW

telephone: 0171 637 9111
fax: 01372 227007

contact: Stephen Darvill
product: Complete Systems
services: Audit, Data, Implementation, Management Services, Procurement

LOGICA has implemented successful systems (large and small) incorporating a variety of GIS products in different markets.
Services include :
• Consultancy
  * where and how to apply GIS
  * GIS product selection
• System design, implementation and integration with other systems.
• Ongoing application support
LOGICA is an independent international system integrator with >2600 staff.

Logitrans

34 av Aristide Briand
94110 Arcueil
France

telephone: +331 498 51516
fax: +331 498 51550

contact: William Gorter
product: Hardware, Complete Systems
services: Data, Media, Procurement, Training

London Research Centre

 Parliament House
81 Black Prince Road
London
SE1 7SZ

telephone: 0171 735 4250
fax: 0171 627 9606

contact: John Hollis
product: Data, Publications
services: Data, Training

Longdin & Browning

50 Sketly Road
Swansea
SA2 0LH

telephone: 01792 202244
fax: 01792 203333
email: 100271.2433@compuserve.com

contact: Nick Eales
product: Software, Complete Systems
services: Data, Media

Longman Group Ltd

Longman House
Burnt Mill
Harlow
Essex CM20 2JE
telephone: 01279 623623
fax: 01279 431059

contact: Nila Patel
product: Publications

Lovell Johns Ltd

10 Hanborough Business Park
Long Hanborough
Witney
Oxfordshire OX8 8LH
telephone: 01993 883161
fax: 01993 883096

contact: Richard Hewish
product: Data, Software
services: Data, Implementation, Management Services

LOY Surveys Ltd

 1 Paisley Road
Renfrew PA4 8JH
telephone: 0141 885 0800
fax: 0141 885 1202

contact: Jim Loy
services: Data, Media
DATA CAPTURE
LOY Surveys offer the GIS Market surveying and mapping services using the latest technology such as GPS, Penmap, etc. A Nationwide Service.

LTG Services

Page Street
Mill Hill
London NW7 2ER
telephone: 0181 906 5559
fax: 0181 906 5248

contact: John Callaghan
product: Software
services: Audit, Data, Media, Management Services, Procurement, Training

Macaulay Land Use Research Institute

 Craigiebuckler
Aberdeen AB9 2QJ
telephone: 01224 318611
fax: 01224 311556
email: r.aspinall@uk.ac.sari.mluri

contact: Dr Richard J.Aspinall
product: Data, Publications, Software
services: Management Services

Magdala Sociedade

Magdala House
Azinhaga da Eira
Alcoitao
2765 Estoril
Portugal

telephone: +351 1 460 0684

contact: Peter Wallace
services: Audit,
Implementation, Management
Services, Procurement,
Training

Magellan Systems Corporation

960 Overland Court
San Dimas
CA 91773
USA

telephone: +1909 394 5000
fax: +1909 394 7050

contact: Jim White
product: Hardware, Software,
Complete Systems
Magellan systems offers a variety of
GPS positioning and data collection
products for mapping, survey and
GIS.

Manchester Metropolitan University

 John Dalton Building
Chester Street
Manchester
M1 5GD

telephone: 0161 247 1581
fax: 0161 247 6344
email: b.heyworth@mmu.ac.uk

contact: Ms Beverly
Heyworth
services: Training
The Universities of Huddersfield,
Salford and Manchester
Metropolitan offers a joint
Postgraduate Diploma/MSc in GIS
by distance learning. For further
details contact Ms Beverly
Heyworth Telephone 0161 247 1581

Map Data Management Ltd

Townend
Farleton
Camforth
Lancashire LA6 1PB
telephone: 015395 67431
fax: 015395 67837

contact: Philip Storey
product: Data, Publications,
Software, Complete Systems
services: Data,
Implementation, Management
Services, Training

MapInfo Ltd

Centennial Court
Easthampstead Road
Bracknell
Berkshire
RG12 1YQ

telephone: 01344 482888
fax: 01344 482777
email: matthew.spencer@mapinfo.com

contact: Matthew Spencer
product: Data, Publications,
Software, Complete Systems
Markets: All organisations
including: Telecommunications,
insurance, transportation, healthcare
and the business functions of sales,
marketing, customer service and
operations

MapInfo, MapBasic
Platform: Windows, Macintosh,
Sun, HP UNIX
MapInfo Corporation is the leading
developer of easy-to-use, multiple
platform (Windows, Macintosh, Sun,
Hewlett-Packard) desktop mapping
software for data visualisation and
geographic analysis. Remote client/
server data access via SQL DataLink
and customised vertical applications
with the MapBasic Development
Environment make MapInfo
accessible to the enterprise,
workgroups and end-users. A
comprehensive range of
international data is available.

MAPIT Limited

1 Lawrence Road
Ramsey
Huttingdon
Cambridgeshire
PE17 1UY

telephone: 01487 813745
fax: 01487 813745

contact: Douglas Cross
product: Software
services: Implementation

MAPS geosystems

Corniche Plaza 1
Sharjah
United Arab Emirates

telephone: +9716 356411
fax: +9716 354057

contact: Alistair MacKenzie
product: Data, Hardware,
Software, Complete Systems
services: Data,
Implementation, Procurement,
Recruitment, Training

Mason Land Surveys

 Dickson Street
Dunfermline
Fife KY12 7SL
telephone: 01383 727261
fax: 01383 739480

contact: Robert Owen
product: Data, Hardware,
Publications, Software,
Complete Systems
services: Audit, Data,
Implementation, Media,
Training

Mathshop

127 Middlebridge Street
Romsey
Hampshire SO51 8HH
telephone: 01794 523423

contact: Timothy J.Coffey
services: Data, Procurement

McLintock Limited

244 Barns Road
Cowley
Oxford OX4 3RW
telephone: 01865 749957
fax: 01865 749434

contact: Chris Hallos-
Johnson
product: Data, Hardware,
Publications, Software,
Complete Systems
services: Audit, Data,
Implementation, Media,
Management Services,
Procurement, Recruitment,
Training

MEGRIN Group

abs Institut Geographique
National
2 avenue Pasteur
BP68, 94160 St. Mande
France

telephone: +331 43 988440
fax: +331 43 988443
email: megrin@megrin.ign.fr

contact: Frangois Salge
product: Data

Mentis Management Consultants Ltd

9 Westminster Court
Hipley Street
Woking
Surrey
GU22 9LQ
telephone: 01483 776717
fax: 01483 747337
email: 100646.3576@compuserve.com

contact: Mike Gunner
services: Audit, Data, Implementation, Management Services, Procurement, Recruitment, Training
The Mentis GIS Advisory Centre offers an independent advice service on GIS products, development of business solutions, project management, and integration into your corporate strategy.

Methods Applications Ltd

39 King Street
Covent Garden
London
WC2E 8JS
telephone: 0171 240 1121
fax: 0171 379 8561
email: mal@methods.win-uk.net

contact: Peter Rowlins
services: Data, Implementation, Management Services, Procurement

Midlands Regional Research Laboratory

Bennett Building
University of Leicester
Leicester
LE1 7RH

telephone: 0116 252 3825
fax: 0116 252 3854
email: ajs@le.ac.uk

contact: Dr.Alan.J.Strachan
product: Publications, Software
services: Audit, Data, Implementation, Management Services, Training

Midsummer Computing

Midsummer House
429 Midsummer Boulevard
Milton Keynes
Buckinghamshire
MK9 2HE

telephone: 01908 668866
fax: 01908 667623

contact: Richard Avery
product: Data, Hardware, Software, Complete Systems
services: Data, Implementation, Management Services, Procurement, Recruitment, Training

Modern Maps

Ashmoore House
62 Craven Road
Newbury
Berkshire RG14 5NJ

telephone: 01635 34251
fax: 01635 34251

contact: Tony Vickers
services: Audit, Data, Implementation, Management Services, Procurement, Training
* manager, communicator, geographer, surveyor, analyst
* strategic level projects involving envisioning and lateral thinking
* alone or in association with partners
* independently going where GIS is going…

Morgan Collis Group Ltd

The Mill
Millbrook Close
St James Mill Road
Northampton
NN5 5JF

telephone: 01604 580980
fax: 01604 589208
email: ian@mcgltd.demon.co.uk

contact: Ian Holroyd
product: Software, Complete Systems
services: Data, Implementation, Procurement
Morgan Collis provides an integrated range of engineering specific software solutions suitable for all organisations involved with water and wastewater network information and its use.

MOSS Systems Limited

 MOSS House
North Heath Lane
Horsham
West Sussex
RH12 5QE

telephone: 01403 259511
fax: 01403 217746

contact: Carol Heaton
product: Software

Mott MacDonald Ltd

20-26 Wellesley Road
Croydon
Surrey
CR9 2UL

telephone: 0181 686 5041
fax: 0181 681 5706
email: gsb@mm-croy.mottmac.co.uk

contact: Graham Bugler
product: Software, Complete Systems
services: Audit, Data, Implementation, Procurement

MPSI Systems Ltd

Castlemead
Lower Castle Street
Bristol
BS1 3AG

telephone: 0117 927 9653
fax: 0117 929 2056

contact: Mr.A.Renshaw
product: Data
services: Data, Implementation

MR Data Graphics

Unit 12, The Pines Trading Estate
Broad Street
Guildford
Surrey
GU3 3BH

telephone: 01483 575312
fax: 01483 300167

contact: Pat Jarvis
product: Data
services: Data, Management Services, Procurement
MR Data Graphics GIS Services, from pilot to completion and implementation. Consultancy, conversion, digitising, colour and binary scanning. Supply of Ordnance Survey raster maps 1:10000 and 1:50000

Munro Garratt International

4 Albert Street
Aberdeen
AB1 2AH

telephone: 01224 622888
fax: 01224 622229

contact: David Field
product: Complete Systems
services: Data, Implementation, Training

MVA Systematica

MVA House
Victoria Way
Woking
Surrey
GU21 1DD

telephone: 01483 728051
fax: 01483 755207

contact: Andy Heath
product: Data, Complete Systems
services: Audit, Data, Procurement, Training

MVM Consultants plc

 MVM House
Oakfield Road
Clifton
Bristol
BS8 2AL

telephone: 0117 974 4477
fax: 0117 970 6897

contact: David Rix
product: Software
services: Audit,
Implementation, Management
Services, Procurement

Since 1987 MVM have been providing database and GIS business solutions to a range of national and international private and public sector clients. The MVM Group offers a range of services designed to assist clients with development and implementation of GIS. Our professional services are provided within the framework of an organisation's objectives and overall business development plan. Services include strategic planning for GIS; feasibility studies; systems evaluation, procurement, design and implementation; database design and development; systems integration; and application development.
MVM are also able to supply a range of local government applications within a range of database and GIS environments.

NAG Ltd

Wilkinson House
Jordan Hill Road
Oxford
OX2 8DR

telephone: 01865 511245
fax: 01865 310139
email: infodesk@nag.co.uk

contact: Dr Terry Burgess
product: Software
services: Data,
Implementation, Management
Services

National Remote Sensing Centre Ltd (NRSC)

 Delta House
Southwood Crescent
Southwood
Farnborough
Hampshire GU14 0NL
telephone: 01252 541464
fax: 01252 375016

contact: Peter Beaumont
product: Data, Complete
Systems
services: Data,
Implementation, Management
Services, Procurement,
Training

NRSC is a world leader in the supply, management and applications of remotely sensed data from satellites and aircraft. Data is supplied from all the major commercial satellites and stereo colour aerial photography of the UK at scales of 1:25 000, 1:10,000, 1:5000 and 1:2500 as hardcopy and digital orthophotos.
Products include:
* digital orthophotos
* image maps
* digital elevation models generated from stereo aerial photography and satellite imagery
* derived vector datasets; e.g. road centre lines
* thematic datasets; e.g. land cover maps, signal clutter
NRSC offers an independent GIS Consultancy providing a range of professional services including: applications development, feasibility studies, benchmark evaluation, database design, system implementation, data conversion, project management, training.
The Airphoto Group based at the Barwell Office provide additional expertise in the application of aerial photography and GIS technologies for town planning, terrain clutter, wildlife habitat mapping, new road development, woodland inventories, forest management, mapping, archaeology, geology – soils and geomorphology, erosion and flood defence, river corridor studies, insurance evaluation and legal disputes.
Headquarter, GIS Consultancy
NRSC, Farnborough, see above
DATA Sales, Airphoto Group
NRSC, Arthur Street, Barwell, Leicestershire, LE9 8GZ
Tel : 01455 844513, FAX:01455 841785

Natural Environment Research Council

Holbrook House
Station Road
Swindon
Wiltshire
SN15 4ES

telephone: 01793 411996
fax: 01793 411959
email: g.darwall@nss.nerc.ac.uk

contact: Mr.G.H.D.Darwell
product: Data, Publications
services: Data,
Implementation, Management
Services

Navstar Systems Ltd

 Mansard Close
Westgate
Northampton
NN5 5DL

telephone: 01604 585588
fax: 01604 585599

contact: Clive de la Fuente
product: Hardware, Software,
Complete Systems

NCC Blackwell Ltd

108 Cowley Road
Oxford
OX4 1JF

telephone: 01865 791100
fax: 01865 798210
email: akitson@cix.compulink.co.uk

contact: Anne Kitson
product: Publications

NERC

 Polaris House
North Star Avenue
Swindon
Wiltshire
SN2 1EU

telephone: 01793 411683
fax: 01793 411610
email: ghdd@nss.nerc.ac.uk

contact: George H D Darwall
product: Data, Publications
services: Data,
Implementation, Management
Services

Nestor International Ltd

MacMillan House
96 Kensington High Street
London
W8 4SG

telephone: 0171 937 4434
fax: 0171 937 4180

contact: Jill Farmer
services: Recruitment

NOMIS

Unit 3P
Mountjoy Research Centre
Durham
DH1 3SW

telephone: 0191 374 2468
fax: 0191 384 4971
email:
michael.blakemore@uk.ac.durham

contact: Michael Blakemore
product: Data, Publications
services: Data, Management
Services, Training

Numonics UK (Division of Telmtek Ltd)

Alfreton Road
Derby
DE21 4AD

telephone: 01332 298480
fax: 01332 290667

contact: Clive Rowe
product: Hardware
services: Data,
Implementation

Oaklands I.T.

2 Oaklands
Bunce Common Road
Leigh
Surrey
RH2 8NS

telephone: 01306 611590
fax: 01306 611590

contact: Robert Coote
services: Data,
Implementation, Management
Services

An experienced Independent
Consultancy providing practical
support on a range of PC/Windows
database and GIS Software.
Currently engaged in address based
analysis from existing data.

Office of Population Censuses and Surveys (OPCS)

 Segensworth Road
Titchfield
Fareham
Hampshire
PO15 5RR

telephone: 01329 813536
fax: 01329 813532

contact: Alan Taylor
product: Data

Optimal Software Ltd

Highbank
Halton Street
Hyde
Cheshire
SK14 2NY

telephone: 0161 367 8715
fax: 0161 367 9328

contact: D Brady
product: Data, Hardware,
Publications, Software,
Complete Systems
services: Data,
Implementation, Management
Services, Training

Ordnance Survey

 Ordnance Survey

 Romsey Road
Southampton
Hampshire
SO16 4GU

telephone: 01703 792773
fax: 01703 792324
email: ordsvy.govt.uk

contact: Digital Sales
product: Data, Publications
services: Data,
Implementation, Training

Ordnance Survey is the national
mapping organisation for Great
Britain, providing definitive, up-to-
date digital map data to a wide range
of organisations for business and
professional use.

The **LAND-LINE™** family of
products covers Urban, Rural and
Moorland areas at 1:1250, 1:2500
and 1:10 000 scale respectively.
Land-Line provides a detailed
representation of the real world for
applications requiring accurate large
scale mapping. Land-Line is
continually updated with the latest
information.

ADDRESS-POINT™ is the
definitive tool for the identification
and use of precisely located
residential, business and public
postal addresses in Great Britain.
ADDRESS-POINT can be used
within a database environment, in
command and control systems and in
GIS.

Land-Form PANORAMA™ is the
digital representation of the contours
from 1:50 000 Landranger® maps,
available as digitised contours or as
a Digital Terrain Model. Land-Form
PANORAMA has applications for
the telephone and cable industries
and for environmental impact
analysis.

Land-Form PROFILE™ is the most
accurate and definitive height
dataset available for Great Britain. It
provides digital representation of the
contours from 1:10 000 scale maps,
available as digitised contours or as
a Digital Terrain Model. It is
particularly suitable for planning
applications, civil engineering
schemes, flood risk assessment and
visual impact analysis.

The **OSCAR®** family of products
consists of four road data products
derived from a single definitive
source.

OSCAR Asset Manager provides a
comprehensive, detailed
representation of the public and
private road network, including
pedestrianised streets in major city
centres. It is designed to meet the
needs of those involved with road
infrastructure at a detailed level.

OSCAR Traffic Manager mirrors
the driveable road network
providing a digital road template
suitable for in-car navigation
applications.

OSCAR Route-Manager represents
a simplified road network using
filtered road geometry and **Oscar
Network Manager** represents a
further simplification of the road
network and elimination of geometry
from the road centre line links.

Boundary-Line™ is the definitive
maintained dataset of electoral and
administrative boundaries in
England and Wales derived from
1:10 000 scale Boundary Records.

**1:10 000 Scale Black and White
Raster Data** provides all the detail,
excluding contours, from the 1:10
000 scale published map series. The
data can be displayed on a computer
screen as a map image, or as a
backdrop to vector data.

1:50 000 Scale Colour Raster Data
provides all the map detail from the
Landranger series. The data can be
displayed on a computer screen as a
map image, or as a backdrop to
vector data.

Meridian™ is Ordnance Survey's
new mid-scale vector data product
created by selecting and customising
features from a variety of existing
and maintained databases. It
contains comprehensive road detail,
railway features and developed land
use and settlement names to provide
a high quality entry-level product for
GIS users.

Strategi™ is a geometrically
structured 1:250 000 vector database
that performs as a strategic planning
and analytical tool or can be used for
backdrop display.

BaseData.GB™ is a structured
1:625 000 vector database which
provides suitable entry-level data for
a GIS, or can be used for backdrop
display, giving an overview of Great
Britain.

All digital map data products can be
manipulated, enhanced or simplified
to meet the precise needs of
customers. For further information
about **Customised Data** telephone
01703 792076.

The OS Initiative is a planned,
sustained programme committed to
providing solutions to end users of
geographic information by the
involvement of approved Value
Added Resellers. Ordnance Survey
welcomes enquiries from potential
VARs.

**For more information telephone
01703 792789.**

Ordnance Survey of Northern Ireland

 Colby House
Stranmillis Court
Belfast
BT9 5BJ

telephone: 01232 255755
fax: 01232 255700

contact: Michael Brand
product: Data, Publications
services: Data,
Implementation, Media,
Management Services,
Training

The Northern Ireland GIS centre, based at Colby House, offers a unique opportunity for organisations to explore and evaluate, by hands on experience, a wide range of Geographic Information Systems. OSNI also offer GIS Consultancy, Integration and Implementation and Data Capture facilities, proving a cost effective solution for customers needs.

Organisation Management Systems

Old Butchers Shop
North Street
Islip
Oxford
OX5 2SQ

telephone: 01865 372161
fax: 01865 842664

contact: Jon J Gibbons
product: Data, Software,
Complete Systems
services: Data,
Implementation, Management
Services, Procurement,
Training

Oscar Faber

 Marlborough House
Upper Marlborough Road
St Albans
Hertfordshire
AL1 3UTXE

telephone: 0181 784 5784
fax: 0181 784 5700
email:
marketing@oscarfab.demon.co.uk

contact: Shaun Everett
product: Software, Complete
Systems
services: Audit, Data,
Implementation, Training

Oscar Faber, founded in 1921, provides GIS Consultancy, services and products, within the framework of an established engineering Consultancy.

Clients include Utilities, Rail and Road Transportation, Local and Central Government, industrials and Commerce.
Oscar Faber's THESIS GIS Product is an engineering applications development environment, and is also used as a fast-track development tool to initiate management and user feedback.
Oscar Faber has extensive experience of Microsoft Windows and Windows NT development, and the Map32 product provides budget-priced 32-bit cartographic facilities for these platforms
To complement GIS skills, Oscar Faber also provides data modelling, telecoms, networking, facilities management and data services.

Ove Arup & Partners

Cambrian Buildings
Mount Stuart Square
Cardiff
CF1 6QP

telephone: 01222 473727
fax: 01222 472277

contact: Simon Power
services: Data,
Implementation, Management
Services, Procurement

Oxford Institute of Retail Management

Templeton College
Kennington
Oxford
Oxfordshire
OX1 5NY

telephone: 01865 735422
fax: 01865 736374
email: oxirm@uk.ac.ox.temcol

contact: Christopher Talbot
services: Data, Procurement,
Training

P&L Engineering Surveys Ltd

10 Eaglesham Road
Clarkston
Glasgow
G76 7BT

telephone: 0141 644 1690
fax: 0141 644 5361

contact: Richard Lennox
services: Data

PA Consulting Group

 123 Buckingham Palace Road
London
SW1W 9SR
telephone: 0171 730 9000
fax: 0171 333 5050
email: phil.jeanes@pa-consulting.com

contact: Phil Jeanes
services: Audit,
Implementation, Management
Services, Procurement,
Recruitment

PAFEC Ltd

Strelley Hall
Nottingham
NG8 6PE

telephone: 0115 935 7055
fax: 0115 935 7057
email: appwmm@pafec.co.uk

contact: Amanda Ward
product: Software, Complete
Systems
services: Data,
Implementation, Management
Services, Training

Panda

48 Impington Lane
Impington
Cambridge
CB4 4NJ

telephone: 01223 233577
fax: 01223 233577

contact: Denis.W.Payne
services: Implementation,
Management Services,
Procurement

Paul Clasper & Associates Ltd

Kelso Villa
Upper Bristol Road
Bath
BA1 3AU

telephone: 01225 444561
fax: 01225 422187
email: clasper@csm.uwe.ac.uk

contact: Paul Clasper
product: Software
services: Data,
Implementation, Management
Services, Procurement

Paul Clasper & Associates Ltd is an independent survey Consultancy providing a range of services, these include :
* Bespoke data capture
* Data QC software
* GIS/GPS integration

PAX Technology

Hither Ascension House
Lambridge Wood Road
Henley-on-Thames
Oxfordshire
RG9 3BS

telephone: 01491 572282
fax: 01491 411040

contact: David J Bethel
product: Data, Software,
Complete Systems
services: Implementation,
Procurement

PD Computing Ltd

63 Cornhill Road
Davyhulme
Manchester
M41 5SZ

telephone: 0161 747 7110

contact: Mr.N.Hoe
product: Software, Complete
Systems
services: Audit, Data,
Implementation, Procurement

Pear Technology Services Ltd

24 First Avenue
Havant
Hampshire
PO9 2QN

telephone: 01705 499689
fax: 01705 499689
email: 100554.2753@compuserve.com

contact: John Cowling
product: Data, Hardware,
Software, Complete Systems
ESS MAPS is a low-cost, highly
functional package equally capable
as a self-contained application or as
the mapping 'engine' for specific
GIS/GPS solutions.

Peter Thorpe Consulting

18 Mercia Avenue
Kenilworth
Warwickshire
CV8 1EU

telephone: 0926 52799
fax: 0926 52799

contact: Peter Thorpe
services: Audit, Data,
Implementation, Management
Services, Procurement,
Training

Photarc Surveys Ltd

Beech House
Beech Avenue
Harrogate
North Yorkshire
HG5 0NA

telephone: 01423 871629
fax: 01423 871639

contact: Rory M.Stanbridge
services: Data

Photoair

Photoair House
191A Main Street
Yaxley
Peterborough
PE7 3LD

telephone: 01733 241850
fax: 01733 242964

contact: Richard Young
product: Data, Publications
services: Data

Photogrammetric Data Services Ltd

 1 Ham Business Centre
Shoreham by Sea
West Sussex
BN43 6PA

telephone: 01273 464883
fax: 01273 454238

contact: Robert Finch
product: Data
services: Data, Management
Services
PDS offers the full range of aerial
survey related services specialising
in digital data capture using
analytical photogrammetry.
Topographic, terrestrial and
volumetric surveys undertaken.

Pinpoint Digitising Services

5 Boleyn Court
Tudor Road
Manor Park
Runcorn
Cheshire
WA7 1SR

telephone: 01928 579148
fax: 01928 579192

contact: Suzanne Soper
product: Data
services: Data

PlanGraphics Inc

202 West Main Street, Ste.200
Frankfort
Kentucky 40601
USA

telephone: +1502 223 1501
fax: +1502 223 1235
email: plang@ipx.netcom.com

contact: Dennis Kunkle
product: Publications
services: Audit, Data,
Implementation, Management
Services, Procurement,
Recruitment, Training

Planning & Mapping Ltd

17 Huffwood Estate
Billingshurst
West Sussex
RH14 9UR

telephone: 01403 783314
fax: 01403 784596

contact: Tony Leeds
product: Data
services: Data, Media

PLANTECH Ltd

South Bank Technopark
99 London Road
London
SE1 6LN

telephone: 0171 922 8825
fax: 0171 928 8066

contact: Sylvie Temperley
product: Publications,
Software
services: Data,
Implementation, Management
Services, Procurement,
Training

Plowman Craven & Associates Ltd

 141 Lower Luton Road
Harpenden
Hertfordshire
AL5 5EQ

telephone: 01582 765566
fax: 01582 765370
email: mark@plowcrav.demon.co.uk

contact: Mark Phillips
services: Data,
Implementation, Media
Wide range of experience enables
PCA to provide a high quality
service, supplying digital spatial
data. Mapping in 2D and 3D,
measured surveys and Consultancy.

Posford Duvivier

Rightwell House
Bretton Centre
Peterborough
Cambridgeshire
PE3 8DW

telephone: 01733 334455
fax: 01733 262243

contact: Tim Jeffries-Harris
services: Data,
Implementation, Management
Services, Procurement

Positioning Resources Limited

64 Commerce Street
Aberdeen
AB2 1BP

telephone: 01224 581502
fax: 01224 574354

contact: Judith Collier
product: Hardware, Software,
Complete Systems
services: Training

Primagraphics Ltd

Melbourn Science Park
Melbourn
Royston
Hertfordshire
SG8 6EJ

telephone: 01763 262041
fax: 01763 262551

contact: Dr Glynn Wright
product: Hardware, Complete
Systems

Procis Software Ltd

 Alexander House
19 Fleming Way
Swindon
Wiltshire
SN1 2NG

telephone: 01793 541200
fax: 01793 541025

contact: Pammi Panesar
product: Hardware,
Publications, Complete
Systems
services: Implementation,
Procurement, Training
Suppliers of the AGI award winning
Mobile-GIS application, which
extends corporate GIS to field
operators in an easy to use but
comprehensive package for portable
computers.

Progis GmbH

Italienerstrasse 3
9500 Villach
Austria

telephone: +43 4242 26332
fax: +43 4242 263327

contact: Di Walter H Mayer
product: Data, Publications,
Software, Complete Systems
services: Data,
Implementation, Training

Property Intelligence plc

 Ingram House
13/15 John Adam Street
London
WC2N 6LD

telephone: 0171 839 7684
fax: 0171 839 1060

contact: Michael Nicholson
product: Data, Publications,
Software, Complete Systems
services: Data

QAS Systems Ltd

 7 Old Town
London
SW4 0JT

telephone: 0171 498 7777
fax: 0171 498 0303

contact: Catherine Meader
product: Data, Software,
Complete Systems
services: Implementation
Addressing software which
generates an address from either a
postcode or partial address. Related
data, including Ordnance Survey's
Address Point data can also be used.

QC Data (Ireland) Limited

QC House
Cork Business & Technology
Park
Model Farm Road
Cork
Ireland
telephone: +353 21 341700
fax: +353 21 343645

contact: Brendan Walshe
services: Data

Quail Map Company

31 Lincoln Road
Exeter
EX4 2DZ

telephone: 01392 430277
fax: 01392 430277

contact: John Yonge
product: Publications

Quorum Information Services

Victoria House
Victoria Road
Aldershot
Hampshire
GU11 1DB

telephone: 01252 318884
fax: 01252 313120

contact: Malcolm Orchard
services: Data
Quorum have 26 years experience in
document management including
document image processing.
Scanning services include paper
conversion and scanning from all
microforms.

R.W.A. Dallas FRICS

23 East Mount Road
York
YO2 2BD

telephone: 01904 652408
fax: 01904 652408

contact: R.W.A.Dallas
services: Data,
Implementation, Management
Services, Procurement

Racal Survey (UK) Ltd

Greenwell Road
East Tullos
Aberdeen
AB9 1DA

telephone: 01224 249700
fax: 01224 249446

contact: D.Inglis
product: Software, Complete
Systems
services: Data, Training
Racal Survey is an international
organisation which has specialised
for more than 10 years in the
development and supply of
Differential GPS products and
services, from short range radio
systems to satellite based systems
such as the Global Skyfix DGPS
system.
In addition the company provides
various DGPS data capture systems
for use on land where portability is
the key, and for use in marine
operations.
The company also maintains a well
trained pool of field staff to provide
a complete GPS data capture service
to clients. DGPS training is offered.
DGPS Products :
Deltafix LR and SR
Skyfix
Landstar
DeltaLink
MultiFix

Raindrop Information Systems Ltd

8 Golden Square
London
W1R 3AF
telephone: 0171 734 1091
fax: 0171 734 1095

contact: Catherine Flannigan
product: Hardware,
Publications, Software,
Complete Systems
services: Implementation,
Media, Procurement, Training

Recruit Media Ltd

20 Colebrooke Row
London
N1 8AP

telephone: 0171 704 1227
fax: 0171 704 1370
email: 100536,3201

contact: Melissa Coxon
services: Recruitment,
Training

regioplanDATA GmbH

Paulsborner Strasse 3
D-10709 Berlin
Germany

telephone: +49 30 896704 18
fax: +49 30 896704 10

contact: Georg Egger
product: Data, Publications,
Software, Complete Systems
services: Audit, Data,
Implementation, Management
Services, Procurement,
Training
regioplan DATA offers Consultancy
and development services for
regional and city planning.
AREAL integrates any kind of geo-
related information (raster, vector,
attributes) under Windows.

Remote Sensing Applications Consultants

4B Mansfield Park
Medstead
Alton
Hampshire
GU34 5PZ

telephone: 01420 561377
fax: 01420 561388
email: consultants@rsac.demon.co.uk

contact: Peter Fletcher
product: Data, Software
services: Data,
Implementation, Management
Services, Training
RSAC has particular expertise in the
use of optical and radar remote

sensing imagery for applications in
agriculture, land cover map updating
and the environment.

RICS Books

 12 Great George Street
London SW1P 3AD
telephone: 0171 222 7000
fax: 0171 222 9430

contact: Diane Williams
product: Publications
Activities
Booksellers & Publishers to
the Royal Institution of
Chartered Surveyors.
Products
RICS books supply a wide range of
books and papers on GIS, LIS and
other related subjects. A fast,
efficient mail order service is
available.

Robert Walker Consultants

64 Histon Road
Cottenham
Cambridgeshire CB4 4UD
telephone: 01954 251003
fax: 01954 251003

contact: Rob Walker
services: Implementation,
Management Services,
Procurement

Royal Commission on the Historical Monuments of England

 Kemble Drive
Swindon
Wiltshire SN2 2GZ
telephone: 01793 414727
fax: 01793 414770

contact: Neil Lang
product: Publications

Royal Geographical Society (with The Institute of British Geographers)

 1 Kensington Gore
London SW7 2AR
telephone: 0171 589 5466
fax: 0171 584 4447
email: info@rgs.org

contact: Dr A F Tatham
product: Data, Publications
services: Management
Services, Training

Royal Institute of Navigation

1 Kensington Gore
London
SW7 2AT
telephone: 0171 589 5021
fax: 0171 823 8671

contact: Group Capt
D.W.Broughton
product: Publications
services: Training

Royal Mail – Address Management Centre

 4 St Georges Business Centre
Portsmouth
Hampshire
PO1 3AX

telephone: 01705 838518
fax: 01705 838518

contact: Bill Haken
product: Data, Publications
services: Audit, Data, Media

Royal Town Planning Institute

 26 Portland Place
London
W1N 4BE

telephone: 0171 636 9107
fax: 0171 323 1582

contact: Michael Napier
product: Publications

Salford University Business Services Ltd

Technology House
Lissadel Street
Salford
M6 6AP

telephone: 0161 957 0012
fax: 0161 737 7700

contact: Paul Coward
services: Audit, Data,
Implementation, Management
Services, Procurement,
Training

SAS Institute

Wittington House
Henley Road
Medmenham
Marlow
Buckinghamshire
SL7 2EB

telephone: 01628 486933
fax: 01628 483203

contact: Alastair Sim
product: Publications, Software
services: Implementation, Management Services, Training

SAS/GIS software from SAS Institute

Geographical data visualisation with SAS/GIS software can now be an integral part of SAS applications for EIS, MIS and DSS.

The SAS System is a fully integrated, hardware-independent system of modular component products providing integrated tools to access, manage, analyse and present all types of business and research data.

SAS/GIS software can be supplied to a wide range of reporting and analysis applications, and can easily be utilised in existing SAS applications, running any analyses directly against geographically selected data. The SAS system's ability to directly access data in relational DBMS's and in non-relational formats allows maximum use of your data assets on platforms from laptop PC's to mainframes. The analysis at the user's command ranges from simple statistics and tabular report-writing to complex multivariate techniques and interactive exploratory analysis, while 4GL and GUI application-building features, enable the SAS System to deliver general purpose or customised applications for users of all kinds.

Markets served by SAS/GIS software include Banking, Insurance, Finance, Marketing and Research, plus Management Information applications in Retail, Distribution, Government, Emergency Services and Health.

Saztec Europe Limited

Tangier Lane
Eton
Windsor
Berkshire
SL6 4BB

telephone: 01753 833131
fax: 01753 832454

contact: Michael McLellan
services: Data, Media

SAZTEC specialises in digitisation (vectoring) and the capture of attribute data, especially of backfiles (legacy data) for all applications. e.g. Utilities.

SAZTEC Philippines Inc

54 Prospect Place
Wapping Wall
London
E1 9TJ

telephone: 0171 702 2906
fax: 0171 702 4507
email: compuserve 100535,2213

contact: Conrad Lealand
services: Data, Implementation, Management Services

Scan Group Limited

P.O. Box 10525
Haifa Bay
Israel
26114
telephone: +9724 410339
fax: +9724 413924

contact: Samuel Levin
product: Data, Hardware, Software, Complete Systems

Scientific Software Limited

Highlands Farm
Highlands Lane
Henley-on-Thames
Oxfordshire
RG9 4PR

telephone: 01491 411727
fax: 01491 411627

contact: David Finlay
product: Software
services: Data

SCOT Conseil

Parc Technologique Du Canal
1 Rue Hermes
31 526 Ramonville
France

telephone: +33 613 94600
fax: +33 613 94610

contact: S Gobin
product: Data, Software, Complete Systems
services: Data, Implementation, Procurement, Training

Scott Wilson Kirkpatrick

Scott House
Basing View
Basingstoke
Hampshire
RG21 2JG

telephone: 01256 461161
fax: 01256 460582
email: 100561,3337

contact: Richard Metcalfe
services: Data, Implementation, Media, Management Services, Procurement

Scott Wilson Kirkpatrick are leading International Consulting Engineers, Transportation and Environmental Planners, with 18 offices in the UK, and over 30 worldwide.

SWK's Information Systems Division focuses on the practical application and integration of information systems, both as a part of engineering and earth resource projects and as independent GIS consultants.

Specialist staff provide comprehensive services in:
* Project management, implementation
* Feasibility studies
* System evaluation, design, implementation
* Data management, integration
* Image processing
* Application development
* IS training

SWK's recent applications areas :
* Road maintenance MIS
* Transportation planning
* EIA
* Groundwater management
* Military operations
* Urban planning
* Agriculture development
* Forest inventory
* Natural resource mapping
* Coastal engineering

SDI Ltd

Kinetic Centre
Borehamwood
Hertfordshire
WD6 4SE

telephone: 0181 207 5474
fax: 0181 207 2755

contact: Steve Morris
product: Hardware, Software,
Complete Systems
services: Data,
Implementation, Management
Services, Procurement,
Training

Sector (UK) Limited

Southbank House
Black Prince Road
London
SE1 7SJ

telephone: 0171 582 9982
fax: 0171 587 1908

contact: Stephen Price
product: Hardware, Software,
Complete Systems
services: Audit,
Implementation, Management
Services, Procurement,
Recruitment, Training

SHL Vision* Solutions Limited

Yorktown
House
8 Frimley Road
Camberley
Surrey GU15 3HS
telephone: 01276 677707
fax: 01276 676567
email: info@vision.solns.co.uk

contact: Kevin Challen
product: Software
services: Implementation,
Training

Spatial Data Management
The SHL VISION* Solutions
approach to Spatial Data
Management (SDM) differs
dramatically from traditional GIS.
Proprietary approaches to GIS have
made it an isolated and specialised
information sub-system, unable to
readily integrate with an enterprise
data model. VISION* breaks
through the proprietary technology
barrier by the addition of a spatial
dimension to data contained within
ORACLE, thus providing a new
information context that can be
levered effectively for existing
corporate databases, or used in
strategic, enterprise-scale
applications through the provision of
geo-processing functions, through a
wide range of platforms. SHL
VISION* Solutions supports a
client./server environment, operating
on a wide range of hardware and
software platforms, including
Microsoft and UNIX.

Move beyond GIS
Within VISION*, commercial
database technology may be fully
utilised, while implementation
methods are enhanced through the
VISION*Express development
platform. VISION*Express is a
multi-user development environment
supporting both Object Modelling
methods and Rapid Application
Development. The use of
VISION*Express protects the
organisation's investments by future
proofing both data and applications.
Many of the world's largest telecoms
and utilities have already made the
move to VISION*, implementing
business focused, mission
applications. VISION* clients
include BAA, BT, Eastern
Electricity, North West Water,
Electricite de France, Hydro Quebec,
Ohio Edison, TELERBAS, and US
WEST Communications.

VISION* Services
VISION* SDM service
professionals work with our clients
to get the most out of their enterprise
SDM and achieve sustained business
benefits. Together with our partners
we provide the unique services,
expertise, and quality management
skills needed to build and service
mission-critical applications, and to
implement enterprise solutions.
VISION* Partners include such
organisations as Anderson
Consulting, Bull, IBM, M3i and
Unisys.

Company Background
SHL VISION* Solutions is an
operating division of SHL
Systemhouse Inc., a recognized
leader in business transformation
through client/server computing.
With headquarters in Ottawa, SHL
employs more than 5000
professionals worldwide.

SIA Limited

Edbury Gate
23 Lower Belgrave Street
London
SW1W 0NW
telephone: 0171 730 4544
fax: 0171 730 6762

contact: Janice Mabert
product: Data, Hardware,
Software, Complete Systems
services: Audit, Data,
Implementation, Management
Services, Training

SIAS Limited

37 Manor Place
Edinburgh
EH3 7EB

telephone: 0131 225 7900
fax: 0131 225 9229
email: lucy@sias.demon.co.uk

contact: Sarah E.Muirhead
product: Software
services: Data,
Implementation, Procurement

Siemens Nixdorf

Siemens House
Oldbury
Bracknell
Berkshire
RG12 8FZ

telephone: 01344 850829
fax: 01344 850943

contact: Miss
C.A.J.Twentyman
product: Hardware, Software,
Complete Systems
services: Implementation,
Management Services,
Training

Silicon Graphics Limited

1530 Arlington Business Park
Theale
Reading
Berkshire
RG7 4SB

telephone: 01734 257500
fax: 01734 257569
email: cresci@reading.sgi.com

contact: Andrew Cresci
product: Hardware,
Publications, Software
services: Implementation

Silsoe College, Cranfield University

 Silsoe
Bedfordshire
MK45 4DT

telephone: 01525 863060
fax: 01525 863099
email: c.bird@cranfield.ac.uk

contact: Chris Bird
services: Data,
Implementation, Management
Services, Training

Simmons Survey Partnership Limited

5 West Street
Axbridge
Somerset
BS26 1AA

telephone: 01934 732122
fax: 01934 732938

contact: Grel Simmons
product: Data
services: Data, Media

Sir Alexander Gibb & Partners Ltd

Earley House
London Road
Reading
Berkshire
RG6 1BL

telephone: 01734 261061
fax: 01734 491054

contact: Dr R S Steedman
services: Data, Procurement

Sir William Halcrow and Partners

 Burderop Park
Swindon
Wiltshire
SN3 4LW

telephone: 01793 812479
fax: 01793 812089

contact: Robert Deakin
product: Data, Software
services: Data,
Implementation, Management
Services

Smallworld Systems Ltd

 Brunswick House
61 – 69 Newmarket Road
Cambridge
CB5 8EG

telephone: 01223 460199
fax: 01223 460210
email: smallworld.@smallworld.co.uk

contact: Joy Haigh
product: Software, Complete
Systems
Smallworld is the world's leading
supplier of operational GIS. It is
established on over 300 sites,
primarily in the Utilities, Cable TV
and Telecommunications sectors.

Smartscan Inc

2344 Spruce Street
Boulder
CO 80302
USA

telephone: +1303 443 7226
fax: +1303 443 2997

contact: Rebecca J Culp
services: Audit, Data,
Implementation

Smith System Engineering Ltd

 Surrey Research Park
Guildford
Surrey
GU2 5YP

telephone: 01483 442000
fax: 01483 442144
email: mwoboyle@smithsys.co.uk

contact: Mike O'Boyle
services: Audit, Data,
Implementation, Management
Services, Procurement

Sokkia Ltd

Datum House
Electra Way
Crewe
Cheshire
CW1 1ZT

telephone: 01270 250525
fax: 01270 250533
email: cix@nobbey

contact: Mark Harper
product: Hardware, Software,
Complete Systems

Solent Mapping and Charting (SMAC)

Buckingham Building
Lion Terrace
Portsmouth
Hampshire
PO1 3HE

telephone: 01705 842477
fax: 01705 842512

contact: Dominic Fontana
services: Data

Southbank Systems PLC

Compass Centre North
Chatham Maritime
Kent ME4 4YG
telephone: 01634 880141
fax: 01634 880383

contact: Christopher Megan
product: Hardware, Complete
Systems
services: Implementation,
Training

Sovereign C.S. Ltd

39 Cuckoo Hill Road
Pinner
Middlesex
HA5 1AS

telephone: 0181 866 0713
fax: 0181 429 0959

contact: Roger Werry
product: Hardware, Software
services: Data,
Implementation, Media,
Procurement, Training

Spacesense consultants

19c Kew Gardens Road
Kew
Richmond
Surrey
TW9 3HD

telephone: 0181 940 6290
fax: 0181 332 2786

contact: Dr Adrian Lloyd-
Lawrence
services: Data

Spatial Data Limited

96 Stricklandgate
Kendal
Cumbria
LA9 5BT

telephone: 01539 721070

contact: Mr.N.A.Richardson
product: Data, Complete
Systems
services: Data, Media

Spatial Geographic Services & Applications Ltd

 18 Falstaff Close
Eynsham
Witney
Oxfordshire   OX8 1QA
telephone: 01865 881753
fax: 01865 881753

contact: Ron Linton
product: Software
services: Data,
Implementation, Management
Services
SGSA provides GIS and Digital
Mapping Consultancy, technical and
project support including field data
acquisition for all users of
geographic information.
SGSA distributes, installs and
supports Comgrafix's MapGrafix, a
powerful, versatile and cost-
effective GIS system for Macintosh
and PC computers, in the UK and
Ireland.

MapGrafix comprises:

MapGrafix, the data capture, editing, analysis, and output toolbox;

MapLink providing data conversion between MapGrafix and other GIS and Mapping Systems; and

MapView, the map projection or coordinate system transformation routine.

Spatial Information Services Ltd (SIS)

77 Hill Head Road
Fareham
Hampshire PO14 3JP
telephone: 01329 662891
fax: 01329 668440

contact: Peter D.Lever
product: Software, Complete Systems
services: Data

SPSS UK Ltd

SPSS House
5 London Street
Chertsey
Surrey KT16 8AP
telephone: 01932 566262
fax: 01932 567020

contact: John Davies
product: Publications, Software
services: Data, Implementation, Management Services

Star Informatic S.A.

Parc Scientifique du Sart-Tilman
Avenue du Pre Aily, 24
B-4031 Angleur-Liege
Belgium

telephone: +32 41 675313
fax: +32 41 671711

contact: Manuel Pallage
product: Software, Complete Systems
services: Implementation

StatSci Europe

Osney House
Mill Street
Oxford OX2 0JX
telephone: 01865 200952
fax: 01865 200953
email: sales@statsci.co.uk

contact: Matthew Eagle
product: Software
services: Data, Implementation, Training

Structural Technologies Ltd (STL)

 Woodside
The Slough
Studley
Warwickshire
B80 7EN

telephone: 01527 854819
fax: 01527 854819

contact: David Schindler
product: Data, Software, Complete Systems
services: Data, Implementation, Management Services, Procurement, Training

STL's STRUMAP is a leader in pc Windows-based GIS. Specialist services offered, from data capture to fully implemented GIS.

Summagraphics Europe N.V.

Keiberg III, Paviljoen 409
B-1930 Zaventem
Brussels
Belgium

telephone: +32 2 721 5033
fax: +32 2 721 5289

contact: Carina Govaert
product: Hardware
services: Media

Sun Microsystems Ltd

Bagshot Manor
Green Lane
Bagshot
Surrey
GU19 5NL

telephone: 01276 451440
fax: 01276 472114
email: colint@sun.co.uk

contact: Colin Taylor
product: Hardware, Publications
services: Implementation, Training

Survey Control Services

Old Co-operative Buildings
Hookergate Lane
Rowlands Gill
Tyne & Wear
NE39 2AJ

telephone: 01207 544996

contact: C.Mills
services: Data

Survey & Development Services Ltd

 3 Hope Street
Bo'ness
West Lothian
EH51 0AA

telephone: 01506 825121
fax: 01506 822629

contact: Elspeth Rodger
product: Software
services: Data, Implementation, Media, Management Services, Training

Digital data capture and database compilation from Land/Aerial surveys, Orthophotography or existing documentation. Data output to all major formats including NTF, ArcINFO, Smallworld, DXF.

Survey Supplies Ltd

Blundellsands House
34 – 44 Mersey View
Liverpool
L22 6QB
telephone: 0151 931 3161
fax: 0151 931 2838

contact: Bob Wells
product: Software, Complete Systems

Svitzer Limited

Morton Peto Road
Great Yarmouth
Norfolk
NR3 0LT
telephone: 01493 440320
fax: 01493 440319

contact: Bob Blow
product: Data, Complete Systems
services: Data, Management Services

Symology Limited

Millfield Lane
Caddington
Bedfordshire
LU1 4AJ
telephone: 01582 842626
fax: 01582 842600

contact: Edgar Blazier
product: Data, Hardware, Software, Complete Systems
services: Audit, Implementation, Management Services, Procurement, Training

Sysdeco (UK) Ltd

 Cambridge Science Park
Milton Road
Cambridge
CB4 4WG

telephone: 01223 420464
fax: 01223 420324

contact: Brian Dixon
product: Hardware, Software,
Complete Systems
services: Implementation,
Management Services,
Training
Sysdeco GIS: state of the art full
GIS: data conversion (automatic).
User configurable applications
builder with a wide range of
standard applications and
comprehensive 4GL development
capability.

System Options Limited

45 Victoria Road
Aldershot
Hampshire
GU11 1SJ

telephone: 01252 334383
fax: 01252 28779

contact: Jim Pedroza
product: Data, Hardware,
Publications, Software,
Complete Systems
services: Data,
Implementation, Media,
Management Services,
Procurement

TACTICIAN UK

14 – 15 Regent Parade
Harrogate
North Yorkshire
HG1 5AW

telephone: 01423 560064
fax: 01423 525545

contact: Richard Ives
product: Data, Software,
Complete Systems
services: Data,
Implementation, Management
Services, Training

Tangent Technology Design Associates Ltd

Chesterton House
Rectory Place
Loughborough
Leicestershire LE11 1UW
telephone: 01509 610910
fax: 01509 610403

contact: Carl Billson
product: Software

Target Market Consultancy

Woodlands
Woodlands Close
Holmer Green
High Wycombe
Buckinghamshire
HP15 6QG

telephone: 01494 712371
fax: 01494 714203
email: compuserve 100130 1723

contact: Peter Sleight
services: Data, Management
Services
T.M.C. offers :
- Independent Consultancy in
customer targeting,
geodemographics & Lifestyle
marketing;
- impartial advice on systems, data,
methodologies;
- data sourcing, project management;
Retail location a speciality.

Taylor & Francis Ltd

4 John Street
London
WC1N 2ET
telephone: 0171 400 3500
fax: 0171 831 2035
email: rsteele@tandf.co.uk

contact: Richard Steele
product: Publications

TDS-CAD Graphics Ltd

TDS House
Lower Philips Road
Blackburn
Lancashire
BB1 5TH

telephone: 01254 676921
fax: 01254 581574

contact: Karen Bamford
product: Hardware, Software,
Complete Systems

TEAMS (Taylor Woodrow Electronics Asset Mapping Survey)

 The Old Brewery
High Court
The Calls
Leeds
LS2 7ES

telephone: 0113 242 3802
fax: 0113 242 5172

contact: Keith Ricardo
product: Software, Complete
Systems
services: Data,
Implementation, Management
Services, Procurement
TEAMS have developed a Mobile
GIS on a pen-based computer. It is
used primarily for data capture and
asset record maintenance across
several industrial sectors.

Tekla Oy

Koronakatu 1
02210 Espoo
Finland

telephone: +358 0 803 7722
fax: +358 0 803 9489
email: rs@tekla.fi

contact: Risto Sajaniemi
product: Software, Complete
Systems
services: Data

Tele Atlas

Moutstraat 92
B-9000 Gent
Belgium

telephone: +32 9 222 5658
fax: +32 9 222 7412

contact: Ad Bastiaansen
product: Data, Publications
services: Audit, Data,
Implementation, Media,
Management Services

Tendron Systems Ltd

61 Park Street
Bristol
BS1 5NU

telephone: 0117 929 4759
fax: 0117 922 1320

contact: R.A.Brown
product: Publications,
Software, Complete Systems
services: Data,
Implementation, Management
Services, Procurement

Tenet Systems Ltd

North Heath Lane
Horsham
West Sussex
RH12 5UX

telephone: 01403 273173
fax: 01403 273123

contact: Dr Sharon Cooper
product: Software
services: Implementation,
Training

Terrafix Ltd

23e Newfield Industrial Estate
High Street
Tunstall
Stoke-on-Trent
Staffordshire
ST6 5PD

telephone: 01782 577015
fax: 01782 835667

contact: Mr.J.B.Rosson
product: Data, Hardware,
Publications, Software,
Complete Systems
services: Data,
Implementation

TerraHunt GeoScience Ltd

Herts Business Centre
London Colney
St Albans
Hertfordshire
AL2 1JG

telephone: 01727 822287
fax: 01727 826570

contact: Derek Morris
services: Media

TerraQuest Group Limited

 Burnt Tree House
Burnt Tree Island
Tipton
West Midlands
DY4 7UW

telephone: 0121 520 0111
fax: 0121 520 5800

contact: Alan Ross
services: Audit, Data,
Management Services,
Procurement

TerraQuest is uniquely positioned to work with both public and private sector organisations to arrive at solutions to land and property information management problems. Over twenty years experience gained on over 150 projects in this arena has provided TerraQuest with an in-depth understanding of major land and property functions.

The British Computer Society

1 Sanford Street
Swindon
Wiltshire
SN1 1HJ

telephone: 01743 417417
fax: 01793 480270

contact: Andrew Wilkes
product: Publications

The Business Database from Yellow Pages

8 Wasterside Drive
Lanley
Slough
Berkshire
SL3 6EZ

telephone: 01753 583311
fax: 01753 594001

contact: Simon Green
product: Data
services: Data

The Data Consultancy

 7 Southern Court
South Street
Reading
Berkshire
RG1 4QS

telephone: 01734 588181
fax: 01734 597637

contact: Margaret H.Smee
product: Data, Publications,
Software, Complete Systems
services: Data,
Implementation, Management
Services, Training

THE DATA CONSULTANCY
The Data Consultancy, part of the Reading based URPI Group, is the leading supplier of digital data sets and software for marketing and planning. Established in 1975 The Data Consultancy has a data catalogue of over 1000 off-the-shelf data sets available in a wide range of formats to suit most GIS and applications software. Software products include MapInfo, Illumine for Windows, Drivetime, Markets and a range of applications software.

Data Sets
The Data Consultancy specialises in the provision of spatially referenced data sets for marketing and planning applications. Analysis of data is undertaken together with a full and comprehensive service of help and advice on the sources, collection and use of data. Survey research and cross analysis of data is also undertaken. Most data sets are spatially referenced to the national grid, administrative and postal geographics. Major data sets include:
* **MAP FILES**: roads, rivers, railways, town street maps, central London street map, European and world maps all in digital form.
* **NETWORKS AND DRIVE TIME MATRICES**: link and node road networks for routing applications and matrices of shortest drive times between a wide variety of origins and destinations.
* **BOUNDARY FILES**: UK administrative areas, counties, districts, local authority wards, health authorities, enumeration districts, postcode areas, districts and sectors; drive time isochrones, TV regions, parliamentary constituencies and many others.
* **POINT FILES**: 1981 and 1991 enumeration district centroids, postcode sector centroids, local authority ward centroids, shopping centres, retail stores, banks, building societies, garages, unit postcodes and full gazetteers.
* **DEMOGRAPHIC DATA**: census small area statistics of enumeration district, ward, postcode sector and other levels. Population updates and population projections at enumeration districts, ward and postcode sector levels. Super Profile neighbourhood classification and lifestyle data.
* **CONSUMER EXPENDITURE AND INCOME**: consumer retail expenditure, retail business turnover potential and household income available at any spatial level down to enumeration district.
* **STORES**: UK hypermarket and superstores, grocery stores, car dealers, multiple retailers, restaurants and fast food outlets plus many other service businesses.
* **SHOPPING CENTRE AND SCHEMES**: details of all large managed shopping schemes together with proposed schemes and those with planning permission. Locations of all major shopping centres including town and city centres with cross references to Goad's and Newman's data.
Data and analyses can be supplied with or without associated GIS software.
See our full catalogue of Geographically Referenced Data Sets for more details.

Software
As an established provider of GIS and related systems The Data Consultancy provides drive times and market analysis systems and is an authorised value added reseller of MapInfo the leading desktop GIS and mapping system. The Data Consultancy also provide a range of other software products designed for market analysis and site assessment studies. Although available as separate systems each system interfaces to the others to provide a full range of GIS, market analysis and mapping capabilities. Full support, training and customisation is available.

The software can be purchased on its own or fully configured with maps, boundaries and data. A wide range of databases, boundary files, maps and analyses can be supplied.

* **MAPINFO**: the leading desktop GIS and mapping system which sets the standard for cost effectiveness. It is particularly suitable for marketing, site assessment, planning and business applications. Includes display, editing, overlay, analysis, query, searching, geocoding and digitising in one easy to use package. Fast display, analysis and querying of user data files in dBASE, Lotus and Excel format. Ideal for mapping and analysing Census data at all spatial levels, including enumeration districts.

* **MAPINFO APPLICATIONS**: off-the-shelf MapInfo applications include Macros, Polyline editing, Penetration for market penetration, Tessellate for creating Thiessen polygons, Gridsum for grid square analysis and others.

* **DRIVETIME**: an extremely powerful and flexible interactive system to generate accurate drive time isochrones. Outputs maps and plots and generates boundary files suitable for input to mapping, GIS and other software.

* **ILLUMINE**: a full market analysis and reporting system giving direct access to the Census of Populations small area statistics at all geographic levels together with users' own client/customer data. Analyses can be undertaken for any area however defined. Areas can be compared, component area reports produced, areas selected and targeted to match required criteria. The modular design allows users to start with a simple entry level system with a few variables and add additional variables, population updates, population projections, neighbourhood profiling, expenditure estimates, retail business turnover potential, drive time isochrones as and when required.

* **MARKETS**: our widely acclaimed and widely used store location model which simulates and predicts shopping patterns and expenditure flows. Model results include store or scheme turnover levels, turnover to floorspace ratios, trade draw rates and market penetration by drive time bands or zones.

* **TRANSLATOR**: our modular digital map file translator for OS NTFv2.0, MapInfo, Atlas, Gimms, ArcInfo and other formats. Translates any number of tiles in a single pass to produce a single map coverage. User definable line, symbols, text and fill pattern and layer separation supported.

Support and Training
All products and services are fully supported by experienced professional staff involved in product use and development. Training courses and seminars held every week at our Reading training facility. Full software customisation available where required.

The LGMB

 Arndale House
The Arndale Centre
Luton
Bedfordshire
LU1 2TS
telephone: 01582 451166
fax: 01582 412524

contact: A.K.Black
product: Publications, Software
services: Training

The Marketing Information Consultancy (MIC)

Causeway House
The Causeway
Teddington
TW11 0JR
telephone: 0181 213 5500
fax: 0181 213 5599
email: mic@hyperlink.com

contact: Richard Bandell
product: Data, Publications, Software, Complete Systems
services: Data

Provide complete and innovative market analysis solution to retail, automotive, financial and leisure companies using the most comprehensive lifestyle database covering 13 million UK households.

The NPA Group Ltd

1 Fircroft Way
Edenbridge
Kent
TN8 6HS
telephone: 01732 865023
fax: 01732 866521
email: info@npagroup.co.uk

contact: Ren Capes
product: Data, Hardware, Publications, Software, Complete Systems
services: Data, Media, Management Services, Procurement

The Severn Partnership

The Maltings
59 Lythwood Road
Bayston Hill
Shrewsbury SY3 0NA
telephone: 01743 874135
fax: 01743 874716

contact: Nigel Atkinson
product: Data
services: Data, Procurement

The Survey Centre

Waterworks Road
Worcester
Worcestershire WR1 3EZ
telephone: 01905 21073
fax: 01905 29085

contact: R.M.Whitfield
product: Data
services: Data, Media, Training

Trac Consultancy

Silverdale House
64 Pepy's Road
New Cross
London SE14 5SD
telephone: 0171 639 9825
fax: 0171 639 9825

contact: David Griffiths
product: Data
services: Data, Management Services

Trident Map Services

70 High Street
Houghton Regis
Dunstable
Bedfordshire LU5 5BJ
telephone: 01582 867211
fax: 01582 867689

contact: Paul Shingfield
product: Publications

Trimble Navigation Europe Ltd

Trimble House
Meridian Office Park
Osborn Way
Hook
Hampshire RG27 9HX
telephone: 01256 760150
fax: 01256 760148

contact: Helen Knight
product: Hardware, Software, Complete Systems
services: Implementation, Training

TYDAC Technologies Ltd

Chilworth Research Centre
2 Venture Road
Southampton
Hampshire
SO16 7NP

telephone: 01703 760824
fax: 01703 760944
email: sales@tydac.demon.co.uk

contact: James Oliver
product: Data, Software,
Complete Systems
services: Data,
Implementation, Management
Services, Training

Unistride Sewer Technology

Red Lion House
Bentley
Near Farnham
Surrey
GU10 5HY

telephone: 01420 23456
fax: 01420 22712

contact: Paul Marvin
services: Data

UNISYS Ltd

 Bakers Court
Bakers Road
Uxbridge
Middlesex
UB8 1RG

telephone: 01895 237137
fax: 01895 862984
email: ecobe@uxbpo1.gb.unisys.com

contact: Edwin Ecob
product: Hardware, Software
services: Audit, Data,
Implementation, Management
Services, Procurement,
Training

Unisys provides information
services, technology, and software to
clients throughout the world. The
company specialises in business-
critical solutions for organisations
that operate in transaction-intensive
environments. These solutions are
used by 1,600 government agencies
worldwide, 41 of the world's largest
banks, 140 airlines worldwide, and
35 of the world's largest
telecommunications companies.
Unisys has been delivering GIS
systems in Europe since the late
1980s. The GIS programme
combines specialised information
services with advanced GIS
technology, and is aimed particularly
at organisations requiring enterprise-
wide access to, and the management
of geographic business data. These
organisations specifically include
telecommunications companies, the
utilities, and land and infrastructure
management agencies.
A key component of the Unisys GIS
programme is the provision of
information services (IS). These
planning, analysis, and
implementation services have been
designed to ensure that the systems
which are delivered to an
organisation properly reflect the
needs of that organisation, in terms
of efficiency, responsiveness, and
competitiveness. To this end, Unisys
provides information services which
address the major phases and tasks
of a GIS project. These include
executive planning, needs analysis,
solution delivery, and post-delivery
review. Unisys can provide overall
GIS project management services
from start to finish.
Unisys supports over one hundred
clients who use GIS solutions in
private and public agencies. The
company is actively involved in one
of the largest GIS projects in Europe,
helping a British
telecommunications company
replace its existing plant records
across more than 5,000 exchanges
throughout the UK. Unisys
consultants are providing executive
planning services and GIS feasibility
studies to the largest airport
authority in Europe. In Switzerland,
Unisys GIS technology is being used
by local government in 80% of the
French-speaking cantons. And a
major Spanish electrical distribution
company relies on Unisys GIS to
ensure the reliable delivery of
electricity to 8 million people.

Universal Systems Ltd

270 Rookwood Avenue
Fredericton
NB
Canada
E3B 2M2
telephone: +1506 458 8533
fax: +1506 459 3849
email: nyarady@unb.ca

contact: Rick Nyarady
product: Hardware, Software,
Complete Systems
services: Implementation,
Training

University of East London

 Longbridge Road
Dagenham
Essex
RM8 2AS

telephone: 0181 849 3618
fax: 0181 849 3514

contact: Andrew Larner
services: Training

University of Edinburgh

 Drummond Street
Edinburgh
EH8 9XP

telephone: 0131 650 2565/
2543
fax: 0131 650 2524
email: gisadmin@geovax.ed.ac.uk

contact: Bruce M.Gittings
product: Publications
services: Data, Management
Services, Procurement,
Training
Edinburgh is the premier GIS
education and research (Grade
5) establishment in the U.K.
Unrivalled equipment. M.Sc.
funding available from NERC
and ESRC. RICS
Accreditation.

University of Greenwich

Pembroke
Chatham Maritime
Kent ME4 4AW
telephone: 0181 331 9800
fax: 0181 331 9805

contact: Dr Gesche Schmid
services: Training
BSc and MSc in Geographic
Information Systems and Remote
Sensing
Fulltime/Parttime/Sandwich
The BSc Courses in GIS or Remote
Sensing (3 years fulltime) offers
training in
• relevant information technology,
technical principles, functionality
and applications in GIS and Remote
Sensing
• combined with a basic knowledge
in one of the pathways offered in the
/School of Earth Sciences (Geography, Geology, Environmental
Sciences)
The MSc Course in GIS or Remote
Sensing (12 months fulltime) is
aimed at applications- with a degree
in Earth/Environmental Sciences
• or with a relevant professional
experience
• who wish to expand their
knowledge in the technical
disciplines of GIS or Remote
Sensing.

University of Leicester

University Road
Leicester
LE1 7RH

telephone: 01533 523839
fax: 01533 523854
email: pffi@le.ac.uk

contact: Dr.Peter.Fisher
services: Training

University of Luton

 Park Square
Luton
Bedfordshire
LU1 3JU

telephone: 01582 489264
fax: 01582 489212
email: rbeard@vax2.luton.ac.uk

contact: Ron Beard
services: Training
Courses include B.Sc Mapping
Science (Full time, part time, work
based learning modes). B.Sc (Minor)
GIS. HND Geographical techniques
(Sandwich course). M.Sc. Waste
Management.

University of Newcastle Upon Tyne

 Mapping Information
Technology Unit
Newcastle-upon-Tyne
NE1 7RU

telephone: 0191 222 6445
fax: 0191 222 8691
email: david.parker@ncl.ac.uk

contact: Dr David Parker
product: Software
services: Implementation,
Management Services,
Training

Walker Ladd Surveys

5 Gas Ferry Road
Bristol
BS1 6UN

telephone: 0117 925 1251
fax: 0117 925 7500

contact: Phil Thompson
services: Data

Wallingford Software Limited

Howbery Park
Wallingford
Oxfordshire
OX10 8BA

telephone: 01491 824777
fax: 01491 826392

contact: Adrian Turner
product: Software
services: Training

WDV (UK)

Mansley Centre
Timothy's Bridge Road
Stratford Upon Avon
Warwickshire
CV37 9NQ

telephone: 01789 297000
fax: 01789 298056

contact: John Lees
product: Hardware, Complete
Systems

WRc (Water Research centre)

 Frankland Road
Blagrove
Swindon
Wiltshire
SN5 8YF

telephone: 01793 511711
fax: 01793 511712

contact: John Cima
product: Hardware,
Publications
services: Audit, Data,
Implementation, Media,
Management Services,
Procurement, Training
WRc specialises in the application of
GIS and data visualisation for water,
waste water, and environmental
applications.
WRc is a "MapInfo"™ Authorised
Partner

WS Atkins Planning & Management Consultants

Woodcote Grove
Epsom
Surrey
KT18 5BW

telephone: 01372 726140
fax: 01372 740055
email: deastwood@wsatkins.co.uk

contact: David Eastwood
product: Software, Complete
Systems
services: Audit, Data,
Implementation, Management
Services, Procurement,
Training

Xcon Data

PO Box 4723
Sofienberg
N-0506 Oslo
Norway

contact: Jon Apneseth
product: Complete Systems
services: Data

directory 2:

product trade names

1991 Census	General Register Office for Scotland 0131 314 4254	**Aerofilms**	Hunting Aerofilms Limited 0181 207 0666
1991 Census Outputs	Office of Population Censuses and Surveys (OPCS) 01329 813536	**AIMI**	Dataview Solutions Ltd 0171 404 0640
3-D Mapper	Graphical Data Capture Ltd (GDC) 0181 349 2151	**ALEX**	A.L.Downloading Services 0181 994 5471
3DMapper	Dataview Solutions Ltd 0171 404 0640	**Alliance**	AP³ Imaging Services Limited 01787 378242
4Base	CAD – Capture Limited 01254 583534	**ALPHAAXP**	Digital Equipment Corporation 01734 203546
4View	CAD – Capture Limited 01254 583534	**ALPHAREL**	CAD – Capture Limited 01254 583534
ACADEMY	Dataflow Information Systems 0117 927 2466	**ALSCAN™**	ALTEK Corporation +1301 572 2555
ACORN	CACI Limited 0171 602 6000	**Alteck**	ITS : Intertrade Scientific Ltd 01908 676633
ACORN*MAPS	CACI Limited 0171 602 6000	**AM**	CAM – Centre for Analysis & Modelling Limited 0171 232 1111
ACORN*PROFILER	CACI Limited 0171 602 6000	**APIC**	APIC Systems 01279 466966
Action	Action Information Management Ltd 01225 777288	**ARC-GIS**	ARC Systems Pty Ltd +612 290 2400
Action Map One	Action Information Management Ltd 01225 777288	**ARC-NET/Gao**	ARC Systems Pty Ltd +612 290 2400
Action Map Two	Action Information Management Ltd 01225 777288	**ARC-NET/Power**	ARC Systems Pty Ltd +612 290 2400
Action PLAN	Action Information Management Ltd 01225 777288	**ARC-NET/Telco**	ARC Systems Pty Ltd +612 290 2400
Active Censtat	Active Software Ltd 01256 56629	**ARC-NET/Water**	ARC Systems Pty Ltd +612 290 2400
Active Provider	Active Software Ltd 01256 56629	**ARC/INFO**	ESRI (UK) Ltd 01923 210450
Active Purchaser	Active Software Ltd 01256 56629	**ARC/INFO**	Eurosense Technologies N.V. +32 2 460 7000
ADDMAPS	Cobham Digital Services Limited 01932 868133	**ARC/INFO**	GEO-Marketing Systems Ltd (GMSL) 01635 872382
ADDRESS-POINT™	Ordnance Survey 01703 792773	**ARC/INFO**	MAPS geosystems +9716 356411
		ArcCAD	Eurosense Technologies N.V. +32 2 460 7000
		ArcCAD	Graphite Management Services Ltd 0115 969 1114

ArcCAD	Survey & Development Services Ltd 01506 825121
ArcCAD	MAPS geosystems +9716 356411
ArcCAD	ESRI (UK) Ltd 01923 210450
ArcCAD	Glen Computing Ltd 01689 875577
ArcCAD	SDI Ltd 0181 207 5474
ArcCATEL	SDI Ltd 0181 207 5474
ArcLIGHT	SDI Ltd 0181 207 5474
ArcView	MAPS geosystems +9716 356411
ArcView	ESRI (UK) Ltd 01923 210450
ArcView	GEO-Marketing Systems Ltd (GMSL) 01635 872382
ArcView	Eurosense Technologies N.V. +32 2 460 7000
ArcView	Graphite Management Services Ltd 0115 969 1114
ArcVIew	SDI Ltd 0181 207 5474
ArcView2	Glen Computing Ltd 01689 875577
AREALR	regioplanDATA GmbH +49 30 896704 18
ARGIS-4GE	UNISYS Ltd 01895 237137
Argus	Munro Garratt International 01224 622888
Ascodes-AudiGIS	Audifilm Girona S.L. +34 72 242611
Ashtech Z-12	Ashtech Europe Limited 01993 883533
ATHENE	A.L.Downloading Services 0181 994 5471
Atlas GIS	Dotted Eyes 0121 445 6150
Atlas GIS	Adept Scientific Micro Systems Ltd 01462 480055
AutoCAD	Autodesk Ltd 01483 300077
AutoCAD	Datatechnology Datech Ltd 0181 308 1800
AutoCAD	Graphite Management Services Ltd 0115 969 1114
AutoCAD	L.E.S. (Computer Services) Limited 01737 223899

AutoCAD	SDI Ltd 0181 207 5474
AutoCAD	DAT/EM Systems International +1907 274 3681
AutoCAD	Glen Computing Ltd 01689 875577
AutoCAD	Geo/SQL (UK) 01672 562012
AutoCAD LT	Autodesk Ltd 01483 300077
AutoCIVIL	L.E.S. (Computer Services) Ltd 01737 223899
AutoDesk	C.A.Design Services Ltd 01493 440444
AutoFM	SDI Ltd 0181 207 5474
AutoMAPS	MAPS geosystems +9716 356411
AutoNTF	University of Newcastle Upon Tyne 0191 222 6445
Autoplant	WS Atkins Planning & Management Consultants 01372 726140
Autoview	CAD – Capture Limited 01254 583534
AUTOWARP	ERDAS (UK) Ltd 01223 880802
B-GUL	Scientific Software Limited 01491 411727
BADGER	Design Computer Aids Limited (DeCAL) 0131 553 3159
BC.AID	PLANTECH Ltd 0171 922 8825
BDHA	Paul Clasper & Associates Ltd 01225 444561
Bentley	Mason Land Surveys 01383 727261
Bibliography of Economic Geology	Geosystems 01235 813913
Binaer	Geodelta +31 1515 8188
Boundary-Line™	Ordnance Survey 01703 792773
British Borehole Catalogue	Geoinformation International 01223 423020
BROWSER	SIAS Limited 0131 225 7900
BUTTONS	TACTICIAN UK 01423 560064
Byers One-Step Conversion	Byers Engineering Company +1404 843 1000
C-GUL	Scientific Software Limited 01491 411727

CableCad	Enghouse (UK) Limited		**CHAMPS**	Hunting Engineering Ltd
	01703 615228			01525 841000
CADCore™	ALTEK Corporation		**Cities Revealed**	Geoinformation International
	+1301 572 2555			01223 423020
CADCORP Map Editor	Computer Aided		**Cityview**	Citywise
	Development (CADCORP)			0171 636 5448
	Limited		**CIVILCAD**	Survey Supplies Ltd
	01438 747996			0151 931 3161
CADCORP Map Manager	Computer Aided		**ColourPrinter™**	ALTEK Corporation
	Development (CADCORP)			+1301 572 2555
	Limited		**ColourScanner™**	ALTEK Corporation
	01438 747996			+1301 572 2555
CADCORP SIS	Computer Aided		**CONNEX**	Tenet Systems Ltd
	Development (CADCORP)			01403 273173
	Limited		**Consumerscan**	The Marketing Information
	01438 747996			Consultancy (MIC)
CADCORP Space Manager	Computer Aided			0181 213 5500
	Development (CADCORP)		**Contex**	TDS-CAD Graphics Ltd
	Limited			01254 676921
	01438 747996		**Contex**	Geo/SQL (UK)
CADdy Architecture	Aneberie CAD			01672 562012
	01206 331215		**Contour**	TDS-CAD Graphics Ltd
CADdy Civil Engineering	Aneberie CAD			01254 676921
	01206 331215		**Coordinate**	PD Computing Ltd
CADdy Surveying	Aneberie CAD			0161 747 7110
	01206 331215		**Countryside Information**	
CalComp	CalComp Limited		**System**	NERC
	01734 320032			01793 411683
CALSITE	H R Wallingford Ltd		**D.I.A.L.**	Pinpoint Digitising Services
	01491 835381			01928 579148
Capturebase	CAD – Capture Limited		**DARTS**	Geo/SQL (UK)
	01254 583534			01672 562012
CaptureFlow	CAD – Capture Limited		**dataMAP**	SIA Limited
	01254 583534			0171 730 4544
CARIS	Universal Systems Ltd		**Dataquest**	Dataquest Europe Ltd
	+1506 458 8533			01494 422722
CARTograph GRMS	CARTograph Ltd		**DATATABR**	ALTEK Corporation
	01223 67818			+1301 572 2555
Cartology	FastCAD GIS Ltd		**DC.AID**	PLANTECH Ltd
	0117 942 8195			0171 922 8825
CARTOLOGY	FastCAD GIS Ltd		**DCW Data**	DMAP Ltd
	01923 240216			01480 497673
CARTOLOGY LITE	FastCAD GIS Ltd		**DecisioMAP**	Infolink Decision Services
	01923 240216			Limited
Cartology Lite	FastCAD GIS Ltd			0181 686 7777
	0117 942 8195		**Define**	Infolink Decision Services
Cartomap	Foto Res			Limited
	+34 27 216455			0181 686 7777
CDGRAB	A.L.Downloading Services		**Delfy**	Geodelta
	0181 994 5471			+31 1515 8188
CENMAN	Gamma Ltd		**DeltaFix LR and SR**	Racal Survey (UK) Ltd
	+3531 6713066			01224 249700
Central Postcode Directories	Office of Population		**DeltaLink**	Racal Survey (UK) Ltd
	Censuses and Surveys			01224 249700
	(OPCS)		**DemoGraf***	EuroDirect Database
	01329 813536			Marketing Ltd
CFONTS	CAD – Capture Limited			01274 737144
	01254 583534		**DeskMapper**	IME (UK) Ltd
				01259 210210

DeskMapper	Longdin & Browning
	01792 202244
DeskMapper	University of Newcastle
	Upon Tyne
	0191 222 6445
DeskMapper	Autodesk Ltd
	01483 300077
DeskMapper	Geo/SQL (UK)
	01672 562012
DGAS	Paul Clasper &
	Associates Ltd
	01225 444561
DGN/CAPTURE	DAT/EM Systems
	International
	+1907 274 3681
DIDACTIM	AP3 Imaging Services
	Limited
	01787 378242
DIGIT-II	GIMMS (GIS) Ltd
	0131 668 3046
DIGITUS	DAT/EM Systems
	International
	+1907 274 3681
DIGITUS	HJM Imaging Systems
	01604 39792
DIMENSION	Ashtech Europe Limited
	01993 883533
DINIS	ICL Ltd
	0121 456 1111
DRAWBASE	Survey & Development
	Services Ltd
	01506 825121
DriverGuide	European Geographic
	Technologies BV
	+31 4998 93385
Drivetime	The Data Consultancy
	01734 588181
DRY	Kirstol Ltd
	0161 338 7512
DVP	Leica UK Ltd
	01908 666663
DVP	Foto Res
	+34 27 216455
DWG/CAPTURE	DAT/EM Systems
	International
	+1907 274 3681
E-GUL	Scientific Software Limited
	01491 411727
EAI	Environment & Planning
	Library
	0161 708 9799
EASI/PACE	I.S. Ltd
	0117 925 0553
ECLIPSE	Intera Information
	Technologies
	+1613 226 5442
ECoS	Natural Environment
	Research Council
	01793 411996

ECOS	NERC
	01793 411683
ED-LINE	London Research Centre
	0171 735 4250
ED-LINE	MVA Systematica
	01483 728051
ED-Line	The Data Consultancy
	01734 588181
ED91	Graphical Data Capture Ltd
	(GDC)
	0181 349 2151
Empress 4GL	Empress Software UK
	01483 861990
Empress RDBMS	Empress Software UK
	01483 861990
En Route	European Geographic
	Technologies BV
	+31 4998 93385
Environ Analytica™	Environment & Planning
	Library
	0161 708 9799
EOSAT Notes	EOSAT
	+1301 552 0525
Erdas	Geo-Perfect TWI B.V.
	+31 1828 30477
ERGOvista	Remote Sensing
	Applications Consultants
	01420 561377
ERMapper	GISL Limited
	01956 285077
EURIPIDES	Geoinformation International
	01223 423020
EWO System	Byers Engineering Company
	+1404 843 1000
Explorer	TYDAC Technologies Ltd
	01703 760824
Fasset	FastCAD GIS Ltd
	0117 942 8195
FASSET	FastCAD GIS Ltd
	01923 240216
FASTCAD	Survey Supplies Ltd
	0151 931 3161
FASTMAP	Survey Supplies Ltd
	0151 931 3161
FIELDNOTES	Southbank Systems PLC
	01634 880141
FINANCIAL*ACORN	CACI Limited
	0171 602 6000
FINDER	Gamma Ltd
	+3531 6713066
FRAMME	Intergraph (UK) Ltd
	01793 619999
Garmin Srvy II	Effective Solutions (Data
	Products)
	01794 514233
GAS	IBM UK Ltd
	01926 464336

GDS	Graphic Data Systems Corporation (GDS) 01483 725225	**Geostat**	Geographic Management Solutions Ltd 01454 281802
GEMS	Graphite Management Services Ltd 0115 969 1114	**Geotitles**	Geosystems 01235 813913
GENAMAP	Genasys II Limited 0161 232 9444	**GEOTRACER**	Geotronics Limited 01480 433555
GEO EXPLORER	Trimble Navigation Europe Ltd 01256 760150	**GGP**	GGP Systems Limited 0181 656 8562
Geo/Navigator	PAX Technology 01491 572282	**GIMMS**	GIMMS (GIS) Ltd 0131 668 3046
Geo/SQL	Elstree Computing Ltd 0181 906 5656	**GINO**	Bradly Associates Ltd 01344 779381
Geo/SQL	Geo/SQL (UK) 01672 562012	**GINOSURF**	Bradly Associates Ltd 01344 779381
Geo/SQL	Glen Computing Ltd 01689 875577	**GIS Guides for Planners**	Royal Town Planning Institute 0171 636 9107
Geo/SQL	LTG Services 0181 906 5559	**GIS Market in Europe**	Dr Stanley Port 01483 421970
Geo/SQL	Longdin & Browning 01792 202244	**GISELLE**	Corena A/S +47 66 794500
Geo/SQL	Geo-Perfect TWI B.V. +31 1828 30477	**GIST**	Action Information Management Ltd 01225 777288
GeoArchive	Geosystems 01235 813913	**GISTUTOR2**	Geoinformation International 01223 423020
GEODIMETER	Geotronics Limited 01480 433555	**GLOBAL GAZETTEER**	ALLM Systems & Marketing 01923 230150
geoGPG	IBM UK Ltd 01926 464336	**GMAP**	GMAP Ltd 0113 244 6164
Geographical Journal	Royal Geographical Society (with The Institute of British Geographers) 0171 589 5466	**GMS**	Glen Computing Ltd 01689 875577
Geolink	CAM – Centre for Analysis & Modelling Limited 0171 232 1111	**GMSL Service Station Database**	GEO-Marketing Systems Ltd (GMSL) 01635 872382
GeoLink	Geo-Perfect TWI B.V. +31 1828 30477	**GOTHIC ADE**	Laser-Scan Ltd 01223 420414
geoManager	IBM UK Ltd 01926 464336	**GPG**	IBM UK Ltd 01926 464336
geoManager/6000	IBM UK Ltd 01926 464336	**GPHIGS**	Scientific Software Limited 01491 411727
Geomax	ACDS Graphic System Inc +1819 770 9631	**GPS Total Station**	Trimble Navigation Europe Ltd 01256 760150
GEOMEDIA	Intergraph (UK) Ltd 01793 619999	**GR.AID**	PLANTECH Ltd 0171 922 8825
GeoNet	Enghouse (UK) Limited 01703 615228	**GreenLine**	Geoview Systems Kft +361 269 2099
GeoPlan	Geoplan (UK) Ltd 01423 569538	**Grid Point**	Pinpoint Digitising Services 01928 579148
Geopoint	CAM – Centre for Analysis & Modelling Limited 0171 232 1111	**GSQL**	Tenet Systems Ltd 01403 273173
Geoscan	Geodelta +31 1515 8188	**GTX OSR**	TDS-CAD Graphics Ltd 01254 676921
Geoscience documentation	Geosystems 01235 813913	**GTX OSR Contour**	GTX Europe Ltd 01256 843555

GyPSy	Geodelta		**JDL**	Sovereign C.S. Ltd
	+31 1515 8188			0181 866 0713
HEALTH*INSITE	CACI Limited		**JetPro Series**	Summagraphics Europe N.V.
	0171 602 6000			+32 2 721 5033
Helava	Leica UK Ltd		**KALIDOR**	Kalidor Europe (ALPS
	01908 666663			Electric (Ireland) Ltd)
HIGHWAY ONE	Symology Limited			+353 29 21212
	01582 842626		**Kork**	Eurosense Technologies N.V.
HiPlot 7000	Summagraphics Europe N.V.			+32 2 460 7000
	+32 2 721 5033		**LAMPS**	Laser-Scan Ltd
HORIZON	Laser-Scan Ltd			01223 420414
	01223 420414		**Land-Form PANORAMA™**	Ordnance Survey
HP	Geo/SQL (UK)			01703 792773
	01672 562012		**Land-Form PROFILE™**	Ordnance Survey
Husky Field-Base	Husky Computing Limited			01703 792773
	01203 604040		**Land-Line®**	Ordnance Survey
Husky FS/2	Husky Computing Limited			01703 792773
	01203 604040		**LandCADD**	Datatechnology Datech Ltd
Husky Hunter 16	Husky Computing Limited			0181 308 1800
	01203 604040		**LandFlow**	Dowling Associates Limited
Husky Hunter 16/80	Husky Computing Limited			01943 880332
	01203 604040		**LANDSAT**	EOSAT
HYDATA	NERC			+1301 552 0525
	01793 411683		**LANDSCAPE**	Geotronics Limited
Hydrotitles	Geosystems			01480 433555
	01235 813913		**Landstar**	Racal Survey (UK) Ltd
HYTRAN	Institute of Hydrology			01224 249700
	01491 838800		**LARNET**	Hunting Engineering Ltd
IGIS	Laser-Scan Ltd			01525 841000
	01223 420414		**LB.AID**	PLANTECH Ltd
Illumine	The Data Consultancy			0171 922 8825
	01734 588181		**LC.AID**	PLANTECH Ltd
ILWIS	GISL Limited			0171 922 8825
	01956 285077		**LEICA**	Leica UK Ltd
ILWIS	ITC			01908 666663
	+31 53 874444		**Lightening Location**	EA Technology
IMAGE CATALOG	ERDAS (UK) Ltd			0151 347 2451
	01223 880802		**LKDesigner**	EA Technology
Imager Document	Advent Imaging Ltd			0151 347 2451
	01491 411566		**LOCATOR GIS**	Sokkia Ltd
Imager Edit	Advent Imaging Ltd			01270 250525
	01491 411566		**LOCATORGIS**	Conic Systems
Imager View	Advent Imaging Ltd			0131 667 2728
	01491 411566		**LSP2012**	Kirstol Ltd
Imager Wide-Format	Advent Imaging Ltd			0161 338 7512
	01491 411566		**Magellan**	Magellan Systems
IMAGINE	ERDAS (UK) Ltd			Corporation
	01223 880802			+1909 394 5000
INSITE	CACI Limited		**MANHATTAN II**	Raindrop Information
	0171 602 6000			Systems Ltd
Intergraph	C.A.Design Services Ltd			0171 734 1091
	01493 440444		**Map Attribute**	GEOSOFT Ltd
Intergraph	Mason Land Surveys			0113 234 4000
	01383 727261		**Map Control 2**	GEOSOFT Ltd
IRIS EXPLORER	NAG Ltd			0113 234 4000
	01865 511245		**Map Edit Control**	GEOSOFT Ltd
IRSDATA	EOSAT			0113 234 4000
	+1301 552 0525		**Map Editor 2**	GEOSOFT Ltd
				0113 234 4000

Map Processor	GEOSOFT Ltd 0113 234 4000	**Mapping Awareness**	Geoinformation International 01223 423020
Map Server	WS Atkins Planning & Management Consultants 01372 726140	**MAPPING OFFICE**	Intergraph (UK) Ltd 01793 619999
Map Server 2	GEOSOFT Ltd 0113 234 4000	**MAPS**	Pear Technology Services Ltd 01705 499689
Map Viewer	Byers Engineering Company +1404 843 1000	**MAPS**	European Business Mapping 01273 702957
MAP32	Oscar Faber 0181 784 5784	**Maps in Action**	Action Information Management Ltd
MapAccess	Geographic Management Solutions Ltd 01454 281802	**Maps in Action Europe**	01225 777288 Action Information Management Ltd
MapBasic	MapInfo Ltd 01344 482888	**Maps in Action GB**	01225 777288 Action Information
MAPDATA	Map Data Management Ltd 015395 67431		Management Ltd 01225 777288
MAPGEN	Tendron Systems Ltd 0117 929 4759	**MAPSCAPE**	HollyBush Software Limited 01443 482785
MAPGRAFIX	Spatial Geographic Services & Applications Ltd 01865 881753	**MAPSYS**	Kirstol Ltd 0161 338 7512
MapInfo	Dataview Solutions Ltd	**MapViewer**	Dotted Eyes 0121 445 6150
	0171 404 0640	**MapWise**	CDR Group
MapInfo	Glen Computing Ltd 01689 875577	**MARKET ANALYSIS**	01433 621282 Laser-Scan Ltd
MapInfo	Graphical Data Capture Ltd (GDC) 0181 349 2151	**MARKET*MASTER**	01223 420414 CACI Limited 0171 602 6000
MapInfo	MapInfo Ltd 01344 482888	**Market Planner**	CCN Marketing 0115 941 0888
MapInfo	Lantmatenet GIS-centrum +46 8402 1700	**Markets**	The Data Consultancy 01734 588181
Mapinfo	Dotted Eyes 0121 445 6150	**MATPAC**	MVA Systematica 01483 728051
MapInfo	The Data Consultancy 01734 588181	**Meridian™**	Ordnance Survey 01703 792773
MapInfo	Foto Res +34 27 216455	**Metamap**	MAPIT Limited 01487 813745
MapInfo	CDR Group 01433 621282	**MetaMap**	IME (UK) Ltd 01259 210210
MapInfo	Gamma Ltd +3531 6713066	**MGE**	Intergraph (UK) Ltd 01793 619999
MapInfo	Graphics Online Limited 0114 279 7972	**MGIMS**	Procis Software Ltd 01793 541200
MapInfo	Geo-Perfect TWI B.V. +31 1828 30477	**MGIS**	Procis Software Ltd 01793 541200
MapIT!	CDR Group 01433 621282	**Micor Low Flows**	Institute of Hydrology 01491 838800
MAPLEX	HollyBush Software Limited 01443 482785	**Micro-Marketing Machine**	TACTICIAN UK 01423 560064
MAPLINK	Tenet Systems Ltd 01403 273173	**MicroGDS**	Graphic Data Systems Corporation (GDS)
MAPMAN	Cobham Digital Services Limited 01932 868133	**MicroGrid III**	01483 725225 Summagraphics Europe N.V. +32 2 721 5033
MAPPER	SDI Ltd 0181 207 5474	**MicroMAPS**	MAPS geosystems +9716 356411

Microsoft	C.A.Design Services Ltd 01493 440444
Microstation	Bentley Systems UK Ltd 01344 412233
Microstation Field	Bentley Systems UK Ltd 01344 412233
Microstation Mapper	Bentley Systems UK Ltd 01344 412233
Microstation Review	Bentley Systems UK Ltd 01344 412233
Microtek	Sovereign C.S. Ltd 0181 866 0713
MIDAS II	Navstar Systems Ltd 01604 585588
Milemaster	WS Atkins Planning & Management Consultants 01372 726140
MiniMaps	European Business Mapping 01273 702957
MISS	Sovereign C.S. Ltd 0181 866 0713
Mobile Work Manager	Byers Engineering Company +1404 843 1000
Monarch	Royal Commission on the Historical Monuments of England 01793 414727
MOSAIC Systems	CCN Marketing 0115 941 0888
MOSS	MOSS Systems Limited 01403 259511
MultiFix	Racal Survey (UK) Ltd 01224 249700
Multimap	Tele Atlas +32 9 222 5658
Mutoh	Geo/SQL (UK) 01672 562012
Neighbours	EuroDirect Database Marketing Ltd 01274 737144
NetMaster	WS Atkins Planning & Management Consultants 01372 726140
NOMIS	NOMIS 0191 374 2468
NoveJet	TDS-CAD Graphics Ltd 01254 676921
NTF for AutoCAD	Longdin & Browning 01792 202244
NV.AID	PLANTECH Ltd 0171 922 8825
O2	Tenet Systems Ltd 01403 273173
OASIS Mapping Solution	Derek Hunter & Partners Ltd 01629 822100
OMEGA	Geoinformation International 01223 423020
OPEN INVENTOR	NAG Ltd 01865 511245
ORTHOMAX	ERDAS (UK) Ltd 01223 880802
OS Superplan®	Trident Map Services 01582 867211
OSCARR Asset-Manager	Ordnance Survey 01703 792773
OSF/1	Digital Equipment Corporation 01734 203546
PAF	Royal Mail – Address Management Centre 01705 838518
PAF on CD	Royal Mail – Address Management Centre 01705 838518
PAFEC GIS	PAFEC Ltd 0115 935 7055
PAMAP GIS™	EPS – Essential Planning Systems Limited +1604 652 8895
Parish-Line	The Data Consultancy 01734 588181
PARKMAN	Cobham Digital Services Limited 01932 868133
PARTNER II	Raindrop Information Systems Ltd 0171 734 1091
PATHFINDER	Trimble Navigation Europe Ltd 01256 760150
Pathfinder	Hunting Engineering Ltd 01525 841000
PC ARC/INFO	ESRI (UK) Ltd 01923 210450
PC TIN	AP3 Imaging Services Limited 01787 378242
PCI-EASI/PACE	Intera Information Technologies +1613 226 5442
PDS 2	LTG Services 0181 906 5559
PDS2	Elstree Computing Ltd 0181 906 5656
PERSPECTIVE VIEW	ERDAS (UK) Ltd 01223 880802
PHIGURE	Scientific Software Limited 01491 411727
Photonet	Royal Commission on the Historical Monuments of England 01793 414727
Phoxy	Geodelta +31 1515 8188
PlanBASE	Hunting Engineering Ltd 01525 841000

PlanCare	Graphical Data Capture Ltd (GDC) 0181 349 2151	**Quick Address**	The Data Consultancy 01734 588181
PlanChest	Graphical Data Capture Ltd (GDC) 0181 349 2151	**QUICKADDRESS**	QAS Systems Ltd 0171 498 7777
Planning Week	Royal Town Planning Institute 0171 636 9107	**RAMINA**	Spatial Information Services Ltd (SIS) 01329 662891
Plot Station	Byers Engineering Company +1404 843 1000	**RAMINA**	Xcon Data
PLT2TIF	CAD – Capture Limited 01254 583534	**RECLAIM**	SDI Ltd 0181 207 5474
PoleCAD	Optimal Software Ltd 0161 367 8715	**REGIOMAP**	Geoinformation International 01223 423020
PoleLOG	Optimal Software Ltd 0161 367 8715	**RESTORATION**	ERDAS (UK) Ltd 01223 880802
Polyview	Geo-Perfect TWI B.V. +31 1828 30477	**RETAIL*INSITE**	CACI Limited 0171 602 6000
Postcode Boundary File	General Register Office for Scotland 0131 314 4254	**RFDesigner**	EA Technology 0151 347 2451
Postcode Index file	General Register Office for Scotland 0131 314 4254	**RIMNET**	Hunting Engineering Ltd 01525 841000
		RISS	MAPS geosystems +9716 356411
PostGrid	Graphics Online Limited 0114 279 7972	**RM-GIS**	COMSULT 01234 342401
PostView	Graphics Online Limited 0114 279 7972	**RolleiFoToTechnic**	CZ Scientific Instruments Ltd 0181 953 1688
POWERMAP	TACTICIAN UK 01423 560064	**RouteAdviser**	Graphical Data Capture Ltd (GDC) 0181 349 2151
PR.AID	PLANTECH Ltd 0171 922 8825	**ROUTEMASTER**	Logitrans +331 498 51516
PREMIS	Organisation Management Systems 01865 372161	**RouteMeter**	Graphical Data Capture Ltd (GDC) 0181 349 2151
Primascan	Primagraphics Ltd 01763 262041	**RouteView**	Dataview Solutions Ltd 0171 404 0640
PRIME MERIDIAN™	EPS – Essential Planning Systems Limited +1604 652 8895	**RPS**	MPSI Systems Ltd 0117 927 9653
		S-GKS	Scientific Software Limited 01491 411727
Princess	WS Atkins Planning & Management Consultants 01372 726140	**S-Plus**	StatSci Europe 01865 200952
PRISM	Ashtech Europe Limited 01993 883533	**S-Plus for ARC/INFO**	StatSci Europe 01865 200952
PRO SCAN II	ICL Ltd 0121 456 1111	**SAMI**	Universal Systems Ltd +1506 458 8533
PROPAC	MVA Systematica 01483 728051	**SAS/GIS**	SAS Institute 01628 486933
Prospects	EuroDirect Database Marketing Ltd 01274 737144	**SASPAC**	London Research Centre 0171 735 4250
ProSpex for Windows	Beacon Dodsworth Limited 01904 638997	**SASPAC**	MVA Systematica 01483 728051
Provec	Foto Res +34 27 216455	**Saturn**	WS Atkins Planning & Management Consultants 01372 726140
QSign	WS Atkins Planning & Management Consultants 01372 726140	**SD2000**	Leica UK Ltd 01908 666663
		SDBM	Ashtech Europe Limited 01993 883533

SDE	ESRI (UK) Ltd
	01923 210450
Sensor Aircraft	Ashtech Europe Limited
	01993 883533
Sensor Marine	Ashtech Europe Limited
	01993 883533
Sercel GPS	CZ Scientific Instruments
	Ltd
	0181 953 1688
SG906 Scanner	Scan Group Limited
	+9724 410339
SICAD	Siemens Nixdorf
	01344 850829
SICAD/open	Siemens Nixdorf
	01344 850829
SICAD/PC-View	Siemens Nixdorf
	01344 850829
SICAD/WinCAT	Siemens Nixdorf
	01344 850829
SkyFix	Racal Survey (UK) Ltd
	01224 249700
SL-GMS	Tenet Systems Ltd
	01403 273173
SLIMPAC	University of Newcastle
	Upon Tyne
	0191 222 6445
Smallworld	WS Atkins Planning &
	Management Consultants
	01372 726140
Smallworld GIS	Smallworld Systems Ltd
	01223 460199
Smallworld View	Smallworld Systems Ltd
	01223 460199
SNAP	Ashtech Europe Limited
	01993 883533
Softdesk	Datatechnology Datech Ltd
	0181 308 1800
SOFTPLOTTER	ERDAS (UK) Ltd
	01223 880802
Solaris	Sun Microsystems Ltd
	01276 451440
SPACE/Motif	APIC Systems
	01279 466966
SPACE/Windows	APIC Systems
	01279 466966
SPANS GIS	TYDAC Technologies Ltd
	01703 760824
SPANS MAP	TYDAC Technologies Ltd
	01703 760824
SPARC	Sun Microsystems Ltd
	01276 451440
Spatial Analyst	Geo/SQL (UK)
	01672 562012
SPECTRUM	Sokkia Ltd
	01270 250525
STAKE OUT	Ashtech Europe Limited
	01993 883533
STANDARD	TACTICIAN UK
	01423 560064

STAR ARCHI	CODEC Facilities Limited
	01926 330112
STAR ARCHI	Star Informatic S.A.
	+32 41 675313
STAR CAD	CODEC Facilities Limited
	01926 330112
STAR CAD	Star Informatic S.A.
	+32 41 675313
STAR CARTO	CODEC Facilities Limited
	01926 330112
STAR CARTO	Star Informatic S.A.
	+32 41 675313
STAR INFRA	CODEC Facilities Limited
	01926 330112
STAR INFRA	Star Informatic S.A.
	+32 41 675313
STAR TECHNO	CODEC Facilities Limited
	01926 330112
STAR TECHNO	Star Informatic S.A.
	+32 41 675313
STAR VIEWER	CODEC Facilities Limited
	01926 330112
STAR VIEWER	Star Informatic S.A.
	+32 41 675313
STC25	CDR Group
	01433 621282
STDesigner	EA Technology
	0151 347 2451
Strategi™	Ordnance Survey
	01703 792773
StreetMAP	Tele Atlas
	+32 9 222 5658
StreetNET	Tele Atlas
	+32 9 222 5658
Streetpad	Geographic Management
	Solutions Ltd
	01454 281802
STRUMAP	Structural Technologies Ltd
	(STL)
	01527 854819
SummaGrid IV	Summagraphics Europe N.V.
	+32 2 721 5033
Super/Imposition	DAT/EM Systems
	International
	+1907 274 3681
Surveyor	Effective Solutions (Data
	Products)
	01794 514233
SUS2	Morgan Collis Group Ltd
	01604 580980
SUS25	Morgan Collis Group Ltd
	01604 580980
System 200	Leica UK Ltd
	01908 666663
System 9	UNISYS Ltd
	01895 237137
SYSTEM 9	ICL Ltd
	0121 456 1111

T-FLIGHT	MAPS geosystems +9716 356411	VIEWSURF	Bradly Associates Ltd 01344 779381
TEAMS Data Dictionary Manager	TEAMS (Taylor Woodrow Electronics Asset Mapping Survey) 0113 242 3802	VISION*	SHL Vision* Solutions Limited 01276 677707
		Vision	Geo-Perfect TWI B.V. +31 1828 30477
TEAMS Mobile GIS	TEAMS (Taylor Woodrow Electronics Asset Mapping Survey) 0113 242 3802	VISION*EXPRESS	SHL Vision* Solutions Limited 01276 677707
		VISION*MapMaker	SHL Vision* Solutions Limited 01276 677707
TELECOMMS ANALYSIS	Laser-Scan Ltd 01223 420414		
The NTF Toolkit	GID Ltd 01491 671964	Visual WorkFlo	FileNet Ltd 0181 944 5111
The SAS System	SAS Institute 01628 486933	VMS	Digital Equipment Corporation 01734 203546
THESIS	Oscar Faber 0181 784 5784	VTRAK	Laser-Scan Ltd 01223 420414
Thessian_P	Graphical Data Capture Ltd (GDC) 0181 349 2151	Ward-Line	The Data Consultancy 01734 588181
topoLogic	Geometria GIS Systems House Ltd +361 250 0989	WaterPoint	Carl Bro Group 0113 2620000
		WDV Laser Plotters	WDV (UK) 01789 297000
TowerCAD	Optimal Software Ltd 0161 367 8715	WILD	Leica UK Ltd 01908 666663
TowerLOG	Optimal Software Ltd 0161 367 8715	Win3D	Progis GmbH +43 4242 26332
TP.AID	PLANTECH Ltd 0171 922 8825	WinGIS	Progis GmbH +43 4242 26332
TRAKSYS	Structural Technologies Ltd (STL) 01527 854819	WINGIS	Midsummer Computing 01908 668866
Trimble	Geo-Perfect TWI B.V. +31 1828 30477	WINGS	System Options Limited 01252 334383
TRIPS	MVA Systematica 01483 728051	WinMAP	Progis GmbH +43 4242 26332
UK Digital Marine Atlas	NERC 01793 411683	WINMAP	Midsummer Computing 01908 668866
UNIBASE	Raindrop Information Systems Ltd 0171 734 1091	WinMAPLT	Progis GmbH +43 4242 26332
UTILMAN	Cobham Digital Services Limited 01932 868133	WINSAT	Midsummer Computing 01908 668866
		WinSAT	Progis GmbH +43 4242 26332
Varsity	Primagraphics Ltd 01763 262041	WIS	Institute of Hydrology 01491 838800
VAX	Digital Equipment Corporation 01734 203546	WIS	NERC 01793 411683
		WIS	ICL Ltd 0121 456 1111
VECTOR	ERDAS (UK) Ltd 01223 880802		
Vehicle Track	ARC Systems Pty Ltd +612 290 2400		

directory 3:
product suppliers

	data products	hardware products	publications	software products	complete systems
A.L.Downloading Services	–	◆	–	–	◆
ACDS Graphic System Inc	–	–	–	◆	◆
Action Information Management Ltd	◆	◆	◆	◆	◆
Active Software Ltd	◆	–	–	◆	◆
Adept Scientific Micro Systems Ltd	◆	–	◆	◆	◆
Advent Imaging Ltd	–	–	–	◆	–
AGFA UK	–	◆	–	◆	◆
ALLM Systems & Marketing	◆	–	◆	–	–
ALTEK Corporation	–	◆	–	–	–
AM/FM International - European Division	–	–	◆	–	–
AND Mapping B.V.	◆	–	–	–	–
Aneberie CAD	–	–	–	◆	–
AP³ Imaging Services Limited	◆	◆	◆	◆	◆
APIC Systems	–	–	–	◆	◆
ARC Systems Pty Ltd	–	–	–	◆	–
Ashtech Europe Limited	–	◆	–	◆	◆
Assist Applications Limited	–	◆	–	◆	◆
Audifilm Girona S.L.	–	–	–	◆	◆
Autodesk Ltd	–	–	–	◆	–
Bartholomew	◆	–	◆	–	◆
Beacon Dodsworth Limited	◆	–	–	◆	◆
Bentley Systems UK Ltd	–	–	–	◆	–
Birkbeck College London	◆	–	◆	–	–
BKS Surveys Ltd	◆	–	–	–	–
Bradly Associates Ltd	–	–	–	◆	–
British Geological Survey	◆	–	◆	◆	◆
Byers Engineering Company	–	–	–	◆	◆
C.A.Design Services Ltd	–	◆	–	◆	◆
CACI Limited	◆	–	◆	◆	◆
CAD - Capture Limited	–	◆	◆	◆	◆
CAD R&D Centre Limited	◆	–	–	◆	◆
CADAC Ltd	◆	◆	–	◆	◆
CalComp Limited	–	◆	–	–	–
CAM - Centre for Analysis & Modelling Limited	◆	–	–	◆	◆
Cambridge Market Intelligence	–	–	◆	–	–
Carl Bro Group	◆	–	–	◆	◆
Carl Zeiss Limited	–	◆	–	◆	◆
CARTograph Ltd	◆	–	–	◆	–
Cartographical Services (Southampton) Limited	–	◆	–	–	◆
Cartwright Associates	◆	–	–	◆	◆
CATALIST	◆	–	–	◆	◆
CCN Marketing	◆	–	◆	◆	◆
CCTA - The Government Centre for Information Systems	–	–	◆	–	–
CDR Group	–	–	–	◆	◆
Chiltern Digitising Services	–	–	–	–	◆
Citywise	◆	–	–	–	–

	data products	hardware products	publications	software products	complete systems
Cobham Digital Services Limited	◆	◆	–	◆	◆
CODEC Facilities Limited	–	–	–	◆	◆
Colorgraph (UK) Ltd	–	◆	–	◆	–
ColourMap Scanning Ltd	◆	–	–	–	–
Computer Aided Development (CADCORP) Ltd	–	–	–	◆	◆
Computer Graphic Suppliers Association	–	–	◆	–	–
COMSULT	–	–	–	◆	◆
Conic Systems	–	–	–	◆	◆
Consensus Information Technology Ltd	–	–	–	◆	◆
Construction Industry Computing Association	–	–	◆	–	–
Corena A/S	–	–	–	◆	–
Council of European Professional Informatics Societies	–	–	◆	–	–
Cray Systems	◆	◆	–	◆	◆
CSI	–	◆	–	◆	◆
CZ Scientific Instruments Ltd	–	◆	–	◆	◆
DAT/EM Systems International	–	◆	–	◆	◆
Dataman Computer Solutions UK Ltd	–	◆	–	–	–
Dataquest Europe Ltd	–	–	◆	–	–
Datatechnology Datech Ltd	–	◆	–	◆	◆
Dataview Solutions Ltd	◆	◆	–	◆	◆
Derek Hunter & Partners Ltd	–	–	–	◆	◆
Design Computer Aids Limited (DeCAL)	◆	◆	◆	◆	◆
Digital Equipment Corporation	–	◆	◆	–	–
DMAP Ltd	◆	–	–	◆	–
DMV Consultants BV	◆	–	–	◆	◆
Dolphin Consulting Group	◆	–	–	–	–
Dotted Eyes	◆	◆	◆	◆	◆
Dowling Associates Limited	–	–	–	◆	–
Dr Stanley Port	–	–	◆	–	–
EA Technology	◆	–	◆	◆	–
Earth Observation Sciences Ltd	–	◆	◆	◆	◆
Earth Resource Mapping	–	–	–	◆	–
Effective Solutions (Data Products)	◆	◆	◆	◆	◆
Elstree Computing Ltd	–	◆	–	◆	◆
Empress Software UK	–	–	–	◆	◆
Enghouse (UK) Limited	◆	–	–	◆	◆
Environment & Planning Library	–	–	◆	◆	–
EOSAT	◆	–	◆	–	–
EPS - Essential Planning Systems Limited	–	–	–	◆	◆
ERA Technology Ltd	–	–	–	–	◆
ERA-Maptec Ltd	◆	◆	–	–	–
ERDAS (UK) Ltd	◆	–	–	◆	◆
ESR Cartographers Ltd	◆	–	◆	–	◆
ESRI (UK) Ltd	◆	◆	–	◆	◆
EuroDirect Database Marketing Ltd	◆	–	–	◆	◆

	data products	hardware products	publications	software products	complete systems
European Business Mapping	◆	–	–	◆	–
European Geographic Technologies BV	◆	◆	◆	◆	–
Eurosense Technologies N.V.	◆	◆	–	◆	◆
Evox Facilities Ltd	◆	–	–	–	–
FastCAD GIS Ltd	◆	◆	–	◆	◆
FastCAD GIS Ltd	–	◆	–	◆	◆
FileNet Ltd	–	–	–	◆	◆
Foto Res	◆	◆	◆	◆	◆
Gamma Ltd	◆	–	–	◆	–
Gardline Infotech	–	–	–	◆	◆
GEC Marconi Research Centre	◆	–	–	◆	◆
Genasys II Limited	–	–	–	◆	◆
General Register Office for Scotland	◆	–	◆	–	–
GEO-Marketing Systems Ltd (GMSL)	◆	◆	–	◆	◆
Geo-Perfect TWI B.V.	◆	◆	–	◆	◆
Geo/SQL (UK)	–	◆	–	◆	◆
GeoData Institute	–	–	◆	–	–
Geodelta	–	–	–	◆	◆
Geografix Limited	◆	◆	–	◆	◆
Geographic Management Solutions Ltd	–	–	–	◆	◆
Geoinformation International	◆	–	◆	–	–
GeoMEM Software	◆	◆	–	◆	–
Geometria GIS Systems House Ltd	◆	–	–	◆	–
Geoplan (UK) Ltd	◆	–	◆	◆	◆
Geops BV	–	–	–	◆	–
GEOSOFT Ltd	–	–	–	◆	–
Geosystems	◆	–	◆	–	–
Geotronics Limited	–	◆	–	◆	◆
Geoview Systems Kft	–	–	–	◆	◆
GGP Systems Limited	–	–	–	◆	–
GID Ltd	–	–	–	◆	–
GIMMS (GIS) Ltd	–	–	–	◆	◆
GIS Services Ltd	–	◆	–	–	–
GISDATA	–	–	◆	–	–
GISL Limited	◆	–	–	◆	–
Glen Computing Ltd	◆	◆	–	◆	◆
Global Surveys Ltd	–	–	–	–	◆
GMAP Ltd	◆	–	–	◆	◆
Graphic Data Systems Corporation (GDS)	–	–	–	◆	–
Graphical Data Capture Ltd (GDC)	◆	◆	–	◆	◆
Graphics Online Limited	◆	–	–	◆	◆
Graphite Management Services Ltd	–	◆	–	◆	◆
Graphtec (UK) Ltd	–	◆	–	–	–
GTCO Corporation	–	◆	–	–	–
GTX Europe Ltd	–	–	–	–	◆

	data products	hardware products	publications	software products	complete systems
H R Wallingford Ltd	–	–	–	◆	–
Hall & Watts Systems Limited	◆	◆	–	◆	◆
Hansa Luftbild	◆	–	–	◆	–
Hitachi Home Electronics (Europe) Limited	–	◆	–	–	–
HJM Imaging Systems	–	◆	–	–	◆
HMSO Books	–	–	◆	–	–
HollyBush Software Limited	–	–	–	◆	–
Hunting Aerofilms Limited	◆	–	◆	–	◆
Hunting Engineering Ltd	–	–	–	◆	◆
Husky Computing Limited	–	◆	–	–	◆
I.S. Ltd	–	–	–	◆	◆
IBM UK Ltd	–	◆	◆	◆	◆
ICL Ltd	–	◆	◆	◆	◆
IMASS Limited	◆	◆	–	◆	◆
IME (UK) Ltd	–	◆	–	◆	◆
Infolink Decision Services Limited	◆	–	–	◆	–
Institute of Hydrology	◆	–	◆	◆	◆
Institute of Terrestrial Ecology	◆	–	◆	–	–
Intera Information Technologies	◆	–	–	◆	◆
Intergraph (UK) Ltd	–	◆	◆	◆	◆
International Map Trade Association	–	–	◆	–	–
International Products	◆	◆	–	◆	◆
IT Southern Ltd	–	◆	–	◆	◆
ITC	–	–	◆	◆	◆
ITS : Intertrade Scientific Ltd	–	◆	◆	–	◆
Kalidor Europe (ALPS Electric (Ireland) Ltd)	–	◆	–	–	–
Kirstol Ltd	–	◆	–	◆	◆
L.E.S. (Computer Services) Ltd	–	◆	–	◆	◆
Lancaster University	–	–	◆	◆	–
Land Aspects Consultancy Ltd (Parkman Group)	◆	–	–	–	–
Lantmatenet GIS-centrum	◆	–	–	◆	◆
Laser Technology International Ltd	–	◆	–	–	–
Laser-Scan Ltd	–	◆	–	◆	◆
Leica UK Ltd	–	◆	–	◆	◆
LiveChart	◆	◆	–	◆	◆
Logica UK Limited	–	–	–	–	◆
Logitrans	–	◆	–	–	◆
London Research Centre	◆	–	◆	–	–
Longdin & Browning	–	–	–	◆	◆
Longman Group Ltd	–	–	◆	–	–
Lovell Johns Ltd	◆	–	–	◆	–
LTG Services	–	–	–	◆	–
Macaulay Land Use Research Institute	◆	–	◆	◆	–
Magellan Systems Corporation	–	◆	–	◆	◆
Map Data Management Ltd	◆	–	◆	◆	◆
MapInfo Ltd	◆	–	◆	◆	◆

	data products	hardware products	publications	software products	complete systems
MAPIT Limited	–	–	–	♦	–
MAPS geosystems	♦	♦	–	♦	♦
Mason Land Surveys	♦	♦	♦	♦	♦
McLintock Limited	♦	♦	♦	♦	♦
MEGRIN Group	♦	–	–	–	–
Midlands Regional Research Laboratory	–	–	♦	♦	–
Midsummer Computing	♦	♦	–	♦	♦
Morgan Collis Group Ltd	–	–	–	♦	♦
MOSS Systems Limited	–	–	–	♦	–
Mott MacDonald Ltd	–	–	–	♦	♦
MPSI Systems Ltd	♦	–	–	–	–
MR Data Graphics	♦	–	–	–	–
Munro Garratt International	–	–	–	–	♦
MVA Systematica	♦	–	–	–	♦
MVM Consultants plc	–	–	–	♦	–
NAG Ltd	–	–	–	♦	–
National Remote Sensing Centre Ltd (NRSC)	♦	–	–	–	♦
Natural Environment Research Council	♦	–	♦	–	–
Navstar Systems Ltd	–	♦	–	♦	♦
NCC Blackwell Ltd	–	–	♦	–	–
NERC	♦	–	♦	–	–
NOMIS	♦	–	♦	–	–
Numonics UK (Division of Telmtek Ltd)	–	♦	–	–	–
Office of Population Censuses and Surveys (OPCS)	♦	–	–	–	–
Optimal Software Ltd	♦	♦	♦	♦	♦
Ordnance Survey	♦	–	♦	–	–
Ordnance Survey of Northern Ireland	♦	–	♦	–	–
Organisation Management Systems	♦	–	–	♦	♦
Oscar Faber	–	–	–	♦	♦
PAFEC Ltd	–	–	–	♦	♦
Paul Clasper & Associates Ltd	–	–	–	♦	–
PAX Technology	♦	–	–	♦	♦
PD Computing Ltd	–	–	–	♦	♦
Pear Technology Services Ltd	♦	♦	–	♦	♦
Photoair	♦	–	♦	–	–
Photogrammetric Data Services Ltd	♦	–	–	–	–
Pinpoint Digitising Services	♦	–	–	–	–
PlanGraphics Inc	–	–	♦	–	–
Planning & Mapping Ltd	♦	–	–	–	–
PLANTECH Ltd	–	–	♦	♦	–
Positioning Resources Limited	–	♦	–	♦	♦
Primagraphics Ltd	–	♦	–	–	♦
Procis Software Ltd	–	♦	–	–	♦
Progis GmbH	♦	–	♦	♦	♦
Property Intelligence plc	♦	–	♦	♦	♦

	data products	hardware products	publications	software products	complete systems
QAS Systems Ltd	◆	–	–	◆	◆
Quail Map Company	–	–	◆	–	–
Racal Survey (UK) Ltd	–	–	–	◆	◆
Raindrop Information Systems Ltd	–	◆	◆	◆	◆
regioplanDATA GmbH	◆	–	◆	◆	◆
Remote Sensing Applications Consultants	◆	–	–	◆	–
RICS Books	–	–	◆	–	–
Royal Commission on the Historical Monuments of England	–	–	◆	–	–
Royal Geographical Society (with The Institute of British Geographers)	◆	–	◆	–	–
Royal Institute of Navigation	–	–	◆	–	–
Royal Mail - Address Management Centre	◆	–	◆	–	–
Royal Town Planning Institute	–	–	◆	–	–
SAS Institute	–	–	◆	◆	–
Scan Group Limited	◆	◆	–	◆	◆
Scientific Software Limited	–	–	–	◆	–
SCOT Conseil	◆	–	–	◆	◆
SDI Ltd	–	◆	–	◆	◆
Sector (UK) Limited	–	◆	–	◆	◆
SHL Vision* Solutions Limited	–	–	–	◆	–
SIA Limited	◆	◆	–	◆	◆
SIAS Limited	–	–	–	◆	–
Siemens Nixdorf	–	◆	–	◆	◆
Silicon Graphics Limited	–	◆	◆	◆	–
Simmons Survey Partnership Limited	◆	–	–	–	–
Sir William Halcrow and Partners	◆	–	–	◆	–
Smallworld Systems Ltd	–	–	–	◆	◆
Sokkia Ltd	–	◆	–	◆	◆
Southbank Systems PLC	–	◆	–	–	◆
Sovereign C.S. Ltd	–	◆	–	◆	–
Spatial Data Limited	◆	–	–	–	◆
Spatial Geographic Services & Applications Ltd	–	–	–	◆	–
Spatial Information Services Ltd (SIS)	–	–	–	◆	◆
SPSS UK Ltd	–	–	◆	◆	–
Star Informatic S.A.	–	–	–	◆	◆
StatSci Europe	–	–	–	◆	–
Structural Technologies Ltd (STL)	◆	–	–	◆	◆
Summagraphics Europe N.V.	–	◆	–	–	–
Sun Microsystems Ltd	–	◆	◆	–	–
Survey & Development Services Ltd	–	–	–	◆	–
Survey Supplies Ltd	–	–	–	◆	◆
Svitzer Limited	◆	–	–	–	◆
Symology Limited	◆	◆	–	◆	◆
Sysdeco (UK) Ltd	–	◆	–	◆	◆
System Options Limited	◆	◆	◆	◆	◆

	data products	hardware products	publications	software products	complete systems
TACTICIAN UK	◆	–	–	◆	◆
Tangent Technology Design Associates Ltd	–	–	–	◆	–
Taylor & Francis Ltd	–	–	◆	–	–
TDS-CAD Graphics Ltd	–	◆	–	◆	◆
TEAMS (Taylor Woodrow Electronics Asset Mapping Survey)	–	–	–	◆	◆
Tekla Oy	–	–	–	◆	◆
Tele Atlas	◆	–	◆	–	–
Tendron Systems Ltd	–	–	◆	◆	◆
Tenet Systems Ltd	–	–	–	◆	–
Terrafix Ltd	◆	◆	◆	◆	◆
The British Computer Society	–	–	◆	–	–
The Business Database from Yellow Pages	◆	–	–	–	–
The Data Consultancy	◆	–	◆	◆	◆
The LGMB	–	–	◆	◆	–
The Marketing Information Consultancy (MIC)	◆	–	◆	◆	◆
The NPA Group Ltd	◆	◆	◆	◆	◆
The Severn Partnership	◆	–	–	–	–
The Survey Centre	◆	–	–	–	–
Trac Consultancy	◆	–	–	–	–
Trident Map Services	–	–	◆	–	–
Trimble Navigation Europe Ltd	–	◆	–	◆	◆
TYDAC Technologies Ltd	◆	–	–	◆	◆
UNISYS Ltd	–	◆	–	◆	–
Universal Systems Ltd	–	◆	–	◆	◆
University of Edinburgh	–	–	◆	–	–
University of Newcastle Upon Tyne	–	–	–	◆	–
Wallingford Software Limited	–	–	–	◆	–
WDV (UK)	–	◆	–	–	◆
WRc (Water Research centre)	–	◆	◆	–	–
WS Atkins Planning & Management Consultants	–	–	–	◆	◆
Xcon Data	–	–	–	–	◆

directory 4:
suppliers of data sets
group 1

	geological data products	land cover data products	land use data products	river data products	soil data products	terrain data products
Action Information Management Ltd	–	◆	◆	–	–	–
AP³ Imaging Services Limited	–	◆	–	–	–	–
Bartholomew	–	–	◆	◆	–	◆
Birkbeck College London	–	◆	–	–	–	–
BKS Surveys Ltd	–	–	◆	–	–	–
British Geological Survey	–	–	◆	–	◆	–
CAD R&D Centre Limited	–	◆	◆	◆	◆	–
Cartwright Associates	–	◆	◆	◆	◆	◆
Citywise	–	–	◆	–	–	–
Cray Systems	–	◆	◆	–	–	–
DMV Consultants BV	–	◆	–	–	◆	–
Dotted Eyes	–	–	–	◆	–	◆
Effective Solutions (Data Products)	–	◆	–	–	–	–
EOSAT	–	◆	–	–	–	–
ERA-Maptec Ltd	–	◆	–	–	–	–
ESRI (UK) Ltd	–	–	–	–	–	◆
European Geographic Technologies BV	–	–	◆	–	–	–
Eurosense Technologies N.V.	–	◆	–	–	–	–
FastCAD GIS Ltd	–	◆	◆	–	–	–
Gamma Ltd	–	◆	–	–	–	–
Geo-Perfect TWI B.V.	–	◆	◆	–	–	◆
Geoinformation International	–	–	–	–	–	◆
GeoMEM Software	–	–	–	◆	–	–
Geometria GIS Systems House Ltd	–	–	◆	–	–	–
GISL Limited	–	◆	◆	–	–	–
Glen Computing Ltd	–	◆	◆	◆	◆	◆
Hall & Watts Systems Limited	–	–	–	–	–	◆
Hansa Luftbild	–	◆	–	–	◆	–
Hunting Aerofilms Limited	–	◆	–	–	–	–
Infolink Decision Services Limited	–	◆	–	–	–	–
Institute of Hydrology	–	◆	–	◆	◆	–
Institute of Terrestrial Ecology	–	◆	–	–	–	–
Intera Information Technologies	◆	–	–	–	–	–
Land Aspects Consultancy Ltd (Parkman Group)	–	◆	◆	◆	◆	◆
Lantmatenet GIS-centrum	–	◆	–	–	–	–
LiveChart	–	◆	◆	–	–	–
Macaulay Land Use Research Institute	–	◆	–	–	◆	–
MapInfo Ltd	–	◆	◆	◆	–	–
National Remote Sensing Centre Ltd (NRSC)	◆	◆	–	–	–	–
Natural Environment Research Council	◆	◆	–	–	◆	–
NERC	–	◆	◆	◆	–	–
Ordnance Survey	–	◆	–	◆	–	◆
Ordnance Survey of Northern Ireland	–	◆	◆	–	◆	◆
Property Intelligence plc	–	–	◆	–	–	–
Scan Group Limited	–	◆	–	–	–	–
SIA Limited	–	–	◆	–	–	–
Sir William Halcrow and Partners	–	–	–	◆	–	◆
System Options Limited	–	◆	–	–	–	–
The Data Consultancy	–	◆	◆	◆	–	–
The NPA Group Ltd	–	◆	◆	–	–	◆

directory 5:
suppliers of data sets
group 2

	digital map products	digital photographs	remote sensed data products
Action Information Management Ltd	◆	◆	◆
Active Software Ltd	◆	–	–
Adept Scientific Micro Systems Ltd	◆	–	–
AND Mapping B.V.	◆	–	–
AP³ Imaging Services Limited	◆	–	◆
Bartholomew	◆	–	–
Birkbeck College London	◆	–	◆
BKS Surveys Ltd	◆	◆	–
British Geological Survey	◆	–	◆
CACI Limited	◆	–	–
CAD R&D Centre Limited	◆	–	–
CADAC Ltd	◆	–	–
CAM - Centre for Analysis & Modelling Limited	◆	–	–
Carl Bro Group	◆	–	–
Cartwright Associates	◆	–	◆
CATALIST	◆	–	–
CCN Marketing	◆	–	–
Citywise	◆	–	–
Cobham Digital Services Limited	◆	–	–
ColourMap Scanning Ltd	◆	–	–
Cray Systems	–	–	◆
Dataview Solutions Ltd	◆	◆	–
Design Computer Aids Limited (DeCAL)	◆	–	–
DMAP Ltd	◆	–	–
DMV Consultants BV	◆	–	◆
Dolphin Consulting Group	◆	–	–
Dotted Eyes	◆	◆	–
Enghouse (UK) Limited	◆	–	–
ERA-Maptec Ltd	◆	–	◆
ESR Cartographers Ltd	◆	–	◆
ESRI (UK) Ltd	◆	–	◆
European Geographic Technologies BV	◆	–	–
Eurosense Technologies N.V.	◆	–	◆
FastCAD GIS Ltd	◆	–	–
Foto Res	◆	◆	◆
GEC Marconi Research Centre	–	–	◆
General Register Office for Scotland	◆	–	–
GEO-Marketing Systems Ltd (GMSL)	◆	–	–
Geo-Perfect TWI B.V.	◆	–	◆
Geografix Limited	◆	–	–
Geoinformation International	◆	◆	–
GeoMEM Software	◆	◆	–
Geometria GIS Systems House Ltd	◆	–	–
Geoplan (UK) Ltd	◆	–	–
GISL Limited	–	–	◆
Glen Computing Ltd	◆	◆	◆

	digital map products	digital photographs	remote sensed data products
Graphical Data Capture Ltd (GDC)	◆	–	–
Hansa Luftbild	◆	–	◆
Hunting Aerofilms Limited	–	–	◆
Infolink Decision Services Limited	◆	–	◆
Institute of Hydrology	◆	–	–
Institute of Terrestrial Ecology	◆	–	◆
Intera Information Technologies	–	–	◆
International Products	◆	–	–
Land Aspects Consultancy Ltd (Parkman Group)	◆	–	–
Lantmatenet GIS-centrum	◆	–	–
LiveChart	◆	–	–
Lovell Johns Ltd	◆	–	–
Macaulay Land Use Research Institute	◆	–	◆
Map Data Management Ltd	◆	–	–
MapInfo Ltd	◆	–	–
MAPS geosystems	◆	◆	◆
Midsummer Computing	◆	–	–
MPSI Systems Ltd	◆	–	–
MR Data Graphics	◆	–	–
National Remote Sensing Centre Ltd (NRSC)	◆	◆	◆
Natural Environment Research Council	◆	–	◆
NERC	◆	–	◆
Optimal Software Ltd	◆	–	–
Ordnance Survey	◆	–	–
Ordnance Survey of Northern Ireland	◆	–	◆
Pear Technology Services Ltd	◆	–	–
Photoair	–	◆	◆
Photogrammetric Data Services Ltd	◆	◆	–
Planning & Mapping Ltd	◆	–	–
Progis GmbH	◆	◆	–
Property Intelligence plc	◆	–	–
regioplanDATA GmbH	◆	–	–
Remote Sensing Applications Consultants	–	–	◆
Royal Geographical Society (with The Institute of British Geographers)	◆	–	–
Scan Group Limited	◆	–	–
SCOT Conseil	–	–	◆
SIA Limited	◆	◆	–
Simmons Survey Partnership Limited	◆	–	–
Sir William Halcrow and Partners	◆	–	–
Spatial Data Limited	◆	–	◆
Structural Technologies Ltd (STL)	◆	◆	◆
Svitzer Limited	◆	–	–
System Options Limited	◆	–	–
TACTICIAN UK	◆	–	–
Tele Atlas	◆	–	–
Terrafix Ltd	◆	–	◆

	digital map products	digital photographs	remote sensed data products
The Data Consultancy	◆	◆	–
The NPA Group Ltd	◆	–	◆
The Severn Partnership	◆	–	–
The Survey Centre	◆	–	–
TYDAC Technologies Ltd	◆	–	◆

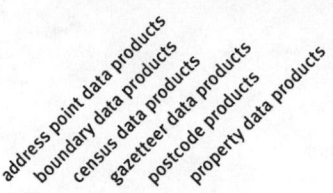

	address point data products	boundary data products	census data products	gazetteer data products	postcode products	property data products
Action Information Management Ltd	◆	◆	–	◆	◆	–
Active Software Ltd	◆	◆	◆	◆	◆	–
Adept Scientific Micro Systems Ltd	–	◆	◆	◆	◆	–
ALLM Systems & Marketing	–	–	◆	◆	◆	–
AND Mapping B.V.	–	◆	–	◆	◆	–
Bartholomew	–	◆	–	◆	◆	–
Beacon Dodsworth Limited	–	–	–	◆	◆	–
Birkbeck College London	–	◆	◆	◆	◆	–
CACI Limited	◆	◆	◆	◆	◆	◆
CAD R&D Centre Limited	–	◆	–	–	–	◆
CAM - Centre for Analysis & Modelling Limited	◆	◆	◆	◆	◆	–
Carl Bro Group	–	–	–	◆	–	–
Cartwright Associates	–	◆	–	◆	–	–
CATALIST	–	–	–	–	–	◆
CCN Marketing	–	◆	◆	–	◆	–
Citywise	◆	◆	–	◆	◆	◆
Cobham Digital Services Limited	◆	◆	–	◆	◆	–
Dataview Solutions Ltd	◆	◆	◆	◆	◆	–
Design Computer Aids Limited (DeCAL)	◆	–	–	◆	◆	◆
DMAP Ltd	–	◆	–	–	–	–
Dolphin Consulting Group	–	◆	◆	–	◆	–
Dotted Eyes	◆	◆	–	◆	◆	◆
ERA-Maptec Ltd	–	–	–	◆	–	–
ESRI (UK) Ltd	–	–	–	–	◆	◆
EuroDirect Database Marketing Ltd	–	–	◆	–	–	–
European Business Mapping	–	◆	–	–	–	–
European Geographic Technologies BV	◆	◆	–	◆	◆	–
Evox Facilities Ltd	–	–	–	◆	◆	–
FastCAD GIS Ltd	◆	◆	–	◆	◆	–
Foto Res	–	◆	–	–	–	–
Gamma Ltd	–	–	◆	–	–	–
General Register Office for Scotland	–	◆	◆	–	◆	–
GEO-Marketing Systems Ltd (GMSL)	–	–	◆	–	–	–
Geo-Perfect TWI B.V.	–	◆	–	–	◆	–
Geoinformation International	◆	◆	◆	–	–	–
GeoMEM Software	◆	◆	◆	◆	◆	◆
Geoplan (UK) Ltd	–	◆	◆	◆	◆	–
Glen Computing Ltd	◆	◆	◆	◆	◆	◆
GMAP Ltd	–	–	◆	–	–	–
Graphical Data Capture Ltd (GDC)	–	◆	–	◆	◆	–
Graphics Online Limited	–	◆	–	–	◆	–
IMASS Limited	◆	–	–	◆	–	◆
Infolink Decision Services Limited	–	–	–	◆	◆	–
Institute of Hydrology	–	◆	–	–	–	–
Land Aspects Consultancy Ltd (Parkman Group)	–	◆	–	◆	–	◆

	address point data products	boundary data products	census data products	gazetteer data products	postcode products	property data products
Lantmatenet GIS-centrum	◆	–	–	–	◆	–
London Research Centre	–	◆	◆	–	–	–
MapInfo Ltd	◆	◆	◆	◆	◆	–
McLintock Limited	◆	–	–	◆	–	◆
MEGRIN Group	–	◆	–	–	–	–
Midsummer Computing	–	◆	◆	–	–	–
MPSI Systems Ltd	◆	–	–	–	–	◆
MVA Systematica	◆	◆	◆	–	–	–
NOMIS	–	–	◆	–	–	–
Office of Population Censuses and Surveys (OPCS)	–	◆	◆	◆	◆	–
Ordnance Survey	◆	◆	–	◆	–	–
Ordnance Survey of Northern Ireland	◆	◆	◆	◆	◆	–
Pinpoint Digitising Services	–	–	–	–	◆	◆
Planning & Mapping Ltd	–	◆	–	–	–	–
Progis GmbH	–	◆	–	–	◆	–
Property Intelligence plc	◆	◆	◆	◆	◆	◆
QAS Systems Ltd	◆	–	–	◆	◆	–
regioplanDATA GmbH	◆	–	–	◆	–	–
Royal Geographical Society (with The Institute of British Geographers)	–	–	–	◆	–	–
Royal Mail - Address Management Centre	–	–	–	–	◆	–
Scan Group Limited	–	◆	–	–	–	–
SIA Limited	◆	◆	◆	◆	◆	◆
Structural Technologies Ltd (STL)	◆	–	–	◆	◆	–
Symology Limited	–	–	–	◆	–	–
System Options Limited	◆	◆	◆	◆	◆	–
TACTICIAN UK	–	◆	◆	◆	◆	–
Tele Atlas	◆	◆	–	–	◆	–
The Data Consultancy	–	◆	◆	◆	◆	◆
The Severn Partnership	–	◆	–	–	–	–
The Survey Centre	–	◆	–	–	–	–
Trac Consultancy	–	◆	–	–	◆	–
TYDAC Technologies Ltd	–	◆	–	–	◆	–

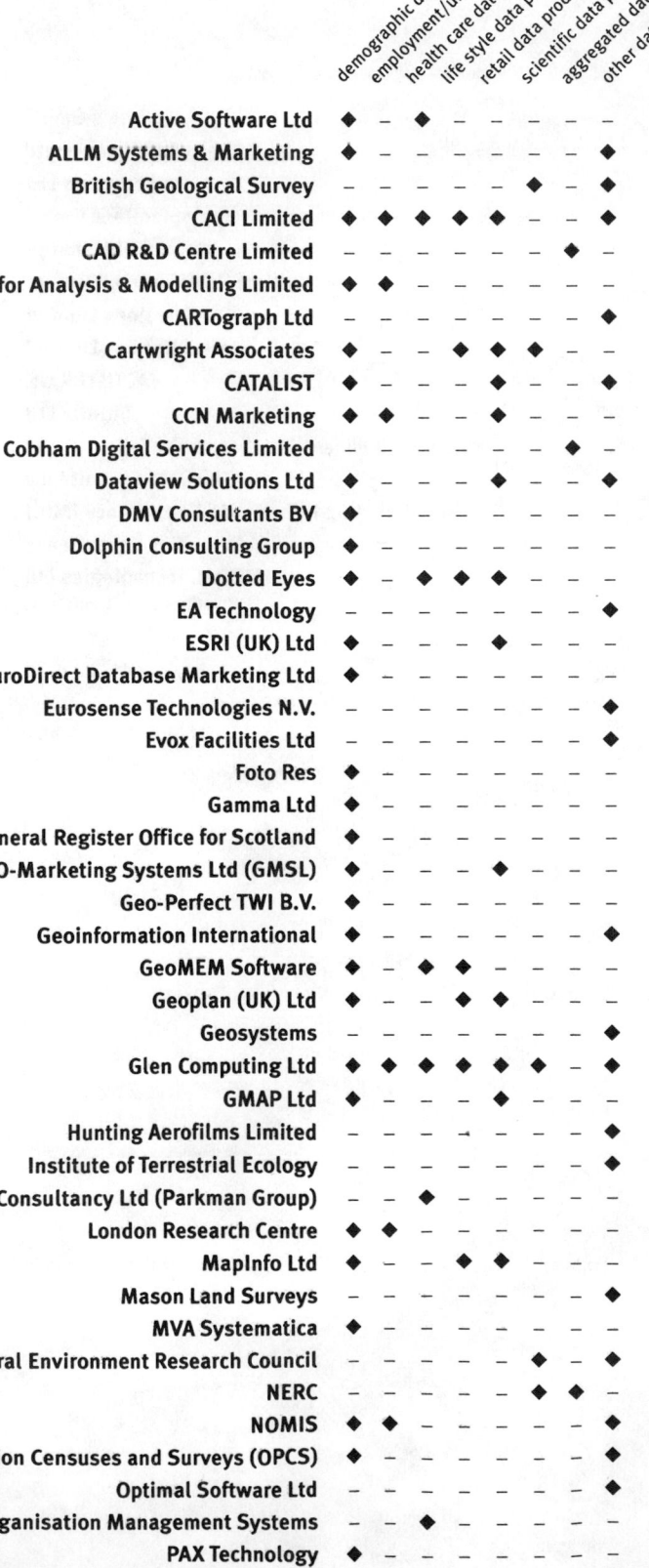

	demographic data products	employment/unemployment data products	health care data products	life style data products	retail data products	scientific data products	aggregated data products	other data products
Active Software Ltd	◆	–	◆	–	–	–	–	–
ALLM Systems & Marketing	◆	–	–	–	–	–	–	◆
British Geological Survey	–	–	–	–	–	◆	–	◆
CACI Limited	◆	◆	◆	◆	◆	–	–	◆
CAD R&D Centre Limited	–	–	–	–	–	–	◆	–
CAM - Centre for Analysis & Modelling Limited	◆	◆	–	–	–	–	–	–
CARTograph Ltd	–	–	–	–	–	–	–	◆
Cartwright Associates	◆	–	–	◆	◆	◆	–	–
CATALIST	◆	–	–	–	◆	–	–	◆
CCN Marketing	◆	◆	–	–	◆	–	–	–
Cobham Digital Services Limited	–	–	–	–	–	–	◆	–
Dataview Solutions Ltd	◆	–	–	◆	–	–	–	◆
DMV Consultants BV	◆	–	–	–	–	–	–	–
Dolphin Consulting Group	◆	–	–	–	–	–	–	–
Dotted Eyes	◆	–	◆	◆	◆	–	–	–
EA Technology	–	–	–	–	–	–	–	◆
ESRI (UK) Ltd	◆	–	–	–	◆	–	–	–
EuroDirect Database Marketing Ltd	◆	–	–	–	–	–	–	–
Eurosense Technologies N.V.	–	–	–	–	–	–	–	◆
Evox Facilities Ltd	–	–	–	–	–	–	–	◆
Foto Res	◆	–	–	–	–	–	–	–
Gamma Ltd	◆	–	–	–	–	–	–	–
General Register Office for Scotland	◆	–	–	–	–	–	–	–
GEO-Marketing Systems Ltd (GMSL)	◆	–	–	–	◆	–	–	–
Geo-Perfect TWI B.V.	◆	–	–	–	–	–	–	–
Geoinformation International	◆	–	–	–	–	–	–	◆
GeoMEM Software	◆	–	◆	◆	–	–	–	–
Geoplan (UK) Ltd	◆	–	–	◆	◆	–	–	–
Geosystems	–	–	–	–	–	–	–	◆
Glen Computing Ltd	◆	◆	◆	◆	◆	◆	–	◆
GMAP Ltd	◆	–	–	◆	◆	–	–	–
Hunting Aerofilms Limited	–	–	–	–	–	–	–	◆
Institute of Terrestrial Ecology	–	–	–	–	–	–	–	◆
Land Aspects Consultancy Ltd (Parkman Group)	–	–	◆	–	–	–	–	–
London Research Centre	◆	◆	–	–	–	–	–	–
MapInfo Ltd	◆	–	–	◆	◆	–	–	–
Mason Land Surveys	–	–	–	–	–	–	–	◆
MVA Systematica	◆	–	–	–	–	–	–	–
Natural Environment Research Council	–	–	–	–	–	◆	–	◆
NERC	–	–	–	–	–	◆	◆	–
NOMIS	◆	◆	–	–	–	–	–	◆
Office of Population Censuses and Surveys (OPCS)	◆	–	–	–	–	–	–	◆
Optimal Software Ltd	–	–	–	–	–	–	–	◆
Organisation Management Systems	–	–	◆	–	–	–	–	–
PAX Technology	◆	–	–	–	–	–	–	–

	demographic data products	employment/unemployment data products	health care data products	life style data products	retail data products	scientific data products	aggregated data products	other data products
Progis GmbH	◆	–	–	–	–	–	–	–
Property Intelligence plc	◆	–	–	–	◆	–	◆	–
QAS Systems Ltd	◆	–	–	–	–	–	–	–
regioplanDATA GmbH	–	–	–	–	–	–	◆	–
SIA Limited	◆	–	–	◆	–	–	–	–
Sir William Halcrow and Partners	–	–	–	–	–	◆	–	◆
Symology Limited	–	–	–	–	–	–	–	◆
System Options Limited	◆	–	–	–	–	–	–	–
TACTICIAN UK	◆	–	–	◆	◆	–	–	–
Terrafix Ltd	◆	–	–	–	–	–	–	–
The Business Database from Yellow Pages	–	–	–	–	–	–	–	◆
The Data Consultancy	◆	◆	◆	◆	◆	–	–	–
The Marketing Information Consultancy (MIC)	◆	–	–	◆	–	–	◆	◆
Trac Consultancy	◆	◆	–	–	–	–	◆	–
TYDAC Technologies Ltd	◆	–	–	–	–	–	–	–

directory 8:

suppliers of hardware products

	computer products	digitiser products	plotter products	printer products	scanner products	data communication products	GPS products	other peripheral products	photogrammetry products	portable computer products	surveying equipment products
A.L.Downloading Services	◆	–	–	–	–	◆	–	–	–	–	–
Action Information Management Ltd	◆	–	◆	◆	◆	◆	–	–	–	◆	–
AGFA UK	–	–	–	–	◆	–	–	–	–	–	–
ALTEK Corporation	–	◆	◆	◆	◆	–	–	–	–	–	–
AP³ Imaging Services Limited	◆	◆	◆	◆	◆	◆	–	–	–	–	–
Ashtech Europe Limited	–	–	–	–	–	◆	–	–	–	–	–
Assist Applications Limited	◆	◆	◆	◆	◆	–	–	–	–	–	–
C.A.Design Services Ltd	◆	◆	◆	◆	◆	–	–	–	–	–	–
CAD - Capture Limited	◆	–	◆	–	◆	–	–	–	–	◆	–
CADAC Ltd	◆	◆	◆	◆	–	–	◆	–	–	◆	◆
CalComp Limited	–	◆	◆	◆	◆	–	–	–	–	–	–
Carl Zeiss Limited	–	–	–	–	◆	–	◆	–	◆	–	◆
Cartographical Services (Southampton) Limited	–	–	–	–	–	–	–	–	◆	–	–
Cobham Digital Services Limited	◆	◆	◆	◆	◆	–	–	–	–	–	–
Colorgraph (UK) Ltd	–	–	–	◆	◆	–	–	–	–	–	–
Cray Systems	◆	◆	◆	◆	◆	◆	–	–	◆	–	–
CSI	–	–	◆	–	◆	–	–	–	–	–	–
CZ Scientific Instruments Ltd	–	–	–	–	◆	–	◆	–	◆	–	◆
DAT/EM Systems International	◆	–	–	–	–	–	–	–	◆	–	–
Dataman Computer Solutions UK Ltd	◆	–	–	–	–	–	–	–	–	–	–
Datatechnology Datech Ltd	◆	◆	◆	◆	◆	–	–	–	–	–	–
Dataview Solutions Ltd	–	–	–	–	–	◆	–	–	–	–	–
Design Computer Aids Limited (DeCAL)	◆	◆	◆	◆	◆	–	–	–	–	–	–
Digital Equipment Corporation	◆	–	–	–	–	◆	–	–	–	◆	–
Dotted Eyes	–	–	–	–	–	–	–	–	–	–	–
Earth Observation Sciences Ltd	◆	–	–	–	–	–	–	–	–	–	–
Effective Solutions (Data Products)	–	–	◆	–	–	–	◆	–	–	–	◆
Elstree Computing Ltd	◆	◆	◆	◆	◆	◆	–	◆	–	◆	–
ESRI (UK) Ltd	◆	◆	◆	◆	◆	–	–	–	–	–	–
European Geographic Technologies BV	–	–	–	–	–	–	–	–	–	–	–
Eurosense Technologies N.V.	◆	◆	◆	◆	◆	–	–	–	–	–	–
FastCAD GIS Ltd	◆	◆	◆	◆	◆	–	–	–	–	◆	–
FastCAD GIS Ltd	◆	◆	◆	◆	◆	–	–	–	–	◆	–
Foto Res	–	–	–	–	–	–	–	–	◆	–	–
GEO-Marketing Systems Ltd (GMSL)	◆	–	–	–	–	–	–	–	–	–	–
Geo-Perfect TWI B.V.	◆	◆	◆	◆	◆	–	◆	–	◆	–	–
Geo/SQL (UK)	◆	◆	◆	◆	◆	–	◆	–	–	–	–
Geografix Limited	–	–	–	–	–	–	–	◆	–	–	◆
GeoMEM Software	–	–	–	–	–	–	–	–	–	◆	–
Geotronics Limited	◆	–	–	–	–	–	◆	–	◆	–	◆
GIS Services Ltd	◆	◆	◆	◆	◆	–	–	–	–	–	–
Glen Computing Ltd	◆	◆	◆	◆	◆	–	–	–	–	◆	–
Graphical Data Capture Ltd (GDC)	◆	◆	◆	◆	◆	–	–	–	–	–	–
Graphite Management Services Ltd	◆	◆	◆	◆	◆	◆	–	–	–	◆	–
Graphtec (UK) Ltd	◆	◆	◆	–	–	–	–	–	–	◆	–
GTCO Corporation	–	◆	–	–	–	–	–	–	–	–	–

	computer products	digitiser products	plotter products	printer products	scanner products	data communication products	GPS products	other peripheral products	photogrammetry products	portable computer products	surveying equipment
Hall & Watts Systems Limited	–	–	–	–	–	–	–	–	–	–	◆
Hitachi Home Electronics (Europe) Limited	–	–	–	–	–	–	◆	–	–	–	–
HJM Imaging Systems	◆	–	◆	–	–	–	–	–	◆	–	–
Husky Computing Limited	◆	–	–	–	◆	–	◆	–	◆	–	–
IBM UK Ltd	◆	–	–	◆	–	◆	–	–	◆	–	–
ICL Ltd	◆	◆	◆	◆	◆	◆	–	◆	–	–	–
IMASS Limited	◆	◆	◆	◆	◆	–	–	–	–	–	–
IME (UK) Ltd	◆	◆	◆	–	◆	–	–	–	–	–	–
Intergraph (UK) Ltd	◆	◆	◆	◆	◆	◆	–	◆	–	–	–
International Products	◆	–	◆	◆	◆	–	–	–	–	–	–
IT Southern Ltd	◆	◆	◆	◆	◆	◆	–	◆	–	–	–
ITS : Intertrade Scientific Ltd	–	◆	–	–	◆	–	–	–	–	–	–
Kalidor Europe (ALPS Electric (Ireland) Ltd)	–	–	–	–	–	–	–	–	◆	–	–
Kirstol Ltd	◆	–	◆	◆	◆	–	–	◆	–	–	–
L.E.S. (Computer Services) Ltd	◆	◆	◆	◆	–	–	–	–	–	–	–
Laser Technology International Ltd	–	–	–	–	–	–	–	–	–	–	◆
Laser-Scan Ltd	◆	◆	◆	◆	◆	–	–	–	–	–	–
Leica UK Ltd	–	–	–	–	–	◆	–	◆	–	–	◆
LiveChart	–	◆	–	–	◆	–	–	–	–	–	–
Logitrans	–	–	–	–	–	–	–	–	–	–	◆
Magellan Systems Corporation	–	–	–	–	–	–	◆	–	–	–	–
MAPS geosystems	–	◆	–	–	–	–	–	–	–	–	–
Mason Land Surveys	◆	–	–	–	–	–	–	–	–	–	–
McLintock Limited	◆	◆	◆	◆	◆	–	–	–	–	–	–
Midsummer Computing	◆	◆	◆	◆	◆	–	–	–	–	◆	–
Navstar Systems Ltd	–	–	–	–	–	–	◆	–	–	–	◆
Numonics UK (Division of Telmtek Ltd)	–	◆	–	–	–	–	–	–	–	–	–
Optimal Software Ltd	◆	◆	◆	◆	◆	◆	–	◆	–	◆	◆
Pear Technology Services Ltd	–	◆	◆	◆	◆	–	–	◆	–	–	–
Positioning Resources Limited	–	–	–	–	–	–	◆	–	–	–	–
Primagraphics Ltd	◆	◆	–	–	◆	–	–	–	–	–	–
Procis Software Ltd	◆	–	–	◆	–	–	–	–	–	◆	–
Raindrop Information Systems Ltd	◆	–	◆	◆	◆	◆	–	◆	–	–	–
Scan Group Limited	–	–	–	–	◆	–	–	–	–	–	–
SDI Ltd	◆	◆	◆	◆	◆	–	–	–	–	–	–
Sector (UK) Limited	◆	–	–	–	–	◆	–	–	–	–	–
SIA Limited	◆	◆	◆	◆	◆	◆	◆	–	◆	–	–
Siemens Nixdorf	◆	–	–	◆	–	◆	–	–	◆	–	–
Silicon Graphics Limited	◆	–	–	–	–	–	–	–	–	–	–
Sokkia Ltd	–	–	–	–	–	–	–	–	◆	–	◆
Southbank Systems PLC	◆	◆	◆	◆	◆	–	–	◆	–	–	–
Sovereign C.S. Ltd	–	–	◆	◆	◆	–	–	–	–	–	–
Summagraphics Europe N.V.	–	◆	–	◆	–	–	–	–	–	–	–
Sun Microsystems Ltd	◆	–	–	–	–	–	–	–	–	–	–
Symology Limited	–	–	–	–	–	–	–	–	–	◆	–

	computer products	digitiser products	plotter products	printer products	scanner products	data communication products	GPS products	other peripheral products	photogrammetry products	portable computer products	surveying equipment products
Sysdeco (UK) Ltd	♦	♦	♦	♦	♦	♦	–	–	–	–	–
System Options Limited	♦	–	–	♦	♦	♦	♦	–	–	–	–
TDS-CAD Graphics Ltd	–	♦	♦	–	♦	–	–	–	–	–	–
Terrafix Ltd	♦	♦	♦	♦	–	♦	♦	♦	–	♦	–
The NPA Group Ltd	–	–	–	–	–	–	–	–	–	–	–
Trimble Navigation Europe Ltd	–	–	–	–	–	–	♦	♦	–	–	–
UNISYS Ltd	♦	–	–	♦	–	♦	♦	–	–	♦	–
Universal Systems Ltd	♦	♦	♦	♦	–	–	–	–	–	♦	–
WDV (UK)	–	–	♦	–	–	–	–	–	–	–	–

directory 9:
publications suppliers

	book publisher	guide publisher	magazine publisher	map publisher	newsletter publisher	technical report publisher	research publisher	technical manual publisher	CDROM publisher	WWW publisher	audio/video tape publisher
Action Information Management Ltd	–	–	–	◆	–	–	–	–	◆	–	–
Adept Scientific Micro Systems Ltd	–	–	◆	–	–	–	–	–	–	–	–
ALLM Systems & Marketing	◆	–	–	–	–	–	–	–	–	–	–
AM/FM International - European Division	–	–	◆	–	–	–	–	–	–	–	–
AP³ Imaging Services Limited	–	–	–	–	–	–	◆	–	–	–	–
Bartholomew	◆	◆	–	◆	◆	–	–	–	◆	–	◆
Birkbeck College London	◆	–	–	–	–	◆	–	–	–	–	–
British Geological Survey	◆	◆	◆	◆	◆	◆	–	–	◆	–	–
CACI Limited	–	–	–	–	–	◆	–	–	–	–	–
CAD - Capture Limited	–	–	–	–	–	–	◆	–	–	–	–
Cambridge Market Intelligence	–	–	–	–	◆	–	–	–	–	–	–
CCN Marketing	–	–	–	–	–	–	◆	–	–	–	–
CCTA - The Government Centre for Information Systems	◆	–	–	–	–	–	–	–	–	–	–
Computer Graphic Suppliers Association	–	–	–	–	◆	–	–	–	–	–	–
Construction Industry Computing Association	◆	◆	–	–	◆	◆	–	–	◆	–	–
Council of European Professional Informatics Societies	–	–	–	–	◆	–	–	–	–	–	–
Dataquest Europe Ltd	–	–	–	–	◆	–	–	–	◆	–	–
Design Computer Aids Limited (DeCAL)	–	–	◆	–	–	–	–	–	–	–	–
Digital Equipment Corporation	–	–	–	–	–	–	–	–	–	◆	–
Dotted Eyes	–	–	–	–	–	–	–	◆	◆	–	–
Dr Stanley Port	–	–	–	–	–	–	–	–	–	–	–
EA Technology	–	–	–	–	–	–	–	◆	–	–	–
Earth Observation Sciences Ltd	–	–	–	–	–	–	–	–	◆	–	–
Effective Solutions (Data Products)	–	–	◆	–	–	–	–	–	◆	–	–
Environment & Planning Library	–	–	–	–	◆	–	–	–	–	–	–
EOSAT	–	–	◆	–	–	–	–	–	–	◆	–
ERA-Maptec Ltd	–	–	–	◆	–	–	–	–	–	–	–
ESR Cartographers Ltd	–	–	–	◆	–	–	–	–	–	–	–
European Geographic Technologies BV	–	–	–	–	–	–	–	–	◆	–	–
Foto Res	–	–	◆	–	–	–	–	–	–	–	–
General Register Office for Scotland	◆	◆	–	–	◆	–	–	–	◆	–	◆
GeoData Institute	◆	◆	–	–	–	–	–	–	◆	–	–
Geoinformation International	◆	◆	◆	◆	◆	–	–	–	◆	–	–
Geoplan (UK) Ltd	–	–	–	◆	–	–	–	–	–	–	–
Geosystems	–	–	–	–	–	–	–	–	◆	–	–
GISDATA	–	–	–	–	–	–	–	–	–	◆	–
HMSO Books	◆	–	–	–	–	–	–	–	–	–	–
Hunting Aerofilms Limited	–	◆	–	–	–	–	–	–	–	–	–
IBM UK Ltd	–	–	–	–	–	–	–	–	–	◆	–
ICL Ltd	–	–	–	–	–	–	–	–	–	◆	–
Institute of Hydrology	–	–	–	–	–	–	–	◆	–	–	–
Institute of Terrestrial Ecology	◆	–	◆	–	–	–	◆	–	–	–	–
Intergraph (UK) Ltd	–	–	–	–	–	–	◆	◆	◆	◆	◆
International Map Trade Association	–	–	–	–	–	–	–	–	–	–	–
ITC	–	◆	–	–	◆	–	–	◆	◆	–	–
Lancaster University	◆	–	–	–	◆	–	–	–	–	–	–
London Research Centre	◆	–	–	◆	◆	◆	–	–	◆	–	–

	book publisher	guide publisher	magazine publisher	map publisher	newsletter publisher	technical report publisher	research publisher	technical manual publisher	CDROM publisher	WWW publisher	audio/video tape publisher
Longman Group Ltd	♦	♦	–	–	–	–	♦	–	–	–	–
Macaulay Land Use Research Institute	–	–	–	–	–	–	♦	–	–	–	–
Map Data Management Ltd	–	–	–	♦	–	–	–	–	–	–	–
MapInfo Ltd	–	–	–	–	♦	–	♦	–	–	–	–
Mason Land Surveys	–	–	–	♦	–	–	–	–	–	–	–
McLintock Limited	–	–	–	♦	–	–	–	–	–	–	–
Midlands Regional Research Laboratory	♦	–	–	–	♦	♦	–	–	–	–	–
Natural Environment Research Council	♦	–	–	–	–	–	–	–	–	–	–
NCC Blackwell Ltd	♦	♦	–	–	–	♦	♦	–	–	–	–
NERC	–	–	–	♦	–	♦	–	–	♦	♦	–
NOMIS	–	–	–	–	–	–	♦	–	–	–	–
Optimal Software Ltd	–	–	–	–	–	–	–	♦	–	–	–
Ordnance Survey	♦	♦	–	♦	♦	♦	–	♦	–	–	–
Ordnance Survey of Northern Ireland	–	–	–	♦	–	–	–	–	–	–	–
Photoair	–	–	–	–	–	–	–	♦	–	–	–
PlanGraphics Inc	♦	–	–	–	–	–	–	–	–	–	–
PLANTECH Ltd	–	–	–	–	♦	–	♦	–	–	–	–
Procis Software Ltd	–	–	–	–	♦	–	–	–	–	–	–
Progis GmbH	–	–	–	–	♦	–	–	–	–	–	–
Property Intelligence plc	–	–	–	–	♦	–	–	–	–	–	–
Quail Map Company	♦	–	–	♦	–	–	–	–	–	–	–
Raindrop Information Systems Ltd	–	–	–	–	–	–	–	♦	–	–	–
regioplanDATA GmbH	–	–	–	–	–	–	–	♦	–	–	–
RICS Books	♦	–	♦	–	–	–	♦	–	–	–	♦
Royal Commission on the Historical Monuments of England	–	♦	–	–	–	–	–	–	–	–	–
Royal Geographical Society (with The Institute of British Geographers)	♦	–	♦	♦	–	–	–	–	–	–	–
Royal Institute of Navigation	–	–	♦	–	♦	♦	♦	–	–	–	–
Royal Mail - Address Management Centre	♦	–	–	–	–	–	♦	–	–	–	–
Royal Town Planning Institute	–	–	♦	–	–	–	♦	–	–	–	–
SAS Institute	–	–	–	–	♦	♦	–	–	–	–	–
Silicon Graphics Limited	–	–	–	–	–	–	–	–	–	♦	–
SPSS UK Ltd	♦	–	–	–	–	–	–	–	–	–	–
Sun Microsystems Ltd	–	–	–	–	–	–	–	–	–	♦	–
System Options Limited	–	–	–	–	–	–	–	♦	–	–	–
Taylor & Francis Ltd	♦	–	–	–	–	–	♦	–	♦	–	–
Tele Atlas	–	–	–	♦	–	–	–	–	♦	–	–
Tendron Systems Ltd	–	–	–	–	–	–	–	♦	–	–	–
Terrafix Ltd	–	–	–	–	–	–	–	♦	–	–	–
The British Computer Society	–	–	♦	–	♦	–	–	♦	–	–	–
The Data Consultancy	–	–	–	–	♦	–	–	♦	–	–	–
The LGMB	♦	–	–	–	–	–	♦	–	–	–	–
The Marketing Information Consultancy (MIC)	–	–	–	–	♦	–	–	♦	–	♦	–
The NPA Group Ltd	–	–	–	♦	–	–	–	–	♦	–	–
Trident Map Services	–	–	–	♦	–	–	–	–	–	–	–
University of Edinburgh	–	–	–	–	–	–	–	–	–	♦	–
WRc (Water Research centre)	–	–	–	♦	♦	♦	♦	♦	♦	–	–

directory 9a:

publications and supplier details on the World Wide Web

British Geological Survey	–http://www.nkw.ac.uk.bgs
Dataquest Europe Ltd	–http://www.dataquest.com
Digital Equipment Corporation	–http://www.digital.com
EOSAT	–http://www.bsenet.com/GEO/geomain.htm
General Register Office for Scotland	–http://www.open.gov.uk/gros/groshome.htm
GeoData Institute	–http://www.geodata.ac.uk
GISDATA	–http://www.shef.ac.uk/uni/academic/D-H/gis/gisdata.html
IBM UK Ltd	–http://www.ibm.com
ICL Ltd	–http://www.icl.com
Intergraph (UK) Ltd	–http://www.ingr.com
ITC	–http://www.nez.com/ilwis/ilwis.htme
NERC	–http://www.nerc.ac.uk/
Silicon Graphics Limited	–http://www.sgi.com & http://www.europe.sgi.com
Sun Microsystems Ltd	–http://www.sun.com
Taylor & Francis Ltd	–http://www.tandf.co.uk/
The Marketing Information Consultancy (MIC)	–http://www.hyperlink.com/mic
University of Edinburgh	–http://www.geo.ed.ac.uk/

directory 10:
software products
complete systems

	CAD software	cartographic design software	document imaging software	GIS software	GPS software
ACDS Graphic System Inc	–	–	–	◆	–
Action Information Management Ltd	–	–	–	◆	◆
Active Software Ltd	–	–	–	◆	–
Adept Scientific Micro Systems Ltd	–	–	–	◆	–
AGFA UK	–	–	◆	–	–
Aneberie CAD	◆	–	–	◆	–
AP³ Imaging Services Limited	–	–	–	◆	–
APIC Systems	–	–	–	◆	–
ARC Systems Pty Ltd	◆	–	–	◆	◆
Ashtech Europe Limited	–	–	–	–	◆
Assist Applications Limited	–	–	–	◆	–
Audifilm Girona S.L.	◆	–	–	◆	–
Autodesk Ltd	◆	–	–	◆	–
Beacon Dodsworth Limited	–	–	–	◆	–
Bentley Systems UK Ltd	◆	–	–	◆	–
British Geological Survey	◆	–	–	◆	–
C.A.Design Services Ltd	◆	–	–	◆	–
CACI Limited	–	–	–	◆	–
CAD - Capture Limited	–	–	◆	◆	–
CAD R&D Centre Limited	◆	–	–	◆	–
CADAC Ltd	◆	–	–	◆	◆
CAM - Centre for Analysis & Modelling Limited	–	–	–	◆	–
Carl Bro Group	◆	–	–	◆	–
CARTograph Ltd	–	–	–	◆	–
Cartwright Associates	–	–	–	◆	–
CATALIST	–	–	–	◆	–
CCN Marketing	–	–	–	◆	–
CDR Group	–	–	–	◆	–
Citywise	–	–	–	◆	–
Cobham Digital Services Limited	–	–	–	◆	–
CODEC Facilities Limited	◆	–	–	◆	–
Computer Aided Development (CADCORP) Ltd	–	–	–	◆	–
COMSULT	–	–	–	◆	–
Conic Systems	–	–	–	◆	–
Consensus Information Technology Ltd	–	–	◆	◆	–
Cray Systems	◆	–	–	◆	–
CZ Scientific Instruments Ltd	–	–	–	–	◆
Datatechnology Datech Ltd	◆	–	◆	◆	–
Dataview Solutions Ltd	–	–	–	◆	◆
Derek Hunter & Partners Ltd	–	–	–	◆	◆
Design Computer Aids Limited (DeCAL)	◆	–	–	◆	–
DMAP Ltd	–	–	◆	◆	–
Dotted Eyes	–	–	–	◆	–
Dowling Associates Limited	–	–	–	◆	–
EA Technology	◆	–	–	–	–

	CAD software	cartographic design software	document imaging software	GIS software	GPS software
Earth Resource Mapping	–	–	◆	–	–
Effective Solutions (Data Products)	–	–	–	◆	◆
Elstree Computing Ltd	◆	–	–	◆	–
Empress Software UK	–	–	–	◆	–
Enghouse (UK) Limited	–	–	–	◆	–
EPS - Essential Planning Systems Limited	–	–	–	◆	–
ERDAS (UK) Ltd	–	–	◆	◆	–
ESRI (UK) Ltd	–	–	–	◆	–
EuroDirect Database Marketing Ltd	–	–	–	◆	–
European Business Mapping	–	–	–	◆	–
Eurosense Technologies N.V.	–	–	–	◆	–
FastCAD GIS Ltd	◆	–	–	◆	–
FastCAD GIS Ltd	◆	–	–	◆	–
FileNet Ltd	–	–	◆	–	–
Foto Res	–	–	–	◆	–
Gamma Ltd	–	–	–	◆	–
Gardline Infotech	–	–	–	◆	–
Genasys II Limited	–	–	–	◆	–
GEO-Marketing Systems Ltd (GMSL)	–	–	–	◆	–
Geo-Perfect TWI B.V.	–	–	–	◆	◆
Geo/SQL (UK)	◆	–	◆	◆	–
Geodelta	◆	–	◆	◆	◆
Geografix Limited	–	–	–	◆	◆
Geographic Management Solutions Ltd	–	–	–	◆	◆
GeoMEM Software	◆	–	–	◆	–
Geometria GIS Systems House Ltd	–	–	–	◆	–
Geoplan (UK) Ltd	–	–	–	◆	–
Geops BV	–	–	–	◆	–
GEOSOFT Ltd	◆	–	–	◆	–
Geotronics Limited	◆	–	–	–	◆
Geoview Systems Kft	–	–	–	◆	–
GGP Systems Limited	–	–	–	◆	–
GIMMS (GIS) Ltd	–	–	–	◆	–
GISL Limited	–	–	–	◆	–
Glen Computing Ltd	◆	–	–	◆	–
GMAP Ltd	–	–	–	◆	–
Graphic Data Systems Corporation (GDS)	◆	–	–	◆	–
Graphical Data Capture Ltd (GDC)	–	–	–	◆	–
Graphics Online Limited	–	–	–	◆	–
Graphite Management Services Ltd	◆	–	◆	◆	–
HollyBush Software Limited	–	◆	–	–	–
Hunting Engineering Ltd	–	–	–	◆	–
I.S. Ltd	–	–	–	◆	–
IBM UK Ltd	◆	–	–	◆	–
ICL Ltd	◆	–	◆	◆	–

	CAD software	cartographic design software	document imaging software	GIS software	GPS software
IMASS Limited	♦	–	–	♦	–
IME (UK) Ltd	–	–	–	♦	–
Infolink Decision Services Limited	–	–	–	♦	–
Institute of Hydrology	–	–	–	♦	–
Intergraph (UK) Ltd	♦	–	♦	♦	♦
International Products	♦	–	♦	♦	–
IT Southern Ltd	♦	–	–	♦	–
ITC	–	–	♦	♦	–
L.E.S. (Computer Services) Ltd	♦	–	–	–	–
Lantmatenet GIS-centrum	–	–	–	♦	–
Laser-Scan Ltd	–	–	–	♦	–
Leica UK Ltd	–	–	–	–	♦
Longdin & Browning	♦	–	–	♦	–
LTG Services	♦	–	–	♦	–
Magellan Systems Corporation	–	–	–	–	♦
Map Data Management Ltd	–	–	–	♦	–
MapInfo Ltd	–	–	–	♦	–
MAPIT Limited	–	–	–	♦	–
MAPS geosystems	♦	–	–	♦	♦
Mason Land Surveys	♦	–	–	♦	–
McLintock Limited	–	–	–	♦	–
Midlands Regional Research Laboratory	–	–	–	♦	–
Midsummer Computing	♦	–	–	♦	–
Morgan Collis Group Ltd	–	–	–	♦	–
MOSS Systems Limited	♦	–	–	–	–
Mott MacDonald Ltd	♦	–	–	♦	–
MVM Consultants plc	–	–	–	♦	–
NAG Ltd	–	–	–	♦	–
Navstar Systems Ltd	–	–	–	–	♦
Optimal Software Ltd	♦	–	–	♦	♦
Organisation Management Systems	–	–	–	♦	–
Oscar Faber	–	–	–	♦	–
PAFEC Ltd	♦	–	♦	♦	–
Paul Clasper & Associates Ltd	–	–	–	–	♦
PAX Technology	♦	–	♦	♦	–
PD Computing Ltd	–	–	–	♦	–
Pear Technology Services Ltd	–	–	–	♦	♦
PLANTECH Ltd	–	–	♦	–	–
Positioning Resources Limited	–	–	–	–	♦
Progis GmbH	–	–	–	♦	–
Property Intelligence plc	–	–	–	♦	–
QAS Systems Ltd	–	–	–	♦	–
Racal Survey (UK) Ltd	–	–	–	–	♦
Raindrop Information Systems Ltd	♦	–	–	–	–
regioplanDATA GmbH	–	–	–	♦	–
Remote Sensing Applications Consultants	–	–	–	–	♦

	CAD software	cartographic design software	document imaging software	GIS software	GPS software
SAS Institute	–	–	–	◆	–
SCOT Conseil	–	–	–	◆	–
SDI Ltd	◆	–	–	◆	–
Sector (UK) Limited	–	–	–	◆	◆
SHL Vision* Solutions Limited	–	–	–	◆	–
SIA Limited	◆	–	–	◆	◆
SIAS Limited	–	–	–	◆	–
Siemens Nixdorf	–	–	–	◆	–
Smallworld Systems Ltd	–	–	–	◆	–
Sokkia Ltd	–	–	–	◆	–
Sovereign C.S. Ltd	–	–	–	◆	–
Spatial Geographic Services & Applications Ltd	–	–	–	◆	–
Star Informatic S.A.	◆	–	–	◆	–
Structural Technologies Ltd (STL)	–	–	–	◆	◆
Survey & Development Services Ltd	◆	–	–	◆	–
Survey Supplies Ltd	◆	–	–	◆	–
Symology Limited	–	–	–	◆	–
Sysdeco (UK) Ltd	–	–	–	◆	–
System Options Limited	–	–	–	◆	◆
TACTICIAN UK	–	–	–	◆	–
TDS-CAD Graphics Ltd	◆	–	–	–	–
TEAMS (Taylor Woodrow Electronics Asset Mapping Survey)	–	–	–	◆	–
Tekla Oy	–	–	–	◆	–
Tendron Systems Ltd	–	–	–	◆	–
Tenet Systems Ltd	–	–	–	◆	–
Terrafix Ltd	◆	–	–	◆	◆
The Data Consultancy	–	–	–	◆	–
The Marketing Information Consultancy (MIC)	–	–	–	◆	–
The NPA Group Ltd	◆	–	–	◆	–
Trimble Navigation Europe Ltd	–	–	–	◆	◆
TYDAC Technologies Ltd	–	–	–	◆	–
UNISYS Ltd	–	–	◆	◆	–
Universal Systems Ltd	–	–	–	◆	–
University of Newcastle Upon Tyne	◆	–	–	◆	–
WS Atkins Planning & Management Consultants	–	–	–	◆	–

software products
software libraries

	coordinate transformation software	data transformation software	remote sensing software	scanning software	spatial analysis software	statistical software	surface modelling software	terrain modelling software
ACDS Graphic System Inc	–	–	–	–	–	–	–	◆
Action Information Management Ltd	◆	◆	–	◆	◆	–	–	–
Active Software Ltd	–	–	–	–	◆	◆	–	–
Adept Scientific Micro Systems Ltd	–	–	–	–	◆	◆	–	–
Advent Imaging Ltd	–	–	–	◆	–	–	–	–
AGFA UK	–	–	–	◆	–	–	–	–
Aneberie CAD	–	–	–	–	–	–	◆	◆
AP³ Imaging Services Limited	–	–	◆	–	◆	–	◆	◆
Assist Applications Limited	–	–	–	–	–	–	◆	◆
Autodesk Ltd	–	–	–	–	–	–	◆	–
Bradly Associates Ltd	–	–	–	–	–	–	◆	–
British Geological Survey	◆	◆	◆	◆	◆	◆	◆	◆
Byers Engineering Company	–	◆	–	–	–	–	–	–
CAD - Capture Limited	–	◆	–	◆	–	–	–	–
CAD R&D Centre Limited	◆	–	–	–	–	–	–	–
CADAC Ltd	–	–	–	–	–	–	◆	◆
Carl Zeiss Limited	–	–	–	◆	–	–	–	–
Cartwright Associates	◆	◆	–	–	◆	–	◆	◆
CCN Marketing	–	–	–	–	◆	◆	–	–
Cobham Digital Services Limited	–	◆	–	–	–	–	–	–
CODEC Facilities Limited	–	◆	–	◆	◆	◆	◆	◆
Colorgraph (UK) Ltd	–	–	–	◆	–	–	–	–
Cray Systems	◆	◆	–	–	◆	–	◆	◆
CSI	–	–	◆	◆	–	–	–	–
Datatechnology Datech Ltd	–	–	–	◆	–	–	◆	◆
Dataview Solutions Ltd	–	◆	–	–	–	–	–	◆
Derek Hunter & Partners Ltd	–	◆	–	–	–	–	–	–
DMV Consultants BV	–	–	◆	–	–	–	–	–
Dotted Eyes	–	◆	–	–	–	–	–	–
EA Technology	–	–	–	–	–	–	–	◆
Earth Observation Sciences Ltd	–	–	◆	–	–	◆	–	–
Earth Resource Mapping	–	–	◆	–	◆	–	–	–
Elstree Computing Ltd	–	–	–	◆	–	–	◆	◆
Enghouse (UK) Limited	–	–	–	–	◆	◆	–	–
EPS - Essential Planning Systems Limited	◆	◆	–	–	◆	–	◆	◆
ERDAS (UK) Ltd	–	–	◆	–	◆	–	◆	◆
ESRI (UK) Ltd	–	◆	–	–	◆	◆	◆	◆
FastCAD GIS Ltd	◆	◆	–	–	◆	–	–	–
FastCAD GIS Ltd	–	◆	–	◆	◆	–	–	–
Foto Res	–	–	◆	–	–	–	–	–
GEC Marconi Research Centre	–	–	◆	–	–	–	–	–
Genasys II Limited	–	–	◆	–	◆	–	–	◆
Geo-Perfect TWI B.V.	–	–	◆	–	◆	–	◆	◆
Geo/SQL (UK)	–	–	–	◆	◆	–	◆	◆
Geodelta	–	–	–	–	–	◆	–	–

	coordinate transformation software	data transformation software	remote sensing software	scanning software	spatial analysis software	statistical software	surface modelling software	terrain modelling software
Geografix Limited	–	◆	◆	–	–	–	–	◆
GeoMEM Software	◆	◆	–	–	–	–	◆	◆
Geotronics Limited	–	–	–	–	–	–	◆	◆
GISL Limited	–	–	◆	–	–	–	–	–
Glen Computing Ltd	◆	◆	◆	◆	◆	◆	◆	◆
GMAP Ltd	–	◆	–	–	◆	–	–	–
Graphical Data Capture Ltd (GDC)	–	◆	–	–	–	–	–	–
Graphite Management Services Ltd	–	–	◆	–	–	–	◆	◆
Hall & Watts Systems Limited	–	–	–	–	–	–	–	◆
Hunting Engineering Ltd	–	–	◆	–	–	◆	◆	◆
I.S. Ltd	–	–	◆	–	–	–	–	◆
ICL Ltd	–	–	–	–	◆	–	◆	◆
IMASS Limited	–	◆	–	–	◆	◆	◆	◆
IME (UK) Ltd	–	◆	–	–	–	–	–	–
Infolink Decision Services Limited	–	–	–	–	◆	◆	–	–
Institute of Hydrology	–	–	◆	–	◆	◆	◆	◆
Intera Information Technologies	–	–	◆	–	–	–	–	–
Intergraph (UK) Ltd	◆	◆	◆	◆	◆	◆	◆	◆
International Products	–	–	–	◆	–	–	–	–
IT Southern Ltd	–	◆	–	◆	◆	–	◆	◆
ITC	◆	◆	◆	–	◆	◆	◆	◆
ITS : Intertrade Scientific Ltd	–	◆	–	–	–	–	–	–
Kirstol Ltd	–	–	–	◆	–	–	–	–
L.E.S. (Computer Services) Ltd	–	–	–	–	–	–	◆	◆
Laser-Scan Ltd	◆	–	◆	◆	◆	–	◆	◆
Leica UK Ltd	–	–	–	–	–	–	◆	◆
LiveChart	–	◆	–	◆	–	–	–	–
Lovell Johns Ltd	–	◆	–	–	–	–	–	–
Macaulay Land Use Research Institute	–	–	◆	–	◆	–	–	◆
Map Data Management Ltd	◆	◆	–	–	–	–	–	◆
MapInfo Ltd	–	◆	–	–	◆	–	–	–
MAPS geosystems	–	–	–	–	◆	–	◆	◆
Midsummer Computing	–	–	◆	–	◆	◆	◆	◆
MOSS Systems Limited	–	–	–	–	–	–	◆	◆
Mott MacDonald Ltd	–	–	–	–	◆	–	◆	◆
NAG Ltd	–	–	–	–	–	◆	◆	◆
Optimal Software Ltd	–	–	–	–	–	–	–	◆
Oscar Faber	–	◆	–	–	◆	–	–	–
PAX Technology	–	–	–	–	◆	◆	–	–
Progis GmbH	◆	–	–	–	–	–	–	–
Raindrop Information Systems Ltd	–	–	–	◆	–	–	–	–
regioplanDATA GmbH	◆	◆	–	–	–	–	–	–
SAS Institute	–	–	–	–	–	◆	–	–
Scan Group Limited	–	–	–	◆	–	–	–	–
SDI Ltd	–	–	–	–	◆	–	◆	◆

	coordinate transformation software	data transformation software	remote sensing software	scanning software	spatial analysis software	statistical software	surface modelling software	terrain modelling software
SIA Limited	–	–	–	◆	◆	◆	–	–
Sir William Halcrow and Partners	–	–	–	–	–	◆	–	–
Spatial Information Services Ltd (SIS)	◆	◆	–	◆	–	–	–	–
SPSS UK Ltd	–	–	–	◆	–	◆	–	–
Star Informatic S.A.	–	–	–	◆	◆	◆	◆	◆
StatSci Europe	–	–	–	–	–	◆	–	–
Structural Technologies Ltd (STL)	◆	◆	◆	–	◆	◆	–	–
Survey Supplies Ltd	◆	◆	–	–	◆	◆	◆	◆
Sysdeco (UK) Ltd	◆	◆	–	◆	◆	–	–	–
System Options Limited	–	–	◆	◆	◆	◆	◆	◆
TDS-CAD Graphics Ltd	–	–	–	◆	–	–	–	–
TEAMS (Taylor Woodrow Electronics Asset Mapping Survey)	–	◆	–	–	–	–	–	–
Tekla Oy	–	–	–	–	–	–	–	◆
Tendron Systems Ltd	–	–	◆	–	–	◆	◆	◆
Terrafix Ltd	–	–	◆	–	–	–	–	–
The Data Consultancy	◆	◆	–	–	◆	◆	–	–
The NPA Group Ltd	–	–	◆	◆	◆	–	◆	◆
Trimble Navigation Europe Ltd	–	–	–	–	–	–	◆	◆
TYDAC Technologies Ltd	–	–	◆	–	◆	–	◆	◆
UNISYS Ltd	◆	◆	–	–	◆	◆	◆	◆
Universal Systems Ltd	–	–	◆	–	◆	–	◆	◆

directory 12:
software products
system software

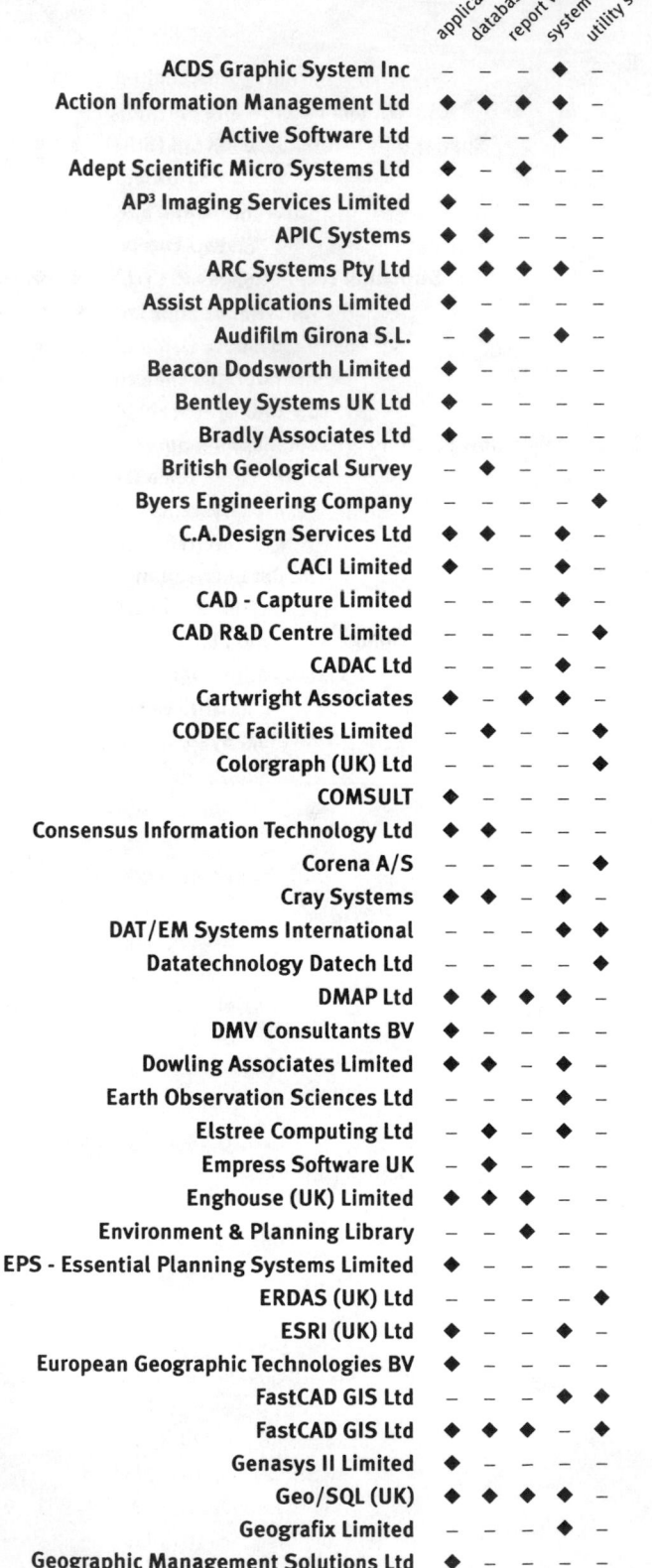

	application builder software	database management system software	report writer software	system software	utility software
ACDS Graphic System Inc	–	–	–	◆	–
Action Information Management Ltd	◆	◆	◆	◆	–
Active Software Ltd	–	–	–	◆	–
Adept Scientific Micro Systems Ltd	◆	–	◆	–	–
AP³ Imaging Services Limited	◆	–	–	–	–
APIC Systems	◆	◆	–	–	–
ARC Systems Pty Ltd	◆	◆	◆	◆	–
Assist Applications Limited	◆	–	–	–	–
Audifilm Girona S.L.	–	◆	–	◆	–
Beacon Dodsworth Limited	◆	–	–	–	–
Bentley Systems UK Ltd	◆	–	–	–	–
Bradly Associates Ltd	◆	–	–	–	–
British Geological Survey	–	◆	–	–	–
Byers Engineering Company	–	–	–	–	◆
C.A.Design Services Ltd	◆	◆	–	◆	–
CACI Limited	◆	–	–	◆	–
CAD - Capture Limited	–	–	–	◆	–
CAD R&D Centre Limited	–	–	–	–	◆
CADAC Ltd	–	–	–	◆	–
Cartwright Associates	◆	–	◆	◆	–
CODEC Facilities Limited	–	◆	–	–	◆
Colorgraph (UK) Ltd	–	–	–	–	◆
COMSULT	◆	–	–	–	–
Consensus Information Technology Ltd	◆	◆	–	–	–
Corena A/S	–	–	–	–	◆
Cray Systems	◆	◆	–	◆	–
DAT/EM Systems International	–	–	–	◆	◆
Datatechnology Datech Ltd	–	–	–	–	◆
DMAP Ltd	◆	◆	◆	◆	–
DMV Consultants BV	◆	–	–	–	–
Dowling Associates Limited	◆	◆	–	◆	–
Earth Observation Sciences Ltd	–	–	–	◆	–
Elstree Computing Ltd	–	◆	–	◆	–
Empress Software UK	–	◆	–	–	–
Enghouse (UK) Limited	◆	◆	◆	–	–
Environment & Planning Library	–	–	◆	◆	–
EPS - Essential Planning Systems Limited	◆	–	–	–	–
ERDAS (UK) Ltd	–	–	–	–	◆
ESRI (UK) Ltd	◆	–	–	◆	–
European Geographic Technologies BV	◆	–	–	–	–
FastCAD GIS Ltd	–	–	–	◆	◆
FastCAD GIS Ltd	◆	◆	◆	–	◆
Genasys II Limited	◆	–	–	–	–
Geo/SQL (UK)	◆	◆	◆	◆	–
Geografix Limited	–	–	–	◆	–
Geographic Management Solutions Ltd	◆	–	–	–	–

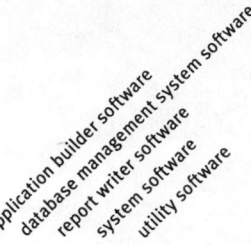

	application builder software	database management system software	report writer software	system software	utility software
Geometria GIS Systems House Ltd	–	–	–	–	♦
GEOSOFT Ltd	♦	–	–	–	♦
Geoview Systems Kft	–	♦	♦	–	–
GID Ltd	–	–	–	–	♦
Glen Computing Ltd	♦	♦	–	♦	–
Graphic Data Systems Corporation (GDS)	–	–	–	–	♦
Graphite Management Services Ltd	–	♦	–	–	–
Hansa Luftbild	♦	–	–	–	–
Hunting Engineering Ltd	♦	♦	–	♦	–
IBM UK Ltd	–	–	–	♦	–
ICL Ltd	♦	♦	♦	♦	♦
IMASS Limited	♦	♦	–	–	♦
Institute of Hydrology	–	–	–	♦	–
Intera Information Technologies	–	–	–	♦	–
Intergraph (UK) Ltd	♦	♦	♦	♦	♦
IT Southern Ltd	–	♦	–	♦	–
ITC	–	♦	–	–	–
Kirstol Ltd	–	–	–	♦	–
Laser-Scan Ltd	♦	–	–	–	–
MapInfo Ltd	♦	–	–	–	–
MAPIT Limited	♦	–	–	–	–
Midsummer Computing	–	–	–	♦	–
Mott MacDonald Ltd	♦	♦	♦	♦	–
NAG Ltd	♦	–	–	♦	–
Optimal Software Ltd	♦	♦	–	♦	–
Organisation Management Systems	♦	–	–	–	–
Oscar Faber	♦	–	–	♦	–
PAFEC Ltd	♦	♦	–	♦	–
PAX Technology	♦	♦	♦	♦	–
PLANTECH Ltd	–	–	♦	–	–
Progis GmbH	♦	♦	–	–	–
Raindrop Information Systems Ltd	♦	♦	♦	♦	–
SAS Institute	♦	–	♦	–	–
Scientific Software Limited	♦	–	–	–	–
SDI Ltd	–	–	–	♦	–
Sector (UK) Limited	–	–	♦	♦	–
SIA Limited	♦	♦	♦	♦	–
Siemens Nixdorf	♦	–	–	–	–
Silicon Graphics Limited	♦	–	–	–	–
Sovereign C.S. Ltd	♦	–	–	–	–
Star Informatic S.A.	–	♦	–	–	♦
Structural Technologies Ltd (STL)	♦	–	–	♦	–
Survey Supplies Ltd	♦	–	–	♦	♦
Symology Limited	–	–	♦	♦	–
Sysdeco (UK) Ltd	♦	–	♦	♦	–

	application builder software	database management system software	report writer software	system software	utility software
System Options Limited	◆	◆	◆	◆	–
Tangent Technology Design Associates Ltd	–	–	–	–	◆
TDS-CAD Graphics Ltd	–	–	–	–	◆
TEAMS (Taylor Woodrow Electronics Asset Mapping Survey)	◆	–	–	◆	–
Tekla Oy	–	–	–	–	◆
Tendron Systems Ltd	–	◆	–	◆	–
Tenet Systems Ltd	◆	◆	–	–	–
Terrafix Ltd	◆	–	–	◆	–
The Data Consultancy	◆	–	◆	–	–
Trimble Navigation Europe Ltd	–	–	–	◆	–
UNISYS Ltd	◆	◆	◆	–	–
Universal Systems Ltd	◆	–	–	–	–
Wallingford Software Limited	–	–	–	–	◆
WS Atkins Planning & Management Consultants	◆	–	–	–	–

suppliers of complete systems

	data capture systems	data conversion systems	document imaging systems	geographic information systems	global positioning systems	digital mapping systems	photogrammetry systems	plotting systems	scanning systems	surveying systems	other systems
A.L.Downloading Services	◆	◆	◆	–	–	–	–	–	◆	–	–
ACDS Graphic System Inc	◆	◆	–	◆	–	◆	–	◆	–	◆	–
Action Information Management Ltd	–	–	–	◆	–	◆	–	◆	–	–	–
Active Software Ltd	–	–	–	◆	–	–	–	–	–	–	–
Adept Scientific Micro Systems Ltd	–	–	–	◆	–	◆	–	–	–	–	–
AGFA UK	–	–	–	–	–	–	–	◆	–	–	–
AP³ Imaging Services Limited	◆	◆	–	◆	–	◆	–	◆	–	–	–
APIC Systems	–	–	–	◆	–	–	–	◆	–	–	–
Ashtech Europe Limited	–	–	–	–	◆	–	–	–	–	◆	–
Assist Applications Limited	◆	◆	◆	◆	–	◆	–	◆	–	◆	–
Audifilm Girona S.L.	–	–	–	◆	–	–	–	–	–	–	–
Bartholomew	◆	◆	–	–	–	◆	–	–	–	–	–
Beacon Dodsworth Limited	–	–	–	◆	–	–	–	–	–	–	–
British Geological Survey	–	–	◆	◆	–	◆	–	◆	–	–	–
Byers Engineering Company	–	–	–	◆	–	◆	–	◆	–	–	–
C.A.Design Services Ltd	◆	◆	◆	◆	–	◆	–	◆	◆	◆	–
CACI Limited	–	–	–	◆	–	◆	–	–	–	–	–
CAD - Capture Limited	◆	◆	◆	◆	–	–	–	–	–	–	–
CAD R&D Centre Limited	–	–	–	◆	–	◆	–	◆	◆	–	–
CADAC Ltd	◆	–	–	◆	◆	◆	–	–	◆	–	–
CAM - Centre for Analysis & Modelling Limited	◆	◆	–	◆	–	–	–	–	–	–	–
Carl Bro Group	–	–	–	◆	–	–	–	–	–	–	–
Carl Zeiss Limited	◆	◆	◆	◆	–	◆	◆	–	–	–	–
Cartographical Services (Southampton) Limited	–	–	–	–	–	–	◆	–	–	–	–
Cartwright Associates	◆	◆	–	–	–	–	–	–	–	–	–
CATALIST	–	–	–	◆	–	–	–	–	–	–	–
CCN Marketing	–	–	–	◆	–	◆	–	◆	–	–	–
CDR Group	◆	◆	–	◆	–	◆	–	–	◆	–	–
Chiltern Digitising Services	◆	◆	–	–	–	–	–	–	–	–	–
Cobham Digital Services Limited	◆	◆	–	◆	–	◆	–	–	–	–	–
CODEC Facilities Limited	◆	◆	◆	◆	–	◆	–	◆	◆	–	–
Computer Aided Development (CADCORP) Ltd	–	–	–	◆	–	◆	–	–	–	–	–
COMSULT	–	–	–	◆	–	◆	–	–	–	–	–
Conic Systems	◆	–	–	–	–	–	–	–	–	–	–
Consensus Information Technology Ltd	–	–	–	◆	–	–	–	–	–	–	–
Cray Systems	◆	◆	◆	◆	–	–	–	◆	–	–	–
CSI	–	–	◆	–	–	–	–	◆	◆	–	–
CZ Scientific Instruments Ltd	–	–	–	–	◆	–	◆	–	–	–	–
DAT/EM Systems International	◆	–	–	–	◆	–	◆	–	–	–	–
Datatechnology Datech Ltd	–	–	◆	◆	–	–	◆	◆	◆	–	–
Dataview Solutions Ltd	–	–	–	◆	◆	◆	–	–	–	–	–
Derek Hunter & Partners Ltd	–	–	–	–	–	◆	–	–	–	–	–
Design Computer Aids Limited (DeCAL)	–	–	–	◆	–	◆	–	–	–	–	–
DMV Consultants BV	–	–	–	◆	–	◆	–	–	–	–	–
Dotted Eyes	–	–	–	◆	–	◆	–	–	–	–	–

	data capture systems	data conversion systems	document imaging systems	geographic information systems	global positioning systems	digital mapping systems	photogrammetry systems	plotting systems	scanning systems	surveying systems	other systems
Earth Observation Sciences Ltd	–	–	♦	–	–	–	–	–	–	–	–
Effective Solutions (Data Products)	–	–	–	–	♦	–	–	♦	–	♦	–
Elstree Computing Ltd	–	–	–	♦	–	–	–	–	–	–	–
Enghouse (UK) Limited	♦	♦	–	♦	–	♦	–	♦	–	–	–
EPS - Essential Planning Systems Limited	–	–	–	♦	–	–	–	–	–	–	–
ERA Technology Ltd	–	–	–	–	–	–	–	–	–	–	♦
ERDAS (UK) Ltd	♦	♦	♦	♦	–	♦	♦	–	♦	–	–
ESR Cartographers Ltd	–	–	–	–	–	♦	–	–	♦	–	–
ESRI (UK) Ltd	♦	♦	–	♦	–	♦	♦	♦	–	–	–
EuroDirect Database Marketing Ltd	–	–	–	♦	–	–	–	–	–	–	–
Eurosense Technologies N.V.	♦	–	–	♦	–	♦	–	♦	–	–	–
FastCAD GIS Ltd	–	–	–	♦	–	♦	–	♦	–	–	–
FastCAD GIS Ltd	–	♦	–	♦	–	♦	–	–	–	–	–
FileNet Ltd	–	–	♦	–	–	–	–	–	–	–	–
Foto Res	♦	♦	–	♦	–	♦	–	–	♦	♦	–
Gardline Infotech	–	–	–	♦	–	–	–	–	–	–	–
GEC Marconi Research Centre	–	–	♦	–	–	–	–	–	–	–	–
Genasys II Limited	–	–	–	♦	–	–	–	–	–	–	–
GEO-Marketing Systems Ltd (GMSL)	–	–	–	♦	–	–	–	–	–	–	–
Geo-Perfect TWI B.V.	♦	–	♦	♦	♦	♦	♦	–	–	–	–
Geo/SQL (UK)	♦	♦	♦	♦	–	♦	–	♦	♦	–	–
Geodelta	–	–	♦	♦	–	♦	–	–	–	–	–
Geografix Limited	♦	♦	♦	♦	♦	♦	–	–	♦	♦	–
Geographic Management Solutions Ltd	♦	–	–	♦	♦	–	–	–	–	–	–
GeoMEM Software	♦	–	♦	–	–	♦	–	–	–	–	–
Geoplan (UK) Ltd	–	–	–	♦	–	♦	–	–	–	–	–
Geotronics Limited	–	–	–	–	♦	♦	–	♦	–	♦	–
Geoview Systems Kft	–	–	–	♦	–	–	–	–	–	–	–
GIMMS (GIS) Ltd	♦	♦	–	♦	–	♦	–	–	–	–	–
Glen Computing Ltd	♦	♦	♦	♦	–	♦	–	♦	♦	–	–
Global Surveys Ltd	–	–	–	–	♦	–	–	–	–	♦	–
GMAP Ltd	–	♦	–	♦	–	♦	–	–	–	–	–
Graphical Data Capture Ltd (GDC)	♦	–	–	♦	–	–	–	–	–	–	–
Graphics Online Limited	–	–	–	♦	–	–	–	–	–	–	–
Graphite Management Services Ltd	♦	♦	♦	♦	–	♦	–	–	♦	♦	–
GTX Europe Ltd	–	♦	–	–	–	–	–	–	–	–	–
Hall & Watts Systems Limited	–	–	–	–	–	♦	–	–	♦	–	–
HJM Imaging Systems	♦	–	–	–	–	♦	♦	–	–	–	–
Hunting Aerofilms Limited	–	–	–	–	–	–	♦	♦	–	–	–
Hunting Engineering Ltd	♦	♦	–	♦	♦	♦	–	♦	♦	♦	–
Husky Computing Limited	♦	–	–	–	–	–	–	–	–	–	–
I.S. Ltd	–	–	♦	♦	–	♦	♦	–	–	–	–
IBM UK Ltd	♦	♦	♦	♦	–	–	–	–	–	–	–
ICL Ltd	♦	♦	♦	♦	–	♦	–	♦	♦	–	–
IMASS Limited	♦	♦	–	♦	–	♦	–	♦	♦	–	–

	data capture systems	data conversion systems	document imaging systems	geographic information systems	global positioning systems	digital mapping systems	photogrammetry systems	plotting systems	scanning systems	surveying systems	other systems
IME (UK) Ltd	◆	◆	–	◆	–	◆	–	–	–	–	–
Institute of Hydrology	◆	◆	–	◆	–	–	–	–	–	–	–
Intera Information Technologies	◆	◆	–	◆	–	◆	–	◆	◆	–	–
Intergraph (UK) Ltd	◆	◆	◆	◆	–	◆	◆	◆	◆	–	–
International Products	◆	◆	◆	◆	–	–	◆	◆	◆	–	–
IT Southern Ltd	◆	◆	–	◆	–	◆	–	◆	–	–	–
ITC	◆	◆	◆	◆	–	–	–	◆	–	–	–
Kirstol Ltd	◆	◆	–	–	–	–	–	◆	–	–	–
L.E.S. (Computer Services) Ltd	◆	–	–	–	–	–	–	◆	–	◆	–
Lantmatenet GIS-centrum	–	–	–	◆	–	–	–	–	–	–	–
Laser-Scan Ltd	◆	◆	◆	◆	◆	–	◆	–	◆	◆	–
Leica UK Ltd	–	–	–	–	◆	–	◆	–	–	◆	–
LiveChart	–	◆	–	–	–	◆	–	◆	–	–	–
Logica UK Limited	–	–	–	◆	–	–	–	–	–	–	–
Logitrans	–	–	–	–	–	–	–	–	–	–	◆
Longdin & Browning	–	–	–	◆	–	◆	–	–	–	–	–
Magellan Systems Corporation	–	–	–	–	◆	–	–	–	–	–	–
Map Data Management Ltd	◆	◆	–	◆	–	◆	◆	◆	–	–	–
MapInfo Ltd	–	◆	–	◆	–	◆	–	–	–	–	–
MAPS geosystems	◆	◆	–	◆	◆	◆	–	–	◆	–	–
Mason Land Surveys	◆	–	–	◆	–	◆	–	◆	–	–	–
McLintock Limited	–	–	–	◆	–	◆	–	–	–	–	–
Midsummer Computing	◆	◆	–	◆	–	–	–	◆	◆	–	–
Morgan Collis Group Ltd	◆	–	–	◆	–	◆	–	–	–	–	–
Mott MacDonald Ltd	–	◆	–	◆	–	◆	–	–	–	–	–
Munroe Garrett International	–	–	–	◆	–	◆	–	◆	–	–	–
MVA Systematica	–	–	–	◆	–	◆	–	–	–	–	–
National Remote Sensing Centre Ltd (NRSC)	–	–	–	◆	–	◆	–	–	–	–	–
Navstar Systems Ltd	–	◆	–	–	◆	–	–	–	–	–	–
Optimal Software Ltd	◆	◆	–	◆	◆	◆	–	◆	◆	◆	–
Organisation Management Systems	◆	–	–	◆	–	–	–	◆	–	–	–
Oscar Faber	–	–	–	◆	–	–	–	–	–	–	–
PAFEC Ltd	◆	◆	◆	◆	–	◆	–	◆	◆	–	–
PAX Technology	–	–	–	◆	–	–	–	–	–	–	–
PD Computing Ltd	–	◆	–	◆	–	–	–	–	–	–	–
Pear Technology Services Ltd	–	–	◆	◆	◆	◆	–	–	–	–	–
Positioning Resources Limited	–	–	–	–	◆	–	–	–	–	–	–
Primagraphics Ltd	–	–	–	–	–	–	–	–	◆	–	–
Procis Software Ltd	◆	–	–	–	◆	◆	–	–	–	–	–
Progis GmbH	◆	◆	◆	◆	–	◆	–	◆	–	◆	–
Property Intelligence plc	◆	–	–	◆	–	–	–	–	–	–	–
QAS Systems Ltd	◆	–	–	◆	–	–	–	–	–	–	–
Racal Survey (UK) Ltd	◆	–	–	–	◆	◆	–	–	–	◆	–
Raindrop Information Systems Ltd	◆	◆	–	–	–	–	–	–	◆	–	–
regioplanDATA GmbH	–	–	–	◆	–	◆	–	◆	◆	–	–

	data capture systems	data conversion systems	document imaging systems	geographic information systems	global positioning systems	digital mapping systems	photogrammetry systems	plotting systems	scanning systems	surveying systems	other systems
Scan Group Limited	◆	◆	–	–	–	–	–	◆	–	–	–
SCOT Conseil	–	–	–	◆	–	◆	–	–	–	◆	–
SDI Ltd	◆	–	–	◆	–	–	–	–	–	–	–
Sector (UK) Limited	◆	◆	◆	◆	◆	◆	–	–	–	◆	–
SIA Limited	◆	◆	–	◆	◆	◆	–	◆	◆	–	–
Siemens Nixdorf	–	–	–	–	–	–	–	–	–	–	–
Smallworld Systems Ltd	–	–	–	◆	–	–	–	–	–	–	–
Sokkia Ltd	◆	–	–	–	◆	–	–	–	◆	–	–
Southbank Systems PLC	◆	–	–	–	–	◆	–	–	◆	–	–
Spatial Data Limited	◆	◆	–	◆	◆	◆	–	◆	–	◆	–
Spatial Information Services Ltd (SIS)	◆	◆	–	◆	–	◆	–	–	–	–	–
Star Informatic S.A.	◆	◆	◆	◆	–	◆	–	◆	◆	◆	–
Structural Technologies Ltd (STL)	◆	◆	–	◆	◆	◆	–	◆	–	–	–
Survey Supplies Ltd	◆	◆	–	◆	–	◆	–	–	–	◆	–
Svitzer Limited	–	–	–	–	–	◆	–	◆	–	◆	–
Symology Limited	◆	–	–	◆	–	◆	–	–	–	–	–
Sysdeco (UK) Ltd	–	–	–	◆	–	◆	–	◆	–	–	–
System Options Limited	–	◆	–	◆	◆	◆	–	◆	◆	–	–
TACTICIAN UK	–	–	–	◆	–	◆	–	–	–	–	–
TDS-CAD Graphics Ltd	–	◆	–	◆	–	–	–	◆	◆	–	–
TEAMS (Taylor Woodrow Electronics Asset Mapping Survey)	◆	◆	–	–	–	–	–	–	–	◆	–
Tekla Oy	–	–	–	◆	–	◆	–	–	–	–	–
Tendron Systems Ltd	–	–	–	–	–	◆	–	–	–	–	–
Terrafix Ltd	◆	◆	–	◆	◆	◆	–	◆	–	–	–
The Data Consultancy	–	◆	–	◆	–	◆	–	–	–	–	–
The Marketing Information Consultancy (MIC)	–	–	–	◆	–	◆	–	–	–	–	–
The NPA Group Ltd	◆	◆	–	◆	–	◆	◆	◆	◆	–	–
Trimble Navigation Europe Ltd	◆	◆	–	◆	◆	◆	–	–	–	◆	–
TYDAC Technologies Ltd	–	–	–	◆	–	◆	–	–	–	–	–
✳niversal Systems Ltd	◆	◆	–	◆	–	◆	–	◆	◆	–	–
WDV (UK)	–	–	–	–	–	–	–	◆	–	–	–
WS Atkins Planning & Management Consultants	–	–	–	◆	–	◆	–	–	–	–	–
Xcon Data	◆	◆	–	–	–	–	–	–	–	–	–

directory 14:

providers of services

	audit services	data services	implementation services	management services	media services	procurement services	staff recruitment services	training services
A.L.Downloading Services	–	◆	–	–	–	–	–	–
A.Rutherford Ltd	◆	◆	◆	◆	–	–	–	–
ACDS Graphic System Inc	–	◆	◆	◆	◆	–	–	–
Action Information Management Ltd	–	◆	◆	◆	–	–	–	–
Active Software Ltd	–	◆	◆	◆	–	◆	–	◆
Adept Scientific Micro Systems Ltd	–	◆	◆	–	–	–	–	–
AGFA UK	–	–	◆	–	◆	◆	–	◆
AiC Analysts	–	◆	–	◆	–	◆	◆	◆
ALLM Systems & Marketing	–	◆	–	–	–	–	–	–
AM/FM International - European Division	–	–	–	–	–	–	–	◆
AND Mapping B.V.	◆	◆	–	–	–	–	–	–
Anglian Engineering & International Consultancy	–	–	–	◆	–	◆	–	–
AP³ Imaging Services Limited	–	◆	◆	–	◆	◆	–	–
APIC Systems	–	–	–	◆	–	◆	–	◆
ARC Systems Pty Ltd	–	◆	–	–	–	–	–	◆
Ashtech Europe Limited	–	–	◆	–	–	–	–	◆
Assist Applications Limited	–	◆	◆	–	–	–	–	◆
Audifilm Girona S.L.	–	◆	◆	◆	–	–	◆	◆
Babtie Shaw & Morton Limited	–	◆	–	◆	–	◆	–	–
Bartholomew	–	◆	–	–	–	–	–	–
Baymont Technologies Inc	–	◆	–	–	–	–	–	–
Beacon Dodsworth Limited	–	◆	–	–	–	–	–	–
Bentley Systems UK Ltd	–	–	◆	–	–	–	–	–
Binnie Black & Veatch	–	◆	–	◆	–	–	–	–
Birkbeck College London	–	◆	◆	◆	–	–	–	◆
BKS Surveys Ltd	–	◆	–	◆	–	–	–	–
British Geological Survey	–	◆	◆	◆	◆	–	–	◆
BS International Consultants	◆	◆	◆	◆	–	◆	–	–
Bull Information Systems Limited	◆	◆	◆	◆	–	–	–	◆
Business Information Management	◆	◆	◆	◆	–	◆	◆	◆
Byers Engineering Company	◆	◆	◆	◆	–	◆	◆	–
C.A.Design Services Ltd	◆	◆	◆	◆	–	◆	◆	◆
CACI Limited	–	◆	◆	◆	–	◆	–	◆
CAD - Capture Limited	–	◆	◆	◆	◆	◆	–	–
CAD R&D Centre Limited	◆	◆	◆	◆	◆	◆	◆	◆
CADAC Ltd	–	◆	–	–	◆	–	◆	◆
CAM - Centre for Analysis & Modelling Limited	◆	◆	◆	◆	◆	◆	◆	◆
Cambashi Ltd	–	–	–	◆	–	–	◆	–
Cambridge Computer Consultants (UK) Ltd	–	◆	◆	◆	–	–	–	–
Carl Bro Group	–	◆	–	–	◆	–	–	◆
Carl Zeiss Limited	–	–	◆	–	–	–	–	◆
CARTograph Ltd	–	◆	◆	–	–	–	–	–
Cartographical Services (Southampton) Limited	–	◆	◆	–	–	–	–	–
Cartwright Associates	◆	◆	◆	◆	◆	◆	–	◆
CATALIST	–	◆	–	◆	–	–	–	◆
CCN Marketing	–	◆	–	◆	–	–	–	◆

	audit services	data services	implementation services	management services	media services	procurement services	staff recruitment services	training services
CDD Ltd	–	◆	◆	–	–	◆	◆	–
CDR Group	–	◆	◆	◆	–	–	–	◆
Chiltern Digitising Services	–	◆	–	–	–	–	–	–
Chroson Ltd	–	–	◆	–	–	◆	–	◆
Citywise	–	◆	◆	–	–	–	–	◆
Cliffe House Associates	◆	◆	◆	◆	–	◆	–	–
CMG Computer Management Group (UK) Ltd	–	◆	◆	◆	–	◆	–	–
Cobham Digital Services Limited	–	◆	◆	◆	–	◆	–	◆
CODEC Facilities Limited	–	–	◆	–	–	–	–	–
ColourMap Scanning Ltd	–	◆	–	–	◆	–	–	–
Computer Aided Development (CADCORP) Ltd	–	–	◆	◆	–	–	–	◆
COMSULT	–	–	◆	◆	–	◆	–	–
Concurrent Appointments International	–	–	–	–	–	–	◆	–
Consensus Information Technology Ltd	–	–	◆	◆	–	–	–	–
Construction Industry Computing Association	◆	◆	–	◆	–	◆	–	◆
Coopers & Lybrand	–	–	◆	◆	–	◆	–	–
Corbins Consultancy	◆	◆	◆	◆	–	◆	–	–
Corena A/S	–	–	◆	–	–	–	–	–
Cray Systems	◆	◆	◆	◆	–	◆	◆	◆
Cromwell House Technical Services	–	◆	–	–	–	–	–	–
Data Base Builders	◆	–	◆	◆	–	◆	–	–
Data Collection Ltd	–	◆	–	–	–	–	–	–
Data Dictionary Systems Limited	–	◆	–	◆	–	–	–	–
Dataflow Information Systems	–	◆	◆	–	–	–	–	◆
Dataquest Europe Ltd	–	–	–	◆	–	◆	–	–
Datatechnology Datech Ltd	–	–	◆	◆	–	–	–	◆
Dataview Solutions Ltd	–	◆	◆	◆	–	–	–	◆
DCL Consulting	–	◆	◆	–	–	◆	–	◆
Derek Hunter & Partners Ltd	–	–	◆	◆	–	–	◆	–
Design Computer Aids Limited (DeCAL)	–	–	◆	–	–	–	–	–
Digital Equipment Corporation	–	–	◆	–	–	◆	–	–
DM Management Consultants Ltd (DMMC)	◆	◆	◆	◆	–	–	◆	◆
DMAP Ltd	–	◆	◆	–	◆	–	–	–
DMV Consultants BV	–	◆	◆	–	◆	◆	◆	◆
Dolphin Consulting Group	–	◆	◆	◆	◆	◆	–	–
Dotted Eyes	◆	◆	◆	◆	◆	◆	–	◆
Dowling Associates Limited	–	–	◆	–	–	–	–	◆
Dr Stanley Port	–	–	–	◆	–	◆	–	◆
EA Technology	–	◆	◆	–	–	–	–	–
Earth Observation Sciences Ltd	–	◆	◆	◆	–	◆	–	–
Earth Resource Mapping	–	–	–	–	–	–	–	◆
ECM Selection Limited	–	–	–	–	–	–	◆	–
Effective Solutions (Data Products)	–	◆	–	–	–	–	–	–
Elstree Computing Ltd	–	◆	◆	–	◆	◆	–	◆
Enghouse (UK) Limited	–	◆	◆	–	–	◆	–	◆
Environment & Planning Library	–	◆	–	–	–	–	–	–

	audit services	data services	implementation services	management services	media services	procurement services	staff recruitment services	training services
EPS - Essential Planning Systems Limited	–	–	◆	–	–	–	–	◆
ERA Technology Ltd	–	–	◆	–	–	–	–	◆
ERA-Maptec Ltd	–	◆	–	–	◆	–	–	◆
ERDAS (UK) Ltd	–	–	◆	–	–	–	–	◆
ERTEC	–	◆	–	–	–	–	–	◆
ESR Cartographers Ltd	–	–	–	–	◆	–	–	–
ESRI (UK) Ltd	–	◆	◆	◆	–	◆	–	◆
European Business Mapping	–	–	–	◆	–	–	–	–
European Geographic Technologies BV	–	◆	◆	–	–	–	–	–
Eurosense Technologies N.V.	–	◆	◆	◆	◆	–	–	◆
Evox Facilities Ltd	–	◆	–	◆	–	◆	–	–
Fairbairn Services Limited	◆	◆	◆	◆	◆	◆	◆	◆
FastCAD GIS Ltd	–	◆	◆	◆	–	–	–	–
FastCAD GIS Ltd	–	◆	◆	–	–	–	–	◆
FileNet Ltd	–	◆	◆	◆	–	◆	–	◆
Flynn & Rothwell	–	◆	◆	◆	–	–	–	–
Foto Res	–	◆	–	–	–	–	◆	◆
G.L. Consulting Ltd	◆	◆	◆	–	–	◆	–	–
Gamma Ltd	–	◆	◆	◆	–	–	–	◆
Gardline Infotech	◆	◆	◆	◆	◆	–	◆	◆
GEC Marconi Research Centre	–	◆	◆	–	–	–	–	–
Genasys II Limited	–	–	◆	◆	–	–	–	◆
GEO-Marketing Systems Ltd (GMSL)	–	◆	◆	◆	–	–	–	–
Geo-Perfect TWI B.V.	–	◆	◆	–	–	–	–	–
GEO-UK Ltd	–	–	–	◆	–	–	–	–
Geo/SQL (UK)	◆	◆	◆	◆	◆	◆	–	◆
Geo2 Consulting	◆	–	◆	◆	–	◆	–	◆
GEOBASE Consultants Ltd	◆	◆	◆	◆	–	◆	◆	◆
GeoData Institute	–	◆	◆	◆	–	◆	–	◆
Geodelta	–	◆	–	–	–	◆	–	◆
Geografix Limited	–	◆	–	–	–	–	–	–
Geographic Management Solutions Ltd	–	–	–	–	–	◆	–	◆
Geoinformation International	–	◆	–	◆	–	–	–	◆
GeoMEM Software	◆	◆	–	–	–	–	–	◆
Geometria GIS Systems House Ltd	–	◆	◆	◆	–	–	–	◆
Geoplan (UK) Ltd	–	◆	◆	◆	◆	–	–	–
Geops BV	–	◆	◆	–	–	–	–	◆
GEOSOFT Ltd	–	◆	–	◆	–	◆	–	–
Geosystems	◆	◆	–	–	–	–	–	–
Geoview Systems Kft	–	–	◆	–	–	–	–	–
GID Ltd	–	◆	–	–	–	◆	–	–
GIMMS (GIS) Ltd	–	–	–	◆	–	◆	–	–
GIS Services Ltd	◆	◆	–	◆	–	◆	◆	◆
GISDATA	–	–	◆	◆	–	–	–	◆
GISL Limited	–	◆	◆	◆	◆	◆	◆	◆

	audit services	data services	implementation services	management services	media services	procurement services	staff recruitment services	training services
Glen Computing Ltd	◆	◆	◆	◆	◆	◆	–	◆
Global Surveys Ltd	–	◆	◆	–	–	–	–	–
GMAP Ltd	–	◆	◆	◆	–	◆	–	–
Graphic Data Systems Corporation (GDS)	–	–	◆	◆	–	◆	–	◆
Graphical Data Capture Ltd (GDC)	–	◆	◆	◆	◆	–	–	◆
Graphite Management Services Ltd	–	◆	◆	–	–	–	–	–
Greig Fester Limited	–	◆	–	◆	–	◆	–	–
Grove Projects Ltd	◆	◆	◆	◆	–	◆	–	–
H R Wallingford Ltd	–	◆	–	–	–	–	–	◆
Hansa Luftbild	–	◆	◆	–	–	◆	–	–
Hunting Engineering Ltd	◆	◆	◆	–	–	◆	–	–
Hunting Technical Services Ltd	–	◆	◆	–	–	–	–	◆
Husky Computing Limited	–	–	–	–	–	–	–	◆
I.S. Ltd	–	◆	◆	◆	–	–	–	–
IBM UK Ltd	◆	–	–	◆	–	◆	–	–
ICL Ltd	–	◆	◆	◆	–	◆	◆	◆
IMASS Limited	–	◆	◆	◆	–	–	–	◆
IME (UK) Ltd	–	–	◆	◆	–	◆	–	–
Infolink Decision Services Limited	–	◆	–	–	–	–	–	–
Informed Solutions Limited	◆	◆	◆	–	◆	–	◆	◆
Ingecon B.V.	–	◆	◆	◆	–	◆	–	◆
Institute of Hydrology	–	◆	◆	◆	–	–	–	–
Institute of Terrestrial Ecology	–	◆	–	–	–	–	–	–
Intera Information Technologies	–	◆	◆	◆	◆	◆	–	◆
Intergraph (UK) Ltd	◆	◆	◆	–	–	◆	–	◆
International Products	–	◆	–	–	◆	–	–	◆
IT Southern Ltd	◆	◆	◆	–	◆	◆	–	–
ITC	–	–	◆	–	–	–	–	◆
ITS : Intertrade Scientific Ltd	–	–	◆	–	–	–	–	–
J.C.White Chartered Land Surveyors	–	◆	–	–	◆	–	◆	–
John D Leatherdale FRICS	◆	◆	◆	◆	–	◆	◆	◆
Kamyco International	–	◆	◆	–	–	◆	◆	–
Keele University	–	–	–	–	–	–	–	◆
Kingswood Consulting Limited	◆	◆	◆	◆	–	◆	–	◆
Kirstol Ltd	–	◆	–	–	◆	–	–	–
KJB Consulting	–	–	◆	◆	–	◆	–	◆
Know Edge Ltd	◆	–	–	◆	–	◆	◆	◆
KPMG	◆	◆	◆	◆	–	–	◆	◆
L.E.S. (Computer Services) Ltd	◆	◆	–	–	◆	–	–	◆
Lancaster University	–	◆	–	–	–	–	–	◆
Land and Satellite Surveys	–	◆	◆	◆	–	–	–	–
Land Aspects Consultancy Ltd (Parkman Group)	–	◆	–	◆	◆	◆	–	–
Lantmatenet GIS-centrum	–	◆	◆	◆	◆	–	◆	◆
LASCO Ltd	–	◆	–	–	–	–	◆	–
Laser-Scan Ltd	–	◆	◆	◆	–	–	–	◆

	audit services	data services	implementation services	management services	media services	procurement services	staff recruitment services	training services
LiveChart	–	◆	–	◆	–	–	–	–
Logica UK Limited	◆	◆	◆	◆	–	◆	–	–
Logitrans	–	◆	–	–	◆	◆	–	◆
London Research Centre	–	◆	–	–	–	–	–	◆
Longdin & Browning	–	◆	–	–	◆	–	–	–
Lovell Johns Ltd	–	◆	◆	◆	–	–	–	–
LOY Surveys Ltd	–	◆	–	–	◆	–	–	–
LTG Services	◆	◆	–	◆	◆	◆	–	◆
Macaulay Land Use Research Institute	–	–	◆	–	–	–	–	–
Magdala Sociedade	◆	–	◆	◆	–	◆	–	◆
Manchester Metropolitan University	–	–	–	–	–	–	–	◆
Map Data Management Ltd	–	◆	◆	◆	◆	–	–	◆
MAPIT Limited	–	–	◆	–	–	–	–	–
MAPS geosystems	–	◆	◆	–	–	◆	◆	◆
Mason Land Surveys	◆	◆	◆	–	◆	–	–	◆
Mathshop	–	◆	–	–	–	◆	–	–
McLintock Limited	◆	◆	◆	◆	◆	◆	◆	◆
Mentis Management Consultants Ltd	◆	◆	◆	◆	–	◆	◆	◆
Methods Applications Ltd	–	◆	◆	◆	–	◆	–	–
Midlands Regional Research Laboratory	◆	◆	◆	◆	–	–	–	–
Midsummer Computing	–	◆	◆	◆	–	◆	◆	◆
Modern Maps	◆	◆	◆	◆	–	◆	–	◆
Morgan Collis Group Ltd	–	◆	◆	–	–	◆	–	–
Mott MacDonald Ltd	◆	◆	◆	–	–	◆	–	–
MPSI Systems Ltd	–	◆	◆	–	–	–	–	–
MR Data Graphics	–	◆	–	◆	–	◆	–	–
Munro Garratt International	–	◆	◆	–	–	–	–	◆
MVA Systematica	◆	◆	–	–	–	◆	–	◆
MVM Consultants plc	◆	–	◆	◆	–	◆	–	–
NAG Ltd	–	◆	◆	◆	–	–	–	–
National Remote Sensing Centre Ltd (NRSC)	–	◆	◆	◆	–	◆	–	◆
Natural Environment Research Council	–	◆	◆	◆	–	–	–	–
NERC	–	◆	◆	◆	–	–	–	–
Nestor International Ltd	–	–	–	–	–	–	◆	–
NOMIS	–	◆	–	◆	–	–	–	◆
Numonics UK (Division of Telmtek Ltd)	–	◆	◆	–	–	–	–	–
Oaklands I.T.	–	◆	◆	◆	–	–	–	–
Optimal Software Ltd	–	◆	◆	◆	–	–	–	◆
Ordnance Survey	–	◆	◆	–	–	–	–	◆
Ordnance Survey of Northern Ireland	–	◆	◆	◆	◆	–	–	◆
Organisation Management Systems	–	◆	◆	◆	–	◆	–	◆
Oscar Faber	◆	◆	◆	–	–	–	–	◆
Ove Arup & Partners	–	◆	◆	◆	–	◆	–	–
Oxford Institute of Retail Management	–	◆	–	–	–	◆	–	◆
P&L Engineering Surveys Ltd	–	◆	–	–	–	–	–	–

	audit services	data services	implementation services	management services	media services	procurement services	staff recruitment services	training services
PA Consulting Group	◆	–	◆	◆	–	◆	◆	–
PAFEC Ltd	–	◆	◆	◆	–	–	–	◆
Panda	–	–	◆	◆	–	◆	–	–
Paul Clasper & Associates Ltd	–	◆	◆	◆	–	◆	–	–
PAX Technology	–	–	◆	–	–	◆	–	–
PD Computing Ltd	◆	◆	◆	–	–	–	–	–
Peter Thorpe Consulting	◆	◆	◆	◆	–	◆	–	◆
Photarc Surveys Ltd	–	◆	–	◆	–	–	–	–
Photoair	–	◆	–	–	–	–	–	–
Photogrammetric Data Services Ltd	–	◆	–	◆	–	–	–	–
Pinpoint Digitising Services	–	◆	–	–	–	–	–	–
PlanGraphics Inc	◆	◆	◆	◆	–	◆	◆	◆
Planning & Mapping Ltd	–	◆	–	–	◆	–	–	–
PLANTECH Ltd	–	◆	◆	◆	–	◆	–	◆
Plowman Craven & Associates Ltd	–	◆	◆	–	◆	–	–	–
Posford Duvivier	–	◆	◆	◆	–	◆	–	–
Positioning Resources Limited	–	–	–	–	–	–	–	◆
Procis Software Ltd	–	–	◆	◆	–	◆	–	◆
Progis GmbH	–	◆	◆	–	–	–	–	◆
Property Intelligence plc	–	◆	–	–	–	–	–	–
QAS Systems Ltd	–	–	◆	–	–	–	–	–
QC Data (Ireland) Limited	–	◆	–	–	–	–	–	–
Quorum Information Services	–	◆	–	–	–	–	–	–
R.W.A. Dalls FRICS	–	◆	◆	◆	–	◆	–	–
Racal Survey (UK) Ltd	–	◆	–	–	–	–	–	◆
Raindrop Information Systems Ltd	–	–	◆	–	◆	◆	–	◆
Recruit Media Ltd	–	–	–	–	–	–	◆	◆
regioplanDATA GmbH	◆	◆	◆	◆	–	◆	–	◆
Remote Sensing Applications Consultants	–	◆	◆	◆	–	–	–	◆
Robert Walker Consultants	–	–	◆	◆	–	◆	–	–
Royal Geographical Society (with The Institute of British Geographers)	–	–	–	◆	–	–	–	◆
Royal Institute of Navigation	–	–	–	–	–	–	–	◆
Royal Mail - Address Management Centre	◆	◆	–	–	◆	–	–	–
Salford University Business Services Ltd	◆	◆	◆	◆	–	◆	–	◆
SAS Institute	–	–	◆	◆	–	–	–	◆
Saztec Europe Limited	–	◆	–	–	◆	–	–	–
SAZTEC Philippines Inc	–	◆	◆	◆	–	–	–	–
Scientific Software Limited	–	◆	–	–	–	–	–	–
SCOT Conseil	–	◆	◆	◆	–	◆	–	◆
Scott Wilson Kirkpatrick	–	◆	◆	◆	◆	◆	–	–
SDI Ltd	–	◆	◆	◆	–	◆	–	◆
Sector (UK) Limited	◆	–	◆	◆	–	◆	◆	◆
SHL Vision* Solutions Limited	–	–	–	–	–	–	–	◆
SIA Limited	◆	◆	◆	◆	–	–	–	◆
SIAS Limited	–	◆	◆	–	–	◆	–	–

	audit services	data services	implementation services	management services	media services	procurement services	staff recruitment services	training services
Siemens Nixdorf	–	–	◆	◆	–	–	–	◆
Silicon Graphics Limited	–	–	◆	–	–	–	–	–
Silsoe College, Cranfield University	–	◆	◆	◆	–	–	–	◆
Simmons Survey Partnership Limited	–	◆	–	–	◆	–	–	–
Sir Alexander Gibb & Partners Ltd	–	◆	–	–	–	◆	–	–
Sir William Halcrow and Partners	–	◆	◆	◆	–	–	–	–
Smartscan Inc	◆	◆	◆	–	–	–	–	–
Smith System Engineering Ltd	◆	◆	◆	◆	–	◆	–	–
Solent Mapping and Charting (SMAC)	–	◆	–	–	–	–	–	–
Southbank Systems PLC	–	–	◆	–	–	–	–	◆
Sovereign C.S. Ltd	–	◆	◆	–	◆	◆	–	◆
Spacesense consultants	–	◆	–	–	–	–	–	–
Spatial Data Limited	–	◆	–	–	◆	–	–	–
Spatial Geographic Services & Applications Ltd	–	◆	◆	◆	–	–	–	–
Spatial Information Services Ltd (SIS)	–	◆	–	–	–	–	–	–
SPSS UK Ltd	–	◆	◆	◆	–	–	–	–
Star Informatic S.A.	–	–	◆	–	–	–	–	–
StatSci Europe	–	◆	◆	–	–	–	–	◆
Structural Technologies Ltd (STL)	–	◆	◆	◆	–	◆	–	◆
Summagraphics Europe N.V.	–	–	–	–	◆	–	–	–
Sun Microsystems Ltd	–	–	◆	–	–	–	–	◆
Survey Control Services	–	◆	–	–	–	–	–	–
Survey & Development Services Ltd	–	◆	◆	◆	◆	–	–	◆
Svitzer Limited	–	◆	–	◆	–	–	–	–
Symology Limited	◆	–	◆	◆	–	◆	–	◆
Sysdeco (UK) Ltd	–	–	◆	◆	–	–	–	◆
System Options Limited	–	◆	◆	◆	◆	◆	–	–
TACTICIAN UK	–	◆	◆	◆	–	–	–	◆
Target Market Consultancy	–	◆	–	◆	–	–	–	–
TEAMS (Taylor Woodrow Electronics Asset Mapping Survey)	–	◆	◆	◆	–	◆	–	–
Tekla Oy	–	◆	–	–	–	–	–	–
Tele Atlas	◆	◆	◆	◆	◆	–	–	–
Tendron Systems Ltd	–	◆	◆	◆	–	◆	–	–
Tenet Systems Ltd	–	–	◆	–	–	–	–	◆
Terrafix Ltd	–	◆	◆	–	–	–	–	–
TerraHunt GeoScience Ltd	–	–	–	–	◆	–	–	–
TerraQuest Group Limited	◆	◆	–	◆	–	◆	–	–
The Business Database from Yellow Pages	–	◆	–	–	–	–	–	–
The Data Consultancy	–	◆	◆	◆	–	–	–	◆
The LGMB	–	–	–	–	–	–	–	◆
The Marketing Information Consultancy (MIC)	–	◆	–	–	–	–	–	–
The NPA Group Ltd	–	◆	–	◆	◆	◆	–	–
The Severn Partnership	–	◆	–	–	–	◆	–	–
The Survey Centre	–	◆	–	–	◆	–	–	◆
Trac Consultancy	–	◆	–	◆	–	–	–	–

	audit services	data services	implementation services	management services	media services	procurement services	staff recruitment services	training services
Trimble Navigation Europe Ltd	–	–	♦	–	–	–	–	♦
TYDAC Technologies Ltd	–	♦	♦	♦	–	–	–	♦
Unistride Sewer Technology	–	♦	–	–	–	–	–	–
UNISYS Ltd	♦	♦	♦	♦	–	♦	–	♦
Universal Systems Ltd	–	–	♦	–	–	–	–	♦
University of East London	–	–	–	–	–	–	–	♦
University of Edinburgh	–	♦	–	♦	–	♦	–	♦
University of Greenwich	–	–	–	–	–	–	–	♦
University of Leicester	–	–	–	–	–	–	–	♦
University of Luton	–	–	–	–	–	–	–	♦
University of Newcastle Upon Tyne	–	–	♦	♦	–	–	–	♦
Walker Ladd Surveys	–	♦	–	–	–	–	–	–
Wallingford Software Limited	–	–	–	–	–	–	–	♦
WRc (Water Research centre)	♦	♦	♦	♦	♦	♦	–	♦
WS Atkins Planning & Management Consultants	♦	♦	♦	♦	–	♦	–	♦
Xcon Data	–	♦	–	–	–	–	–	–

	audit services	data services	implementation services	management services	media services	procurement services	staff recruitment services	training services
A.Rutherford Ltd	♦	♦	♦	♦	–	–	–	–
AiC Analysts	–	♦	–	♦	–	♦	♦	♦
Anglian Engineering & International Consultancy	–	–	–	♦	♦	–	♦	–
Babtie Shaw & Morton Limited	–	♦	–	♦	–	♦	–	–
Baymont Technologies Inc	–	♦	–	–	–	–	–	–
Binnie Black & Veatch	–	♦	–	–	–	–	–	–
BS International Consultants	♦	♦	♦	♦	–	–	–	–
Bull Information Systems Limited	♦	♦	♦	♦	–	–	–	♦
Business Information Management	♦	♦	♦	♦	–	♦	♦	♦
Cambashi Ltd	–	–	♦	–	–	♦	–	–
Cambridge Computer Consultants (UK) Ltd	–	♦	♦	♦	–	♦	–	–
CDD Ltd	–	♦	♦	–	–	♦	♦	–
Chroson Ltd	–	–	♦	–	–	♦	–	♦
Cliffe House Associates	♦	♦	♦	♦	–	♦	–	–
CMG Computer Management Group (UK) Ltd	–	♦	♦	♦	–	–	–	–
Concurrent Appointments International	–	–	–	–	–	–	♦	–
Coopers & Lybrand	–	–	♦	♦	–	–	–	–
Corbins Consultancy	♦	♦	♦	♦	–	–	–	–
Cromwell House Technical Services	–	♦	–	–	–	–	–	–
Data Base Builders	♦	–	♦	♦	–	♦	–	–
Data Collection Ltd	–	♦	–	–	–	–	–	–
Data Dictionary Systems Limited	–	♦	–	♦	–	–	–	–
Dataflow Information Systems	–	♦	♦	–	–	–	–	♦
DCL Consulting	–	♦	♦	–	–	♦	–	♦
DM Management Consultants Ltd (DMMC)	♦	♦	♦	♦	–	–	♦	♦
ECM Selection Limited	–	–	–	–	–	–	♦	–
ERTEC	–	♦	–	–	–	–	–	♦
Fairbairn Services Limited	♦	♦	♦	♦	♦	♦	♦	♦
Flynn & Rothwell	–	♦	♦	♦	–	–	–	–
G.L. Consulting Ltd	♦	♦	♦	–	–	♦	–	–
GEO-UK Ltd	–	–	–	♦	–	–	–	–
Geo2 Consulting	♦	–	♦	♦	–	♦	–	♦
GEOBASE Consultants Ltd	♦	♦	♦	♦	–	♦	♦	♦
Greig Fester Limited	–	♦	–	♦	–	♦	–	–
Grove Projects Ltd	♦	♦	♦	♦	♦	♦	–	–
Hunting Technical Services Ltd	–	♦	♦	–	–	–	–	♦
Informed Solutions Limited	♦	♦	♦	♦	–	♦	–	♦
Ingecon B.V.	–	♦	♦	♦	–	♦	–	♦
J.C.White Chartered Land Surveyors	–	♦	–	–	♦	–	♦	–
John D Leatherdale FRICS	♦	♦	♦	♦	–	♦	♦	♦
Kamyco International	–	♦	♦	–	–	♦	♦	–
Keele University	–	–	–	–	–	–	–	♦
Kingswood Consulting Limited	♦	♦	♦	♦	–	♦	–	♦
KJB Consulting	–	–	♦	♦	–	♦	–	♦
Know Edge Ltd	♦	–	–	♦	–	♦	♦	♦

	audit services	data services	implementation services	management services	media services	procurement services	staff recruitment services	training services
KPMG	◆	◆	◆	◆	–	–	◆	◆
Land and Satellite Surveys	–	◆	◆	◆	–	–	–	–
LASCO Ltd	–	◆	–	–	–	–	◆	–
LOY Surveys Ltd	–	◆	–	–	◆	–	–	–
Magdala Sociedade	◆	–	◆	◆	–	◆	–	◆
Manchester Metropolitan University	–	–	–	–	–	–	–	◆
Mathshop	–	◆	–	–	–	◆	–	–
Mentis Management Consultants Ltd	◆	◆	◆	◆	–	◆	◆	◆
Methods Applications Ltd	–	◆	◆	◆	–	–	◆	–
Modern Maps	◆	◆	◆	◆	–	–	–	◆
Nestor International Ltd	–	–	–	–	–	–	◆	–
Oaklands I.T.	–	◆	◆	◆	–	–	–	–
Ove Arup & Partners	–	◆	◆	◆	–	◆	–	–
Oxford Institute of Retail Management	–	◆	–	–	–	◆	–	◆
P&L Engineering Surveys Ltd	–	◆	–	–	–	–	–	–
PA Consulting Group	◆	–	◆	◆	–	◆	◆	–
Panda	–	–	◆	◆	–	◆	–	–
Peter Thorpe Consulting	◆	◆	◆	◆	–	◆	–	◆
Photarc Surveys Ltd	–	◆	–	–	–	–	–	–
Plowman Craven & Associates Ltd	–	◆	◆	–	◆	–	–	–
Posford Duvivier	–	◆	◆	◆	–	–	–	–
QC Data (Ireland) Limited	–	◆	–	–	–	–	–	–
Quorum Information Services	–	◆	–	–	–	–	–	–
R.W.A. Dalls FRICS	–	◆	◆	◆	–	◆	–	–
Recruit Media Ltd	–	–	–	–	–	–	◆	◆
Robert Walker Consultants	–	–	◆	◆	–	◆	–	–
Salford University Business Services Ltd	◆	◆	◆	◆	–	◆	–	◆
Saztec Europe Limited	–	◆	–	–	◆	–	–	–
SAZTEC Philippines Inc	–	◆	◆	◆	–	–	–	–
Scott Wilson Kirkpatrick	–	◆	◆	◆	◆	◆	–	–
Silsoe College, Cranfield University	–	◆	◆	◆	–	–	–	◆
Sir Alexander Gibb & Partners Ltd	–	◆	–	–	–	◆	–	–
Smartscan Inc	◆	◆	◆	–	–	–	–	–
Smith System Engineering Ltd	◆	◆	◆	◆	–	◆	–	–
Solent Mapping and Charting (SMAC)	–	◆	–	–	–	–	–	–
Spacesense consultants	–	◆	–	–	–	–	–	–
Survey Control Services	–	◆	–	–	–	–	–	–
Target Market Consultancy	–	◆	–	◆	–	–	–	–
TerraHunt GeoScience Ltd	–	–	–	–	◆	–	–	–
TerraQuest Group Limited	◆	◆	–	◆	–	◆	–	–
Unistride Sewer Technology	–	◆	–	–	–	–	–	–
University of East London	–	–	–	–	–	–	–	◆
University of Greenwich	–	–	–	–	–	–	–	◆
University of Leicester	–	–	–	–	–	–	–	◆
University of Luton	–	–	–	–	–	–	–	◆
Walker Ladd Surveys	–	◆	–	–	–	–	–	–

directory 16:
providers of audit services

	data audit services	post implementation review services	procedure audit services	quality audit services	security audit services	system audit services
A.Rutherford Ltd	◆	–	–	–	–	–
BS International Consultants	–	–	–	◆	–	–
Bull Information Systems Limited	◆	◆	◆	◆	◆	◆
Business Information Management	◆	◆	◆	◆	◆	◆
Byers Engineering Company	◆	–	–	◆	–	–
C.A.Design Services Ltd	◆	◆	–	–	–	–
CAD R&D Centre Limited	–	–	◆	–	–	◆
CAM - Centre for Analysis & Modelling Limited	◆	–	–	–	–	–
Cartwright Associates	◆	◆	◆	–	–	◆
Cliffe House Associates	◆	◆	◆	◆	–	–
Construction Industry Computing Association	–	◆	◆	–	–	–
Corbins Consultancy	◆	◆	◆	–	–	◆
Cray Systems	◆	◆	◆	◆	–	◆
Data Base Builders	–	–	◆	◆	–	–
DM Management Consultants Ltd (DMMC)	–	–	◆	–	–	–
Dotted Eyes	◆	–	–	◆	–	–
Fairbairn Services Limited	◆	◆	◆	◆	◆	◆
G.L. Consulting Ltd	–	–	–	–	–	–
Gardline Infotech	◆	–	–	◆	–	–
Geo/SQL (UK)	◆	◆	◆	–	–	◆
Geo2 Consulting	◆	–	◆	◆	–	–
GEOBASE Consultants Ltd	◆	◆	◆	◆	◆	◆
GeoMEM Software	◆	–	–	–	–	–
Geosystems	◆	–	–	–	–	–
GIS Services Ltd	◆	◆	◆	◆	◆	◆
Glen Computing Ltd	◆	◆	–	–	–	–
Grove Projects Ltd	–	–	–	◆	–	–
Hunting Engineering Ltd	–	–	◆	◆	◆	–
IBM UK Ltd	–	◆	–	–	–	–
Informed Solutions Limited	◆	◆	◆	◆	◆	◆
Intergraph (UK) Ltd	–	◆	–	–	–	◆
IT Southern Ltd	◆	◆	◆	–	◆	◆
John D Leatherdale FRICS	–	◆	◆	◆	–	–
Kingswood Consulting Limited	◆	◆	◆	◆	◆	◆
Know Edge Ltd	◆	◆	◆	–	–	–
KPMG	◆	◆	◆	◆	◆	◆
L.E.S. (Computer Services) Ltd	◆	–	–	–	–	–
Logica UK Limited	◆	◆	◆	◆	◆	◆
LTG Services	–	–	◆	◆	–	◆
Magdala Sociedade	–	–	◆	◆	◆	◆
Mason Land Surveys	◆	–	–	–	–	–
McLintock Limited	–	◆	–	–	–	◆
Mentis Management Consultants Ltd	◆	–	◆	–	–	–
Midlands Regional Research Laboratory	◆	–	–	–	–	–
Modern Maps	◆	–	–	–	–	–

	data audit services	post implementation review services	procedure audit services	quality audit services	security audit services	system audit services
Mott MacDonald Ltd	◆	◆	◆	◆	◆	◆
MVA Systematica	◆	◆	◆	◆	◆	◆
MVM Consultants plc	–	◆	◆	–	–	◆
Oscar Faber	◆	–	–	–	–	–
PA Consulting Group	–	◆	–	–	◆	◆
PD Computing Ltd	◆	◆	◆	◆	◆	◆
Peter Thorpe Consulting	◆	◆	◆	◆	◆	◆
PlanGraphics Inc	◆	◆	◆	–	–	◆
regioplanDATA GmbH	◆	–	–	–	–	–
Royal Mail - Address Management Centre	◆	–	–	–	–	–
Salford University Business Services Ltd	◆	◆	–	–	–	◆
Sector (UK) Limited	◆	◆	◆	◆	◆	◆
SIA Limited	◆	◆	◆	◆	◆	◆
Smartscan Inc	◆	–	◆	◆	–	–
Smith System Engineering Ltd	–	–	–	–	◆	◆
Symology Limited	–	–	–	–	–	◆
Tele Atlas	◆	–	◆	◆	◆	–
TerraQuest Group Limited	◆	–	–	–	–	–
UNISYS Ltd	◆	◆	◆	◆	◆	◆
WRc (Water Research centre)	◆	◆	◆	◆	–	◆
WS Atkins Planning & Management Consultants	◆	◆	◆	◆	◆	◆

	address matching services	data cleansing services	data geopointing services	data matching services	data integration services	data transfer services	data validation services	data verification services
A.L.Downloading Services	◆	◆	–	–	–	◆	–	–
A.Rutherford Ltd	–	–	–	–	◆	–	–	–
ACDS Graphic System Inc	–	–	–	–	◆	–	◆	◆
Action Information Management Ltd	◆	–	–	–	◆	–	–	–
Active Software Ltd	–	–	–	–	◆	–	–	–
Adept Scientific Micro Systems Ltd	–	–	–	–	–	◆	–	–
ALLM Systems & Marketing	–	◆	–	–	–	–	–	–
AP³ Imaging Services Limited	–	◆	–	◆	◆	–	–	–
Assist Applications Limited	–	–	◆	–	–	–	◆	◆
Audifilm Girona S.L.	–	–	◆	–	◆	–	◆	◆
Babtie Shaw & Morton Limited	–	◆	–	◆	◆	◆	◆	◆
Beacon Dodsworth Limited	–	–	–	–	◆	–	–	–
Binnie Black & Veatch	–	–	–	–	◆	–	–	–
Bull Information Systems Limited	◆	◆	◆	◆	◆	◆	◆	◆
Byers Engineering Company	–	◆	–	–	◆	◆	◆	◆
CACI Limited	◆	◆	◆	◆	◆	–	–	–
CAD R&D Centre Limited	–	–	◆	–	◆	–	◆	◆
CADAC Ltd	–	–	–	–	◆	–	–	–
CAM - Centre for Analysis & Modelling Limited	◆	◆	◆	–	–	–	–	–
Carl Bro Group	–	◆	–	–	–	–	–	–
Cartwright Associates	–	◆	◆	–	◆	◆	–	◆
CATALIST	–	–	–	◆	◆	–	–	–
CCN Marketing	◆	◆	–	–	–	–	◆	◆
CDR Group	–	–	–	–	–	–	◆	–
Citywise	◆	–	◆	◆	◆	–	–	–
Cobham Digital Services Limited	◆	–	◆	–	–	–	–	–
Cray Systems	–	–	–	–	◆	–	–	–
Cromwell House Technical Services	–	◆	◆	–	–	◆	◆	◆
Dataflow Information Systems	◆	–	–	–	–	◆	–	–
DMAP Ltd	–	◆	–	–	◆	◆	◆	◆
DMV Consultants BV	–	◆	–	–	–	◆	◆	◆
Dotted Eyes	◆	◆	◆	◆	◆	–	–	–
Earth Observation Sciences Ltd	–	–	–	–	◆	–	–	–
Enghouse (UK) Limited	–	–	–	–	◆	–	–	–
ERA-Maptec Ltd	–	–	–	–	◆	–	◆	–
ESRI (UK) Ltd	◆	◆	–	◆	◆	–	◆	◆
European Geographic Technologies BV	–	–	◆	–	–	–	–	–
Eurosense Technologies N.V.	–	–	◆	–	◆	◆	–	◆
Evox Facilities Ltd	–	◆	–	–	–	–	◆	–
Fairbairn Services Limited	–	◆	–	–	–	◆	◆	◆
FastCAD GIS Ltd	◆	–	–	◆	–	–	–	–
FastCAD GIS Ltd	◆	–	–	◆	–	–	–	–
Flynn & Rothwell	–	–	–	–	–	–	–	◆
Gardline Infotech	◆	◆	◆	◆	◆	◆	◆	◆
GEC Marconi Research Centre	–	–	–	–	◆	–	◆	◆

	address matching services	data cleansing services	data geopointing services	data matching services	data integration services	data transfer services	data validation services	data verification services
GEO-Marketing Systems Ltd (GMSL)	◆	–	–	–	–	–	–	–
Geo/SQL (UK)	–	–	–	–	◆	◆	◆	◆
GEOBASE Consultants Ltd	–	◆	◆	◆	◆	◆	◆	◆
GeoData Institute	–	◆	–	–	◆	◆	–	–
Geodelta	–	–	–	–	–	–	–	◆
Geoplan (UK) Ltd	◆	◆	◆	–	◆	–	–	–
GID Ltd	–	–	–	–	◆	–	◆	◆
GIS Services Ltd	◆	◆	–	◆	◆	–	–	–
Glen Computing Ltd	◆	◆	◆	◆	◆	–	–	–
GMAP Ltd	–	–	–	–	◆	–	◆	◆
Grove Projects Ltd	◆	–	–	–	–	◆	–	–
Hansa Luftbild	–	–	–	–	–	◆	–	–
Hunting Engineering Ltd	–	–	–	◆	◆	–	◆	◆
ICL Ltd	◆	◆	◆	◆	◆	–	◆	◆
IMASS Limited	–	–	–	–	◆	◆	◆	◆
Infolink Decision Services Limited	◆	◆	◆	◆	–	–	◆	◆
Informed Solutions Limited	◆	–	◆	–	◆	◆	◆	◆
International Products	–	◆	–	◆	–	–	–	–
IT Southern Ltd	–	◆	◆	–	◆	–	◆	◆
Kamyco International	–	–	–	–	–	–	◆	◆
Kingswood Consulting Limited	◆	–	–	–	–	–	–	–
KPMG	–	–	–	–	–	–	◆	◆
L.E.S. (Computer Services) Ltd	–	–	–	–	–	–	–	◆
Land Aspects Consultancy Ltd (Parkman Group)	◆	◆	–	◆	◆	◆	◆	◆
Lantmatenet GIS-centrum	◆	–	◆	◆	◆	–	–	–
Laser-Scan Ltd	–	–	–	–	–	◆	–	–
Logica UK Limited	–	–	–	–	◆	–	–	–
Logitrans	◆	–	◆	–	–	–	–	–
Lovell Johns Ltd	–	◆	◆	–	–	–	–	–
LOY Surveys Ltd	–	–	–	–	–	◆	–	–
LTG Services	–	–	–	–	◆	◆	◆	◆
MAPS geosystems	–	◆	◆	◆	◆	–	–	–
Mason Land Surveys	◆	◆	◆	◆	◆	◆	◆	–
McLintock Limited	◆	–	–	◆	◆	–	◆	◆
Mentis Management Consultants Ltd	–	–	–	–	–	–	◆	◆
Midsummer Computing	◆	–	◆	◆	◆	◆	◆	◆
Morgan Collis Group Ltd	–	–	–	–	–	–	◆	–
Mott MacDonald Ltd	–	–	–	◆	◆	◆	◆	–
MVA Systematica	◆	◆	–	–	◆	–	–	–
National Remote Sensing Centre Ltd (NRSC)	◆	◆	◆	◆	◆	◆	◆	◆
NERC	–	–	–	–	◆	–	–	–
NOMIS	–	–	–	◆	◆	–	◆	◆
Optimal Software Ltd	–	–	–	–	◆	–	–	–
Ordnance Survey	◆	–	–	–	◆	◆	–	–
Oscar Faber	–	◆	◆	–	◆	◆	◆	◆
Ove Arup & Partners	–	◆	◆	–	–	◆	–	–

	address matching services	data cleansing services	data geopointing services	data matching services	data integration services	data transfer services	data validation services	data verification services
P&L Engineering Surveys Ltd	–	–	–	–	–	◆	–	–
PD Computing Ltd	◆	–	–	◆	◆	◆	–	–
Peter Thorpe Consulting	–	–	–	–	◆	–	–	–
Photogrammetric Data Services Ltd	–	–	–	–	◆	–	–	–
Pinpoint Digitising Services	◆	◆	–	◆	–	–	◆	◆
PLANTECH Ltd	–	–	–	–	–	◆	–	–
Plowman Craven & Associates Ltd	–	–	◆	–	–	◆	–	–
Progis GmbH	◆	◆	◆	◆	◆	◆	◆	◆
Property Intelligence plc	◆	◆	◆	–	–	–	–	◆
regioplanDATA GmbH	◆	◆	◆	◆	◆	–	◆	◆
Royal Mail - Address Management Centre	–	–	–	–	–	–	–	◆
Salford University Business Services Ltd	–	◆	–	◆	◆	◆	◆	◆
SAZTEC Philippines Inc	–	◆	–	–	◆	◆	◆	◆
Scott Wilson Kirkpatrick	–	◆	–	–	◆	◆	◆	◆
SDI Ltd	–	–	–	◆	–	–	–	–
SIA Limited	–	–	–	–	–	–	◆	◆
SIAS Limited	–	◆	–	–	–	–	◆	◆
Silsoe College, Cranfield University	–	–	–	–	◆	–	–	–
Simmons Survey Partnership Limited	–	–	–	–	–	–	◆	◆
Smartscan Inc	◆	–	◆	◆	◆	–	–	–
Smith System Engineering Ltd	–	–	–	–	–	–	◆	–
Sovereign C.S. Ltd	–	–	–	◆	◆	–	–	–
Spatial Data Limited	–	–	–	◆	◆	◆	◆	◆
Spatial Geographic Services & Applications Ltd	–	–	–	◆	◆	–	–	–
Spatial Information Services Ltd (SIS)	–	–	◆	◆	◆	–	–	–
Structural Technologies Ltd (STL)	–	–	◆	–	◆	◆	◆	–
Survey & Development Services Ltd	◆	–	–	–	–	◆	◆	◆
Svitzer Limited	–	–	–	◆	◆	–	–	–
System Options Limited	–	–	◆	–	–	–	–	–
TACTICIAN UK	◆	–	–	–	–	–	–	–
TEAMS (Taylor Woodrow Electronics Asset Mapping Survey)	–	–	–	–	◆	◆	–	–
Tele Atlas	◆	–	◆	◆	◆	–	◆	◆
Tendron Systems Ltd	–	–	–	–	◆	–	◆	◆
Terrafix Ltd	–	–	–	–	◆	–	–	–
TerraQuest Group Limited	◆	–	◆	–	◆	◆	◆	◆
The Business Database from Yellow Pages	◆	–	–	–	–	–	–	–
The Data Consultancy	–	–	◆	◆	◆	–	–	–
The Marketing Information Consultancy (MIC)	◆	–	–	–	–	–	–	–
The NPA Group Ltd	–	–	◆	–	–	–	–	–
The Severn Partnership	–	–	–	–	–	◆	–	–
Trac Consultancy	◆	◆	–	◆	–	–	–	–
TYDAC Technologies Ltd	◆	–	◆	–	–	–	–	–
UNISYS Ltd	◆	◆	◆	◆	◆	◆	◆	◆
WRc (Water Research centre)	–	–	–	–	◆	–	–	–
WS Atkins Planning & Management Consultants	–	–	–	–	◆	–	◆	◆
Xcon Data	–	–	–	–	◆	–	–	–

directory 18:
data services group 2

	data consultancy services	data analysis services	data modelling services	data capture services	data conversion services	digitising services	photogrammetry services	scanning services	data QA services
A.L.Downloading Services	◆	–	–	◆	◆	–	–	◆	–
A.Rutherford Ltd	◆	◆	◆	–	◆	–	–	–	–
ACDS Graphic System Inc	–	–	–	–	–	–	–	–	◆
Action Information Management Ltd	◆	◆	–	–	◆	–	–	◆	–
Active Software Ltd	◆	◆	–	–	–	◆	–	–	–
Adept Scientific Micro Systems Ltd	◆	–	–	–	◆	◆	–	–	–
AiC Analysts	◆	◆	◆	–	–	–	–	–	–
ALLM Systems & Marketing	◆	–	–	–	–	–	–	–	–
AND Mapping B.V.	–	–	–	–	–	◆	–	–	–
AP³ Imaging Services Limited	◆	◆	◆	◆	◆	◆	◆	◆	–
Assist Applications Limited	◆	–	–	◆	◆	◆	–	◆	–
Audifilm Girona S.L.	◆	◆	◆	◆	◆	◆	◆	◆	◆
Babtie Shaw & Morton Limited	◆	◆	◆	◆	–	◆	–	◆	–
Bartholomew	–	–	–	–	–	◆	–	–	–
Baymont Technologies Inc	◆	–	–	–	◆	◆	–	–	–
Binnie Black & Veatch	◆	◆	◆	◆	–	–	–	–	–
Birkbeck College London	◆	◆	◆	–	–	–	–	–	–
BKS Surveys Ltd	◆	–	◆	◆	◆	◆	◆	◆	–
British Geological Survey	◆	◆	◆	◆	–	◆	–	◆	–
BS International Consultants	◆	–	–	–	–	–	–	–	–
Bull Information Systems Limited	◆	◆	◆	◆	◆	◆	–	◆	◆
Business Information Management	◆	◆	–	◆	◆	–	–	–	–
Byers Engineering Company	◆	◆	–	◆	◆	◆	–	–	◆
C.A.Design Services Ltd	◆	–	–	◆	◆	◆	–	◆	–
CACI Limited	◆	◆	◆	–	–	–	–	–	–
CAD - Capture Limited	◆	◆	–	◆	◆	–	–	◆	–
CAD R&D Centre Limited	◆	–	◆	–	◆	–	◆	◆	◆
CADAC Ltd	–	–	–	◆	◆	◆	–	–	–
CAM - Centre for Analysis & Modelling Limited	◆	◆	◆	–	◆	◆	–	–	–
Cambridge Computer Consultants (UK) Ltd	◆	–	◆	◆	–	◆	–	–	–
Carl Bro Group	◆	–	–	◆	◆	◆	–	–	–
CARTograph Ltd	–	◆	◆	–	–	◆	–	–	–
Cartographical Services (Southampton) Limited	–	–	◆	◆	◆	◆	◆	–	–
Cartwright Associates	◆	◆	◆	◆	◆	◆	–	–	◆
CATALIST	◆	◆	◆	◆	–	–	–	–	–
CCN Marketing	◆	◆	◆	–	–	–	–	–	◆
CDD Ltd	◆	◆	–	◆	◆	◆	–	–	–
CDR Group	◆	–	–	◆	◆	◆	–	◆	–
Chiltern Digitising Services	–	–	–	◆	◆	◆	–	◆	–
Citywise	◆	◆	–	–	–	◆	–	–	–
Cliffe House Associates	◆	–	–	–	–	–	–	–	–
CMG Computer Management Group (UK) Ltd	◆	◆	◆	–	–	–	–	–	–
Cobham Digital Services Limited	◆	–	◆	◆	◆	◆	–	◆	–
ColourMap Scanning Ltd	–	–	–	–	–	◆	–	◆	–
Construction Industry Computing Association	◆	–	–	–	–	–	–	–	–

	data consultancy services	data analysis services	data modelling services	data capture services	data conversion services	digitising services	photogrammetry services	scanning services	data QA services
Corbins Consultancy	◆	–	–	–	–	–	–	–	–
Cray Systems	◆	–	–	–	◆	–	–	–	–
Cromwell House Technical Services	◆	◆	◆	◆	◆	◆	–	◆	◆
Data Collection Ltd	–	–	–	◆	◆	–	–	–	–
Data Dictionary Systems Limited	◆	–	◆	–	–	–	–	–	–
Dataflow Information Systems	◆	◆	–	–	◆	–	–	–	–
Dataview Solutions Ltd	◆	◆	–	–	◆	–	–	–	–
DCL Consulting	◆	–	–	–	–	–	–	–	–
DM Management Consultants Ltd (DMMC)	◆	–	–	–	–	–	–	–	–
DMAP Ltd	◆	◆	–	◆	◆	◆	–	◆	◆
DMV Consultants BV	◆	◆	–	–	◆	◆	–	◆	◆
Dolphin Consulting Group	◆	◆	–	◆	–	–	–	–	–
Dotted Eyes	◆	◆	◆	◆	◆	–	–	–	–
EA Technology	–	◆	◆	◆	◆	–	–	–	–
Earth Observation Sciences Ltd	–	◆	◆	–	–	◆	–	–	–
Elstree Computing Ltd	◆	–	◆	–	–	–	–	–	–
Enghouse (UK) Limited	◆	–	–	–	◆	–	–	–	–
Environment & Planning Library	◆	–	–	–	–	◆	–	◆	–
ERTEC	◆	–	–	–	–	–	–	–	–
ESRI (UK) Ltd	◆	◆	◆	◆	◆	◆	–	–	◆
European Geographic Technologies BV	–	–	–	–	–	◆	–	–	–
Eurosense Technologies N.V.	◆	◆	◆	◆	◆	◆	◆	◆	–
Evox Facilities Ltd	–	◆	–	–	–	–	–	–	–
Fairbairn Services Limited	◆	–	◆	◆	◆	◆	◆	◆	◆
FastCAD GIS Ltd	◆	◆	–	–	◆	◆	–	–	–
FastCAD GIS Ltd	◆	–	–	–	◆	◆	–	–	–
FileNet Ltd	–	–	–	–	–	–	–	◆	–
Flynn & Rothwell	◆	◆	–	◆	–	–	–	–	◆
Foto Res	◆	◆	–	◆	◆	◆	◆	◆	–
G.L. Consulting Ltd	◆	–	–	–	–	–	–	–	◆
Gamma Ltd	◆	–	–	–	–	–	–	–	–
Gardline Infotech	◆	◆	◆	◆	◆	◆	–	◆	◆
GEC Marconi Research Centre	◆	◆	–	–	–	–	–	–	◆
GEO-Marketing Systems Ltd (GMSL)	◆	◆	◆	–	–	–	–	–	–
Geo-Perfect TWI B.V.	–	–	–	◆	–	◆	◆	–	–
Geo/SQL (UK)	◆	◆	◆	◆	◆	◆	–	◆	◆
GEOBASE Consultants Ltd	◆	◆	◆	◆	◆	◆	◆	◆	◆
GeoData Institute	◆	◆	◆	◆	◆	◆	–	–	–
Geodelta	◆	◆	–	–	◆	–	◆	–	◆
Geografix Limited	–	–	–	–	–	◆	–	–	–
Geoinformation International	◆	–	–	–	–	–	–	–	–
GeoMEM Software	–	–	–	◆	–	–	–	–	–
Geometria GIS Systems House Ltd	◆	–	◆	◆	◆	◆	–	◆	◆
Geoplan (UK) Ltd	◆	◆	–	–	◆	◆	◆	–	–
Geops BV	◆	◆	–	◆	◆	–	–	◆	–

	data consultancy services	data analysis services	data modelling services	data capture services	data conversion services	digitising services	photogrammetry services	scanning services	data QA services
GEOSOFT Ltd	◆	–	–	◆	–	–	–	–	–
Geosystems	–	◆	–	–	◆	◆	–	–	–
GID Ltd	◆	–	–	–	–	–	–	–	◆
GIS Services Ltd	◆	◆	◆	◆	◆	◆	–	◆	–
GISL Limited	◆	◆	◆	◆	◆	◆	–	–	–
Glen Computing Ltd	◆	◆	◆	◆	◆	◆	◆	◆	◆
GMAP Ltd	◆	◆	◆	–	◆	–	–	–	–
Graphical Data Capture Ltd (GDC)	◆	–	–	◆	◆	◆	–	–	–
Graphite Management Services Ltd	◆	–	–	–	–	◆	◆	–	–
Greig Fester Limited	–	–	◆	–	–	–	–	–	–
Grove Projects Ltd	◆	–	◆	◆	◆	◆	–	–	–
Hansa Luftbild	◆	–	–	◆	◆	◆	◆	◆	–
Hunting Engineering Ltd	◆	–	◆	–	◆	◆	◆	◆	◆
Hunting Technical Services Ltd	◆	◆	◆	–	–	◆	◆	–	◆
I.S. Ltd	◆	–	–	–	–	◆	–	–	–
ICL Ltd	◆	◆	◆	◆	◆	◆	–	◆	◆
IMASS Limited	◆	–	◆	◆	◆	◆	–	◆	◆
Infolink Decision Services Limited	◆	◆	◆	◆	◆	–	–	–	–
Informed Solutions Limited	◆	–	◆	–	◆	–	–	–	◆
Ingecon B.V.	◆	–	–	–	–	–	–	–	–
Institute of Hydrology	–	◆	◆	–	–	◆	–	–	–
Institute of Terrestrial Ecology	◆	–	◆	–	–	–	–	–	–
Intera Information Technologies	◆	–	–	◆	–	◆	◆	–	◆
Intergraph (UK) Ltd	◆	–	–	–	–	–	–	–	–
International Products	◆	–	–	◆	◆	◆	–	◆	–
IT Southern Ltd	◆	–	–	◆	◆	◆	–	◆	◆
J.C.White Chartered Land Surveyors	–	–	–	–	–	◆	◆	–	–
John D Leatherdale FRICS	◆	–	–	–	–	–	–	–	–
Kamyco International	◆	–	–	–	–	–	–	–	◆
Kingswood Consulting Limited	◆	◆	–	–	–	–	–	–	–
Kirstol Ltd	–	–	–	–	–	–	–	◆	–
KPMG	◆	◆	◆	–	–	–	–	–	–
L.E.S. (Computer Services) Ltd	–	–	◆	◆	–	◆	–	–	–
Lancaster University	◆	◆	◆	–	–	–	–	–	–
Land Aspects Consultancy Ltd (Parkman Group)	◆	◆	◆	◆	◆	◆	–	◆	◆
Lantmatenet GIS-centrum	◆	◆	◆	–	◆	–	–	–	–
LASCO Ltd	–	–	–	◆	–	–	–	–	–
Laser-Scan Ltd	◆	–	–	–	–	◆	–	–	–
LiveChart	–	–	–	◆	–	◆	–	◆	–
Logica UK Limited	◆	–	–	–	–	–	–	–	–
London Research Centre	◆	◆	◆	–	–	–	–	–	–
Longdin & Browning	–	–	–	◆	◆	◆	–	–	–
Lovell Johns Ltd	◆	–	–	◆	◆	◆	–	–	–
LOY Surveys Ltd	◆	–	◆	◆	◆	◆	◆	◆	–
LTG Services	◆	–	◆	◆	◆	◆	◆	◆	◆

	data consultancy services	data analysis services	data modelling services	data capture services	data conversion services	digitising services	photogrammetry services	scanning services	data QA services
Map Data Management Ltd	–	–	–	◆	–	◆	◆	–	–
MAPS geosystems	◆	–	◆	◆	◆	◆	◆	–	–
Mason Land Surveys	◆	◆	◆	◆	◆	◆	◆	◆	◆
Mathshop	◆	–	–	–	–	–	–	–	–
McLintock Limited	◆	–	◆	◆	◆	–	–	–	◆
Mentis Management Consultants Ltd	◆	◆	–	–	–	–	–	–	◆
Methods Applications Ltd	◆	–	–	–	–	–	–	–	–
Midlands Regional Research Laboratory	◆	–	–	–	–	–	–	–	–
Midsummer Computing	◆	◆	◆	◆	◆	◆	◆	◆	◆
Modern Maps	◆	–	–	–	–	–	–	–	–
Morgan Collis Group Ltd	–	–	◆	◆	–	◆	–	–	–
Mott MacDonald Ltd	◆	◆	◆	–	◆	–	–	–	◆
MPSI Systems Ltd	◆	◆	◆	◆	–	◆	–	–	–
MR Data Graphics	◆	–	–	◆	◆	◆	–	◆	–
Munro Garratt International	◆	–	–	–	–	–	–	–	–
MVA Systematica	◆	◆	◆	–	–	–	–	–	◆
NAG Ltd	◆	–	–	–	–	–	–	–	–
National Remote Sensing Centre Ltd (NRSC)	◆	◆	◆	◆	◆	◆	◆	◆	◆
Natural Environment Research Council	◆	–	◆	–	–	–	–	–	–
NERC	◆	–	◆	–	–	–	–	–	–
NOMIS	◆	–	–	–	–	–	–	–	◆
Numonics UK (Division of Telmtek Ltd)	–	–	–	–	–	◆	–	–	–
Oaklands I.T.	◆	–	–	◆	–	–	–	–	–
Optimal Software Ltd	◆	◆	◆	–	◆	◆	◆	◆	–
Ordnance Survey	◆	◆	◆	◆	◆	◆	◆	◆	◆
Ordnance Survey of Northern Ireland	◆	◆	◆	◆	◆	◆	◆	◆	–
Organisation Management Systems	◆	–	◆	◆	–	–	–	–	–
Oscar Faber	◆	◆	◆	◆	◆	◆	–	◆	◆
Ove Arup & Partners	◆	◆	◆	◆	◆	◆	–	–	–
Oxford Institute of Retail Management	◆	–	–	–	–	–	–	–	–
P&L Engineering Surveys Ltd	–	–	◆	◆	–	–	–	–	–
PAFEC Ltd	◆	–	–	–	–	–	–	–	–
PD Computing Ltd	◆	–	◆	◆	–	–	–	–	–
Peter Thorpe Consulting	◆	◆	◆	–	–	–	–	–	◆
Photarc Surveys Ltd	◆	–	–	◆	◆	◆	◆	–	–
Photogrammetric Data Services Ltd	–	–	–	◆	◆	◆	◆	◆	–
Pinpoint Digitising Services	–	–	–	◆	–	◆	–	–	◆
PlanGraphics Inc	◆	–	–	–	–	–	–	–	–
Planning & Mapping Ltd	–	–	◆	◆	–	◆	◆	–	–
PLANTECH Ltd	–	–	–	–	◆	–	–	–	–
Plowman Craven & Associates Ltd	◆	–	–	◆	◆	◆	◆	–	–
Posford Duvivier	◆	◆	–	–	–	–	–	–	–
Progis GmbH	◆	◆	◆	◆	◆	◆	◆	◆	–
Property Intelligence plc	◆	◆	–	–	–	–	–	–	–
QC Data (Ireland) Limited	◆	–	–	◆	◆	◆	–	◆	–

	data consultancy services	data analysis services	data modelling services	data capture services	data conversion services	digitising services	photogrammetry services	scanning services	data QA services
Quorum Information Services	–	–	–	◆	◆	◆	–	◆	–
Racal Survey (UK) Ltd	–	–	–	◆	–	–	–	–	–
regioplanDATA GmbH	◆	◆	◆	◆	◆	◆	–	◆	◆
Salford University Business Services Ltd	◆	◆	◆	◆	◆	◆	–	–	◆
Saztec Europe Limited	–	–	–	◆	◆	◆	–	–	–
SAZTEC Philippines Inc	◆	◆	–	◆	◆	◆	◆	◆	◆
Scientific Software Limited	–	–	–	–	◆	–	–	–	–
SCOT Conseil	◆	◆	–	–	–	◆	–	–	–
Scott Wilson Kirkpatrick	◆	–	◆	◆	◆	◆	–	–	–
SDI Ltd	◆	–	–	–	–	–	–	◆	–
SIA Limited	◆	◆	–	◆	◆	◆	–	◆	◆
SIAS Limited	◆	◆	◆	–	◆	–	–	–	–
Silsoe College, Cranfield University	◆	◆	◆	–	–	–	–	–	–
Simmons Survey Partnership Limited	–	–	◆	◆	◆	◆	◆	◆	–
Sir Alexander Gibb & Partners Ltd	◆	◆	◆	–	–	–	–	–	–
Sir William Halcrow and Partners	◆	◆	◆	◆	◆	–	–	–	–
Smartscan Inc	◆	–	–	◆	◆	◆	–	–	◆
Solent Mapping and Charting (SMAC)	–	–	–	◆	–	◆	◆	–	–
Sovereign C.S. Ltd	◆	–	–	–	◆	–	◆	◆	–
Spacesense consultants	◆	–	–	–	–	–	–	–	–
Spatial Data Limited	–	◆	◆	◆	◆	◆	◆	–	◆
Spatial Geographic Services & Applications Ltd	◆	◆	–	◆	–	◆	–	◆	–
Spatial Information Services Ltd (SIS)	–	–	◆	◆	◆	◆	–	◆	–
SPSS UK Ltd	◆	◆	–	–	–	–	–	–	–
StatSci Europe	–	◆	–	–	–	–	–	–	–
Structural Technologies Ltd (STL)	◆	◆	◆	◆	◆	◆	–	◆	–
Survey & Development Services Ltd	◆	–	◆	◆	◆	◆	◆	◆	◆
Svitzer Limited	–	◆	–	◆	◆	◆	–	–	◆
System Options Limited	◆	◆	◆	◆	◆	◆	◆	◆	◆
TACTICIAN UK	◆	◆	–	–	–	–	–	–	–
Target Market Consultancy	◆	–	–	–	–	–	–	–	–
TEAMS (Taylor Woodrow Electronics Asset Mapping Survey)	◆	◆	–	◆	–	–	–	–	–
Tekla Oy	◆	–	–	–	–	◆	–	◆	–
Tele Atlas	◆	–	–	◆	◆	◆	–	–	◆
Tendron Systems Ltd	◆	◆	◆	–	–	–	–	◆	–
Terrafix Ltd	◆	–	–	◆	◆	◆	–	◆	–
TerraQuest Group Limited	◆	◆	◆	◆	◆	◆	◆	◆	–
The Business Database from Yellow Pages	–	◆	–	–	–	–	–	–	–
The Data Consultancy	◆	◆	◆	–	◆	◆	–	–	–
The Marketing Information Consultancy (MIC)	◆	◆	◆	◆	–	–	–	–	–
The NPA Group Ltd	◆	◆	◆	◆	–	◆	◆	◆	–
The Severn Partnership	–	–	◆	◆	◆	–	◆	–	–
The Survey Centre	–	–	◆	◆	◆	◆	◆	–	◆
Trac Consultancy	◆	◆	◆	–	◆	–	–	–	–
TYDAC Technologies Ltd	◆	◆	◆	◆	◆	◆	–	–	–

	data consultancy services	data analysis services	data modelling services	data capture services	data conversion services	digitising services	photogrammetry services	scanning services	data QA services
Unistride Sewer Technology	–	–	–	◆	–	–	–	–	–
UNISYS Ltd	◆	◆	◆	◆	◆	◆	–	–	◆
University of Edinburgh	◆	–	–	–	–	–	–	–	–
Walker Ladd Surveys	–	–	◆	◆	–	◆	–	–	–
WRc (Water Research centre)	◆	–	◆	–	–	–	–	–	–
WS Atkins Planning & Management Consultants	◆	◆	◆	◆	◆	◆	◆	–	◆
Xcon Data	◆	–	–	◆	◆	–	–	◆	–

directory 19:
providers of surveying services

	aerial surveying services	land surveying services	marine surveying services	remote sensing services
AP³ Imaging Services Limited	◆	◆	–	◆
Audifilm Girona S.L.	–	◆	–	–
Babtie Shaw & Morton Limited	–	◆	–	–
Birkbeck College London	–	–	–	◆
BKS Surveys Ltd	◆	◆	–	–
British Geological Survey	–	◆	◆	◆
BS International Consultants	◆	◆	◆	–
C.A.Design Services Ltd	–	◆	–	–
CAD R&D Centre Limited	–	◆	–	–
Cartographical Services (Southampton) Limited	◆	◆	–	–
CDR Group	–	◆	–	–
Citywise	–	◆	–	–
Cray Systems	–	–	–	◆
Cromwell House Technical Services	–	◆	–	–
Data Collection Ltd	–	◆	–	–
DMV Consultants BV	◆	◆	–	◆
Earth Observation Sciences Ltd	–	–	–	◆
Effective Solutions (Data Products)	◆	◆	◆	–
ERA-Maptec Ltd	–	◆	◆	◆
Eurosense Technologies N.V.	◆	◆	◆	◆
Fairbairn Services Limited	–	◆	–	–
Foto Res	◆	◆	–	◆
Gardline Infotech	–	–	◆	–
GEC Marconi Research Centre	–	–	–	◆
Geo-Perfect TWI B.V.	◆	–	–	◆
GEOBASE Consultants Ltd	◆	◆	◆	◆
GeoData Institute	–	–	–	◆
Geodelta	◆	–	–	–
GISL Limited	◆	◆	–	◆
Glen Computing Ltd	◆	◆	◆	◆
Global Surveys Ltd	–	◆	–	–
Grove Projects Ltd	–	◆	–	–
H R Wallingford Ltd	–	–	◆	–
Hansa Luftbild	◆	–	–	◆
Hunting Engineering Ltd	◆	◆	◆	◆
Hunting Technical Services Ltd	◆	◆	◆	◆
I.S. Ltd	◆	–	–	◆
Institute of Hydrology	◆	–	–	◆
Institute of Terrestrial Ecology	–	◆	–	–
Intera Information Technologies	◆	–	–	◆
J.C.White Chartered Land Surveyors	◆	◆	◆	–
L.E.S. (Computer Services) Ltd	–	◆	–	–
Land and Satellite Surveys	–	◆	–	–
Land Aspects Consultancy Ltd (Parkman Group)	–	◆	–	–
LASCO Ltd	–	◆	–	–

	aerial surveying services	land surveying services	marine surveying services	remote sensing services
Longdin & Browning	–	◆	◆	–
LOY Surveys Ltd	◆	◆	–	–
LTG Services	◆	◆	–	–
Map Data Management Ltd	◆	–	–	–
MAPS geosystems	◆	◆	◆	◆
Mason Land Surveys	◆	◆	◆	◆
Midsummer Computing	◆	◆	◆	◆
Morgan Collis Group Ltd	–	◆	–	–
National Remote Sensing Centre Ltd (NRSC)	◆	◆	◆	◆
Natural Environment Research Council	◆	◆	◆	◆
NERC	◆	◆	◆	◆
Optimal Software Ltd	◆	◆	–	–
Ordnance Survey	◆	◆	–	–
Ordnance Survey of Northern Ireland	◆	◆	–	◆
P&L Engineering Surveys Ltd	–	◆	–	–
Paul Clasper & Associates Ltd	–	◆	◆	–
Photarc Surveys Ltd	◆	◆	–	◆
Photoair	◆	–	–	–
Photogrammetric Data Services Ltd	◆	–	–	–
Planning & Mapping Ltd	◆	◆	–	–
Plowman Craven & Associates Ltd	◆	◆	–	–
Progis GmbH	◆	◆	◆	◆
R.W.A. Dalls FRICS	–	◆	–	–
Racal Survey (UK) Ltd	–	◆	◆	–
Remote Sensing Applications Consultants	–	–	–	◆
SCOT Conseil	–	◆	–	◆
Scott Wilson Kirkpatrick	◆	◆	◆	◆
SIA Limited	◆	–	–	–
Silsoe College, Cranfield University	–	–	–	◆
Simmons Survey Partnership Limited	◆	◆	–	–
Sir Alexander Gibb & Partners Ltd	–	–	–	◆
Sir William Halcrow and Partners	–	◆	–	–
Solent Mapping and Charting (SMAC)	–	◆	◆	–
Sovereign C.S. Ltd	◆	◆	◆	◆
Spacesense consultants	–	–	–	◆
Spatial Data Limited	–	◆	◆	◆
Structural Technologies Ltd (STL)	–	–	–	◆
Survey Control Services	–	◆	–	–
Survey & Development Services Ltd	◆	◆	–	–
Svitzer Limited	–	◆	◆	–
System Options Limited	◆	–	–	–
Terrafix Ltd	–	–	–	◆
TerraQuest Group Limited	◆	◆	–	–
The NPA Group Ltd	–	–	–	◆
The Severn Partnership	–	◆	–	–

	aerial surveying services	land surveying services	marine surveying services	remote sensing services
The Survey Centre	◆	◆	–	–
TYDAC Technologies Ltd	–	–	–	◆
Unistride Sewer Technology	–	◆	–	–
Walker Ladd Surveys	–	◆	–	–
WRc (Water Research centre)	–	◆	◆	–
WS Atkins Planning & Management Consultants	◆	◆	–	–

directory 20:
providers of procurement services

	conformance to standards services	cost benefit analysis studies	feasibility services	business analysis services	detail design services	software development services	project management services	risk analysis services	tender process services	testing services	documentation services
Active Software Ltd	–	–	–	–	◆	◆	–	–	–	–	–
AGFA UK	◆	–	◆	–	◆	◆	◆	–	–	–	–
AiC Analysts	–	–	–	–	◆	◆	–	–	–	–	–
Anglian Engineering & International Consultancy	–	–	–	–	–	◆	–	–	–	–	–
AP³ Imaging Services Limited	–	–	–	–	◆	–	–	–	–	–	–
APIC Systems	◆	–	–	–	–	–	–	–	–	–	–
Babtie Shaw & Morton Limited	–	–	◆	◆	◆	◆	◆	–	◆	–	–
BS International Consultants	–	◆	◆	◆	–	–	◆	–	–	◆	–
Bull Information Systems Limited	◆	◆	◆	◆	◆	◆	◆	◆	◆	◆	◆
Business Information Management	◆	◆	◆	◆	◆	◆	◆	◆	◆	◆	–
Byers Engineering Company	–	◆	–	–	◆	◆	◆	–	–	◆	◆
C.A.Design Services Ltd	–	◆	–	◆	◆	–	◆	–	–	–	–
CACI Limited	–	◆	–	◆	◆	–	◆	–	–	–	–
CAD - Capture Limited	–	◆	–	◆	◆	◆	◆	–	–	–	–
CAD R&D Centre Limited	–	◆	◆	◆	◆	◆	◆	◆	◆	–	◆
CAM - Centre for Analysis & Modelling Limited	–	–	◆	–	–	–	–	–	–	–	–
Cambridge Computer Consultants (UK) Ltd	–	◆	◆	◆	◆	–	◆	◆	◆	◆	–
Cartwright Associates	◆	◆	◆	◆	◆	◆	◆	◆	◆	◆	◆
CDD Ltd	–	–	◆	◆	–	◆	◆	–	–	–	–
Chroson Ltd	–	◆	◆	◆	–	◆	–	◆	–	–	–
Cliffe House Associates	◆	◆	◆	◆	–	–	◆	–	–	◆	◆
CMG Computer Management Group (UK) Ltd	◆	◆	◆	◆	◆	◆	◆	◆	◆	◆	◆
Cobham Digital Services Limited	–	◆	◆	–	◆	◆	–	–	–	–	–
COMSULT	◆	–	◆	–	◆	◆	◆	◆	◆	◆	◆
Construction Industry Computing Association	◆	◆	◆	◆	–	–	◆	◆	◆	◆	–
Coopers & Lybrand	◆	◆	◆	◆	◆	◆	◆	◆	◆	◆	◆
Corbins Consultancy	–	◆	◆	◆	–	–	◆	◆	◆	◆	◆
Cray Systems	◆	◆	◆	◆	◆	◆	◆	◆	◆	◆	◆
Data Base Builders	–	◆	◆	◆	–	–	–	–	◆	–	◆
DCL Consulting	–	–	◆	–	–	◆	–	–	–	–	◆
Digital Equipment Corporation	–	◆	–	–	–	–	–	◆	◆	–	–
DMV Consultants BV	–	◆	◆	◆	◆	◆	◆	–	◆	–	◆
Dolphin Consulting Group	–	◆	◆	–	–	◆	◆	–	–	–	◆
Dotted Eyes	◆	◆	◆	◆	◆	◆	◆	–	◆	◆	◆
Dr Stanley Port	–	◆	◆	◆	–	–	–	–	–	–	–
Earth Observation Sciences Ltd	–	–	◆	–	◆	◆	◆	–	–	◆	◆
Elstree Computing Ltd	◆	◆	◆	◆	–	◆	–	–	–	–	–
Enghouse (UK) Limited	–	◆	◆	◆	–	◆	◆	–	◆	–	◆
ESRI (UK) Ltd	–	◆	◆	◆	◆	◆	◆	–	–	–	–
Evox Facilities Ltd	–	–	–	–	◆	◆	–	–	–	–	◆
Fairbairn Services Limited	–	–	◆	◆	◆	◆	◆	–	–	–	◆
FileNet Ltd	–	◆	◆	◆	–	◆	◆	–	◆	–	◆
G.L. Consulting Ltd	–	◆	◆	◆	–	–	◆	◆	◆	–	–
Geo/SQL (UK)	–	◆	◆	◆	◆	◆	◆	◆	◆	◆	◆
Geo2 Consulting	◆	◆	◆	◆	–	–	◆	–	–	–	–

	conformance to standards services	cost benefit analysis studies	feasibility services	business analysis services	detail design services	software development services	project management services	risk analysis services	tender process services	testing services	documentation services
GEOBASE Consultants Ltd	◆	◆	◆	◆	◆	◆	◆	◆	◆	◆	◆
GeoData Institute	–	◆	◆	◆	–	–	–	–	◆	–	–
Geodelta	–	–	◆	–	–	◆	–	–	–	–	–
Geographic Management Solutions Ltd	–	–	–	◆	–	◆	◆	–	–	–	–
GEOSOFT Ltd	–	–	◆	–	◆	◆	◆	–	–	–	–
GID Ltd	–	–	–	◆	◆	◆	◆	–	–	–	◆
GIMMS (GIS) Ltd	–	–	–	–	◆	–	◆	–	–	–	–
GIS Services Ltd	–	◆	◆	◆	◆	◆	◆	–	◆	–	◆
GISL Limited	–	–	◆	–	–	–	–	–	–	–	–
Glen Computing Ltd	–	◆	◆	◆	–	◆	◆	–	–	–	–
GMAP Ltd	–	–	◆	◆	–	–	–	–	–	–	–
Greig Fester Limited	–	–	–	–	–	–	–	◆	–	–	–
Grove Projects Ltd	–	–	–	◆	◆	◆	◆	◆	◆	◆	–
Hansa Luftbild	–	◆	◆	–	◆	◆	◆	–	◆	–	◆
Hunting Engineering Ltd	◆	◆	◆	◆	◆	◆	◆	◆	◆	◆	◆
IBM UK Ltd	–	◆	◆	◆	◆	◆	◆	◆	◆	–	–
ICL Ltd	–	–	◆	◆	◆	◆	◆	◆	◆	◆	◆
IME (UK) Ltd	–	–	◆	–	–	–	–	–	–	–	–
Informed Solutions Limited	◆	◆	◆	◆	◆	◆	◆	◆	◆	◆	◆
Ingecon B.V.	–	◆	◆	◆	–	◆	◆	–	◆	–	–
Intera Information Technologies	–	–	◆	–	–	–	–	–	–	–	–
Intergraph (UK) Ltd	–	–	–	◆	◆	◆	–	–	–	–	–
IT Southern Ltd	◆	◆	◆	◆	◆	◆	◆	◆	◆	◆	◆
John D Leatherdale FRICS	◆	◆	◆	◆	◆	◆	◆	◆	◆	–	–
Kamyco International	–	◆	◆	◆	–	–	◆	◆	–	–	–
Kingswood Consulting Limited	◆	◆	◆	◆	◆	◆	◆	◆	◆	◆	◆
KJB Consulting	–	–	◆	◆	–	–	◆	–	–	◆	–
Know Edge Ltd	◆	◆	◆	◆	◆	–	◆	–	◆	◆	◆
Land Aspects Consultancy Ltd (Parkman Group)	◆	◆	◆	◆	◆	◆	◆	–	–	–	◆
Logica UK Limited	◆	–	◆	◆	◆	◆	◆	–	–	–	◆
Logitrans	–	◆	◆	–	◆	◆	◆	–	◆	◆	◆
LTG Services	◆	◆	◆	◆	–	◆	◆	◆	◆	◆	◆
Magdala Sociedade	◆	◆	◆	◆	◆	◆	◆	◆	◆	–	◆
MAPS geosystems	–	–	◆	◆	◆	–	–	◆	–	◆	–
Mathshop	–	–	–	–	–	◆	–	–	–	–	–
McLintock Limited	–	◆	–	–	–	–	–	◆	–	–	–
Mentis Management Consultants Ltd	◆	◆	◆	◆	◆	–	◆	◆	◆	◆	◆
Methods Applications Ltd	◆	◆	◆	–	◆	–	◆	–	◆	◆	–
Midsummer Computing	◆	◆	◆	◆	◆	◆	◆	◆	◆	◆	◆
Modern Maps	–	◆	◆	◆	–	–	–	–	–	–	◆
Morgan Collis Group Ltd	–	–	–	–	◆	–	–	◆	–	–	–
Mott MacDonald Ltd	◆	◆	◆	◆	◆	◆	◆	◆	◆	◆	◆
MR Data Graphics	–	–	◆	–	–	◆	–	–	–	–	–
MVA Systematica	–	◆	◆	◆	◆	◆	◆	–	–	–	◆
MVM Consultants plc	◆	◆	◆	◆	◆	◆	◆	◆	◆	◆	–

	conformance to standards services	cost benefit analysis studies	feasibility services	business analysis services	detail design services	software development services	project management services	risk analysis services	tender process services	testing services	documentation services
National Remote Sensing Centre Ltd (NRSC)	◆	◆	◆	◆	◆	◆	◆	◆	◆	◆	◆
Organisation Management Systems	–	–	◆	◆	◆	◆	◆	–	–	◆	◆
Ove Arup & Partners	–	◆	◆	◆	◆	◆	◆	◆	◆	–	–
Oxford Institute of Retail Management	–	–	–	◆	–	–	–	–	–	–	–
PA Consulting Group	–	◆	◆	◆	◆	◆	◆	◆	◆	◆	–
Panda	–	–	–	–	–	–	–	–	–	–	–
Paul Clasper & Associates Ltd	–	–	◆	–	–	–	–	–	–	–	–
PAX Technology	–	–	◆	◆	◆	–	◆	–	–	–	–
PD Computing Ltd	–	◆	◆	◆	◆	–	–	–	–	–	–
Peter Thorpe Consulting	◆	◆	◆	◆	◆	◆	◆	◆	◆	◆	◆
PlanGraphics Inc	◆	◆	◆	◆	◆	◆	◆	◆	◆	◆	◆
PLANTECH Ltd	–	–	◆	◆	◆	–	–	–	–	–	–
Posford Duvivier	–	◆	◆	–	◆	◆	–	–	–	–	–
Procis Software Ltd	–	◆	◆	◆	◆	◆	◆	–	–	◆	◆
R.W.A. Dalls FRICS	–	◆	◆	–	–	–	–	–	◆	–	–
Raindrop Information Systems Ltd	◆	◆	◆	◆	–	◆	◆	◆	◆	◆	◆
regioplanDATA GmbH	–	–	◆	◆	◆	◆	◆	–	◆	◆	◆
Robert Walker Consultants	◆	◆	◆	◆	◆	–	◆	◆	◆	◆	◆
Salford University Business Services Ltd	◆	◆	◆	◆	◆	◆	◆	–	◆	◆	◆
SCOT Conseil	–	–	◆	–	–	◆	◆	–	–	–	–
Scott Wilson Kirkpatrick	◆	◆	◆	◆	◆	◆	◆	◆	◆	◆	–
SDI Ltd	–	–	◆	–	◆	◆	–	–	–	–	–
Sector (UK) Limited	◆	◆	◆	◆	◆	◆	◆	◆	◆	◆	◆
SIAS Limited	–	–	◆	–	◆	◆	–	–	–	–	–
Sir Alexander Gibb & Partners Ltd	–	◆	◆	–	–	◆	◆	◆	–	–	–
Smith System Engineering Ltd	◆	◆	◆	◆	◆	◆	◆	◆	◆	◆	◆
Sovereign C.S. Ltd	–	–	◆	–	–	–	–	◆	–	–	–
Structural Technologies Ltd (STL)	–	–	–	–	–	◆	◆	–	–	–	–
Symology Limited	–	–	◆	◆	–	–	–	–	–	–	–
System Options Limited	–	◆	◆	◆	◆	◆	–	◆	◆	◆	◆
TEAMS (Taylor Woodrow Electronics Asset Mapping Survey)	–	–	–	◆	◆	◆	◆	–	–	–	◆
Tendron Systems Ltd	◆	◆	◆	◆	◆	◆	◆	◆	◆	◆	◆
TerraQuest Group Limited	–	◆	◆	◆	–	–	◆	◆	–	–	–
The NPA Group Ltd	–	–	◆	–	–	–	◆	◆	–	–	–
The Severn Partnership	–	–	–	–	◆	–	◆	–	–	–	–
UNISYS Ltd	◆	◆	◆	◆	◆	◆	◆	◆	◆	◆	◆
University of Edinburgh	–	–	–	–	◆	–	–	–	–	–	–
WRc (Water Research centre)	◆	◆	◆	◆	◆	◆	◆	◆	◆	◆	◆
WS Atkins Planning & Management Consultants	◆	◆	◆	◆	◆	◆	◆	◆	◆	◆	◆

	maintenance services	management services	system implementation services	training services	project management services	application configuration services
A.Rutherford Ltd	–	◆	–	–	–	–
ACDS Graphic System Inc	◆	◆	◆	◆	◆	–
Action Information Management Ltd	◆	◆	◆	◆	◆	–
Active Software Ltd	◆	◆	◆	–	–	◆
Adept Scientific Micro Systems Ltd	–	–	–	◆	–	–
AGFA UK	–	–	◆	◆	–	◆
AP³ Imaging Services Limited	–	–	–	◆	–	–
ARC Systems Pty Ltd	◆	◆	◆	◆	◆	◆
Ashtech Europe Limited	◆	–	–	◆	–	◆
Assist Applications Limited	◆	–	◆	◆	◆	◆
Audifilm Girona S.L.	◆	◆	◆	◆	◆	–
Bentley Systems UK Ltd	–	–	–	–	–	◆
Birkbeck College London	–	–	–	◆	–	–
British Geological Survey	–	–	–	◆	◆	–
BS International Consultants	–	◆	–	–	◆	–
Bull Information Systems Limited	◆	◆	◆	◆	◆	◆
Business Information Management	–	–	–	◆	◆	–
Byers Engineering Company	◆	–	–	–	–	–
C.A.Design Services Ltd	–	◆	–	◆	◆	◆
CACI Limited	–	–	◆	◆	–	–
CAD - Capture Limited	◆	◆	–	–	◆	–
CAD R&D Centre Limited	◆	◆	◆	◆	◆	◆
CAM - Centre for Analysis & Modelling Limited	–	–	–	◆	–	–
Cambridge Computer Consultants (UK) Ltd	–	◆	◆	◆	◆	◆
Carl Zeiss Limited	◆	–	◆	◆	–	◆
CARTograph Ltd	–	–	◆	◆	–	–
Cartographical Services (Southampton) Limited	–	–	–	◆	–	–
Cartwright Associates	◆	◆	◆	◆	◆	◆
CDD Ltd	◆	–	◆	◆	◆	◆
CDR Group	–	–	–	◆	–	◆
Chroson Ltd	–	–	–	–	◆	◆
Citywise	–	–	◆	–	–	–
Cliffe House Associates	–	–	–	–	◆	–
CMG Computer Management Group (UK) Ltd	◆	◆	◆	◆	◆	◆
Cobham Digital Services Limited	–	◆	◆	◆	◆	◆
CODEC Facilities Limited	◆	–	◆	–	◆	◆
Computer Aided Development (CADCORP) Ltd	◆	–	◆	◆	–	◆
COMSULT	◆	◆	◆	◆	◆	◆
Consensus Information Technology Ltd	–	–	◆	◆	◆	◆
Coopers & Lybrand	–	–	–	–	◆	–
Corbins Consultancy	–	–	–	–	◆	–
Corena A/S	–	–	◆	◆	–	–
Cray Systems	◆	◆	◆	◆	◆	◆
Data Base Builders	–	◆	◆	–	◆	◆
Dataflow Information Systems	◆	–	–	◆	◆	◆

	maintenance services	management services	system implementation services	training services	project management services	application configuration services
Datatechnology Datech Ltd	–	–	◆	–	–	–
Dataview Solutions Ltd	–	–	◆	◆	◆	◆
DCL Consulting	–	◆	–	–	◆	–
Derek Hunter & Partners Ltd	◆	–	◆	–	–	◆
Digital Equipment Corporation	–	◆	◆	–	◆	–
DM Management Consultants Ltd (DMMC)	–	–	–	–	◆	–
DMAP Ltd	–	–	◆	◆	◆	–
DMV Consultants BV	–	◆	◆	◆	◆	◆
Dolphin Consulting Group	–	–	–	–	◆	–
Dotted Eyes	–	◆	◆	◆	◆	–
Dowling Associates Limited	◆	◆	◆	◆	◆	◆
EA Technology	◆	–	–	◆	–	◆
Earth Observation Sciences Ltd	–	◆	◆	–	◆	–
Effective Solutions (Data Products)	◆	–	–	◆	–	–
Elstree Computing Ltd	◆	◆	◆	◆	–	–
Enghouse (UK) Limited	◆	◆	◆	◆	◆	◆
EPS - Essential Planning Systems Limited	◆	–	–	◆	–	◆
ERA Technology Ltd	–	–	–	◆	◆	◆
ERDAS (UK) Ltd	–	–	◆	◆	–	–
ESRI (UK) Ltd	◆	◆	◆	◆	◆	◆
European Geographic Technologies BV	◆	–	–	–	–	–
Eurosense Technologies N.V.	◆	–	◆	◆	◆	◆
Fairbairn Services Limited	◆	◆	◆	◆	◆	◆
FastCAD GIS Ltd	◆	–	◆	◆	◆	◆
FastCAD GIS Ltd	◆	◆	◆	◆	–	–
FileNet Ltd	◆	◆	◆	◆	◆	◆
Flynn & Rothwell	–	–	–	–	◆	–
G.L. Consulting Ltd	–	–	–	–	◆	–
Gamma Ltd	◆	◆	◆	◆	◆	◆
Gardline Infotech	◆	◆	◆	◆	◆	◆
GEC Marconi Research Centre	–	–	◆	–	–	–
Genasys II Limited	◆	–	–	◆	–	–
GEO-Marketing Systems Ltd (GMSL)	–	–	–	–	–	–
Geo-Perfect TWI B.V.	◆	–	◆	◆	–	◆
Geo/SQL (UK)	◆	◆	◆	◆	◆	◆
Geo2 Consulting	–	◆	–	–	◆	–
GEOBASE Consultants Ltd	◆	◆	◆	◆	◆	◆
GeoData Institute	–	–	◆	◆	–	–
Geometria GIS Systems House Ltd	–	–	◆	–	–	◆
Geoplan (UK) Ltd	◆	◆	◆	–	–	–
Geops BV	–	–	◆	◆	◆	◆
Geoview Systems Kft	–	–	◆	–	◆	–
GISDATA	–	–	–	–	–	–
GISL Limited	–	–	–	◆	–	–
Glen Computing Ltd	◆	◆	◆	◆	◆	◆

	maintenance services	management services	system implementation services	training services	project management services	application configuration services
Global Surveys Ltd	–	–	–	◆	–	–
GMAP Ltd	◆	◆	◆	◆	◆	◆
Graphic Data Systems Corporation (GDS)	–	◆	◆	◆	◆	◆
Graphical Data Capture Ltd (GDC)	–	–	◆	◆	–	◆
Graphite Management Services Ltd	◆	◆	◆	◆	–	◆
Grove Projects Ltd	–	–	–	–	◆	–
Hansa Luftbild	–	–	◆	◆	◆	–
Hunting Engineering Ltd	◆	◆	◆	◆	◆	◆
Hunting Technical Services Ltd	–	–	–	◆	◆	◆
I.S. Ltd	–	–	–	◆	–	–
IBM UK Ltd	◆	–	◆	–	◆	◆
ICL Ltd	◆	◆	◆	◆	◆	◆
IMASS Limited	◆	◆	◆	◆	◆	◆
IME (UK) Ltd	◆	–	–	–	◆	◆
Informed Solutions Limited	◆	◆	◆	◆	◆	◆
Ingecon B.V.	◆	◆	◆	◆	◆	◆
Institute of Hydrology	–	–	–	–	–	◆
Intera Information Technologies	–	–	◆	◆	◆	◆
Intergraph (UK) Ltd	◆	◆	◆	◆	◆	◆
IT Southern Ltd	◆	◆	◆	◆	◆	◆
ITC	–	–	–	◆	–	–
ITS : Intertrade Scientific Ltd	◆	–	–	–	–	–
John D Leatherdale FRICS	–	◆	–	–	◆	◆
Kamyco International	–	–	–	–	◆	–
Kingswood Consulting Limited	◆	◆	◆	◆	◆	◆
KJB Consulting	–	◆	–	◆	◆	–
KPMG	–	◆	◆	◆	◆	◆
Land and Satellite Surveys	–	–	–	–	◆	–
Lantmatenet GIS-centrum	–	–	–	◆	–	–
Laser-Scan Ltd	◆	–	◆	◆	◆	◆
Logica UK Limited	◆	◆	◆	◆	◆	◆
Lovell Johns Ltd	–	–	–	◆	◆	–
Magdala Sociedade	–	–	◆	◆	◆	◆
Map Data Management Ltd	◆	◆	◆	◆	◆	◆
MAPIT Limited	◆	◆	◆	◆	◆	◆
MAPS geosystems	◆	–	◆	◆	◆	–
Mason Land Surveys	◆	–	–	◆	–	–
McLintock Limited	◆	◆	–	–	◆	–
Mentis Management Consultants Ltd	–	–	–	◆	◆	◆
Methods Applications Ltd	–	–	–	–	◆	–
Midlands Regional Research Laboratory	–	–	–	–	◆	–
Midsummer Computing	◆	◆	◆	◆	◆	◆
Modern Maps	–	–	–	–	◆	–
Morgan Collis Group Ltd	–	–	–	◆	◆	–
Mott MacDonald Ltd	–	◆	◆	◆	◆	◆

	maintenance services	management services	system implementation services	training services	project management services	application configuration services
MPSI Systems Ltd	–	–	◆	–	◆	–
Munro Garratt International	◆	–	–	◆	◆	–
MVM Consultants plc	–	◆	◆	–	◆	◆
NAG Ltd	–	–	◆	–	–	–
National Remote Sensing Centre Ltd (NRSC)	◆	◆	◆	◆	◆	◆
Natural Environment Research Council	–	–	–	–	–	◆
NERC	–	◆	–	–	◆	–
Numonics UK (Division of Telmtek Ltd)	–	–	–	–	–	◆
Oaklands I.T.	–	–	–	–	◆	–
Optimal Software Ltd	◆	–	◆	◆	–	◆
Ordnance Survey	–	◆	–	◆	◆	–
Ordnance Survey of Northern Ireland	◆	◆	◆	◆	◆	◆
Organisation Management Systems	–	◆	–	◆	–	◆
Oscar Faber	◆	◆	◆	◆	◆	◆
Ove Arup & Partners	–	◆	◆	◆	◆	◆
PA Consulting Group	–	◆	◆	–	◆	–
PAFEC Ltd	◆	◆	◆	◆	◆	◆
Panda	–	◆	–	–	◆	–
Paul Clasper & Associates Ltd	–	–	◆	–	–	–
PAX Technology	–	–	◆	◆	◆	◆
PD Computing Ltd	–	◆	◆	◆	◆	–
Peter Thorpe Consulting	–	◆	–	◆	◆	–
PlanGraphics Inc	–	◆	◆	◆	◆	◆
PLANTECH Ltd	◆	–	◆	◆	◆	◆
Plowman Craven & Associates Ltd	◆	◆	–	–	◆	◆
Posford Duvivier	–	–	–	–	–	◆
Procis Software Ltd	–	◆	◆	◆	–	◆
Progis GmbH	◆	◆	–	◆	–	◆
QAS Systems Ltd	–	◆	◆	–	–	◆
R.W.A. Dalls FRICS	–	–	–	–	◆	–
Raindrop Information Systems Ltd	◆	◆	◆	◆	◆	◆
regioplanDATA GmbH	◆	◆	◆	◆	◆	◆
Remote Sensing Applications Consultants	–	–	–	–	–	◆
Robert Walker Consultants	–	◆	–	–	◆	–
Salford University Business Services Ltd	◆	◆	–	–	◆	–
SAS Institute	–	–	◆	–	–	◆
SAZTEC Philippines Inc	◆	◆	–	–	◆	–
SCOT Conseil	–	–	–	◆	◆	–
Scott Wilson Kirkpatrick	◆	◆	◆	◆	◆	◆
SDI Ltd	–	–	◆	–	◆	◆
Sector (UK) Limited	◆	◆	◆	◆	◆	◆
SHL Vision* Solutions Limited	◆	◆	◆	◆	◆	◆
SIA Limited	◆	–	◆	◆	◆	◆
SIAS Limited	◆	◆	–	–	–	–
Siemens Nixdorf	◆	◆	◆	◆	◆	◆

	maintenance services	management services	system implementation services	training services	project management services	application configuration services
Silicon Graphics Limited	◆	–	–	◆	–	–
Silsoe College, Cranfield University	–	–	–	◆	–	–
Sir William Halcrow and Partners	◆	◆	–	◆	◆	–
Smartscan Inc	◆	◆	◆	–	–	–
Smith System Engineering Ltd	–	–	◆	–	◆	◆
Southbank Systems PLC	◆	◆	◆	◆	◆	◆
Sovereign C.S. Ltd	–	–	–	–	◆	–
Spatial Geographic Services & Applications Ltd	–	–	–	◆	◆	–
SPSS UK Ltd	–	–	◆	◆	◆	–
Star Informatic S.A.	◆	–	◆	◆	–	◆
StatSci Europe	–	–	–	◆	◆	–
Structural Technologies Ltd (STL)	–	–	◆	◆	◆	◆
Sun Microsystems Ltd	◆	–	◆	◆	–	–
Survey & Development Services Ltd	–	–	–	◆	–	–
Symology Limited	◆	–	◆	◆	–	–
Sysdeco (UK) Ltd	◆	–	◆	◆	◆	◆
System Options Limited	◆	◆	◆	◆	◆	◆
TACTICIAN UK	◆	–	◆	◆	–	◆
TEAMS (Taylor Woodrow Electronics Asset Mapping Survey)	◆	◆	◆	◆	◆	◆
Tele Atlas	◆	◆	–	◆	◆	◆
Tendron Systems Ltd	–	◆	◆	–	◆	◆
Tenet Systems Ltd	–	–	–	◆	–	–
Terrafix Ltd	◆	◆	◆	◆	◆	◆
The Data Consultancy	◆	◆	◆	◆	◆	◆
Trimble Navigation Europe Ltd	–	–	–	◆	–	◆
TYDAC Technologies Ltd	◆	◆	–	◆	◆	◆
UNISYS Ltd	◆	◆	◆	◆	◆	◆
Universal Systems Ltd	◆	–	◆	◆	–	◆
University of Newcastle Upon Tyne	–	–	◆	–	–	◆
WRc (Water Research centre)	–	◆	◆	◆	◆	◆
WS Atkins Planning & Management Consultants	◆	◆	◆	◆	◆	◆

directory 22:
providers of management services

	general consultancy services	facilities management services	legal services	process re-engineering services	professional development services	general project management services	quality accreditation services	other management services
A.Rutherford Ltd	◆	–	–	–	–	–	–	–
ACDS Graphic System Inc	–	◆	–	–	–	–	–	–
Action Information Management Ltd	◆	–	–	–	–	–	–	–
Active Software Ltd	◆	–	–	–	–	–	–	–
AiC Analysts	–	–	–	◆	◆	–	–	–
Anglian Engineering & International Consultancy	–	–	–	–	–	◆	–	–
APIC Systems	◆	–	–	–	–	–	–	–
Audifilm Girona S.L.	◆	–	–	–	◆	◆	–	–
Babtie Shaw & Morton Limited	◆	–	–	–	–	–	–	–
Binnie Black & Veatch	–	–	–	–	–	◆	–	–
Birkbeck College London	◆	–	–	–	–	–	–	–
BKS Surveys Ltd	◆	–	–	–	–	–	–	–
British Geological Survey	–	–	–	–	–	◆	–	–
BS International Consultants	◆	–	–	◆	◆	◆	–	–
Bull Information Systems Limited	◆	◆	–	◆	–	◆	–	–
Business Information Management	◆	–	–	–	–	–	–	–
Byers Engineering Company	◆	–	–	◆	–	–	–	–
C.A.Design Services Ltd	◆	◆	–	–	–	◆	–	–
CACI Limited	◆	–	–	–	–	–	–	–
CAD - Capture Limited	◆	◆	–	–	–	◆	–	–
CAD R&D Centre Limited	◆	–	–	◆	◆	◆	–	–
CAM - Centre for Analysis & Modelling Limited	◆	–	–	–	–	–	–	–
Cambashi Ltd	◆	–	–	◆	–	◆	–	–
Cambridge Computer Consultants (UK) Ltd	◆	–	–	◆	–	◆	–	–
Cartwright Associates	◆	–	–	◆	◆	◆	–	–
CATALIST	◆	–	–	–	–	–	–	–
CCN Marketing	◆	–	–	–	–	–	–	–
CDR Group	◆	–	–	–	–	–	–	–
Cliffe House Associates	◆	–	–	–	–	–	–	–
CMG Computer Management Group (UK) Ltd	◆	◆	–	◆	–	◆	–	–
Cobham Digital Services Limited	◆	–	–	–	–	–	–	–
COMSULT	◆	–	–	–	–	–	–	–
Consensus Information Technology Ltd	◆	–	–	◆	–	–	–	–
Construction Industry Computing Association	◆	–	–	◆	◆	◆	–	–
Coopers & Lybrand	◆	–	–	◆	–	◆	–	–
Corbins Consultancy	◆	–	–	–	–	◆	–	–
Cray Systems	◆	◆	◆	◆	◆	◆	◆	–
Data Base Builders	◆	–	–	–	–	◆	–	–
Data Dictionary Systems Limited	◆	–	–	–	–	–	–	–
Dataquest Europe Ltd	◆	–	–	–	–	–	–	–
Datatechnology Datech Ltd	–	◆	–	–	–	–	–	–
Dataview Solutions Ltd	◆	–	–	–	–	–	–	–
Derek Hunter & Partners Ltd	◆	–	–	–	–	–	–	–
Design Computer Aids Limited (DeCAL)	◆	–	–	–	–	–	–	–
DM Management Consultants Ltd (DMMC)	◆	–	–	–	–	–	–	–

	general consultancy services	facilities management services	legal services	process re-engineering services	professional development services	general project management services	quality accreditation services	other management services
Dolphin Consulting Group	◆	–	–	–	–	◆	–	–
Dotted Eyes	◆	–	–	◆	–	◆	–	–
Dr Stanley Port	◆	–	–	–	–	–	–	–
Earth Observation Sciences Ltd	◆	–	–	–	–	◆	–	–
ESRI (UK) Ltd	◆	–	–	–	–	–	–	–
European Business Mapping	◆	–	–	–	–	–	–	–
Eurosense Technologies N.V.	◆	◆	–	–	–	◆	–	–
Evox Facilities Ltd	◆	–	–	–	–	–	–	–
Fairbairn Services Limited	◆	◆	–	◆	–	◆	–	–
FastCAD GIS Ltd	◆	–	–	–	–	–	–	–
FileNet Ltd	–	–	–	–	◆	◆	–	–
Flynn & Rothwell	◆	–	–	–	–	◆	–	–
Gamma Ltd	◆	–	–	–	–	◆	–	–
Gardline Infotech	◆	–	–	–	–	◆	–	–
Genasys II Limited	◆	–	–	–	–	–	–	–
GEO-Marketing Systems Ltd (GMSL)	◆	–	–	–	–	–	–	–
GEO-UK Ltd	–	–	–	–	–	–	–	◆
Geo/SQL (UK)	◆	◆	–	–	–	–	–	–
Geo2 Consulting	◆	–	–	◆	–	◆	–	–
GEOBASE Consultants Ltd	◆	◆	◆	◆	◆	◆	◆	–
GeoData Institute	◆	–	–	–	–	–	–	–
Geoinformation International	◆	–	–	–	–	–	–	–
Geometria GIS Systems House Ltd	–	–	–	–	–	◆	–	–
Geoplan (UK) Ltd	◆	–	–	–	–	–	–	–
GEOSOFT Ltd	◆	–	–	–	–	–	–	–
GIMMS (GIS) Ltd	◆	–	–	–	–	–	–	–
GIS Services Ltd	◆	–	–	◆	–	◆	◆	–
GISDATA	◆	–	–	–	–	–	–	–
GISL Limited	◆	–	–	–	–	◆	–	–
Glen Computing Ltd	◆	–	–	–	–	–	–	–
GMAP Ltd	◆	–	–	–	–	◆	–	–
Graphic Data Systems Corporation (GDS)	◆	–	–	–	–	–	–	–
Graphical Data Capture Ltd (GDC)	◆	–	–	–	–	–	–	–
Greig Fester Limited	◆	–	–	–	–	–	–	–
Grove Projects Ltd	◆	–	–	–	–	–	–	–
I.S. Ltd	◆	–	–	–	–	–	–	–
ICL Ltd	◆	◆	–	–	–	–	–	–
IMASS Limited	◆	–	–	–	–	◆	–	–
IME (UK) Ltd	◆	–	–	–	–	◆	–	–
Informed Solutions Limited	◆	–	–	◆	◆	◆	–	–
Ingecon B.V.	◆	◆	–	–	◆	◆	–	–
Institute of Hydrology	◆	–	–	–	–	–	–	–
Intera Information Technologies	◆	–	–	–	–	◆	–	–
John D Leatherdale FRICS	◆	–	◆	–	◆	◆	–	–
Kingswood Consulting Limited	◆	◆	–	◆	–	◆	–	–

	general consultancy services	facilities management services	legal services	process re-engineering services	professional development services	general project management services	quality accreditation services	other management services
KJB Consulting	◆	–	–	–	◆	◆	–	◆
Know Edge Ltd	◆	–	–	◆	–	◆	–	–
KPMG	◆	–	◆	◆	◆	◆	◆	–
Land and Satellite Surveys	◆	–	–	–	–	–	–	–
Land Aspects Consultancy Ltd (Parkman Group)	◆	◆	◆	◆	–	◆	–	–
Lantmatenet GIS-centrum	◆	–	–	–	–	–	–	–
Laser-Scan Ltd	◆	–	–	–	–	◆	–	–
LiveChart	–	–	–	–	◆	–	–	–
Logica UK Limited	◆	–	–	◆	◆	◆	–	–
Lovell Johns Ltd	◆	–	–	–	–	–	–	–
LTG Services	◆	◆	–	–	–	◆	–	–
Macaulay Land Use Research Institute	◆	–	–	–	–	–	–	–
Magdala Sociedade	◆	–	–	◆	–	◆	◆	–
Map Data Management Ltd	–	–	–	–	–	◆	–	–
McLintock Limited	–	◆	–	–	–	–	–	–
Mentis Management Consultants Ltd	◆	–	◆	◆	◆	◆	–	–
Methods Applications Ltd	◆	–	–	◆	–	◆	–	–
Midlands Regional Research Laboratory	◆	–	–	–	–	–	–	–
Midsummer Computing	◆	◆	–	–	–	–	–	–
Modern Maps	◆	–	–	–	–	◆	–	–
MR Data Graphics	–	◆	–	–	–	–	–	–
MVM Consultants plc	–	–	–	◆	◆	◆	–	–
NAG Ltd	◆	–	–	–	–	–	–	–
National Remote Sensing Centre Ltd (NRSC)	◆	–	–	–	–	–	–	–
Natural Environment Research Council	◆	–	–	–	–	–	–	–
NERC	◆	◆	–	–	–	◆	–	–
NOMIS	◆	–	–	–	–	–	–	–
Oaklands I.T.	◆	–	–	–	◆	–	–	–
Optimal Software Ltd	◆	◆	–	–	–	–	–	–
Ordnance Survey of Northern Ireland	◆	–	–	–	–	◆	–	–
Organisation Management Systems	◆	–	–	–	–	–	–	–
Ove Arup & Partners	◆	–	–	–	–	◆	–	–
PA Consulting Group	◆	–	–	◆	–	◆	–	–
PAFEC Ltd	–	–	–	–	–	◆	–	–
Panda	◆	–	–	–	–	◆	–	–
Paul Clasper & Associates Ltd	◆	–	–	–	–	–	–	–
Peter Thorpe Consulting	◆	–	–	◆	◆	◆	–	–
Photogrammetric Data Services Ltd	–	–	–	–	–	◆	–	–
PlanGraphics Inc	◆	◆	–	–	–	–	–	–
PLANTECH Ltd	◆	–	–	–	–	–	–	–
Posford Duvivier	◆	–	–	–	–	–	–	–
R.W.A. Dalls FRICS	◆	–	–	–	–	–	–	–
regioplanDATA GmbH	◆	–	–	–	◆	◆	–	–
Remote Sensing Applications Consultants	◆	–	–	–	–	–	–	–
Robert Walker Consultants	◆	–	–	–	–	◆	–	–

	general consultancy services	facilities management services	legal services	process re-engineering services	professional development services	general project management services	quality accreditation services	other management services
Royal Geographical Society (with The Institute of British Geographers)	◆	–	–	–	–	–	–	–
Salford University Business Services Ltd	◆	◆	–	–	–	◆	–	–
SAS Institute	◆	–	–	–	–	–	–	–
SAZTEC Philippines Inc	◆	–	–	–	–	◆	–	–
Scott Wilson Kirkpatrick	◆	–	–	–	–	◆	–	–
SDI Ltd	◆	◆	–	–	–	–	–	–
Sector (UK) Limited	◆	◆	◆	–	–	◆	◆	–
SIA Limited	◆	–	–	–	◆	◆	–	–
Siemens Nixdorf	◆	–	–	–	–	◆	–	–
Silsoe College, Cranfield University	◆	–	–	◆	–	–	–	–
Sir William Halcrow and Partners	◆	–	–	–	–	◆	–	–
Smith System Engineering Ltd	◆	–	–	–	–	–	–	–
Spatial Geographic Services & Applications Ltd	◆	–	–	–	–	–	–	–
SPSS UK Ltd	◆	–	–	–	–	–	–	–
Structural Technologies Ltd (STL)	◆	–	–	–	–	◆	–	–
Survey & Development Services Ltd	–	◆	–	–	–	–	–	–
Svitzer Limited	◆	–	–	–	–	–	–	–
Symology Limited	◆	–	–	–	–	–	–	–
Sysdeco (UK) Ltd	◆	–	–	–	◆	–	–	–
System Options Limited	◆	◆	–	–	–	–	–	–
TACTICIAN UK	◆	–	–	–	–	–	–	–
Target Market Consultancy	◆	–	–	–	–	–	–	–
TEAMS (Taylor Woodrow Electronics Asset Mapping Survey)	◆	◆	–	–	–	◆	–	–
Tele Atlas	◆	–	–	–	–	◆	–	–
Tendron Systems Ltd	◆	◆	–	–	–	–	–	–
TerraQuest Group Limited	◆	–	–	–	–	◆	–	–
The Data Consultancy	◆	–	–	–	–	–	–	–
The NPA Group Ltd	◆	–	–	–	–	–	–	–
Trac Consultancy	◆	–	–	–	–	–	–	–
TYDAC Technologies Ltd	◆	–	–	–	–	–	–	–
UNISYS Ltd	◆	◆	–	◆	◆	◆	◆	–
University of Edinburgh	◆	–	–	–	–	–	–	–
University of Newcastle Upon Tyne	◆	–	–	–	–	–	–	–
WRc (Water Research centre)	◆	–	–	◆	–	◆	–	–
WS Atkins Planning & Management Consultants	◆	◆	–	◆	◆	◆	◆	–

directory 23:
providers of media services

	CDROM authoring/production services	provider of consumables	microfiche production services	microfilm production services	plotting services	printing services
ACDS Graphic System Inc	–	–	–	–	–	♦
AGFA UK	–	♦	♦	–	–	♦
AP³ Imaging Services Limited	–	–	–	–	♦	♦
British Geological Survey	–	–	–	–	♦	♦
CAD - Capture Limited	–	–	–	♦	–	♦
CAD R&D Centre Limited	–	–	–	–	♦	–
CADAC Ltd	–	–	–	–	♦	♦
CAM - Centre for Analysis & Modelling Limited	–	–	–	–	♦	–
Carl Bro Group	–	–	–	–	♦	♦
Cartwright Associates	–	–	–	–	♦	♦
ColourMap Scanning Ltd	–	–	–	–	♦	–
DMAP Ltd	♦	–	–	–	–	–
DMV Consultants BV	–	–	–	–	–	♦
Dolphin Consulting Group	–	–	–	–	♦	–
Dotted Eyes	♦	–	–	–	–	–
Elstree Computing Ltd	–	♦	–	–	–	–
ERA-Maptec Ltd	–	–	–	–	♦	♦
ESR Cartographers Ltd	–	–	–	–	–	♦
Eurosense Technologies N.V.	–	♦	–	–	♦	♦
Fairbairn Services Limited	–	–	–	–	♦	–
Gardline Infotech	–	–	–	–	♦	♦
Geo/SQL (UK)	–	♦	–	–	♦	♦
Geoplan (UK) Ltd	–	–	–	–	♦	♦
GISL Limited	–	♦	–	–	♦	–
Glen Computing Ltd	–	–	–	–	♦	–
Graphical Data Capture Ltd (GDC)	–	–	–	–	♦	–
Grove Projects Ltd	–	–	–	–	♦	–
Intera Information Technologies	–	♦	–	–	♦	♦
International Products	–	♦	–	–	♦	–
IT Southern Ltd	–	♦	♦	♦	♦	♦
J.C.White Chartered Land Surveyors	–	–	–	–	♦	–
Kirstol Ltd	–	–	♦	–	♦	–
L.E.S. (Computer Services) Ltd	–	–	–	–	♦	–
Land Aspects Consultancy Ltd (Parkman Group)	–	–	–	–	♦	–
Lantmatenet GIS-centrum	–	–	–	–	–	♦
Logitrans	–	–	–	–	–	♦
Longdin & Browning	–	–	–	–	♦	–
LOY Surveys Ltd	–	–	–	–	♦	♦
LTG Services	–	♦	–	–	♦	♦
Mason Land Surveys	–	–	–	–	♦	♦
McLintock Limited	–	♦	–	–	♦	–
Ordnance Survey of Northern Ireland	–	–	–	–	♦	♦
Planning & Mapping Ltd	–	–	–	–	♦	♦
Plowman Craven & Associates Ltd	–	–	–	–	♦	–
Raindrop Information Systems Ltd	–	♦	–	–	–	♦

	CDROM authoring/production services	provider of consumables	microfiche production services	microfilm services	plotting services	printing services
Royal Mail - Address Management Centre	–	◆	◆	–	–	–
Saztec Europe Limited	–	–	◆	◆	–	–
Scott Wilson Kirkpatrick	–	–	–	–	◆	◆
Simmons Survey Partnership Limited	–	–	–	–	◆	–
Sovereign C.S. Ltd	–	–	–	–	◆	◆
Spatial Data Limited	–	–	–	–	◆	◆
Summagraphics Europe N.V.	–	◆	–	–	–	–
Survey & Development Services Ltd	–	–	–	–	◆	◆
System Options Limited	–	–	–	–	◆	◆
Tele Atlas	◆	–	–	–	–	–
TerraHunt GeoScience Ltd	–	–	–	–	◆	–
The NPA Group Ltd	–	–	–	–	◆	◆
The Survey Centre	–	◆	–	–	◆	◆
WRc (Water Research centre)	–	–	–	–	◆	–

directory 24:
providers of staff recruitment services

	recruitment consultancy services	contract staff agency	staff selection services	staff search services
AiC Analysts	◆	◆	◆	–
Audifilm Girona S.L.	◆	–	–	–
Business Information Management	–	–	◆	–
Byers Engineering Company	◆	–	–	–
C.A.Design Services Ltd	–	◆	–	–
CAD R&D Centre Limited	◆	–	◆	–
CADAC Ltd	–	◆	–	–
Cambashi Ltd	◆	–	◆	–
CDD Ltd	◆	–	–	–
Concurrent Appointments International	◆	◆	◆	–
Cray Systems	◆	◆	◆	–
Derek Hunter & Partners Ltd	◆	◆	–	–
DM Management Consultants Ltd (DMMC)	◆	–	◆	–
DMV Consultants BV	–	–	◆	–
ECM Selection Limited	◆	–	◆	◆
Fairbairn Services Limited	◆	◆	–	–
Foto Res	◆	–	–	–
Gardline Infotech	◆	◆	◆	–
GEOBASE Consultants Ltd	◆	◆	◆	–
GIS Services Ltd	◆	◆	◆	–
GISL Limited	◆	◆	◆	–
ICL Ltd	◆	◆	◆	–
J.C.White Chartered Land Surveyors	–	◆	–	–
John D Leatherdale FRICS	◆	–	–	–
Kamyco International	◆	–	–	–
Know Edge Ltd	–	–	◆	–
KPMG	◆	–	◆	–
Lantmatenet GIS-centrum	◆	–	–	–
LASCO Ltd	◆	◆	–	–
MAPS geosystems	–	–	◆	–
McLintock Limited	–	–	◆	–
Mentis Management Consultants Ltd	◆	◆	◆	–
Midsummer Computing	–	–	◆	–
Nestor International Ltd	◆	◆	◆	◆
PA Consulting Group	◆	–	–	–
PlanGraphics Inc	◆	◆	◆	–
Recruit Media Ltd	◆	◆	–	–
Sector (UK) Limited	–	◆	–	–

directory 25:
providers of training services

	awareness training	training management services	product training services	training programme services	formal education	masters degree
Active Software Ltd	◆	–	–	–	–	–
Adept Scientific Micro Systems Ltd	–	–	◆	–	–	–
AGFA UK	–	–	◆	◆	–	–
AiC Analysts	◆	◆	–	–	–	–
AM/FM International - European Division	◆	◆	–	–	–	–
APIC Systems	–	–	◆	◆	–	–
ARC Systems Pty Ltd	◆	◆	◆	◆	–	–
Ashtech Europe Limited	–	–	◆	–	–	–
Assist Applications Limited	–	–	◆	–	–	–
Audifilm Girona S.L.	–	◆	◆	–	–	–
Birkbeck College London	◆	–	–	◆	◆	◆
British Geological Survey	–	–	–	◆	–	–
Bull Information Systems Limited	◆	–	–	–	–	–
Business Information Management	◆	◆	–	–	–	–
C.A.Design Services Ltd	–	–	◆	◆	–	–
CACI Limited	–	–	◆	◆	–	–
CAD R&D Centre Limited	–	◆	◆	◆	–	–
CADAC Ltd	–	–	◆	–	–	–
CAM - Centre for Analysis & Modelling Limited	◆	–	◆	–	–	–
Carl Bro Group	–	–	◆	–	–	–
Carl Zeiss Limited	–	–	◆	–	–	–
Cartwright Associates	◆	◆	◆	◆	–	–
CATALIST	–	–	◆	–	–	–
CCN Marketing	–	–	◆	–	–	–
CDR Group	–	–	◆	–	–	–
Chroson Ltd	◆	–	–	–	–	–
Citywise	◆	–	◆	–	–	–
Cobham Digital Services Limited	–	–	◆	◆	–	–
Computer Aided Development (CADCORP) Ltd	–	–	◆	–	–	–
Construction Industry Computing Association	◆	◆	–	–	–	–
Cray Systems	◆	◆	◆	◆	–	–
Dataflow Information Systems	–	–	◆	–	–	–
Datatechnology Datech Ltd	–	–	◆	–	–	–
Dataview Solutions Ltd	–	–	◆	–	–	–
DCL Consulting	◆	–	–	–	–	–
DM Management Consultants Ltd (DMMC)	◆	◆	–	–	–	–
DMV Consultants BV	–	◆	–	–	–	–
Dotted Eyes	◆	◆	◆	◆	–	–
Dowling Associates Limited	–	–	◆	◆	–	–
Dr Stanley Port	◆	◆	–	–	–	–
Earth Resource Mapping	–	–	◆	–	–	–
Elstree Computing Ltd	–	–	◆	–	–	–
Enghouse (UK) Limited	◆	◆	◆	◆	–	–
EPS - Essential Planning Systems Limited	–	–	◆	–	–	–
ERA Technology Ltd	–	–	◆	–	–	–

	awareness training	training management services	product training services	training programme services	formal education	masters degree
ERA-Maptec Ltd	–	–	◆	–	–	–
ERDAS (UK) Ltd	–	–	◆	–	–	–
ERTEC	◆	–	–	–	–	–
ESRI (UK) Ltd	◆	◆	◆	◆	–	–
Eurosense Technologies N.V.	◆	–	◆	–	–	–
Fairbairn Services Limited	◆	–	–	–	–	–
FastCAD GIS Ltd	–	–	◆	–	–	–
FileNet Ltd	◆	◆	◆	◆	–	–
Foto Res	–	–	–	◆	–	–
Gamma Ltd	–	◆	◆	◆	–	–
Gardline Infotech	◆	◆	◆	–	–	–
Genasys II Limited	–	–	–	◆	–	–
Geo/SQL (UK)	◆	◆	◆	◆	–	–
Geo2 Consulting	◆	◆	–	–	–	–
GEOBASE Consultants Ltd	◆	◆	–	–	–	–
GeoData Institute	◆	–	◆	–	–	◆
Geodelta	◆	–	◆	◆	–	–
Geographic Management Solutions Ltd	◆	–	–	–	–	–
Geoinformation International	◆	◆	–	◆	–	–
GeoMEM Software	–	–	◆	–	–	–
Geometria GIS Systems House Ltd	–	–	◆	–	–	–
Geops BV	–	◆	◆	–	–	–
GIS Services Ltd	◆	◆	–	◆	–	–
GISDATA	◆	–	–	–	–	–
GISL Limited	◆	–	◆	◆	–	–
Glen Computing Ltd	◆	◆	◆	◆	–	–
Graphic Data Systems Corporation (GDS)	–	–	◆	–	–	–
Graphical Data Capture Ltd (GDC)	–	–	◆	–	–	–
H R Wallingford Ltd	–	–	◆	–	–	–
Hansa Luftbild	–	–	◆	–	–	–
Hunting Technical Services Ltd	◆	–	–	–	–	–
Husky Computing Limited	–	–	◆	–	–	–
ICL Ltd	◆	◆	◆	◆	–	–
IMASS Limited	◆	◆	–	◆	–	–
Informed Solutions Limited	◆	◆	◆	–	–	–
Ingecon B.V.	◆	◆	◆	–	–	–
Intera Information Technologies	◆	–	◆	–	–	–
Intergraph (UK) Ltd	◆	–	◆	–	–	–
International Products	–	–	◆	–	–	–
ITC	–	–	◆	–	–	–
John D Leatherdale FRICS	◆	◆	–	–	–	–
Keele University	–	–	–	◆	◆	◆
Kingswood Consulting Limited	◆	◆	◆	–	–	–
KJB Consulting	◆	◆	–	–	–	–
Know Edge Ltd	◆	◆	◆	–	–	–

	awareness training	training management services	product training services	training programme services	formal education	masters degree
KPMG	◆	◆	–	–	–	–
L.E.S. (Computer Services) Ltd	–	–	–	◆	–	–
Lancaster University	◆	–	–	–	–	–
Lantmatenet GIS-centrum	–	–	◆	–	–	–
Laser-Scan Ltd	–	–	◆	–	–	–
Logitrans	–	–	◆	–	–	–
London Research Centre	–	–	◆	–	–	–
LTG Services	–	–	◆	–	–	–
Magdala Sociedade	◆	◆	–	–	–	–
Manchester Metropolitan University	–	–	–	–	◆	◆
Map Data Management Ltd	–	–	◆	–	–	–
MAPS geosystems	–	–	◆	◆	–	–
Mason Land Surveys	◆	–	◆	–	–	–
McLintock Limited	–	◆	–	◆	–	–
Mentis Management Consultants Ltd	◆	◆	–	◆	–	–
Midlands Regional Research Laboratory	◆	–	◆	◆	–	–
Midsummer Computing	◆	◆	◆	◆	–	–
Modern Maps	◆	–	–	–	–	–
Munro Garratt International	–	–	◆	–	–	–
MVA Systematica	–	–	◆	–	–	–
National Remote Sensing Centre Ltd (NRSC)	◆	◆	◆	◆	–	–
NOMIS	–	–	◆	◆	–	–
Optimal Software Ltd	◆	–	◆	◆	–	–
Ordnance Survey	◆	–	◆	–	–	–
Ordnance Survey of Northern Ireland	◆	–	–	–	–	–
Organisation Management Systems	–	–	–	◆	–	–
Oscar Faber	–	–	◆	–	–	–
Oxford Institute of Retail Management	◆	◆	–	–	–	–
PAFEC Ltd	◆	◆	◆	–	–	–
Peter Thorpe Consulting	◆	◆	–	–	–	–
PlanGraphics Inc	◆	◆	◆	◆	–	–
PLANTECH Ltd	–	–	◆	–	–	–
Positioning Resources Limited	–	–	◆	–	–	–
Procis Software Ltd	–	–	◆	–	–	–
Progis GmbH	–	–	◆	–	–	–
Racal Survey (UK) Ltd	–	–	◆	–	–	–
Raindrop Information Systems Ltd	–	–	–	◆	–	–
Recruit Media Ltd	–	–	–	◆	–	–
regioplanDATA GmbH	◆	–	◆	–	–	–
Remote Sensing Applications Consultants	–	–	–	◆	–	–
Royal Geographical Society (with The Institute of British Geographers)	◆	–	–	–	–	–
Royal Institute of Navigation	◆	–	–	–	–	–
Salford University Business Services Ltd	◆	◆	◆	◆	–	–
SAS Institute	–	–	◆	–	–	–
SCOT Conseil	◆	–	–	–	–	–

	awareness training	training management services	product training services	training programme services	formal education	masters degree
SDI Ltd	–	–	◆	–	–	–
Sector (UK) Limited	–	◆	◆	–	–	–
SHL Vision* Solutions Limited	◆	◆	◆	◆	–	–
SIA Limited	◆	◆	◆	◆	–	–
Siemens Nixdorf	–	–	◆	◆	–	–
Silsoe College, Cranfield University	◆	–	–	◆	–	–
Southbank Systems PLC	–	–	◆	–	–	–
Sovereign C.S. Ltd	–	–	◆	–	–	–
StatSci Europe	–	–	◆	–	–	–
Structural Technologies Ltd (STL)	–	–	◆	–	–	–
Sun Microsystems Ltd	–	–	◆	–	–	–
Survey & Development Services Ltd	–	–	◆	–	–	–
Symology Limited	–	–	◆	–	–	–
Sysdeco (UK) Ltd	–	–	◆	–	–	–
TACTICIAN UK	–	–	◆	◆	–	–
Tenet Systems Ltd	◆	–	◆	–	–	–
The Data Consultancy	–	–	◆	◆	–	–
The LGMB	◆	–	–	–	–	–
The Survey Centre	–	–	–	◆	–	–
Trimble Navigation Europe Ltd	–	–	–	◆	–	–
TYDAC Technologies Ltd	◆	◆	◆	◆	–	–
UNISYS Ltd	◆	◆	◆	◆	–	–
Universal Systems Ltd	–	–	◆	–	–	–
University of East London	–	–	–	–	◆	–
University of Edinburgh	◆	–	–	–	◆	◆
University of Greenwich	–	–	–	–	◆	◆
University of Leicester	–	–	–	–	◆	◆
University of Luton	–	–	–	◆	–	–
University of Newcastle Upon Tyne	–	–	–	◆	–	–
Wallingford Software Limited	–	–	◆	–	–	–
WRc (Water Research centre)	◆	◆	–	–	–	–
WS Atkins Planning & Management Consultants	◆	◆	◆	◆	–	–

directory 26:

conformance to project management methodology

	PRINCE	SSADM	PRISM	CRAMM
Bull Information Systems Limited	◆	◆	–	–
CACI Limited	◆	◆	–	–
Cambridge Computer Consultants (UK) Ltd	◆	◆	◆	–
Carl Bro Group	–	◆	–	–
CMG Computer Management Group (UK) Ltd	◆	◆	◆	◆
Construction Industry Computing Association	◆	◆	◆	◆
Coopers & Lybrand	◆	◆	–	◆
Corbins Consultancy	◆	–	–	–
Cray Systems	◆	◆	–	–
DM Management Consultants Ltd (DMMC)	◆	–	–	–
DMAP Ltd	◆	◆	–	–
Enghouse (UK) Limited	–	◆	–	–
ESRI (UK) Ltd	◆	–	–	–
Fairbairn Services Limited	◆	◆	–	–
G.L. Consulting Ltd	◆	◆	–	–
Gamma Ltd	–	◆	–	–
Geo/SQL (UK)	–	◆	–	–
GEOBASE Consultants Ltd	◆	◆	–	–
Geoview Systems Kft	–	◆	–	–
GIS Services Ltd	◆	◆	–	–
Hunting Engineering Ltd	◆	◆	–	–
ICL Ltd	◆	◆	–	–
IMASS Limited	◆	◆	–	–
Informed Solutions Limited	◆	◆	–	–
Intergraph (UK) Ltd	◆	◆	–	–
IT Southern Ltd	–	◆	–	–
Kamyco International	–	◆	–	–
Kingswood Consulting Limited	◆	◆	–	–
Know Edge Ltd	◆	◆	–	–
KPMG	–	◆	◆	◆
Land Aspects Consultancy Ltd (Parkman Group)	–	◆	–	–
Logica UK Limited	◆	◆	◆	◆
Longdin & Browning	◆	–	–	–
McLintock Limited	◆	◆	–	–
Mentis Management Consultants Ltd	◆	◆	◆	◆
Methods Applications Ltd	◆	◆	◆	◆
Midsummer Computing	◆	–	–	–
Modern Maps	◆	–	–	–
Mott MacDonald Ltd	–	◆	–	–
Munro Garratt International	–	◆	–	–
MVM Consultants plc	◆	◆	–	–
National Remote Sensing Centre Ltd (NRSC)	◆	◆	–	–
NCC Blackwell Ltd	◆	◆	–	–
NERC	◆	–	–	–
Ordnance Survey	◆	–	–	–

	PRINCE	SSADM	PRISM	CRAMM
Ove Arup & Partners	–	◆	–	–
PA Consulting Group	◆	◆	–	◆
PAFEC Ltd	◆	◆	◆	◆
Scott Wilson Kirkpatrick	–	◆	–	–
Sector (UK) Limited	◆	◆	◆	–
SIAS Limited	◆	◆	–	–
Smith System Engineering Ltd	◆	◆	–	–
Symology Limited	◆	–	–	–
System Options Limited	◆	◆	–	–
Tendron Systems Ltd	–	◆	–	–
TerraQuest Group Limited	◆	◆	–	–
UNISYS Ltd	◆	◆	–	–
WRc (Water Research centre)	◆	◆	–	–
WS Atkins Planning & Management Consultants	◆	◆	–	◆

directory 27:
conformance to standards by suppliers

	BS7567	BS7666 Part 1	BS7666 Part 2	BS7666 Part 3	BS7666 Part 4 (Draft DC95/643511)	DIGEST	EDIGeO	SDTS	NJUG13	other standards
A.Rutherford Ltd	♦	♦	♦	♦	♦	–	–	–	–	–
AGFA UK	♦	–	–	–	–	–	–	–	–	–
APIC Systems	–	–	–	–	–	♦	–	–	–	–
ARC Systems Pty Ltd	–	–	–	–	–	–	–	♦	–	–
Assist Applications Limited	♦	–	–	–	–	–	–	–	–	–
CAD R&D Centre Limited	–	–	–	–	–	–	–	–	–	♦
Carl Bro Group	–	♦	♦	♦	–	–	–	–	–	–
CODEC Facilities Limited	–	–	–	–	–	♦	–	–	–	–
Computer Aided Development (CADCORP) Ltd	♦	–	–	–	–	–	–	–	–	–
Cray Systems	♦	–	–	–	–	–	–	–	–	–
CZ Scientific Instruments Ltd	–	–	–	–	–	–	–	–	–	♦
Dataflow Information Systems	–	♦	–	–	–	–	–	–	–	–
Derek Hunter & Partners Ltd	♦	–	–	–	–	–	–	–	–	–
DMAP Ltd	–	–	–	–	–	♦	–	–	–	–
ESRI (UK) Ltd	♦	♦	♦	♦	–	♦	–	♦	–	–
Eurosense Technologies N.V.	–	–	–	–	–	♦	–	–	–	–
GEOBASE Consultants Ltd	♦	♦	♦	♦	♦	–	–	–	♦	–
GID Ltd	♦	–	–	–	–	–	–	–	–	–
GIS Services Ltd	–	♦	♦	♦	–	–	–	–	♦	–
Grove Projects Ltd	–	–	–	–	–	–	–	–	♦	–
Hitachi Home Electronics (Europe) Limited	–	–	–	–	–	–	–	–	–	♦
Kamyco International	–	–	–	–	–	–	–	♦	–	–
Land Aspects Consultancy Ltd (Parkman Group)	♦	♦	♦	♦	–	–	–	–	–	♦
Longdin & Browning	♦	–	–	–	–	–	–	–	♦	–
LTG Services	♦	♦	–	–	–	–	–	–	♦	–
Manchester Metropolitan University	–	–	–	–	–	–	–	–	–	–
Mott MacDonald Ltd	–	–	–	–	–	–	–	–	–	♦
Numonics UK (Division of Telmtek Ltd)	–	–	–	–	–	–	–	–	–	♦
Ordnance Survey	♦	–	–	–	–	–	–	–	♦	–
Ordnance Survey of Northern Ireland	♦	–	–	–	–	–	–	–	–	–
PD Computing Ltd	♦	–	–	–	–	–	–	–	–	♦
QC Data (Ireland) Limited	♦	–	–	–	–	–	–	–	–	–
regioplanDATA GmbH	–	–	–	–	–	–	–	–	–	♦
Sokkia Ltd	♦	–	–	–	–	–	–	–	–	–
Star Informatic S.A.	–	–	–	–	–	♦	–	–	–	–
Structural Technologies Ltd (STL)	♦	–	–	–	–	–	–	–	–	–
Survey & Development Services Ltd	–	–	–	–	–	–	–	–	♦	–
Symology Limited	–	♦	♦	♦	–	–	–	–	–	–
Sysdeco (UK) Ltd	♦	♦	–	–	–	–	–	–	–	–
System Options Limited	♦	♦	–	–	–	♦	♦	♦	♦	–
Tele Atlas	–	–	–	–	–	–	–	–	–	♦
TerraQuest Group Limited	–	–	–	–	–	–	–	–	–	♦
UNISYS Ltd	♦	–	–	–	–	–	–	♦	♦	–
Universal Systems Ltd	–	–	–	–	–	♦	–	–	–	–
WRc (Water Research centre)	♦	♦	♦	♦	–	–	–	–	–	♦

directory 28:
suppliers quality accredited

	BS5750	ISO9000	TickIT	AQAP	other accreditation
A.L.Downloading Services	◆	◆	–	–	–
AGFA UK	◆	–	–	–	–
ARC Systems Pty Ltd	–	◆	–	–	–
Autodesk Ltd	◆	◆	◆	–	–
Babtie Shaw & Morton Limited	◆	–	–	–	–
Binnie Black & Veatch	◆	–	–	–	–
Birkbeck College London	–	–	–	–	◆
BKS Surveys Ltd	◆	◆	–	–	–
Bull Information Systems Limited	◆	◆	–	–	–
Byers Engineering Company	–	◆	–	–	–
CAD - Capture Limited	◆	◆	–	–	–
CAD R&D Centre Limited	–	–	–	–	◆
CalComp Limited	◆	◆	–	–	–
Carl Bro Group	◆	◆	–	–	–
Carl Zeiss Limited	◆	◆	–	–	–
Cartographical Services (Southampton) Limited	–	◆	–	–	–
CDR Group	–	◆	–	–	–
CMG Computer Management Group (UK) Ltd	◆	◆	◆	–	–
Coopers & Lybrand	◆	–	–	–	–
Cray Systems	–	◆	–	–	–
Datatechnology Datech Ltd	–	◆	–	–	–
Digital Equipment Corporation	◆	◆	–	–	–
DM Management Consultants Ltd (DMMC)	–	–	–	–	◆
DMV Consultants BV	–	◆	–	–	–
Earth Observation Sciences Ltd	–	◆	◆	◆	–
Elstree Computing Ltd	◆	–	–	–	–
FastCAD GIS Ltd	◆	–	–	–	–
Flynn & Rothwell	◆	–	–	–	–
Gardline Infotech	◆	◆	–	–	–
GEC Marconi Research Centre	◆	◆	–	–	–
Genasys II Limited	–	–	◆	–	–
Geotronics Limited	–	◆	–	–	–
Glen Computing Ltd	◆	◆	–	–	◆
GMAP Ltd	–	◆	◆	–	◆
Grove Projects Ltd	◆	–	–	–	–
Hitachi Home Electronics (Europe) Limited	◆	◆	–	–	–
HMSO Books	◆	–	–	–	–
HollyBush Software Limited	◆	–	–	–	–
Hunting Engineering Ltd	–	◆	–	◆	–
Husky Computing Limited	◆	◆	–	◆	–
IBM UK Ltd	◆	◆	–	–	–
ICL Ltd	◆	◆	–	–	–
IMASS Limited	◆	◆	–	–	–
Intergraph (UK) Ltd	◆	◆	◆	–	–
IT Southern Ltd	–	◆	◆	–	–

	BS5750	ISO9000	TickIT	AQAP	other accreditation
Kalidor Europe (ALPS Electric (Ireland) Ltd)	–	◆	–	–	–
Kamyco International	–	◆	–	–	–
KPMG	◆	◆	◆	–	–
Land Aspects Consultancy Ltd (Parkman Group)	◆	–	–	–	–
Laser-Scan Ltd	–	◆	◆	–	–
Leica UK Ltd	◆	◆	–	–	–
Logica UK Limited	–	◆	◆	–	–
Longdin & Browning	◆	–	–	–	–
LOY Surveys Ltd	◆	◆	–	–	–
LTG Services	◆	–	–	–	–
Mason Land Surveys	◆	–	–	–	–
Methods Applications Ltd	–	◆	–	–	–
MOSS Systems Limited	◆	–	◆	–	–
Mott MacDonald Ltd	–	–	–	–	–
MVM Consultants plc	◆	◆	–	–	–
National Remote Sensing Centre Ltd (NRSC)	–	◆	–	–	–
Numonics UK (Division of Telmtek Ltd)	◆	◆	–	–	–
Optimal Software Ltd	◆	◆	–	◆	–
Oscar Faber	◆	–	–	–	–
PA Consulting Group	◆	◆	–	◆	–
Plowman Craven & Associates Ltd	◆	–	–	–	–
Posford Duvivier	◆	–	–	–	–
Primagraphics Ltd	◆	◆	◆	–	–
QC Data (Ireland) Limited	–	◆	–	–	–
SAS Institute	◆	◆	◆	–	–
SAZTEC Philippines Inc	–	◆	–	–	–
Scott Wilson Kirkpatrick	◆	◆	–	–	–
Siemens Nixdorf	◆	◆	–	–	–
Silicon Graphics Limited	◆	◆	–	–	–
Simmons Survey Partnership Limited	◆	–	–	–	–
Sir Alexander Gibb & Partners Ltd	◆	–	–	–	–
Sir William Halcrow and Partners	◆	–	–	–	–
Smith System Engineering Ltd	◆	◆	◆	◆	–
Sokkia Ltd	◆	–	–	–	–
Southbank Systems PLC	◆	–	–	–	–
SPSS UK Ltd	◆	◆	–	–	–
Survey Control Services	◆	–	–	–	–
Survey & Development Services Ltd	◆	◆	–	–	–
Survey Supplies Ltd	◆	–	–	–	–
Svitzer Limited	–	◆	–	–	–
System Options Limited	◆	–	–	–	–
TDS-CAD Graphics Ltd	–	◆	–	–	–
TEAMS (Taylor Woodrow Electronics Asset Mapping Survey)	◆	–	–	–	–
Tekla Oy	–	◆	–	–	–
Tele Atlas	–	◆	–	–	–

	BS5750	ISO9000	TickIT	AQAP	other accreditation
TerraQuest Group Limited	◆	◆	–	–	–
The Severn Partnership	◆	◆	–	–	–
Trident Map Services	◆	–	–	–	–
UNISYS Ltd	◆	◆	–	–	–
University of Edinburgh	–	–	–	–	◆
WRc (Water Research centre)	◆	–	–	–	–
WS Atkins Planning & Management Consultants	◆	–	◆	–	–

directory 29:

off-the-shelf local government applications

	land charges application	planning applications application	building control application	asset management application	estate management application	highway management application	land terrier application	parks and gardens applications	contaminated land applications
AP³ Imaging Services Limited	–	–	–	–	–	–	–	–	◆
ARC Systems Pty Ltd	◆	◆	–	◆	◆	–	–	–	–
Assist Applications Limited	◆	◆	◆	◆	◆	◆	◆	◆	◆
Audifilm Girona S.L.	◆	◆	◆	–	◆	–	–	–	–
CACI Limited	–	◆	–	◆	◆	–	–	–	–
CAD R&D Centre Limited	◆	◆	–	–	–	–	◆	–	◆
CADAC Ltd	–	–	–	◆	◆	–	–	–	–
CAM - Centre for Analysis & Modelling Limited	–	◆	–	◆	–	–	–	–	–
Carl Bro Group	◆	–	–	–	–	–	–	–	–
CDR Group	–	–	–	–	–	–	–	–	–
Citywise	–	–	–	◆	–	–	–	–	–
Cobham Digital Services Limited	–	–	–	◆	◆	–	◆	◆	◆
CODEC Facilities Limited	◆	–	–	◆	◆	◆	◆	–	–
Cray Systems	–	–	–	◆	–	–	◆	◆	–
Dataflow Information Systems	–	◆	◆	–	–	–	–	–	–
Datatechnology Datech Ltd	–	–	–	◆	◆	◆	–	◆	–
Derek Hunter & Partners Ltd	–	–	–	–	–	–	–	◆	–
Design Computer Aids Limited (DeCAL)	–	–	–	◆	◆	–	◆	–	–
DMV Consultants BV	◆	–	–	–	–	–	–	–	–
Dotted Eyes	–	–	–	–	◆	–	–	–	–
Dowling Associates Limited	◆	◆	–	–	–	–	–	–	–
EA Technology	–	–	–	◆	–	–	–	–	–
Earth Resource Mapping	◆	◆	–	–	–	–	–	–	◆
Elstree Computing Ltd	–	–	–	◆	◆	–	–	–	◆
Enghouse (UK) Limited	–	–	–	◆	–	–	–	–	–
ESRI (UK) Ltd	◆	◆	◆	◆	◆	◆	◆	◆	◆
Eurosense Technologies N.V.	◆	◆	–	–	◆	◆	◆	◆	◆
FastCAD GIS Ltd	◆	◆	◆	–	◆	◆	–	◆	◆
FastCAD GIS Ltd	◆	◆	–	–	◆	◆	–	◆	◆
Foto Res	–	–	–	–	–	–	–	◆	–
Geo-Perfect TWI B.V.	–	◆	◆	◆	–	–	–	◆	◆
Geo/SQL (UK)	◆	◆	–	◆	◆	–	◆	–	◆
Geodelta	◆	◆	–	–	–	–	–	–	–
Geoinformation International	–	◆	–	–	–	–	–	–	◆
Geops BV	–	–	–	–	–	–	–	◆	–
GEOSOFT Ltd	–	◆	–	◆	◆	◆	◆	◆	◆
Glen Computing Ltd	◆	◆	–	◆	◆	◆	◆	◆	◆
Graphic Data Systems Corporation (GDS)	◆	◆	–	◆	◆	◆	◆	◆	◆
Graphite Management Services Ltd	–	◆	–	◆	◆	–	◆	◆	◆
Grove Projects Ltd	–	◆	–	◆	◆	◆	◆	◆	◆
Hunting Engineering Ltd	–	–	–	◆	◆	–	–	◆	◆
Husky Computing Limited	–	–	–	◆	◆	–	◆	–	–
I.S. Ltd	–	◆	–	–	◆	◆	–	◆	◆
ICL Ltd	◆	◆	–	◆	◆	◆	◆	◆	–
IMASS Limited	–	–	–	◆	–	–	◆	–	◆
IME (UK) Ltd	–	◆	–	–	◆	–	◆	◆	◆
Intergraph (UK) Ltd	–	–	–	◆	◆	◆	◆	–	◆
Kamyco International	–	◆	–	–	–	◆	◆	–	–

	land charges application	planning applications application	building control application	asset management application	estate management application	highway management application	land terrier application	parks and gardens applications	contaminated land applications
Land Aspects Consultancy Ltd (Parkman Group)	–	–	–	◆	◆	◆	◆	–	◆
Laser-Scan Ltd	◆	◆	–	–	◆	–	◆	–	–
Logica UK Limited	–	–	–	◆	–	–	–	–	–
Logitrans	–	◆	–	–	–	–	–	–	–
Longdin & Browning	–	–	–	–	–	◆	◆	◆	–
MapInfo Ltd	◆	◆	–	◆	◆	◆	◆	–	◆
MAPIT Limited	–	–	–	–	–	–	–	–	◆
McLintock Limited	◆	◆	◆	–	–	–	–	–	–
Midsummer Computing	–	◆	–	◆	◆	–	◆	–	–
Morgan Collis Group Ltd	–	–	–	◆	–	–	–	–	–
Mott MacDonald Ltd	–	–	–	◆	–	◆	–	–	◆
MPSI Systems Ltd	–	◆	–	◆	–	–	–	–	–
Navstar Systems Ltd	–	–	–	◆	◆	–	–	–	–
Optimal Software Ltd	–	–	–	◆	–	–	–	–	–
Organisation Management Systems	–	–	–	◆	◆	–	◆	–	–
Oscar Faber	–	–	–	◆	–	◆	–	–	–
Ove Arup & Partners	–	◆	–	–	–	◆	◆	–	◆
PAFEC Ltd	–	–	–	◆	–	–	–	–	–
PAX Technology	◆	–	–	◆	◆	–	◆	◆	–
PLANTECH Ltd	◆	◆	◆	–	–	–	–	–	◆
Positioning Resources Limited	–	–	–	–	◆	–	–	–	◆
Procis Software Ltd	–	–	–	–	–	◆	–	–	–
Property Intelligence plc	–	◆	–	◆	◆	–	–	–	–
Racal Survey (UK) Ltd	–	–	–	◆	–	◆	–	–	–
Raindrop Information Systems Ltd	–	–	–	◆	◆	–	–	–	–
regioplanDATA GmbH	◆	◆	–	◆	◆	–	–	–	–
SDI Ltd	–	–	–	◆	◆	◆	◆	–	◆
SIA Limited	◆	◆	◆	◆	◆	◆	◆	◆	–
Siemens Nixdorf	◆	◆	–	◆	–	–	◆	◆	–
Simmons Survey Partnership Limited	–	–	–	◆	◆	–	◆	–	–
Sir William Halcrow and Partners	–	–	–	–	–	–	–	–	◆
Southbank Systems PLC	–	–	–	◆	–	◆	–	◆	–
Star Informatic S.A.	◆	–	–	◆	◆	◆	–	–	–
Structural Technologies Ltd (STL)	–	–	–	◆	–	–	–	–	–
Survey Supplies Ltd	–	◆	–	–	◆	◆	◆	–	–
Symology Limited	–	◆	–	–	–	◆	–	◆	–
Sysdeco (UK) Ltd	◆	◆	◆	◆	◆	◆	◆	◆	◆
System Options Limited	–	◆	–	◆	–	◆	◆	◆	◆
Tangent Technology Design Associates Ltd	–	–	–	◆	–	–	◆	–	–
TEAMS (Taylor Woodrow Electronics Asset Mapping Survey)	–	–	–	◆	–	–	◆	–	–
Tekla Oy	–	–	–	–	–	◆	–	–	–
Trimble Navigation Europe Ltd	–	–	–	◆	–	–	–	–	◆
UNISYS Ltd	◆	–	–	–	–	◆	◆	–	◆
Universal Systems Ltd	–	◆	–	◆	◆	◆	–	◆	◆
University of Newcastle Upon Tyne	–	◆	–	◆	–	◆	◆	◆	◆
Wallingford Software Limited	–	–	–	◆	–	–	–	–	◆
WRc (Water Research centre)	–	–	–	–	–	–	–	–	◆
WS Atkins Planning & Management Consultants	–	–	–	◆	◆	◆	◆	–	–

directory 30:
off-the-shelf utility applications

	computerised street works support application	water distribution network application	sewer catchment network application	sludge disposal application	electricity distribution network application	gas distribution network application	telecommunications network application	cable services network application
ACDS Graphic System Inc	–	◆	◆	–	◆	◆	–	◆
Anglian Engineering & International Consultancy	◆	◆	–	–	–	◆	–	–
AP³ Imaging Services Limited	–	–	–	–	–	–	◆	◆
ARC Systems Pty Ltd	–	◆	◆	–	◆	◆	◆	◆
Assist Applications Limited	◆	–	◆	–	–	–	–	–
Beacon Dodsworth Limited	–	–	–	–	◆	–	◆	–
CACI Limited	–	–	–	–	–	◆	◆	◆
CAD R&D Centre Limited	◆	◆	–	–	◆	–	◆	◆
CADAC Ltd	◆	◆	◆	–	◆	◆	◆	◆
Carl Bro Group	–	◆	–	–	–	–	–	–
CDR Group	–	◆	◆	–	◆	◆	◆	◆
CMG Computer Management Group (UK) Ltd	–	–	–	–	◆	–	–	–
Cobham Digital Services Limited	–	◆	◆	–	–	–	–	–
CODEC Facilities Limited	◆	◆	◆	–	◆	◆	◆	◆
Corena A/S	–	–	–	–	–	–	◆	◆
Cray Systems	–	◆	◆	–	–	–	–	–
Dataview Solutions Ltd	–	–	–	–	–	–	◆	◆
Derek Hunter & Partners Ltd	–	–	◆	–	–	–	–	–
DMV Consultants BV	–	◆	–	–	–	–	–	–
EA Technology	–	–	–	–	◆	–	◆	–
Earth Resource Mapping	–	–	–	–	–	–	◆	–
Elstree Computing Ltd	–	–	–	–	–	–	–	◆
Enghouse (UK) Limited	–	◆	◆	◆	◆	◆	◆	◆
ERDAS (UK) Ltd	–	–	–	–	–	–	◆	–
ESRI (UK) Ltd	◆	◆	◆	◆	◆	◆	◆	◆
Eurosense Technologies N.V.	–	◆	◆	–	◆	◆	◆	◆
FastCAD GIS Ltd	◆	◆	◆	◆	–	–	–	–
FastCAD GIS Ltd	◆	–	◆	◆	–	–	–	◆
Geo-Perfect TWI B.V.	–	◆	◆	–	◆	◆	◆	–
Geo/SQL (UK)	–	–	◆	–	–	◆	◆	◆
Geoinformation International	◆	–	–	–	–	–	◆	–
GEOSOFT Ltd	–	◆	◆	◆	◆	◆	◆	◆
Geosystems	–	◆	◆	◆	–	–	–	–
Geoview Systems Kft	–	–	–	–	◆	◆	–	–
Glen Computing Ltd	◆	◆	◆	–	◆	◆	◆	–
Graphic Data Systems Corporation (GDS)	◆	◆	◆	◆	◆	◆	◆	◆
Grove Projects Ltd	–	◆	◆	–	◆	◆	◆	◆
Husky Computing Limited	–	◆	◆	–	◆	◆	◆	–
I.S. Ltd	–	–	–	◆	–	◆	–	–
ICL Ltd	–	–	–	–	–	◆	–	–
IMASS Limited	–	◆	◆	◆	◆	◆	◆	◆
Intergraph (UK) Ltd	◆	◆	◆	–	◆	◆	◆	◆
Land Aspects Consultancy Ltd (Parkman Group)	◆	◆	–	–	–	–	–	–
Laser-Scan Ltd	–	–	–	–	◆	◆	◆	◆
LiveChart	–	–	–	–	–	–	–	◆

	computerised street works support application	water distribution network application	sewer catchment network application	sludge disposal application	electricity distribution network application	gas distribution network application	telecommunications network application	cable services network application
Logica UK Limited	◆	◆	◆	–	◆	◆	◆	–
Longdin & Browning	–	◆	–	–	◆	◆	–	–
Macaulay Land Use Research Institute	–	–	–	◆	–	–	–	–
MapInfo Ltd	◆	◆	◆	◆	◆	◆	◆	◆
Morgan Collis Group Ltd	–	◆	◆	–	–	–	–	–
Mott MacDonald Ltd	–	◆	◆	–	◆	◆	◆	◆
Navstar Systems Ltd	–	–	–	–	◆	◆	◆	–
Optimal Software Ltd	–	–	–	–	◆	–	–	–
Oscar Faber	–	◆	◆	◆	–	–	–	–
Ove Arup & Partners	–	–	◆	–	–	–	–	–
PAX Technology	–	–	–	–	–	–	◆	–
Procis Software Ltd	–	◆	◆	–	◆	◆	–	◆
SDI Ltd	–	–	–	–	–	◆	◆	–
SIA Limited	◆	–	–	–	–	–	–	–
Siemens Nixdorf	–	◆	◆	–	◆	◆	◆	◆
Smallworld Systems Ltd	–	◆	–	–	◆	–	◆	◆
Star Informatic S.A.	◆	◆	◆	–	◆	◆	◆	◆
Structural Technologies Ltd (STL)	–	◆	◆	◆	–	–	–	–
Survey Supplies Ltd	◆	–	–	–	–	–	–	–
Symology Limited	◆	–	–	–	–	–	–	–
Sysdeco (UK) Ltd	–	◆	–	–	◆	◆	–	◆
System Options Limited	–	◆	◆	◆	◆	◆	◆	◆
TACTICIAN UK	–	–	–	–	◆	◆	◆	–
TEAMS (Taylor Woodrow Electronics Asset Mapping Survey)	◆	◆	◆	–	◆	◆	◆	◆
Tekla Oy	◆	–	–	–	◆	–	–	◆
Trimble Navigation Europe Ltd	–	–	◆	–	◆	◆	◆	–
UNISYS Ltd	–	–	–	–	◆	◆	◆	◆
Universal Systems Ltd	–	◆	◆	–	◆	◆	◆	–
Wallingford Software Limited	–	◆	◆	–	–	–	–	–
WRc (Water Research centre)	–	◆	◆	◆	–	–	◆	–
WS Atkins Planning & Management Consultants	–	◆	–	–	–	–	–	–

directory 31:
off-the-shelf demographic applications

	census data application	geo-demographic application	market profiling application	market penetration application	sales forecasting application	site location application	crime analysis application
Action Information Management Ltd	–	–	–	–	–	–	◆
Active Software Ltd	◆	–	◆	–	◆	–	–
Assist Applications Limited	◆	◆	–	–	–	–	–
Beacon Dodsworth Limited	◆	◆	◆	◆	◆	◆	–
CACI Limited	◆	◆	◆	◆	◆	◆	◆
CAD R&D Centre Limited	–	–	–	–	–	–	◆
CAM - Centre for Analysis & Modelling Limited	◆	◆	◆	◆	–	–	–
Cartwright Associates	–	–	–	–	–	◆	–
CCN Marketing	◆	◆	◆	–	–	–	–
CDR Group	–	–	–	–	–	◆	–
Citywise	–	–	◆	–	–	–	–
Cobham Digital Services Limited	–	–	–	–	–	–	◆
Consensus Information Technology Ltd	–	–	◆	◆	◆	–	–
Dataview Solutions Ltd	◆	◆	◆	◆	◆	–	–
Derek Hunter & Partners Ltd	–	–	–	–	–	–	◆
Dolphin Consulting Group	◆	◆	◆	–	–	–	–
Dotted Eyes	–	◆	◆	–	–	–	–
Enghouse (UK) Limited	–	–	–	◆	◆	–	–
ESRI (UK) Ltd	◆	◆	◆	◆	◆	◆	◆
EuroDirect Database Marketing Ltd	◆	◆	–	–	–	–	–
Foto Res	◆	◆	–	–	–	–	–
Gamma Ltd	◆	–	–	–	–	–	–
GEO-Marketing Systems Ltd (GMSL)	◆	◆	◆	◆	◆	◆	–
Geo-Perfect TWI B.V.	–	◆	◆	◆	–	◆	◆
Geoinformation International	–	◆	–	–	–	◆	–
GeoMEM Software	◆	◆	–	◆	◆	–	–
Geoplan (UK) Ltd	◆	◆	◆	◆	◆	◆	◆
GEOSOFT Ltd	◆	◆	◆	◆	◆	◆	–
Glen Computing Ltd	◆	◆	◆	◆	–	◆	◆
GMAP Ltd	◆	–	◆	◆	◆	◆	–
Graphic Data Systems Corporation (GDS)	◆	◆	–	–	–	◆	◆
Graphical Data Capture Ltd (GDC)	◆	–	–	–	–	–	◆
Graphics Online Limited	–	◆	◆	◆	◆	–	–
Hunting Engineering Ltd	◆	–	–	–	–	–	–
IMASS Limited	◆	–	–	–	–	–	◆
Infolink Decision Services Limited	◆	◆	◆	◆	◆	◆	–
Intergraph (UK) Ltd	◆	–	–	–	–	◆	◆
Laser-Scan Ltd	–	◆	◆	◆	◆	◆	–
London Research Centre	◆	◆	–	–	–	–	–
MapInfo Ltd	◆	◆	◆	◆	◆	◆	◆
MAPIT Limited	–	◆	–	–	–	–	–
Midlands Regional Research Laboratory	◆	–	–	–	–	–	–
MPSI Systems Ltd	–	–	–	◆	◆	–	–
MVA Systematica	◆	◆	◆	–	–	–	–
Navstar Systems Ltd	–	–	–	–	–	◆	–

	census data application	geo-demographic application	market profiling application	market penetration application	sales forecasting application	site location application	crime analysis application
Office of Population Censuses and Surveys (OPCS)	♦	♦	–	–	–	–	–
PAFEC Ltd	–	–	–	–	–	–	♦
PLANTECH Ltd	–	–	–	–	–	♦	–
Property Intelligence plc	♦	♦	–	–	–	–	–
Raindrop Information Systems Ltd	–	–	–	♦	♦	–	–
regioplanDATA GmbH	–	♦	–	–	–	♦	–
SIA Limited	♦	–	♦	–	–	–	–
Siemens Nixdorf	♦	♦	–	–	–	♦	–
Smallworld Systems Ltd	–	–	♦	–	–	♦	–
Spatial Geographic Services & Applications Ltd	–	–	♦	–	–	–	–
Sysdeco (UK) Ltd	♦	–	–	–	–	–	–
System Options Limited	♦	♦	♦	♦	♦	–	–
TACTICIAN UK	♦	♦	♦	♦	♦	♦	♦
The Data Consultancy	♦	♦	♦	♦	♦	♦	–
The Marketing Information Consultancy (MIC)	–	♦	♦	♦	♦	♦	–

directory 32:

off-the-shelf transport applications

	navigation application	vehicle location application	vehicle scheduling/routing application
ACDS Graphic System Inc	–	◆	–
Action Information Management Ltd	◆	◆	◆
ARC Systems Pty Ltd	–	◆	–
Ashtech Europe Limited	◆	◆	–
Assist Applications Limited	–	◆	–
CADAC Ltd	–	◆	–
CODEC Facilities Limited	–	◆	–
CZ Scientific Instruments Ltd	–	◆	–
Dataview Solutions Ltd	–	◆	◆
Dotted Eyes	–	–	◆
Effective Solutions (Data Products)	◆	◆	–
Empress Software UK	–	◆	–
ESRI (UK) Ltd	◆	◆	◆
European Geographic Technologies BV	◆	◆	–
Eurosense Technologies N.V.	–	◆	◆
Foto Res	◆	◆	–
Gamma Ltd	–	◆	–
Geo-Perfect TWI B.V.	◆	◆	–
Geodelta	◆	–	–
GEOSOFT Ltd	◆	◆	◆
Graphic Data Systems Corporation (GDS)	◆	◆	◆
HollyBush Software Limited	◆	–	–
Husky Computing Limited	–	–	◆
I.S. Ltd	◆	–	–
Intergraph (UK) Ltd	◆	◆	–
Laser-Scan Ltd	–	◆	–
LiveChart	◆	◆	–
Logitrans	–	◆	◆
MapInfo Ltd	–	◆	◆
MAPIT Limited	◆	–	–
Oscar Faber	–	◆	–
PAFEC Ltd	–	◆	–
Pear Technology Services Ltd	–	◆	–
Positioning Resources Limited	◆	◆	–
Racal Survey (UK) Ltd	◆	◆	–
SIA Limited	–	◆	◆
Star Informatic S.A.	–	◆	–
Structural Technologies Ltd (STL)	–	◆	–
System Options Limited	◆	◆	◆
Tele Atlas	◆	–	–
Trimble Navigation Europe Ltd	◆	◆	–
Universal Systems Ltd	◆	–	–

directory 33:

off-the-shelf health and environmental applications

	river management application	environmental application	health application
ACDS Graphic System Inc	–	◆	–
Action Information Management Ltd	–	◆	–
Active Software Ltd	–	–	◆
Anglian Engineering & International Consultancy	◆	–	–
AP³ Imaging Services Limited	–	◆	–
Beacon Dodsworth Limited	–	–	◆
CACI Limited	–	–	◆
CAD R&D Centre Limited	◆	◆	–
Cartwright Associates	◆	◆	–
CCN Marketing	–	–	◆
CDR Group	–	◆	–
Chiltern Digitising Services	◆	–	–
CODEC Facilities Limited	–	◆	–
DMV Consultants BV	◆	◆	–
Dotted Eyes	–	–	◆
Earth Resource Mapping	◆	◆	–
Enghouse (UK) Limited	◆	–	–
ERDAS (UK) Ltd	◆	◆	–
ESRI (UK) Ltd	◆	◆	◆
Eurosense Technologies N.V.	◆	◆	–
FastCAD GIS Ltd	–	◆	–
FastCAD GIS Ltd	◆	–	–
Foto Res	–	◆	–
Geo-Perfect TWI B.V.	◆	◆	–
Geo/SQL (UK)	◆	◆	–
Geodelta	–	◆	–
GeoMEM Software	–	◆	◆
Geoplan (UK) Ltd	–	–	◆
Geops BV	–	◆	–
GEOSOFT Ltd	◆	◆	–
Geosystems	◆	◆	–
Glen Computing Ltd	◆	◆	–
Graphic Data Systems Corporation (GDS)	◆	◆	–
Graphical Data Capture Ltd (GDC)	–	–	◆
Grove Projects Ltd	–	◆	–
H R Wallingford Ltd	◆	◆	–
Hunting Engineering Ltd	◆	◆	–
Husky Computing Limited	◆	◆	–
I.S. Ltd	◆	◆	–
ICL Ltd	–	◆	–
Institute of Hydrology	◆	◆	–
Intergraph (UK) Ltd	◆	◆	–
Land Aspects Consultancy Ltd (Parkman Group)	◆	◆	◆
Laser-Scan Ltd	–	◆	–
LiveChart	◆	◆	–

	river management application	environmental application	health application
Logica UK Limited	◆	◆	–
London Research Centre	–	–	◆
MapInfo Ltd	◆	◆	◆
McLintock Limited	–	–	◆
Mott MacDonald Ltd	◆	◆	–
Natural Environment Research Council	◆	◆	–
NERC	◆	◆	–
Office of Population Censuses and Surveys (OPCS)	–	–	◆
Organisation Management Systems	–	–	◆
Ove Arup & Partners	◆	◆	–
PAX Technology	–	◆	–
PLANTECH Ltd	–	◆	◆
Racal Survey (UK) Ltd	◆	–	–
regioplanDATA GmbH	–	◆	–
SCOT Conseil	–	◆	–
Siemens Nixdorf	–	◆	–
Simmons Survey Partnership Limited	◆	–	–
Star Informatic S.A.	–	◆	–
System Options Limited	◆	◆	–
TACTICIAN UK	–	–	◆
The Data Consultancy	–	–	◆
Trimble Navigation Europe Ltd	◆	◆	–
Universal Systems Ltd	◆	◆	–
WRc (Water Research centre)	◆	–	–

directory 34:
suppliers markets group 1

	applicable to all markets	european government market	central government market	local government market	government agencies market	education market	emergency services market	military market
A.L.Downloading Services	◆	–	–	–	–	–	–	–
A.Rutherford Ltd	–	–	–	◆	◆	◆	◆	–
ACDS Graphic System Inc	–	–	–	◆	◆	–	–	–
Action Information Management Ltd	–	–	◆	◆	–	–	◆	◆
Adept Scientific Micro Systems Ltd	◆	–	–	–	–	–	–	–
Advent Imaging Ltd	–	◆	–	–	–	◆	–	◆
AGFA UK	–	◆	◆	◆	◆	◆	◆	◆
AiC Analysts	–	–	–	◆	–	–	◆	–
ALLM Systems & Marketing	–	◆	–	–	–	◆	–	–
ALTEK Corporation	–	◆	◆	◆	–	◆	◆	◆
AM/FM International - European Division	–	◆	◆	◆	◆	◆	◆	–
Aneberie CAD	–	–	–	◆	–	–	–	–
Anglian Engineering & International Consultancy	–	–	◆	◆	◆	–	–	–
AP³ Imaging Services Limited	–	–	–	◆	◆	◆	–	–
APIC Systems	–	–	◆	◆	◆	◆	◆	◆
ARC Systems Pty Ltd	–	–	◆	◆	◆	–	–	–
Ashtech Europe Limited	–	–	–	◆	◆	◆	◆	–
Assist Applications Limited	–	–	◆	◆	◆	–	–	–
Audifilm Girona S.L.	–	–	–	◆	–	–	◆	–
Autodesk Ltd	–	◆	◆	◆	◆	◆	–	◆
Babtie Shaw & Morton Limited	–	–	◆	◆	◆	–	–	–
Bartholomew	◆	–	–	–	–	–	–	–
Baymont Technologies Inc	–	–	◆	◆	◆	–	◆	–
Bentley Systems UK Ltd	◆	–	–	–	–	–	–	–
Binnie Black & Veatch	–	◆	–	–	◆	–	–	–
Birkbeck College London	–	–	◆	◆	◆	◆	–	–
BKS Surveys Ltd	–	–	◆	◆	◆	◆	◆	◆
Bradly Associates Ltd	–	–	◆	◆	–	◆	–	◆
British Geological Survey	–	◆	◆	◆	◆	◆	–	◆
BS International Consultants	–	–	–	◆	–	–	–	–
Bull Information Systems Limited	–	–	◆	◆	◆	–	◆	◆
Business Information Management	–	◆	◆	◆	◆	◆	◆	◆
Byers Engineering Company	–	–	◆	◆	◆	–	–	–
C.A.Design Services Ltd	–	◆	◆	◆	–	–	–	–
CACI Limited	–	◆	◆	◆	◆	◆	◆	–
CAD - Capture Limited	–	◆	◆	◆	◆	–	–	◆
CAD R&D Centre Limited	–	–	◆	◆	◆	◆	–	◆
CADAC Ltd	–	–	–	◆	–	◆	–	–
CalComp Limited	–	–	–	◆	–	–	◆	–
CAM - Centre for Analysis & Modelling Limited	–	–	◆	◆	◆	–	–	–
Cambashi Ltd	◆	◆	◆	◆	–	–	◆	–
Cambridge Computer Consultants (UK) Ltd	–	◆	◆	◆	◆	–	◆	◆
Cambridge Market Intelligence	◆	–	–	–	–	–	–	–
Carl Bro Group	–	–	–	◆	–	–	–	–
Carl Zeiss Limited	–	◆	◆	◆	◆	–	–	◆

	applicable to all markets	european government market	central government market	local government market	government agencies market	education market	emergency services market	military market
Cartographical Services (Southampton) Limited	–	–	◆	◆	◆	–	–	◆
Cartwright Associates	–	–	◆	◆	◆	◆	◆	◆
CATALIST	–	–	◆	◆	◆	–	◆	–
CCTA - The Government Centre for Information Systems	–	–	◆	◆	◆	–	–	–
CDD Ltd	–	◆	◆	◆	◆	–	–	◆
CDR Group	–	◆	◆	◆	–	–	◆	◆
Chroson Ltd	–	–	–	–	–	–	–	◆
Cliffe House Associates	–	–	◆	◆	◆	–	◆	–
CMG Computer Management Group (UK) Ltd	–	–	◆	◆	◆	–	◆	–
Cobham Digital Services Limited	–	–	◆	◆	◆	◆	◆	◆
CODEC Facilities Limited	–	◆	◆	◆	◆	◆	◆	◆
Colorgraph (UK) Ltd	◆	–	–	–	–	–	–	–
ColourMap Scanning Ltd	–	–	–	◆	–	–	–	–
Computer Aided Development (CADCORP) Ltd	–	–	◆	◆	◆	◆	◆	–
COMSULT	–	–	◆	◆	◆	◆	◆	◆
Conic Systems	–	–	◆	◆	◆	–	◆	–
Construction Industry Computing Association	–	◆	◆	◆	◆	–	◆	◆
Coopers & Lybrand	–	◆	◆	◆	◆	–	–	◆
Council of European Professional Informatics Societies	–	◆	–	–	–	–	–	–
Cray Systems	–	◆	◆	◆	◆	–	◆	◆
Cromwell House Technical Services	◆	–	–	–	–	–	–	–
CSI	–	◆	–	–	–	–	–	◆
CZ Scientific Instruments Ltd	–	–	◆	◆	◆	–	◆	–
DAT/EM Systems International	–	◆	–	–	◆	◆	–	–
Data Base Builders	–	–	–	◆	◆	◆	–	–
Data Collection Ltd	–	–	–	◆	–	–	–	–
Data Dictionary Systems Limited	–	–	◆	–	–	–	–	◆
Dataflow Information Systems	–	–	–	◆	–	–	–	–
Dataman Computer Solutions UK Ltd	–	◆	◆	◆	◆	◆	–	◆
Datatechnology Datech Ltd	–	◆	◆	◆	◆	◆	◆	◆
Dataview Solutions Ltd	–	–	◆	◆	◆	◆	◆	◆
DCL Consulting	–	–	–	◆	–	–	–	–
Derek Hunter & Partners Ltd	–	–	–	◆	◆	–	◆	–
Design Computer Aids Limited (DeCAL)	–	–	◆	◆	–	◆	–	–
Digital Equipment Corporation	◆	–	–	–	–	–	–	–
DM Management Consultants Ltd (DMMC)	–	–	◆	◆	–	◆	–	–
DMAP Ltd	–	◆	◆	◆	◆	◆	◆	◆
DMV Consultants BV	–	◆	◆	◆	◆	◆	–	–
Dolphin Consulting Group	–	–	◆	◆	◆	–	–	–
Dotted Eyes	◆	–	–	–	–	–	–	–
Dowling Associates Limited	–	–	◆	◆	–	◆	◆	–
Dr Stanley Port	–	◆	◆	◆	–	–	–	–
EA Technology	–	–	–	◆	◆	–	◆	–
Earth Observation Sciences Ltd	–	◆	◆	–	–	–	–	–
Earth Resource Mapping	–	–	◆	◆	–	◆	–	◆

	applicable to all markets	european government market	central government market	local government market	government agencies market	education market	emergency services market	military market
Effective Solutions (Data Products)	–	◆	–	–	–	–	–	◆
Elstree Computing Ltd	–	–	◆	◆	◆	◆	–	◆
Enghouse (UK) Limited	–	–	–	◆	◆	◆	–	–
Environment & Planning Library	–	–	–	◆	–	–	–	–
EOSAT	–	–	–	◆	◆	◆	–	◆
EPS - Essential Planning Systems Limited	–	◆	◆	◆	◆	◆	◆	◆
ERA Technology Ltd	–	◆	◆	◆	◆	–	◆	◆
ERA-Maptec Ltd	–	◆	◆	◆	◆	–	–	–
ERDAS (UK) Ltd	–	◆	◆	◆	–	–	◆	◆
ERTEC	–	–	–	–	–	◆	–	–
ESR Cartographers Ltd	–	–	◆	◆	◆	–	–	◆
ESRI (UK) Ltd	–	◆	◆	◆	◆	◆	◆	◆
European Business Mapping	–	◆	–	◆	◆	◆	–	–
European Geographic Technologies BV	–	–	◆	◆	◆	–	–	–
Eurosense Technologies N.V.	–	◆	◆	◆	◆	◆	–	◆
Evox Facilities Ltd	◆	–	–	–	–	–	–	–
Fairbairn Services Limited	–	–	◆	◆	–	–	◆	◆
FastCAD GIS Ltd	–	–	◆	◆	◆	–	◆	–
FastCAD GIS Ltd	–	◆	◆	◆	◆	◆	◆	–
Flynn & Rothwell	–	–	◆	◆	◆	–	–	–
Foto Res	–	–	–	◆	◆	◆	–	–
G.L. Consulting Ltd	–	–	◆	◆	◆	–	–	◆
Gamma Ltd	–	◆	◆	◆	◆	–	◆	–
Gardline Infotech	–	◆	◆	◆	◆	–	◆	◆
GEC Marconi Research Centre	–	◆	◆	–	◆	–	–	◆
Genasys II Limited	–	–	–	◆	–	–	◆	◆
General Register Office for Scotland	–	◆	◆	◆	◆	◆	◆	–
Geo-Perfect TWI B.V.	–	◆	◆	◆	–	◆	◆	◆
Geo/SQL (UK)	–	◆	◆	◆	◆	◆	◆	◆
Geo2 Consulting	–	◆	◆	◆	◆	–	◆	–
GEOBASE Consultants Ltd	–	◆	◆	◆	◆	–	◆	◆
GeoData Institute	–	–	◆	◆	◆	◆	–	–
Geodelta	–	◆	◆	◆	◆	–	–	–
Geografix Limited	–	◆	◆	◆	◆	–	◆	◆
Geographic Management Solutions Ltd	–	◆	◆	◆	◆	◆	◆	◆
Geographical Association, The	–	–	–	–	–	◆	–	–
Geoinformation International	◆	–	–	–	–	–	–	–
GeoMEM Software	◆	–	–	–	–	–	–	–
Geometria GIS Systems House Ltd	–	–	–	◆	◆	◆	–	–
Geoplan (UK) Ltd	–	◆	◆	◆	◆	◆	◆	–
Geops BV	–	–	–	◆	◆	–	◆	–
GEOSOFT Ltd	–	◆	◆	◆	◆	◆	–	◆
Geosystems	–	◆	◆	◆	◆	◆	–	–
Geotronics Limited	–	–	◆	◆	◆	◆	◆	◆
Geoview Systems Kft	–	–	–	◆	–	–	–	–
GGP Systems Limited	–	–	–	◆	◆	–	◆	–

	applicable to all markets	european government market	central government market	local government market	government agencies market	education market	emergency services market	military market
GID Ltd	◆	–	–	–	–	–	–	–
GIMMS (GIS) Ltd	–	–	◆	◆	◆	–	◆	–
GIS Services Ltd	◆	–	–	–	–	–	–	–
GISDATA	–	–	–	◆	–	◆	–	–
GISL Limited	–	◆	◆	–	◆	–	◆	–
Glen Computing Ltd	◆	–	–	–	–	–	–	–
Global Surveys Ltd	–	–	–	–	–	◆	–	◆
GMAP Ltd	–	–	◆	–	–	–	–	–
Graphic Data Systems Corporation (GDS)	–	◆	◆	◆	◆	–	◆	–
Graphical Data Capture Ltd (GDC)	–	◆	◆	◆	◆	◆	◆	◆
Graphite Management Services Ltd	–	–	–	◆	◆	–	◆	◆
Grove Projects Ltd	–	◆	◆	◆	◆	◆	◆	◆
GTCO Corporation	–	◆	◆	◆	–	–	–	◆
GTX Europe Ltd	–	–	–	◆	–	–	–	◆
Hansa Luftbild	–	–	–	–	◆	–	–	◆
Hitachi Home Electronics (Europe) Limited	◆	–	–	–	–	–	–	–
HJM Imaging Systems	–	◆	–	–	◆	◆	–	–
HMSO Books	◆	–	–	–	–	–	–	–
HollyBush Software Limited	–	–	–	◆	◆	–	–	◆
Hunting Aerofilms Limited	–	◆	◆	◆	◆	–	◆	–
Hunting Engineering Ltd	–	◆	◆	◆	–	–	◆	◆
Hunting Technical Services Ltd	–	◆	◆	◆	◆	◆	◆	◆
Husky Computing Limited	–	◆	◆	◆	◆	–	–	◆
I.S. Ltd	–	–	–	◆	◆	◆	–	◆
IBM UK Ltd	–	–	–	◆	–	–	–	–
ICL Ltd	–	–	◆	◆	◆	–	◆	◆
IMASS Limited	–	–	–	◆	◆	–	◆	◆
IME (UK) Ltd	–	–	◆	◆	◆	◆	–	–
Infolink Decision Services Limited	–	–	–	◆	◆	–	–	–
Informed Solutions Limited	–	◆	◆	◆	◆	–	◆	◆
Ingecon B.V.	–	◆	◆	◆	◆	–	–	–
Institute of Hydrology	–	◆	◆	–	◆	–	–	–
Institute of Terrestrial Ecology	–	◆	◆	◆	◆	◆	–	–
Intera Information Technologies	–	◆	◆	◆	◆	◆	◆	◆
Intergraph (UK) Ltd	◆	–	–	–	–	–	–	–
International Products	–	–	–	◆	◆	–	◆	–
IT Southern Ltd	–	–	◆	◆	–	–	–	–
ITC	–	◆	◆	◆	◆	◆	–	–
ITS : Intertrade Scientific Ltd	–	–	◆	◆	–	◆	–	◆
J.C.White Chartered Land Surveyors	–	–	–	◆	◆	◆	–	–
John D Leatherdale FRICS	–	◆	◆	◆	◆	–	◆	◆
Kalidor Europe (ALPS Electric (Ireland) Ltd)	–	◆	◆	◆	◆	–	◆	◆
Kamyco International	–	–	–	–	–	◆	–	◆
Keele University	–	–	–	–	–	◆	–	–
Kingswood Consulting Limited	–	◆	◆	◆	◆	◆	◆	◆

	applicable to all markets	european government market	central government market	local government market	government agencies market	education market	emergency services market	military market
Kirstol Ltd	–	◆	–	◆	–	–	–	◆
KJB Consulting	–	–	◆	◆	◆	–	◆	–
Know Edge Ltd	–	◆	◆	◆	◆	–	◆	◆
KPMG	–	◆	◆	◆	◆	◆	◆	◆
L.E.S. (Computer Services) Ltd	–	–	◆	◆	◆	–	◆	◆
Lancaster University	–	◆	◆	◆	–	–	–	–
Land and Satellite Surveys	–	–	–	◆	◆	–	–	–
Land Aspects Consultancy Ltd (Parkman Group)	–	◆	◆	◆	–	–	–	–
Lantmatenet GIS-centrum	–	–	◆	◆	◆	◆	◆	◆
LASCO Ltd	–	–	–	◆	–	–	–	–
Laser Technology International Ltd	◆	–	–	–	–	–	–	–
Laser-Scan Ltd	–	–	◆	◆	–	◆	◆	◆
Leica UK Ltd	–	–	◆	◆	–	◆	–	◆
LiveChart	–	–	–	–	–	–	–	◆
Logica UK Limited	–	–	◆	–	◆	–	–	–
Logitrans	–	–	–	◆	–	–	◆	–
London Research Centre	–	◆	◆	◆	◆	◆	◆	–
Longdin & Browning	–	◆	◆	◆	–	–	◆	◆
Longman Group Ltd	–	–	–	–	–	◆	–	–
Lovell Johns Ltd	–	◆	◆	◆	◆	◆	–	◆
LOY Surveys Ltd	–	–	◆	◆	◆	◆	◆	◆
LTG Services	–	–	◆	◆	◆	–	–	◆
Macaulay Land Use Research Institute	–	–	–	–	◆	–	–	–
Magdala Sociedade	–	◆	◆	◆	◆	–	◆	◆
Magellan Systems Corporation	–	◆	◆	◆	◆	◆	◆	◆
Manchester Metropolitan University	◆	–	–	–	–	–	–	–
Map Data Management Ltd	–	–	◆	◆	◆	◆	–	◆
MapInfo Ltd	◆	–	–	–	–	–	–	–
MAPIT Limited	–	–	–	◆	–	–	–	–
MAPS geosystems	–	◆	◆	◆	◆	–	–	◆
Mason Land Surveys	–	–	◆	◆	◆	–	◆	–
Mathshop	–	–	◆	–	◆	–	–	◆
McLintock Limited	–	–	–	◆	–	–	–	–
MEGRIN Group	–	◆	◆	◆	◆	–	–	–
Mentis Management Consultants Ltd	–	◆	◆	◆	◆	◆	◆	◆
Methods Applications Ltd	–	◆	◆	–	◆	◆	◆	◆
Midlands Regional Research Laboratory	–	–	–	◆	◆	◆	–	–
Midsummer Computing	–	–	◆	◆	◆	–	–	◆
Modern Maps	–	◆	◆	◆	◆	–	–	◆
Morgan Collis Group Ltd	–	◆	◆	◆	◆	–	◆	–
MOSS Systems Limited	–	–	◆	◆	◆	–	–	–
Mott MacDonald Ltd	◆	–	–	–	–	–	–	–
MPSI Systems Ltd	–	–	–	–	◆	–	–	–
MR Data Graphics	–	–	◆	◆	◆	–	◆	◆
MVA Systematica	–	◆	◆	◆	◆	–	◆	–

	applicable to all markets	european government market	central government market	local government market	government agencies market	education market	emergency services market	military market
MVM Consultants plc	–	♦	♦	♦	♦	–	–	♦
NAG Ltd	–	–	–	–	–	♦	–	♦
National Remote Sensing Centre Ltd (NRSC)	–	♦	♦	♦	♦	–	–	♦
Natural Environment Research Council	–	♦	♦	♦	♦	–	–	♦
Navstar Systems Ltd	–	–	–	♦	–	♦	♦	♦
NCC Blackwell Ltd	–	–	♦	♦	♦	–	–	–
NERC	–	♦	♦	♦	♦	–	–	–
Nestor International Ltd	♦	–	–	–	–	–	–	–
NOMIS	–	–	♦	♦	♦	♦	–	–
Numonics UK (Division of Telmtek Ltd)	–	♦	♦	♦	♦	♦	♦	♦
Oaklands I.T.	–	–	–	♦	♦	–	♦	–
Office of Population Censuses and Surveys (OPCS)	–	♦	♦	♦	♦	♦	♦	♦
Ordnance Survey	♦	–	–	–	–	–	–	–
Ordnance Survey of Northern Ireland	♦	–	–	–	–	–	–	–
Organisation Management Systems	–	–	♦	♦	♦	–	♦	–
Oscar Faber	–	♦	♦	♦	♦	–	♦	♦
Ove Arup & Partners	–	♦	♦	♦	–	–	–	–
PA Consulting Group	–	♦	♦	♦	♦	–	♦	♦
PAFEC Ltd	–	–	–	–	–	–	♦	♦
Panda	–	–	–	♦	–	♦	♦	–
Paul Clasper & Associates Ltd	–	–	–	♦	–	–	–	–
PAX Technology	–	♦	♦	♦	♦	♦	–	–
PD Computing Ltd	–	–	♦	♦	♦	–	–	–
Pear Technology Services Ltd	–	–	–	♦	–	♦	–	–
Peter Thorpe Consulting	–	–	–	♦	–	–	–	–
Photarc Surveys Ltd	–	♦	♦	♦	♦	–	–	–
Photoair	–	–	♦	♦	♦	♦	♦	–
Photogrammetric Data Services Ltd	–	–	♦	♦	♦	–	♦	♦
Pinpoint Digitising Services	–	–	♦	♦	♦	♦	♦	–
PlanGraphics Inc	♦	–	–	–	–	–	–	–
Planning & Mapping Ltd	–	♦	♦	♦	♦	–	–	–
PLANTECH Ltd	–	–	♦	♦	–	–	–	–
Plowman Craven & Associates Ltd	–	♦	♦	♦	♦	♦	♦	♦
Posford Duvivier	–	–	♦	♦	♦	–	–	–
Positioning Resources Limited	–	–	–	–	–	♦	♦	–
Primagraphics Ltd	–	–	–	–	–	♦	–	♦
Procis Software Ltd	–	–	♦	♦	♦	–	♦	–
Progis GmbH	♦	–	–	–	–	–	–	–
Property Intelligence plc	–	–	♦	♦	♦	–	–	–
QAS Systems Ltd	–	–	♦	♦	♦	♦	♦	♦
QC Data (Ireland) Limited	–	–	♦	♦	♦	–	–	–
Quorum Information Services	♦	–	–	–	–	–	–	–
R.W.A. Dalls FRICS	–	–	♦	♦	♦	♦	–	–
Racal Survey (UK) Ltd	–	♦	♦	♦	–	–	–	♦
Raindrop Information Systems Ltd	–	–	♦	♦	–	♦	–	–

	applicable to all markets	european government market	central government market	local government market	government agencies market	education market	emergency services market	military market
regioplanDATA GmbH	–	–	–	◆	◆	–	◆	–
Remote Sensing Applications Consultants	–	◆	◆	◆	◆	–	–	◆
RICS Books	◆	–	–	–	–	–	–	–
Robert Walker Consultants	–	◆	◆	◆	◆	–	–	–
Royal Geographical Society (with The Institute of British Geographers)	◆	–	–	–	–	–	–	–
Royal Institute of Navigation	–	◆	◆	–	◆	◆	◆	◆
Royal Mail - Address Management Centre	◆	–	–	–	–	–	–	–
Royal Town Planning Institute	–	–	–	◆	–	–	–	–
Salford University Business Services Ltd	–	◆	◆	◆	◆	–	◆	◆
SAS Institute	–	◆	◆	◆	◆	◆	◆	◆
Saztec Europe Limited	◆	–	–	–	–	–	–	–
SAZTEC Philippines Inc	–	◆	◆	◆	◆	–	◆	◆
Scan Group Limited	–	◆	◆	◆	◆	–	◆	◆
Scientific Software Limited	–	–	–	–	◆	–	–	–
SCOT Conseil	–	◆	–	◆	–	–	–	–
Scott Wilson Kirkpatrick	–	◆	◆	◆	◆	–	–	◆
SDI Ltd	–	–	–	◆	◆	◆	–	◆
Sector (UK) Limited	–	–	◆	◆	◆	–	◆	–
SHL Vision* Solutions Limited	–	◆	◆	◆	◆	–	–	◆
SIA Limited	–	–	◆	◆	◆	–	◆	◆
SIAS Limited	–	–	◆	◆	–	–	–	–
Siemens Nixdorf	–	◆	◆	◆	◆	–	–	–
Silicon Graphics Limited	◆	–	–	–	–	–	–	–
Silsoe College, Cranfield University	–	◆	◆	◆	◆	–	–	◆
Simmons Survey Partnership Limited	◆	–	–	–	–	–	–	–
Sir Alexander Gibb & Partners Ltd	–	–	◆	◆	◆	–	–	–
Sir William Halcrow and Partners	–	–	◆	◆	–	–	–	◆
Smallworld Systems Ltd	–	–	–	◆	–	◆	–	–
Smartscan Inc	–	◆	◆	◆	◆	–	◆	◆
Smith System Engineering Ltd	–	◆	◆	◆	◆	–	◆	◆
Sokkia Ltd	–	◆	◆	◆	◆	◆	–	–
Southbank Systems PLC	–	–	◆	◆	◆	–	◆	–
Sovereign C.S. Ltd	–	◆	◆	◆	◆	–	◆	◆
Spacesense consultants	–	◆	◆	◆	◆	–	–	–
Spatial Data Limited	–	◆	◆	◆	◆	–	–	◆
Spatial Geographic Services & Applications Ltd	–	–	–	◆	◆	–	◆	–
Spatial Information Services Ltd (SIS)	–	◆	◆	◆	◆	–	◆	–
SPSS UK Ltd	–	◆	◆	◆	◆	◆	◆	◆
Star Informatic S.A.	–	◆	◆	◆	◆	◆	◆	◆
StatSci Europe	◆	–	–	–	–	–	–	–
Structural Technologies Ltd (STL)	–	–	◆	◆	–	–	–	–
Summagraphics Europe N.V.	–	–	–	–	–	◆	–	◆
Sun Microsystems Ltd	◆	–	–	–	–	–	–	–
Survey & Development Services Ltd	◆	–	–	–	–	–	–	–
Survey Supplies Ltd	–	–	–	◆	–	–	◆	–

	applicable to all markets	european government market	central government market	local government market	government agencies market	education market	emergency services market	military market
Symology Limited	−	−	−	♦	−	−	−	−
Sysdeco (UK) Ltd	−	−	♦	♦	−	−	♦	−
System Options Limited	♦	−	−	−	−	−	−	−
Tangent Technology Design Associates Ltd	−	−	−	♦	−	−	♦	−
Taylor & Francis Ltd	−	−	−	−	−	♦	−	−
TDS-CAD Graphics Ltd	−	−	−	♦	−	−	♦	♦
TEAMS (Taylor Woodrow Electronics Asset Mapping Survey)	−	♦	♦	♦	−	−	−	♦
Tekla Oy	−	−	−	♦	−	−	−	−
Tele Atlas	−	♦	♦	♦	♦	−	♦	♦
Tendron Systems Ltd	−	−	−	♦	♦	−	♦	♦
Tenet Systems Ltd	−	♦	♦	♦	♦	♦	♦	♦
Terrafix Ltd	−	♦	♦	♦	♦	−	♦	♦
TerraQuest Group Limited	−	−	♦	♦	♦	−	♦	−
The Business Database from Yellow Pages	♦	−	−	−	−	−	−	−
The Data Consultancy	−	♦	♦	♦	♦	♦	♦	♦
The LGMB	−	−	−	♦	−	−	♦	−
The Marketing Information Consultancy (MIC)	−	−	−	♦	♦	♦	−	−
The NPA Group Ltd	−	♦	♦	♦	♦	♦	♦	♦
The Severn Partnership	−	−	−	♦	♦	−	−	−
The Survey Centre	−	−	♦	♦	♦	♦	−	−
Trident Map Services	−	−	♦	♦	♦	♦	−	−
Trimble Navigation Europe Ltd	−	♦	♦	♦	♦	♦	♦	♦
TYDAC Technologies Ltd	♦	−	−	−	−	−	−	−
Unistride Sewer Technology	−	−	−	♦	♦	−	−	♦
UNISYS Ltd	−	−	♦	♦	♦	♦	♦	−
Universal Systems Ltd	−	−	−	♦	−	♦	−	♦
University of Edinburgh	♦	−	−	−	−	−	−	−
University of Greenwich	♦	−	−	−	−	−	−	−
University of Luton	−	−	−	−	♦	♦	♦	−
University of Newcastle Upon Tyne	−	−	−	♦	♦	♦	−	−
Walker Ladd Surveys	−	−	−	♦	−	−	−	−
Wallingford Software Limited	−	−	−	♦	♦	♦	−	−
WDV (UK)	♦	−	−	−	−	−	−	−
WRc (Water Research centre)	−	♦	♦	♦	−	−	−	−
WS Atkins Planning & Management Consultants	−	♦	♦	♦	♦	♦	♦	♦
Xcon Data	−	♦	♦	♦	♦	−	♦	−

	distribution markets	retail market	banking market	finance market	insurance market	marketing sector	other market
A.Rutherford Ltd	◆	◆	◆	–	◆	◆	–
Action Information Management Ltd	◆	–	–	–	–	–	–
AGFA UK	◆	◆	◆	◆	◆	◆	–
AiC Analysts	◆	◆	◆	–	–	–	–
ALLM Systems & Marketing	–	–	–	–	–	–	◆
ALTEK Corporation	–	–	–	–	–	–	◆
AM/FM International - European Division	◆	◆	◆	◆	◆	◆	–
AND Mapping B.V.	◆	–	–	–	–	–	–
AP³ Imaging Services Limited	–	◆	–	–	–	–	–
APIC Systems	–	◆	–	–	◆	◆	–
Autodesk Ltd	◆	◆	◆	◆	◆	◆	–
Beacon Dodsworth Limited	◆	◆	◆	◆	◆	◆	–
Birkbeck College London	◆	◆	–	–	–	–	–
Bull Information Systems Limited	◆	–	–	–	–	–	–
Business Information Management	◆	◆	◆	◆	◆	◆	–
CACI Limited	◆	◆	◆	◆	◆	◆	◆
CAD R&D Centre Limited	◆	–	–	–	–	–	–
CalComp Limited	◆	◆	–	–	–	–	–
CAM - Centre for Analysis & Modelling Limited	–	◆	–	◆	–	◆	–
Cambashi Ltd	◆	◆	◆	◆	◆	◆	–
Cambridge Computer Consultants (UK) Ltd	◆	◆	–	–	–	◆	–
Carl Zeiss Limited	–	◆	–	–	–	–	–
CARTograph Ltd	–	–	◆	–	◆	–	–
Cartwright Associates	◆	◆	–	–	◆	◆	–
CATALIST	◆	◆	◆	◆	◆	◆	–
CCN Marketing	–	◆	◆	◆	◆	◆	–
CDD Ltd	–	–	–	–	–	◆	–
CMG Computer Management Group (UK) Ltd	◆	◆	◆	◆	◆	–	◆
Cobham Digital Services Limited	◆	–	–	–	◆	–	–
CODEC Facilities Limited	◆	◆	◆	–	–	–	–
COMSULT	◆	◆	–	–	–	–	–
Consensus Information Technology Ltd	–	◆	–	–	–	◆	–
Construction Industry Computing Association	◆	◆	◆	–	–	–	–
Coopers & Lybrand	◆	◆	◆	◆	◆	–	–
Cray Systems	◆	–	–	–	–	–	–
Data Base Builders	–	–	–	◆	–	–	–
Dataman Computer Solutions UK Ltd	–	–	–	◆	◆	–	–
Dataquest Europe Ltd	–	–	–	–	–	–	◆
Datatechnology Datech Ltd	◆	◆	◆	◆	◆	◆	–
Dataview Solutions Ltd	◆	◆	◆	◆	◆	◆	–
DM Management Consultants Ltd (DMMC)	–	◆	–	◆	–	◆	◆
DMAP Ltd	◆	◆	◆	◆	◆	◆	◆
Dolphin Consulting Group	◆	◆	◆	◆	◆	◆	–
EA Technology	◆	–	–	–	–	–	–
ECM Selection Limited	–	–	–	–	–	–	◆

	distribution markets	retail market	banking market	finance market	insurance market	marketing sector	other market
Elstree Computing Ltd	–	◆	◆	–	–	–	◆
Empress Software UK	–	–	–	–	–	–	◆
EPS - Essential Planning Systems Limited	–	–	–	–	◆	◆	–
ERA Technology Ltd	–	–	–	–	–	–	◆
ERDAS (UK) Ltd	◆	–	–	–	◆	–	◆
ESR Cartographers Ltd	–	◆	–	–	–	◆	–
ESRI (UK) Ltd	◆	◆	◆	◆	◆	◆	–
EuroDirect Database Marketing Ltd	–	◆	◆	◆	◆	◆	–
European Business Mapping	–	◆	◆	◆	◆	◆	–
European Geographic Technologies BV	–	–	–	–	–	–	◆
Eurosense Technologies N.V.	◆	–	–	–	–	◆	–
Fairbairn Services Limited	◆	◆	◆	◆	◆	◆	–
FileNet Ltd	◆	◆	◆	◆	◆	–	–
Foto Res	◆	–	–	–	–	◆	–
Gamma Ltd	◆	◆	◆	◆	◆	–	–
Gardline Infotech	◆	–	–	–	–	–	–
General Register Office for Scotland	◆	◆	◆	◆	◆	◆	–
GEO-Marketing Systems Ltd (GMSL)	–	◆	–	–	–	◆	–
Geo-Perfect TWI B.V.	◆	◆	◆	–	–	◆	–
GEO-UK Ltd	–	–	–	–	–	–	◆
Geo/SQL (UK)	◆	◆	◆	◆	◆	◆	–
Geo2 Consulting	◆	–	–	–	–	–	–
GEOBASE Consultants Ltd	–	–	–	◆	–	◆	–
Geografix Limited	◆	–	–	–	–	–	–
Geographic Management Solutions Ltd	◆	–	–	–	◆	–	–
Geoplan (UK) Ltd	◆	◆	◆	◆	◆	◆	–
GEOSOFT Ltd	◆	◆	–	–	–	◆	◆
GIMMS (GIS) Ltd	–	–	–	–	–	◆	–
GMAP Ltd	◆	◆	◆	◆	◆	◆	–
Graphic Data Systems Corporation (GDS)	–	–	–	–	–	–	◆
Graphical Data Capture Ltd (GDC)	◆	◆	◆	◆	◆	◆	–
Graphics Online Limited	◆	◆	◆	◆	◆	◆	–
Graphite Management Services Ltd	◆	◆	◆	–	–	–	–
Greig Fester Limited	–	–	–	◆	◆	–	–
Grove Projects Ltd	◆	◆	◆	◆	◆	◆	–
GTCO Corporation	–	–	◆	–	–	–	–
Hunting Aerofilms Limited	–	–	–	–	–	◆	–
Hunting Engineering Ltd	–	–	–	–	–	◆	–
Husky Computing Limited	◆	◆	–	–	◆	◆	◆
IBM UK Ltd	–	◆	◆	–	–	–	–
ICL Ltd	◆	◆	◆	◆	◆	–	–
IME (UK) Ltd	–	–	◆	–	–	–	–
Infolink Decision Services Limited	–	◆	◆	◆	◆	◆	–
Informed Solutions Limited	◆	◆	◆	◆	◆	◆	–
International Map Trade Association	◆	◆	–	–	–	–	◆

	distribution markets	retail market	banking market	finance market	insurance market	marketing sector	other market
International Products	–	◆	–	–	–	–	–
IT Southern Ltd	–	–	–	◆	–	–	–
ITS : Intertrade Scientific Ltd	◆	–	–	–	–	–	◆
John D Leatherdale FRICS	–	–	◆	–	–	◆	–
Kamyco International	–	◆	–	–	–	–	–
Kingswood Consulting Limited	◆	◆	◆	◆	◆	◆	–
Know Edge Ltd	◆	◆	◆	◆	◆	◆	–
KPMG	◆	◆	◆	◆	◆	◆	–
L.E.S. (Computer Services) Ltd	–	◆	–	–	–	–	–
Lancaster University	–	◆	–	–	–	◆	–
Lantmatenet GIS-centrum	◆	◆	◆	◆	◆	◆	–
Laser-Scan Ltd	–	◆	–	–	◆	◆	–
LiveChart	◆	◆	–	–	–	–	–
Logitrans	◆	◆	◆	–	–	–	–
LOY Surveys Ltd	–	◆	◆	–	–	–	–
LTG Services	–	◆	◆	◆	◆	◆	–
Magdala Sociedade	–	–	◆	–	◆	◆	–
MAPIT Limited	–	–	–	–	–	◆	–
Mason Land Surveys	◆	◆	◆	–	–	–	–
MEGRIN Group	–	–	–	–	–	◆	–
Mentis Management Consultants Ltd	◆	◆	◆	◆	◆	◆	–
Methods Applications Ltd	◆	◆	◆	◆	◆	◆	–
Midsummer Computing	◆	–	◆	–	–	–	–
MPSI Systems Ltd	–	◆	–	–	–	–	–
MR Data Graphics	◆	◆	–	–	–	–	–
MVA Systematica	◆	◆	◆	◆	◆	◆	–
MVM Consultants plc	–	◆	–	–	–	◆	–
NAG Ltd	–	–	◆	◆	◆	–	–
National Remote Sensing Centre Ltd (NRSC)	–	–	–	–	◆	–	–
Natural Environment Research Council	–	–	–	–	◆	–	–
NERC	–	–	–	–	◆	–	–
NOMIS	–	◆	–	◆	–	◆	–
Numonics UK (Division of Telmtek Ltd)	◆	–	◆	◆	–	◆	–
Office of Population Censuses and Surveys (OPCS)	◆	◆	◆	◆	◆	◆	–
Organisation Management Systems	–	◆	◆	–	–	–	–
Oxford Institute of Retail Management	–	◆	◆	◆	◆	◆	–
PA Consulting Group	◆	–	◆	◆	◆	◆	–
Panda	–	–	–	–	–	◆	◆
Paul Clasper & Associates Ltd	–	–	–	–	–	–	◆
PD Computing Ltd	–	–	–	–	◆	–	–
Pear Technology Services Ltd	–	◆	–	–	◆	◆	–
Pinpoint Digitising Services	–	–	–	–	◆	–	–
Plowman Craven & Associates Ltd	◆	◆	◆	◆	◆	◆	–
Property Intelligence plc	–	◆	◆	◆	–	–	–
QAS Systems Ltd	◆	◆	◆	◆	◆	◆	–

	distribution markets	retail market	banking market	finance market	insurance market	marketing sector	other market
Raindrop Information Systems Ltd	–	♦	♦	♦	♦	♦	–
Recruit Media Ltd	–	–	–	–	–	–	♦
regioplanDATA GmbH	♦	–	♦	♦	♦	♦	–
Remote Sensing Applications Consultants	–	–	–	–	–	–	♦
Royal Institute of Navigation	♦	–	–	–	–	–	–
Royal Town Planning Institute	–	–	–	–	–	–	♦
Salford University Business Services Ltd	♦	♦	♦	♦	♦	♦	–
SAS Institute	♦	♦	♦	♦	♦	♦	–
Scan Group Limited	♦	♦	–	–	–	♦	♦
SDI Ltd	–	–	–	–	–	♦	♦
Sector (UK) Limited	–	–	♦	♦	–	–	–
SIA Limited	♦	♦	♦	–	♦	♦	–
Sir Alexander Gibb & Partners Ltd	–	–	–	♦	♦	–	–
Sir William Halcrow and Partners	–	–	–	–	♦	–	–
Smallworld Systems Ltd	–	♦	♦	–	♦	♦	–
Smartscan Inc	♦	–	–	–	–	–	–
Solent Mapping and Charting (SMAC)	–	–	–	–	–	–	♦
Southbank Systems PLC	–	♦	–	–	♦	–	–
Sovereign C.S. Ltd	–	♦	♦	–	–	–	♦
Spacesense consultants	–	–	–	–	♦	♦	–
Spatial Information Services Ltd (SIS)	♦	♦	–	–	–	–	–
SPSS UK Ltd	♦	♦	♦	♦	♦	♦	–
Star Informatic S.A.	♦	♦	♦	–	–	–	–
Summagraphics Europe N.V.	♦	–	–	–	–	–	–
Survey Control Services	–	♦	–	–	–	–	–
Sysdeco (UK) Ltd	–	♦	–	–	♦	–	–
TACTICIAN UK	♦	♦	♦	♦	♦	♦	♦
Tangent Technology Design Associates Ltd	♦	♦	–	–	–	–	–
Target Market Consultancy	♦	♦	♦	♦	♦	♦	–
Tele Atlas	♦	♦	♦	♦	♦	♦	–
Tenet Systems Ltd	–	♦	♦	♦	♦	–	–
Terrafix Ltd	♦	–	–	–	–	–	–
TerraQuest Group Limited	♦	–	–	–	–	♦	–
The Data Consultancy	♦	♦	♦	♦	♦	♦	–
The Marketing Information Consultancy (MIC)	–	♦	♦	♦	♦	♦	–
The NPA Group Ltd	♦	♦	–	–	♦	–	–
Trac Consultancy	♦	♦	♦	♦	♦	♦	–
Trident Map Services	♦	♦	♦	–	–	–	–
Trimble Navigation Europe Ltd	♦	–	–	–	–	–	–
Universal Systems Ltd	–	–	–	–	♦	–	–
University of Luton	♦	♦	–	–	–	–	–
Walker Ladd Surveys	–	♦	–	–	–	–	–
WS Atkins Planning & Management Consultants	–	♦	♦	–	–	–	–
Xcon Data	♦	♦	–	–	–	–	–

directory 36:
suppliers markets
group 3

	farming markets	forestry market	geology market	mining market	exploration market	oil/petroleum market	marine and coastal market	utility market	civil engineering market
A.Rutherford Ltd	◆	◆	−	−	−	−	−	◆	◆
ACDS Graphic System Inc	−	◆	−	◆	−	−	−	◆	◆
Action Information Management Ltd	−	−	−	−	−	−	−	◆	−
Advent Imaging Ltd	◆	◆	−	◆	◆	◆	◆	◆	−
AGFA UK	◆	◆	−	−	◆	−	−	◆	◆
AiC Analysts	−	−	−	−	−	◆	−	◆	◆
ALTEK Corporation	◆	◆	◆	◆	◆	−	−	◆	◆
AM/FM International - European Division	−	−	−	−	−	−	−	◆	−
Aneberie CAD	−	−	−	−	−	−	−	−	◆
Anglian Engineering & International Consultancy	−	◆	−	−	−	◆	◆	−	◆
AP³ Imaging Services Limited	−	◆	−	◆	−	−	−	◆	−
APIC Systems	−	◆	−	−	−	◆	◆	−	−
ARC Systems Pty Ltd	−	−	−	−	−	−	−	◆	−
Ashtech Europe Limited	◆	◆	◆	−	◆	◆	−	◆	◆
Audifilm Girona S.L.	−	−	−	−	−	−	−	−	◆
Autodesk Ltd	◆	◆	◆	◆	◆	◆	◆	◆	◆
Babtie Shaw & Morton Limited	−	−	−	−	−	−	−	◆	◆
Baymont Technologies Inc	−	−	−	−	−	−	−	◆	−
Beacon Dodsworth Limited	−	−	−	−	−	−	−	◆	−
Binnie Black & Veatch	−	−	−	◆	−	−	◆	◆	◆
BKS Surveys Ltd	−	◆	−	◆	−	−	◆	◆	◆
Bradly Associates Ltd	−	◆	−	◆	◆	−	−	◆	◆
British Geological Survey	◆	−	◆	◆	◆	◆	◆	−	◆
BS International Consultants	−	−	−	−	◆	◆	−	◆	−
Bull Information Systems Limited	−	−	−	−	◆	◆	−	◆	−
Business Information Management	◆	◆	−	◆	−	−	−	−	−
Byers Engineering Company	−	−	−	−	−	◆	−	◆	−
C.A.Design Services Ltd	−	◆	◆	−	◆	◆	−	◆	◆
CACI Limited	−	−	−	−	−	◆	−	◆	−
CAD - Capture Limited	−	−	−	−	−	◆	−	◆	◆
CAD R&D Centre Limited	◆	◆	◆	−	−	◆	◆	◆	◆
CADAC Ltd	−	−	−	◆	−	◆	◆	◆	◆
CalComp Limited	−	−	−	−	−	−	−	◆	◆
CAM - Centre for Analysis & Modelling Limited	−	−	−	−	−	−	−	◆	−
Cambashi Ltd	−	−	−	−	−	◆	−	◆	◆
Cambridge Computer Consultants (UK) Ltd	◆	◆	◆	−	◆	◆	◆	◆	◆
Carl Bro Group	−	−	−	−	−	−	−	◆	◆
Carl Zeiss Limited	−	◆	−	◆	−	−	−	◆	◆
CARTograph Ltd	−	−	−	−	−	−	−	◆	−
Cartographical Services (Southampton) Limited	◆	◆	◆	◆	◆	◆	◆	◆	◆
Cartwright Associates	◆	◆	◆	−	◆	◆	◆	◆	◆
CATALIST	−	−	−	−	−	−	−	−	−
CCN Marketing	−	−	−	−	−	−	−	◆	−
CDD Ltd	−	◆	◆	−	−	◆	◆	−	−
CDR Group	−	◆	−	◆	−	−	−	◆	◆

	farming markets	forestry market	geology market	mining market	exploration market	oil/petroleum market	marine and coastal market	utility market	civil engineering market
Chiltern Digitising Services	–	–	◆	–	–	–	–	◆	–
Chroson Ltd	–	–	–	–	–	–	–	◆	–
Citywise	–	–	–	–	–	–	–	◆	–
CMG Computer Management Group (UK) Ltd	–	–	–	–	◆	◆	–	◆	–
Cobham Digital Services Limited	◆	◆	–	–	–	◆	–	◆	–
CODEC Facilities Limited	–	◆	◆	◆	◆	–	–	◆	–
ColourMap Scanning Ltd	–	–	–	–	–	◆	–	–	–
Computer Aided Development (CADCORP) Ltd	–	–	–	–	–	–	–	◆	–
COMSULT	◆	◆	–	–	◆	◆	–	◆	◆
Conic Systems	–	◆	◆	◆	◆	◆	◆	◆	–
Construction Industry Computing Association	–	–	–	–	–	–	–	◆	–
Coopers & Lybrand	–	–	–	–	–	◆	–	◆	–
Corbins Consultancy	–	–	–	–	–	–	–	◆	–
Corena A/S	–	–	–	–	–	–	–	◆	–
Cray Systems	◆	◆	–	–	–	◆	◆	–	–
CSI	–	◆	◆	◆	◆	◆	–	–	–
CZ Scientific Instruments Ltd	◆	◆	◆	◆	◆	–	–	◆	◆
DAT/EM Systems International	–	◆	–	◆	◆	–	–	◆	◆
Data Base Builders	–	–	–	–	–	–	–	◆	–
Dataman Computer Solutions UK Ltd	–	–	–	–	–	◆	◆	–	◆
Datatechnology Datech Ltd	◆	◆	–	◆	◆	◆	◆	◆	◆
Dataview Solutions Ltd	◆	–	–	◆	◆	◆	◆	◆	–
DCL Consulting	–	–	–	–	–	–	–	◆	◆
Derek Hunter & Partners Ltd	–	–	–	–	–	–	–	◆	–
DMAP Ltd	–	–	–	◆	–	–	–	◆	–
DMV Consultants BV	◆	◆	–	–	–	–	–	◆	◆
Dolphin Consulting Group	–	–	–	–	–	–	–	◆	–
Dr Stanley Port	–	–	–	–	–	◆	–	◆	◆
EA Technology	–	–	–	–	–	–	–	◆	–
Earth Observation Sciences Ltd	–	◆	–	–	◆	◆	–	–	–
Earth Resource Mapping	◆	◆	◆	◆	◆	◆	◆	◆	–
Effective Solutions (Data Products)	–	–	–	–	◆	–	–	–	–
Elstree Computing Ltd	–	–	–	–	–	–	–	◆	–
Enghouse (UK) Limited	–	–	–	–	–	–	–	◆	–
EOSAT	◆	◆	–	◆	◆	–	–	◆	◆
EPS - Essential Planning Systems Limited	◆	◆	◆	◆	◆	◆	◆	–	◆
ERA Technology Ltd	◆	◆	–	◆	◆	–	–	◆	◆
ERA-Maptec Ltd	◆	◆	–	◆	◆	–	–	◆	–
ERDAS (UK) Ltd	◆	◆	◆	◆	◆	◆	◆	◆	◆
ERTEC	–	–	–	–	–	–	–	◆	–
ESR Cartographers Ltd	◆	◆	◆	–	–	–	–	◆	◆
ESRI (UK) Ltd	◆	◆	◆	◆	◆	◆	◆	◆	–
EuroDirect Database Marketing Ltd	–	–	–	–	–	–	–	◆	–
European Business Mapping	–	–	–	–	–	–	–	◆	–
Eurosense Technologies N.V.	◆	◆	–	◆	–	–	–	◆	◆

	farming markets	forestry market	geology market	mining market	exploration market	oil/petroleum market	marine and coastal market	utility market	civil engineering market
Fairbairn Services Limited	–	–	♦	♦	–	♦	–	♦	♦
FastCAD GIS Ltd	–	–	–	–	–	–	–	♦	♦
FastCAD GIS Ltd	–	♦	–	–	–	–	–	♦	♦
FileNet Ltd	–	–	–	–	–	–	–	♦	–
Flynn & Rothwell	–	–	–	–	–	–	–	♦	♦
Foto Res	–	♦	–	♦	–	–	–	♦	♦
Gamma Ltd	–	–	–	–	–	♦	–	–	–
Gardline Infotech	♦	♦	♦	♦	♦	♦	♦	♦	♦
GEC Marconi Research Centre	♦	♦	–	–	♦	♦	♦	–	–
Genasys II Limited	–	–	–	–	–	♦	–	♦	–
Geo-Perfect TWI B.V.	–	♦	♦	–	♦	♦	–	♦	–
Geo/SQL (UK)	♦	♦	–	♦	♦	–	♦	♦	♦
Geo2 Consulting	–	–	–	–	–	♦	♦	♦	♦
GEOBASE Consultants Ltd	–	–	♦	–	–	♦	♦	♦	♦
GeoData Institute	♦	♦	–	♦	♦	♦	♦	♦	♦
Geodelta	–	–	–	–	–	–	–	♦	–
Geografix Limited	–	–	–	–	♦	–	♦	♦	♦
Geographic Management Solutions Ltd	♦	♦	♦	♦	♦	♦	♦	♦	♦
Geometria GIS Systems House Ltd	–	–	–	–	–	–	–	♦	–
Geoplan (UK) Ltd	–	–	–	–	♦	–	♦	♦	–
GEOSOFT Ltd	♦	♦	–	♦	–	♦	–	♦	–
Geosystems	–	♦	–	♦	♦	♦	–	♦	♦
Geotronics Limited	–	♦	–	♦	♦	–	–	♦	♦
Geoview Systems Kft	–	–	–	–	–	–	–	♦	–
GIMMS (GIS) Ltd	–	–	–	–	–	–	–	♦	–
GISL Limited	♦	♦	♦	♦	♦	♦	♦	–	♦
Global Surveys Ltd	–	♦	–	♦	♦	–	♦	♦	♦
GMAP Ltd	–	–	–	–	–	♦	–	–	–
Graphic Data Systems Corporation (GDS)	–	–	–	–	–	–	–	♦	–
Graphical Data Capture Ltd (GDC)	–	♦	–	–	–	–	–	♦	–
Graphics Online Limited	–	–	–	–	–	–	–	♦	–
Graphite Management Services Ltd	♦	♦	♦	♦	♦	♦	♦	♦	♦
Grove Projects Ltd	♦	–	–	♦	–	–	♦	♦	♦
GTCO Corporation	♦	♦	–	♦	–	♦	–	♦	♦
GTX Europe Ltd	–	♦	–	–	♦	♦	–	♦	♦
H R Wallingford Ltd	–	–	–	–	–	–	–	–	♦
Hall & Watts Systems Limited	–	–	–	♦	–	–	–	–	♦
Hansa Luftbild	–	♦	–	♦	♦	–	–	♦	–
HJM Imaging Systems	–	♦	–	♦	♦	–	–	♦	♦
HollyBush Software Limited	–	–	–	–	–	♦	♦	♦	–
Hunting Aerofilms Limited	♦	♦	–	♦	♦	–	♦	♦	♦
Hunting Engineering Ltd	–	–	–	–	–	–	–	–	♦
Hunting Technical Services Ltd	♦	♦	♦	♦	♦	♦	♦	♦	–
Husky Computing Limited	–	♦	–	–	♦	–	–	♦	♦
I.S. Ltd	♦	♦	♦	♦	♦	♦	♦	♦	♦

	farming markets	forestry market	geology market	mining market	exploration market	oil/petroleum market	marine and coastal market	utility market	civil engineering market
IBM UK Ltd	–	–	–	–	–	–	–	◆	–
ICL Ltd	–	–	–	–	–	–	–	◆	–
IMASS Limited	–	◆	–	–	–	–	–	◆	–
IME (UK) Ltd	–	–	–	–	–	–	–	◆	–
Infolink Decision Services Limited	–	–	–	–	–	–	–	◆	–
Informed Solutions Limited	–	◆	–	◆	◆	◆	◆	◆	◆
Ingecon B.V.	–	–	–	–	–	–	–	◆	◆
Institute of Hydrology	–	◆	–	–	–	–	–	◆	–
Institute of Terrestrial Ecology	–	◆	–	–	–	–	–	◆	–
Intera Information Technologies	◆	◆	–	◆	◆	–	◆	◆	◆
International Map Trade Association	–	–	–	–	–	–	◆	–	–
International Products	–	–	–	–	–	–	–	◆	–
IT Southern Ltd	–	–	–	–	–	–	–	◆	–
ITC	◆	◆	◆	◆	–	–	◆	–	–
ITS : Intertrade Scientific Ltd	–	◆	◆	◆	–	◆	◆	◆	–
J.C.White Chartered Land Surveyors	–	–	–	–	–	–	–	◆	◆
John D Leatherdale FRICS	–	◆	–	–	–	◆	–	–	◆
Kalidor Europe (ALPS Electric (Ireland) Ltd)	–	◆	◆	◆	◆	◆	–	◆	◆
Kamyco International	–	–	–	–	–	–	–	◆	◆
Kingswood Consulting Limited	–	–	–	–	–	◆	–	◆	–
Kirstol Ltd	–	◆	–	–	◆	–	–	◆	–
KJB Consulting	–	◆	–	–	–	–	–	–	◆
Know Edge Ltd	◆	◆	–	–	–	–	–	◆	–
KPMG	–	–	–	–	–	◆	–	◆	◆
L.E.S. (Computer Services) Ltd	◆	◆	–	◆	◆	–	–	◆	◆
Land and Satellite Surveys	◆	◆	–	–	◆	–	–	◆	◆
Land Aspects Consultancy Ltd (Parkman Group)	◆	◆	–	◆	◆	–	◆	◆	◆
LASCO Ltd	–	–	–	–	–	–	–	◆	◆
Laser-Scan Ltd	–	◆	◆	◆	–	–	◆	◆	◆
Leica UK Ltd	–	◆	–	◆	◆	–	–	◆	◆
LiveChart	–	–	–	–	–	–	◆	–	–
Logica UK Limited	◆	–	–	–	–	◆	◆	◆	–
London Research Centre	–	–	–	–	–	–	–	◆	–
Longdin & Browning	–	◆	–	◆	–	◆	◆	◆	◆
Lovell Johns Ltd	◆	◆	◆	◆	◆	◆	◆	◆	–
LOY Surveys Ltd	◆	◆	–	◆	◆	◆	◆	◆	◆
LTG Services	–	◆	–	–	–	–	–	◆	◆
Macaulay Land Use Research Institute	◆	◆	–	–	–	–	–	–	–
Magdala Sociedade	◆	◆	–	–	–	–	◆	◆	–
Magellan Systems Corporation	◆	◆	–	◆	◆	◆	◆	–	–
Map Data Management Ltd	◆	◆	–	–	–	–	–	–	–
MAPIT Limited	–	–	–	–	–	–	–	◆	–
MAPS geosystems	–	–	–	–	◆	–	–	◆	◆
Mason Land Surveys	–	◆	–	◆	–	◆	–	◆	◆
Mathshop	–	–	–	–	–	–	–	–	◆

	farming markets	forestry market	geology market	mining market	exploration market	oil/petroleum market	marine and coastal market	utility market	civil engineering market
Mentis Management Consultants Ltd	–	◆	–	–	–	–	–	◆	◆
Methods Applications Ltd	–	–	–	–	–	–	–	◆	–
Midsummer Computing	–	–	–	◆	◆	–	–	◆	◆
Morgan Collis Group Ltd	–	–	–	–	–	–	–	◆	–
MOSS Systems Limited	–	–	–	◆	–	–	◆	–	◆
MPSI Systems Ltd	–	–	–	–	–	◆	–	–	–
MR Data Graphics	◆	◆	–	–	–	–	–	◆	–
Munro Garratt International	–	–	–	–	–	◆	–	–	–
MVA Systematica	–	–	–	–	–	–	–	◆	–
MVM Consultants plc	–	–	–	◆	–	◆	–	–	◆
NAG Ltd	◆	◆	◆	◆	–	◆	–	◆	–
National Remote Sensing Centre Ltd (NRSC)	◆	◆	◆	◆	◆	–	–	◆	◆
Natural Environment Research Council	◆	◆	–	◆	◆	–	–	◆	◆
Navstar Systems Ltd	◆	◆	◆	◆	–	–	–	–	–
NERC	◆	◆	◆	◆	◆	◆	◆	◆	◆
Numonics UK (Division of Telmtek Ltd)	◆	◆	–	◆	◆	◆	◆	◆	◆
Office of Population Censuses and Surveys (OPCS)	◆	◆	–	◆	◆	◆	–	◆	◆
Optimal Software Ltd	–	–	–	–	–	–	–	◆	◆
Oscar Faber	◆	◆	–	–	◆	◆	◆	◆	◆
Ove Arup & Partners	–	–	–	◆	–	–	◆	◆	◆
P&L Engineering Surveys Ltd	–	–	–	–	–	–	–	–	◆
PA Consulting Group	–	–	–	–	–	◆	–	◆	–
Paul Clasper & Associates Ltd	–	–	–	–	–	◆	◆	–	–
PAX Technology	–	◆	–	–	–	–	–	–	◆
Pear Technology Services Ltd	◆	◆	–	–	–	–	–	◆	–
Photarc Surveys Ltd	–	–	◆	◆	–	◆	◆	◆	◆
Photoair	◆	◆	–	◆	–	–	–	◆	◆
Photogrammetric Data Services Ltd	◆	◆	–	–	◆	–	◆	◆	◆
Pinpoint Digitising Services	–	–	–	–	–	–	–	◆	–
Planning & Mapping Ltd	–	◆	–	◆	–	–	–	◆	◆
Plowman Craven & Associates Ltd	◆	◆	–	◆	–	◆	–	◆	◆
Posford Duvivier	–	–	–	–	–	◆	◆	–	◆
Positioning Resources Limited	–	◆	◆	◆	◆	◆	–	◆	◆
Primagraphics Ltd	–	–	–	–	–	–	–	–	◆
Procis Software Ltd	–	◆	–	–	–	–	–	◆	–
Property Intelligence plc	–	–	–	–	–	–	–	◆	–
QAS Systems Ltd	–	–	◆	◆	–	–	–	◆	–
QC Data (Ireland) Limited	–	–	–	–	–	◆	–	◆	–
Racal Survey (UK) Ltd	◆	◆	◆	◆	◆	◆	◆	◆	◆
regioplanDATA GmbH	–	–	–	–	–	–	–	–	◆
Remote Sensing Applications Consultants	◆	◆	–	–	–	–	–	–	–
Robert Walker Consultants	–	–	–	–	–	–	–	–	–
Royal Institute of Navigation	–	–	◆	–	◆	◆	◆	◆	◆
Salford University Business Services Ltd	◆	◆	◆	◆	–	◆	◆	◆	◆
SAS Institute	–	◆	–	–	–	◆	–	◆	–

	farming markets	forestry market	geology market	mining market	exploration market	oil/petroleum market	marine and coastal market	utility market	civil engineering market
SAZTEC Philippines Inc	–	◆	–	–	–	◆	–	◆	◆
Scan Group Limited	◆	◆	–	◆	◆	–	–	◆	◆
Scientific Software Limited	–	–	–	◆	–	–	–	◆	–
SCOT Conseil	◆	◆	–	–	–	–	–	–	–
Scott Wilson Kirkpatrick	◆	◆	◆	◆	–	–	◆	◆	◆
SDI Ltd	–	–	–	–	–	–	–	–	–
Sector (UK) Limited	–	–	–	–	–	–	–	◆	–
SHL Vision* Solutions Limited	–	–	–	–	–	–	–	◆	–
Siemens Nixdorf	–	–	–	–	–	–	–	–	–
Silsoe College, Cranfield University	◆	◆	–	–	–	–	–	–	–
Sir Alexander Gibb & Partners Ltd	–	–	–	–	◆	–	–	◆	◆
Sir William Halcrow and Partners	◆	–	–	–	–	–	◆	–	◆
Smallworld Systems Ltd	–	–	–	–	–	–	–	◆	–
Smartscan Inc	–	◆	–	◆	◆	–	–	–	◆
Smith System Engineering Ltd	–	–	–	–	–	–	◆	◆	–
Sokkia Ltd	–	–	–	–	–	–	–	◆	◆
Solent Mapping and Charting (SMAC)	–	–	–	–	–	–	◆	–	–
Southbank Systems PLC	–	◆	–	–	–	–	–	◆	–
Sovereign C.S. Ltd	◆	◆	◆	◆	◆	◆	◆	◆	◆
Spacesense consultants	–	–	–	◆	◆	–	–	–	◆
Spatial Data Limited	◆	◆	–	◆	◆	–	–	◆	◆
Spatial Information Services Ltd (SIS)	–	–	–	–	–	◆	◆	◆	◆
SPSS UK Ltd	–	◆	–	–	–	◆	◆	◆	–
Star Informatic S.A.	–	◆	◆	◆	◆	–	–	◆	◆
Structural Technologies Ltd (STL)	–	◆	–	–	–	–	◆	◆	–
Summagraphics Europe N.V.	◆	◆	–	◆	◆	–	–	◆	◆
Survey Control Services	◆	–	–	◆	–	–	–	–	◆
Survey Supplies Ltd	–	–	–	–	–	–	–	–	–
Svitzer Limited	–	–	–	◆	–	–	–	–	◆
Sysdeco (UK) Ltd	–	–	–	–	–	–	–	◆	–
Tangent Technology Design Associates Ltd	–	◆	–	–	–	–	–	◆	–
TDS-CAD Graphics Ltd	–	◆	–	◆	◆	–	–	◆	◆
TEAMS (Taylor Woodrow Electronics Asset Mapping Survey)	–	◆	–	–	◆	◆	◆	◆	◆
Tekla Oy	–	◆	–	–	–	–	–	◆	◆
Tele Atlas	–	–	–	–	–	–	–	◆	–
Tendron Systems Ltd	–	–	–	–	–	–	–	◆	–
Tenet Systems Ltd	–	–	–	◆	–	–	–	◆	–
Terrafix Ltd	–	◆	–	–	◆	–	–	◆	–
TerraHunt GeoScience Ltd	–	–	◆	◆	◆	◆	–	◆	◆
TerraQuest Group Limited	–	–	–	–	–	–	–	◆	◆
The Data Consultancy	◆	◆	–	◆	◆	–	◆	◆	◆
The Marketing Information Consultancy (MIC)	–	–	–	–	–	–	–	◆	–
The NPA Group Ltd	◆	◆	◆	◆	◆	–	◆	–	◆
The Severn Partnership	◆	◆	–	–	–	–	–	◆	◆
The Survey Centre	◆	◆	–	◆	–	–	–	◆	◆

	farming markets	forestry market	geology market	mining market	exploration market	oil/petroleum market	marine and coastal market	utility market	civil engineering market
Trident Map Services	◆	◆	–	–	–	◆	–	◆	◆
Trimble Navigation Europe Ltd	◆	◆	–	◆	◆	–	–	◆	◆
Unistride Sewer Technology	–	–	–	–	–	–	–	◆	◆
UNISYS Ltd	–	–	–	–	–	–	–	◆	◆
Universal Systems Ltd	–	◆	–	◆	◆	–	–	◆	–
University of Luton	◆	◆	◆	◆	–	◆	◆	◆	◆
University of Newcastle Upon Tyne	–	–	–	–	–	–	–	–	◆
Walker Ladd Surveys	–	–	–	–	–	–	–	◆	◆
Wallingford Software Limited	–	–	–	–	–	–	–	◆	◆
WRc (Water Research centre)	◆	◆	–	–	–	◆	◆	◆	◆
WS Atkins Planning & Management Consultants	◆	–	◆	–	–	◆	◆	◆	◆
Xcon Data	–	–	–	–	–	–	–	◆	◆

	property markets	health market	environmental market	natural resources market	research market	transport market	navigation market	tourism market	cartography market
A.Rutherford Ltd	♦	♦	♦	♦	♦	♦	–	–	–
ACDS Graphic System Inc	–	–	♦	♦	–	–	–	–	♦
Action Information Management Ltd	–	–	♦	–	–	–	–	–	–
Active Software Ltd	–	♦	–	–	–	–	–	–	–
Advent Imaging Ltd	–	♦	–	–	–	♦	–	–	–
AGFA UK	♦	♦	♦	♦	♦	♦	♦	–	♦
AiC Analysts	–	♦	–	–	♦	–	–	–	♦
ALLM Systems & Marketing	–	–	–	–	♦	–	–	–	♦
ALTEK Corporation	♦	♦	♦	♦	♦	♦	–	–	♦
AM/FM International - European Division	♦	–	♦	–	♦	♦	–	–	♦
AND Mapping B.V.	–	–	–	–	–	♦	♦	–	♦
Anglian Engineering & International Consultancy	–	–	♦	–	–	–	–	♦	–
AP³ Imaging Services Limited	–	–	♦	♦	–	–	–	–	♦
APIC Systems	♦	♦	♦	♦	♦	♦	♦	♦	♦
Ashtech Europe Limited	–	–	♦	♦	♦	♦	♦	–	♦
Audifilm Girona S.L.	–	–	–	–	–	–	–	–	♦
Autodesk Ltd	♦	♦	♦	♦	♦	♦	♦	♦	–
Babtie Shaw & Morton Limited	–	–	♦	♦	–	♦	–	–	–
Beacon Dodsworth Limited	–	♦	–	–	–	–	–	–	–
Binnie Black & Veatch	–	–	♦	–	–	–	–	–	–
Birkbeck College London	–	–	♦	♦	♦	–	–	–	–
BKS Surveys Ltd	–	–	♦	♦	–	♦	–	–	♦
Bradly Associates Ltd	–	♦	♦	♦	♦	–	–	–	–
British Geological Survey	♦	–	♦	♦	♦	–	–	–	♦
BS International Consultants	–	♦	–	–	–	–	♦	–	–
Bull Information Systems Limited	–	♦	♦	–	–	♦	–	–	–
Business Information Management	♦	♦	♦	–	♦	♦	–	♦	♦
Byers Engineering Company	–	–	–	–	–	♦	–	–	–
C.A.Design Services Ltd	♦	–	♦	♦	–	♦	♦	–	♦
CACI Limited	♦	♦	–	–	♦	♦	–	♦	–
CAD - Capture Limited	–	♦	–	–	–	–	–	–	♦
CAD R&D Centre Limited	♦	–	–	–	–	–	–	–	–
CADAC Ltd	–	–	–	♦	–	♦	–	–	♦
CalComp Limited	–	♦	♦	–	♦	–	–	–	–
CAM - Centre for Analysis & Modelling Limited	–	♦	–	–	–	–	–	–	–
Cambashi Ltd	–	–	–	–	–	♦	–	–	–
Cambridge Computer Consultants (UK) Ltd	♦	–	♦	♦	–	♦	♦	♦	♦
Carl Bro Group	♦	–	♦	–	–	–	–	–	♦
Carl Zeiss Limited	–	–	♦	–	–	–	–	–	♦
Cartographical Services (Southampton) Limited	♦	–	♦	♦	–	♦	–	–	♦
Cartwright Associates	♦	♦	♦	♦	♦	♦	–	–	♦
CATALIST	♦	♦	–	–	–	–	–	–	–
CCN Marketing	–	–	–	–	♦	–	–	–	–
CCTA - The Government Centre for Information Systems	–	♦	–	–	–	–	–	–	–
CDD Ltd	–	–	♦	♦	–	–	–	–	♦

	property markets	health market	environmental market	natural resources market	research market	transport market	navigation market	tourism market	cartography market
CDR Group	◆	◆	◆	–	–	◆	–	–	–
Chiltern Digitising Services	–	–	◆	◆	◆	–	–	–	–
Chroson Ltd	–	–	–	–	–	–	–	–	◆
Citywise	◆	–	–	–	◆	–	–	–	–
Cliffe House Associates	–	◆	–	–	–	–	–	–	–
CMG Computer Management Group (UK) Ltd	◆	–	◆	–	–	◆	–	–	◆
Cobham Digital Services Limited	◆	◆	–	–	◆	–	–	–	◆
CODEC Facilities Limited	◆	◆	◆	◆	◆	◆	–	–	◆
ColourMap Scanning Ltd	–	–	◆	◆	–	–	–	–	◆
Computer Aided Development (CADCORP) Ltd	◆	–	◆	–	–	–	–	◆	–
COMSULT	◆	◆	◆	◆	◆	◆	◆	–	◆
Conic Systems	◆	◆	◆	–	–	◆	◆	–	–
Construction Industry Computing Association	◆	◆	–	–	◆	◆	–	–	–
Coopers & Lybrand	◆	◆	–	–	–	–	–	–	–
Cray Systems	–	◆	◆	◆	◆	◆	–	◆	◆
CSI	–	–	◆	◆	◆	–	–	–	◆
CZ Scientific Instruments Ltd	–	–	–	–	–	–	◆	–	–
DAT/EM Systems International	–	–	◆	◆	◆	◆	–	–	◆
Data Base Builders	–	–	◆	–	–	–	◆	–	◆
Data Collection Ltd	–	–	–	–	–	◆	–	–	–
Dataman Computer Solutions UK Ltd	–	◆	◆	◆	◆	◆	–	–	–
Datatechnology Datech Ltd	◆	◆	–	–	◆	◆	◆	–	–
Dataview Solutions Ltd	◆	◆	–	–	◆	◆	◆	◆	–
Design Computer Aids Limited (DeCAL)	◆	–	–	–	–	◆	–	–	–
DM Management Consultants Ltd (DMMC)	–	◆	–	–	–	–	–	–	–
DMAP Ltd	–	◆	◆	◆	◆	–	–	–	◆
DMV Consultants BV	–	–	◆	◆	◆	–	–	–	◆
Dolphin Consulting Group	–	◆	–	–	–	◆	–	–	–
Dr Stanley Port	–	◆	–	–	–	◆	–	–	–
Earth Observation Sciences Ltd	–	–	◆	◆	–	–	–	–	◆
Earth Resource Mapping	–	–	◆	◆	–	–	–	–	◆
ECM Selection Limited	–	–	–	–	◆	–	–	–	–
Effective Solutions (Data Products)	–	–	–	–	–	–	◆	–	–
Elstree Computing Ltd	◆	–	–	–	◆	–	–	–	–
Empress Software UK	–	–	◆	–	◆	–	◆	–	–
Enghouse (UK) Limited	◆	–	–	◆	–	–	–	–	–
Environment & Planning Library	–	–	–	–	◆	–	–	◆	–
EOSAT	–	–	◆	◆	◆	◆	–	–	◆
EPS - Essential Planning Systems Limited	◆	–	◆	◆	–	–	–	◆	◆
ERA Technology Ltd	◆	–	◆	◆	◆	◆	–	–	–
ERA-Maptec Ltd	–	–	◆	◆	◆	–	–	–	◆
ERDAS (UK) Ltd	◆	–	◆	◆	◆	◆	–	–	◆
ERTEC	–	–	◆	–	◆	–	–	–	◆
ESR Cartographers Ltd	–	–	◆	–	–	◆	–	◆	◆
ESRI (UK) Ltd	◆	◆	◆	◆	◆	◆	–	◆	◆

	property markets	health market	environmental market	natural resources market	research market	transport market	navigation market	tourism market	cartography market
EuroDirect Database Marketing Ltd	–	–	–	–	–	–	–	♦	–
European Business Mapping	–	♦	–	–	–	–	–	–	–
European Geographic Technologies BV	–	–	–	–	–	♦	–	–	–
Eurosense Technologies N.V.	–	–	♦	♦	♦	♦	–	–	♦
Fairbairn Services Limited	♦	♦	♦	–	–	♦	–	–	♦
FastCAD GIS Ltd	–	♦	♦	–	–	–	–	–	♦
FastCAD GIS Ltd	–	♦	–	–	–	–	–	–	♦
FileNet Ltd	–	♦	–	–	–	–	–	–	–
Flynn & Rothwell	–	–	♦	–	–	♦	–	–	–
Foto Res	♦	–	♦	♦	–	♦	–	–	♦
G.L. Consulting Ltd	–	♦	–	–	–	–	–	–	–
Gamma Ltd	♦	–	♦	♦	♦	♦	–	–	♦
Gardline Infotech	♦	–	♦	♦	–	♦	♦	–	♦
GEC Marconi Research Centre	–	–	–	–	♦	–	–	–	–
Genasys II Limited	–	–	–	–	–	–	–	–	–
General Register Office for Scotland	–	♦	–	–	♦	♦	–	–	–
Geo-Perfect TWI B.V.	–	–	♦	♦	♦	–	–	–	♦
Geo/SQL (UK)	♦	♦	♦	♦	♦	♦	–	–	♦
Geo2 Consulting	–	–	♦	–	–	♦	♦	–	♦
GEOBASE Consultants Ltd	♦	♦	♦	♦	–	♦	♦	–	♦
GeoData Institute	–	♦	♦	♦	♦	–	–	♦	–
Geodelta	♦	–	♦	–	♦	–	–	–	♦
Geografix Limited	–	–	♦	♦	–	♦	–	–	♦
Geographic Management Solutions Ltd	♦	–	♦	♦	–	♦	–	♦	♦
Geometria GIS Systems House Ltd	♦	–	–	–	–	–	♦	–	♦
Geoplan (UK) Ltd	–	♦	–	–	–	–	–	–	–
GEOSOFT Ltd	–	♦	♦	–	♦	♦	♦	–	–
Geosystems	–	–	♦	♦	♦	♦	–	–	–
Geotronics Limited	–	♦	♦	♦	♦	♦	–	–	–
Geoview Systems Kft	♦	–	–	–	–	–	–	–	–
GISL Limited	–	–	♦	♦	–	♦	–	♦	♦
Global Surveys Ltd	–	–	♦	–	–	–	–	–	–
GMAP Ltd	♦	♦	–	–	–	–	–	–	–
Graphic Data Systems Corporation (GDS)	–	–	–	–	–	♦	–	–	–
Graphical Data Capture Ltd (GDC)	♦	♦	♦	–	–	♦	–	♦	♦
Graphite Management Services Ltd	♦	♦	♦	♦	–	♦	–	♦	♦
Grove Projects Ltd	♦	–	♦	–	–	♦	–	–	♦
GTCO Corporation	♦	–	♦	♦	–	–	–	–	♦
GTX Europe Ltd	–	♦	♦	–	–	–	–	–	–
H R Wallingford Ltd	–	–	♦	♦	–	–	–	–	♦
Hansa Luftbild	–	–	♦	♦	–	♦	–	–	♦
HJM Imaging Systems	–	–	♦	♦	♦	–	–	–	♦
HollyBush Software Limited	–	–	–	♦	♦	♦	♦	–	♦
Hunting Aerofilms Limited	–	–	♦	♦	–	♦	–	–	–
Hunting Engineering Ltd	♦	♦	♦	♦	♦	♦	–	–	–

	property markets	health market	environmental market	natural resources market	research market	transport market	navigation market	tourism market	cartography market
Hunting Technical Services Ltd	–	–	◆	◆	◆	◆	–	–	◆
Husky Computing Limited	–	–	◆	–	◆	◆	–	–	–
I.S. Ltd	–	–	◆	◆	◆	◆	◆	–	◆
IBM UK Ltd	–	–	–	–	–	◆	–	–	–
ICL Ltd	◆	–	◆	◆	◆	–	–	–	–
IMASS Limited	–	◆	◆	–	–	–	–	–	–
IME (UK) Ltd	◆	–	–	–	–	–	◆	–	–
Informed Solutions Limited	◆	–	◆	–	◆	–	◆	–	◆
Ingecon B.V.	–	–	–	–	–	◆	◆	–	◆
Institute of Hydrology	◆	–	◆	◆	◆	–	–	–	–
Institute of Terrestrial Ecology	–	–	◆	◆	◆	–	–	–	–
Intera Information Technologies	◆	–	◆	◆	◆	◆	◆	–	◆
International Map Trade Association	–	–	–	–	–	–	–	◆	◆
ITC	–	◆	◆	◆	◆	–	–	–	–
ITS : Intertrade Scientific Ltd	–	–	–	–	–	–	–	–	◆
J.C.White Chartered Land Surveyors	◆	◆	–	–	–	–	–	–	–
John D Leatherdale FRICS	◆	–	–	–	–	◆	–	–	–
Kalidor Europe (ALPS Electric (Ireland) Ltd)	–	◆	◆	◆	–	◆	◆	–	◆
Kamyco International	–	–	–	–	◆	–	–	–	–
Keele University	–	–	–	–	◆	–	–	–	–
Kingswood Consulting Limited	◆	◆	–	–	–	◆	◆	◆	–
Kirstol Ltd	–	–	–	–	–	◆	–	–	–
KJB Consulting	◆	–	◆	–	–	◆	–	–	◆
Know Edge Ltd	◆	◆	◆	◆	–	◆	–	–	◆
KPMG	◆	◆	◆	◆	◆	–	–	◆	◆
L.E.S. (Computer Services) Ltd	◆	◆	◆	–	–	◆	–	–	◆
Lancaster University	–	◆	◆	–	–	–	–	◆	–
Land and Satellite Surveys	–	–	◆	–	–	–	–	–	–
Land Aspects Consultancy Ltd (Parkman Group)	◆	◆	◆	◆	◆	◆	–	–	◆
Lantmatenet GIS-centrum	–	◆	◆	◆	◆	◆	–	–	◆
Laser-Scan Ltd	–	–	◆	–	◆	–	–	–	◆
Leica UK Ltd	–	–	–	–	–	–	◆	–	◆
LiveChart	–	–	–	–	–	–	◆	–	–
Logica UK Limited	–	–	◆	◆	–	–	–	–	–
Logitrans	–	–	–	–	–	◆	–	–	◆
London Research Centre	–	◆	–	–	◆	◆	–	–	–
Longdin & Browning	◆	–	–	–	–	◆	–	–	–
Lovell Johns Ltd	–	–	◆	◆	–	–	–	◆	◆
LOY Surveys Ltd	◆	◆	◆	◆	–	–	–	◆	–
LTG Services	◆	◆	◆	◆	–	–	–	–	◆
Macaulay Land Use Research Institute	–	–	◆	◆	◆	–	–	–	–
Magdala Sociedade	◆	–	–	–	–	–	–	–	–
Magellan Systems Corporation	–	◆	◆	◆	◆	◆	◆	–	◆
Map Data Management Ltd	◆	–	–	–	–	–	–	◆	◆
MAPIT Limited	◆	–	–	–	–	–	◆	–	–

	property markets	health market	environmental market	natural resources market	research market	transport market	navigation market	tourism market	cartography market
MAPS geosystems	–	–	–	–	–	◆	–	–	◆
Mason Land Surveys	◆	◆	◆	◆	–	◆	–	–	◆
Mathshop	–	–	–	–	◆	◆	–	–	◆
Mentis Management Consultants Ltd	◆	◆	◆	◆	◆	◆	–	–	◆
Methods Applications Ltd	◆	◆	–	–	–	–	–	–	–
Midlands Regional Research Laboratory	–	◆	◆	◆	–	–	–	–	–
Midsummer Computing	◆	◆	◆	◆	◆	◆	◆	–	–
Modern Maps	–	–	◆	–	–	–	◆	◆	–
MOSS Systems Limited	–	–	–	–	–	◆	–	–	–
MR Data Graphics	–	◆	–	–	–	–	–	–	◆
MVA Systematica	◆	◆	◆	–	–	◆	◆	–	–
MVM Consultants plc	◆	–	◆	–	◆	–	–	–	–
NAG Ltd	–	◆	◆	◆	◆	–	–	–	–
National Remote Sensing Centre Ltd (NRSC)	–	–	◆	◆	–	◆	–	–	◆
Natural Environment Research Council	◆	–	◆	◆	◆	–	–	–	◆
Navstar Systems Ltd	–	–	◆	◆	◆	◆	◆	–	◆
NERC	–	–	◆	◆	◆	–	–	–	◆
NOMIS	◆	◆	–	–	◆	–	–	–	–
Numonics UK (Division of Telmtek Ltd)	◆	◆	◆	◆	◆	–	◆	–	◆
Oaklands I.T.	–	–	◆	–	–	–	–	–	◆
Office of Population Censuses and Surveys (OPCS)	◆	◆	◆	–	◆	◆	◆	◆	–
Organisation Management Systems	◆	◆	–	–	–	–	–	–	–
Oscar Faber	◆	◆	◆	–	–	◆	–	–	–
Ove Arup & Partners	◆	–	◆	◆	–	◆	–	◆	–
PA Consulting Group	–	◆	◆	–	–	◆	–	–	–
Panda	–	–	–	–	–	◆	–	–	◆
PAX Technology	◆	◆	◆	–	◆	–	–	–	◆
PD Computing Ltd	◆	◆	–	–	–	–	–	–	–
Pear Technology Services Ltd	–	–	–	–	–	–	◆	–	–
Peter Thorpe Consulting	◆	–	◆	–	–	–	–	–	–
Photarc Surveys Ltd	–	–	–	–	–	–	–	–	◆
Photoair	–	◆	◆	–	–	–	–	–	◆
Photogrammetric Data Services Ltd	◆	–	◆	◆	–	◆	◆	–	◆
Pinpoint Digitising Services	–	◆	–	–	–	–	–	–	–
Planning & Mapping Ltd	–	–	–	–	–	–	–	–	◆
Plowman Craven & Associates Ltd	◆	◆	◆	◆	◆	◆	–	◆	◆
Posford Duvivier	–	–	◆	–	–	◆	◆	–	–
Positioning Resources Limited	–	–	–	◆	–	◆	–	–	–
Procis Software Ltd	–	–	–	–	–	◆	–	–	–
QAS Systems Ltd	◆	◆	◆	–	–	◆	◆	◆	–
Quail Map Company	–	–	–	–	–	◆	–	–	◆
Racal Survey (UK) Ltd	–	–	◆	◆	–	◆	–	–	◆
regioplanDATA GmbH	◆	–	◆	–	–	–	–	–	◆
Remote Sensing Applications Consultants	–	–	◆	◆	◆	–	–	–	–
Royal Institute of Navigation	–	–	◆	◆	◆	◆	◆	◆	◆

	property markets	health market	environmental market	natural resources market	research market	transport market	navigation market	tourism market	cartography market
Royal Town Planning Institute	–	–	♦	–	–	–	–	–	–
Salford University Business Services Ltd	♦	♦	♦	♦	♦	–	–	♦	♦
SAS Institute	–	♦	♦	–	♦	–	–	♦	–
SAZTEC Philippines Inc	♦	–	♦	♦	–	♦	–	–	♦
Scan Group Limited	♦	♦	♦	♦	♦	♦	♦	–	♦
SCOT Conseil	–	–	♦	♦	–	–	–	–	♦
Scott Wilson Kirkpatrick	–	♦	♦	♦	♦	♦	–	♦	♦
SDI Ltd	♦	–	♦	–	–	♦	–	–	–
Sector (UK) Limited	–	–	–	–	–	♦	–	♦	–
SHL Vision* Solutions Limited	–	–	♦	♦	♦	♦	–	–	–
SIA Limited	♦	–	–	–	–	♦	–	–	–
SIAS Limited	–	–	–	–	–	♦	–	–	–
Siemens Nixdorf	–	–	♦	♦	–	–	–	–	–
Silsoe College, Cranfield University	–	–	♦	♦	♦	–	–	–	–
Sir Alexander Gibb & Partners Ltd	♦	–	♦	♦	–	♦	–	–	–
Sir William Halcrow and Partners	–	–	♦	–	–	♦	–	–	–
Smallworld Systems Ltd	–	–	–	–	♦	♦	–	–	♦
Smartscan Inc	–	♦	♦	♦	–	♦	♦	–	–
Smith System Engineering Ltd	–	♦	♦	♦	♦	♦	♦	–	–
Southbank Systems PLC	♦	♦	♦	–	–	–	♦	–	–
Sovereign C.S. Ltd	♦	♦	♦	♦	–	♦	–	♦	♦
Spacesense consultants	–	–	♦	♦	♦	–	–	–	–
Spatial Data Limited	–	♦	♦	–	–	♦	♦	–	–
Spatial Geographic Services & Applications Ltd	–	–	♦	–	–	–	–	–	♦
Spatial Information Services Ltd (SIS)	–	–	♦	♦	–	♦	–	–	♦
SPSS UK Ltd	–	♦	♦	–	♦	♦	–	♦	–
Star Informatic S.A.	♦	♦	♦	♦	♦	♦	–	–	♦
Structural Technologies Ltd (STL)	–	–	♦	–	♦	♦	♦	–	–
Summagraphics Europe N.V.	♦	–	–	–	♦	–	–	–	♦
Survey Control Services	♦	–	♦	–	–	♦	–	–	–
Svitzer Limited	–	–	♦	♦	–	–	♦	–	–
Sysdeco (UK) Ltd	♦	–	–	–	–	–	–	–	♦
TACTICIAN UK	♦	♦	–	–	–	–	–	–	–
Tangent Technology Design Associates Ltd	–	♦	–	–	–	♦	–	–	–
Taylor & Francis Ltd	–	–	–	–	♦	–	–	–	–
TDS-CAD Graphics Ltd	–	♦	–	–	–	–	–	–	–
TEAMS (Taylor Woodrow Electronics Asset Mapping Survey)	♦	–	–	♦	–	♦	–	♦	–
Tele Atlas	–	♦	–	–	♦	♦	♦	♦	♦
Tendron Systems Ltd	–	♦	–	♦	–	♦	♦	–	–
Tenet Systems Ltd	–	♦	–	–	♦	♦	–	–	♦
Terrafix Ltd	–	♦	–	–	♦	♦	♦	–	–
TerraHunt GeoScience Ltd	♦	–	♦	♦	–	–	♦	–	–
TerraQuest Group Limited	♦	♦	♦	♦	–	♦	–	–	♦
The Data Consultancy	♦	♦	♦	♦	♦	♦	♦	♦	–
The Marketing Information Consultancy (MIC)	♦	♦	–	–	♦	–	–	♦	–

	property markets	health market	environmental market	natural resources market	research market	transport market	navigation market	tourism market	cartography market
The NPA Group Ltd	–	–	♦	♦	♦	–	–	♦	♦
The Severn Partnership	♦	–	–	–	–	–	–	–	♦
The Survey Centre	♦	–	♦	♦	–	♦	–	–	♦
Trident Map Services	♦	–	♦	–	–	♦	–	♦	♦
Trimble Navigation Europe Ltd	–	–	♦	♦	–	♦	♦	–	♦
UNISYS Ltd	–	–	♦	–	♦	♦	♦	–	♦
Universal Systems Ltd	♦	–	♦	♦	♦	♦	♦	–	♦
University of Luton	–	♦	♦	♦	–	♦	–	♦	♦
University of Newcastle Upon Tyne	♦	–	–	–	♦	–	–	–	–
Walker Ladd Surveys	♦	♦	–	–	–	–	–	–	–
Wallingford Software Limited	–	–	♦	–	–	–	–	–	–
WRc (Water Research centre)	♦	–	♦	–	♦	–	–	–	–
WS Atkins Planning & Management Consultants	♦	♦	♦	♦	♦	♦	–	♦	♦
Xcon Data	♦	–	♦	♦	–	♦	♦	–	♦

directory 38:
user group information

ACDS Graphic System Inc
contact: David A Olson
President of the MAPS User Groups
1790 Nautilus Street
La Jolla
CA 92037
telephone: +619 456 7861
1 meeting per year

Active Software Ltd
contact: Paul Smith
Datum House
Roentgen Road
Basingstoke
Hampshire
RG24 8NG
telephone: 01256 56629
1 meeting per year

AM/FM International - European Division
contact:
AM/FM International, European
Division
P.O. Box 6
CH-4005 Basel
Switzerland
telephone: +41 61 6915111
1 meeting per year

AP³ Imaging Services Limited
contact: Miss Emma Bailey
Unit 3B
Bull Lane Industrial Estate
Action
Sudbury
Suffolk
CO10 0BD
telephone: 01787 378242
1 meeting per year

APIC Systems
contact: Pierre Pijourlet
Communaute Urbaine de Lyon
Centre de donnees urbaines
20, Rue du Lac
69399 Lyon Cedex 03
France
telephone: +33 16 78 63 43 90
fax: +33 16 78 63 41 99
3 meetings per year

ARC Systems Pty Ltd
contact: Tony Slocombe
Port of Brisbane
Authority
GPO Box 1818
Brisbane
Queensland
Australia
4001
telephone: +617 8330807
12 meetings per year

Assist Applications Limited
contact: Nigel Brown
User Group Secretary
Wolverhampton M.B.C.
Civic Centre
St. Peter's Square
Wolverhampton
WV1 1RP
telephone: 01902 315610
4 meetings per year

Autodesk Ltd
contact: Richard Hart
Mapping & GIS Special
Interest Group
c/- ELS Land Consultants
Ltd
telephone: 01483 751222
fax: 01483 750687
1 meeting per year

Beacon Dodsworth Limited
contact: Peter Bell
CDMS Limited
4th Floor Standbrook
House
Old Bond Street
London W1X 3TB
telephone: 0171 495 4195
fax: 0171 495 4144
1 meeting per year

Bentley Systems UK Ltd
contact: Norman Mitchell
c/- Sheppard Robson
77 Parkway
Camden Town
London NW1 7PU
telephone: 0171 485 4161
fax: 0171 267 3861
1 meeting per year

Bradly Associates Ltd
contact: Dr Monica Dowson
Secretary, GINO User Group
School of Computing &
Mathematical Sciences
De Montfot University
The Gateway
Leicester
LE1 9BH
telephone: 01533 577462
2 meetings per year

CAD R&D Centre Limited
contact: Slavka Mitova
CAD R & D Centre
P.O. Box 112
Sofia 1113
Bulgaria
telephone: 359 2 705257
fax: 359 2 703556
1 meeting per year

CCN Marketing
contact: Debbie Yonge
Talbot House
Talbot Street
Nottingham
NG1 5HF
telephone: 0115 9410888
fax: 0115 9344903
2 meetings per year

CDR Group
contact: Martin Waters
CDR Group
Birchfield Hall
Aston Lane
Hope
Sheffield
S30 2RA
telephone: 01433 621282
fax: 01433 621292
1 meeting per year

CODEC Facilities Limited
contact: Benoit Fettweis
Star Informatic SA
Park Scientifique du Sart-Tilman
Avenue du Pre Aily, 24
B-4031 Angleur-Liege
Belgium
telephone: +32 41 675313
fax: +32 41 671711
2 meetings per year

**Computer Aided Development
(CADCORP) Ltd**
contact: Joe Burroughs
CADCORP
Sterling Court
Norton Road
Stevenage SGI 2JY
telephone: 01438 747996
fax: 01438 747997
4 meetings per year

**Consensus Information
Technology Ltd**
contact: Consensus
The Paddock
Handforth
Cheshire SK9 3HQ
telephone: 01625 537777
fax: 01625 539621
3 meetings per year

Dataview Solutions Ltd
contact: Terry Tysoe
Secretary
MapInfo User Group (MUG UK)
c/- Milton Keynes Borough Council
1 Saxon Gate East
Central Milton Keynes
MK9 3HQ
telephone: 01908 682760
fax: 01908 682319
4 meetings per year

**Design Computer Aids Limited
(DeCAL)**
contact: Hamish Thomson
Property Department
Tayside Regional Council
Crichton Street
Dundee
DD1
telephone: 01382 303623
4 meetings per year

Elstree Computing Ltd
contact: Barry Blake
PDS User Group
Elstree Computing Limited
133 - 139 Page Street
Mill Hill
London
NW7 2ER
telephone: 0181 9067 5656
2 meetings per year

Empress Software UK
contact: Martyn Bright
Technical Resources (Midlands) Ltd
22 Maryland Close
Barwell
Leicester
LE9 8ES
telephone: 01455 840004
2 meetings per year

Enghouse (UK) Limited
contact: Harry Mowat
Enghouse (UK) Ltd
18 The Business Village
Tollgate
Eastleigh
Hampshire SO53 3TG
telephone: 01703 615228
fax: 01703 615253
3 meetings per year

**EPS - Essential Planning Systems
Limited**
contact: Alison Malis
200 - 6772 Oldfield Road
Victoria
BC
Canada
V8M 2A2
telephone: +604 652 8845
fax: +604 652 8896
1 meeting per year

ERDAS (UK) Ltd
contact: Dr Andy Wells
School of Earth Sciences
University of Greenwich
Medway Towns Campus
Chatham Maritime
Kent
ME4 4AW
telephone: 0181 331 9838
fax: 0181 331 9805
2 meetings per year

ESRI (UK) Ltd
contact: Peter Beaumont
NRSC Ltd
Delta House
Southwood Crescent
Southwood
Farnborough
GU14 0NL
telephone: 01252 541464
4 meetings per year

Eurosense Technologies N.V.
contact: ARC/INFO User's Group
Belgium & Luxembourg
p/a ISSep
rue de Chera 200
4000 Liege
Belgium
+32 41 527150
1 meeting per year

FastCAD GIS Ltd
contact: Louise Solomon
10 Cotham Road South
Cotham
Bristol
BS6 5TZ
telephone: 0117 942 8195
fax: 0117 942 8196
2 meetings per year

FastCAD GIS Ltd
contact: Mrs L Solomon
FastCAD GIS Ltd
10 Cotham Road South
Cotham
Bristol
BS6 5TZ
telephone: 01179 428195
fax: 01179 428196
3 meetings per year

FileNet Ltd
contact: Julie Edwards
Britannia Building Society
Britannia House
Cheadle Road
Leel
Staffordshire
ST13 5RG
telephone: 01538 399399
fax: 01538 399149
3 meetings per year

Genasys II Limited
contact: Mr Allen Elston
Central Services Department
Swansea City Council
The Guild hall
Swansea
SA1 4PE
telephone: 01792 302330
fax: 01792 467432
2 meetings per year

**General Register Office for
Scotland**
contact: Peter Jamieson
Secretary Scottish Census Users
Group
c/- General Register Office for
Scotland
Ladywell House
Ladywell House
Edinburgh
EH12 7TF
telephone: 0131 314 4254
fax: 0131 314 4344
4 meetings per year

Geops BV
contact: Geops BV
Agro Business Park 36
6708 PW Wageningen
The Netherlands
+31 8370 79636
2 meetings per year

Geotronics Limited
contact: John Davey
Walker LADD Partnership
9 Park Place
Clifton
Bristol
BS8 1JP
4 meetings per year

Global Surveys Ltd
contact: Inst C.E.S.

Graphic Data Systems Corporation (GDS)
contact: John Warburton
EDS User Group
1 Mill Cottage
Birdsedge
Huddersfield
West Yorkshire
HD8 8XU
telephone: 01484 602197
1 meeting per year

Graphical Data Capture Ltd (GDC)
contact: John Allen
MapInfo User Group
c/- AT&T
P.O. Box 5,
Ravens bank Drive
Redditch
B98 9HB
telephone: 01527 492160
3 meetings per year

HollyBush Software Limited
contact: HollyBush Software Ltd
Innovation House
Dr William Price Business Centre
Pontypridd
MidGlamorgan
CF37 1TJ
UK
telephone: 01443 482785
fax: 01443 482788
2 meetings per year

I.S. Ltd
contact: Neil Quarmby
EASI/PACE User Group
c/- I.S. Ltd
Maggs House
78 Queens Road
Bristol
BS8 1QX
telephone: 01179 250553
fax: 01179 250663
1 meeting per year

IBM UK Ltd
contact: Roger H Morgan
Thames Water Utilities Ltd
Nugent House
Vastern Road
Reading
Berkshire
RG1 8DB
telephone: 01734 593623
1 meeting per year

ICL Ltd
contact: B Hall
c/- ICL
Observatory House
Windsor Road
Slough
SL1 2EY
1 meeting per year

Ingecon B.V.
contact: European AeroSig
P.O. Box 164
6720 AD Bennekom
The Netherlands
telephone: +31 8389 16681
fax: +31 8389 18648
2 meetings per year

Intera Information Technologies
contact: Robert Dams
200, 2 Gurdwana Road
Nepean
Ontario
Canada
K2E 1A2
telephone: +613 226 5442
1 meeting per year

Intergraph (UK) Ltd
contact: Ken Robinson
Commission for New Towns
Saxon Court
502 Avebury Boulevard
Central Milton Keynes MK9 3HS
telephone: 01908 692692
fax: 01908 691333
4 meetings per year

Laser-Scan Ltd
contact: Peter Haywood (Chairman)
Ordnance Survey
Romsey Road
Maybush
Southampton SO16 4GU
telephone: 01703 492675
fax: 01703 792404
1 meeting per year

Logitrans
contact: William Gorter
LOGITRANS
34 av. Aristide Briand
94110 Arcueil
France
telephone: +331 49 851516
1 meeting per year

London Research Centre
contact: Hywel Davies
SASPAC User Group
London Research Centre
81 Black Prince Road
London SE1 7SZ
telephone: 0171 627 9696
fax: 0171 627 9606
1 meeting per year

LTG Services
contact: S.Pittard
LTG Services
Page Street
Mill Hill
London
NW7 2ER
telephone: 0181 906 5047
4 meetings per year

Macaulay Land Use Research Institute
contact: Dr Richard Aspinall
Land Cover Data User Group
Macaulay Land Use Research
Institute
Craigiebuckler
Aberdeen ABE 2QJ
telephone: 01224 318611
fax: 01224 311556
1 meeting per year

MapInfo Ltd
contact: Terry Tysoe
Secretary
MapInfo User Group (MUG UK)
c/- Milton Keynes Borough Council
1 Saxon Gate East
Central Milton Keynes
MK9 3HQ
telephone: 01908 682760
fax: 01908 682319
4 meetings per year

Monroe Garrett International
contact: Linda Stein
c/- Monroe Garrett International
4 Albert Street
Aberdeen AB1 1XQ
telephone: 01224 622888
fax: 01224 622229
1 meeting per year

Morgan Collie Group Ltd
contact: Ian Hollered
Morgan Collie Group Ltd
The Mill
Mill brook Close
St. James Mill Road
Northampton NN5 5JF
telephone: 01604 580980
fax: 01604 589208
2 meetings per year

MOSS Systems Limited
contact: Pam George
MOSS Systems Limited
MOSS House
North Heath Lane
Hiroshima
RH12 5QE
telephone: 01403 272750
fax: 01403 272750
4 meetings per year

MVA Systematica
contact: Martin Bach
TRIPS User Group
MVA House
Victoria Way
Woking
Surrey
GU21 1DD
telephone: 01483 728051
fax: 01483 755207
1 meeting per year

NAG Ltd
contact: Emmanuel Vergison
Solvay SA
310 Rue de Ransbeek
1120 Brussels
Belgium
telephone: +32 2 264 2164
fax: +32 2 264 3061
1 meeting per year

NCC Blackwell Ltd
contact: International SSADM
Users Group
21 Windsor Forest Court
Mill Ride
Ascot
Berkshire
SL5 8LT
telephone: 01344 884415
fax: 01344 884413
1 meeting per year

Office of Population Censuses and Surveys (OPCS)
contact: Alan Taylor
Central Postcode Directory User
Group (Public Sector)
c/- OPCS
Segensworth Road,
Titchfield
Fareham
Hampshire
PO15 5RR
telephone: 01329 813536
fax: 01329 813532
2 meetings per year

Ordnance Survey
contact: Ordnance Survey
Liaison Committees
Mr Tom Leaney
Secretary
Royal Society OS Scientific Committee
Royal Society
6 Charlton House Terrace
London SW1Y 5AG

telephone: 0171 839 5561
fax: 0171 930 2170

Miss Karen Killaspy
Secretary
Standing Committee of Professional Map Users
12 Great George Street
London SW1P 3AD

telephone: 0171 222 7000
fax: 0171 334 3795

Mr A K Black/Ms Rosie Somerville
Secretary
Local Authorities OS Committee
Local Government Management
Board
Arndale House
The Arndale Centre
Luton LU1 2TS
telephone: 01582 451 166
fax: 01582 412 525

Ms Anne Taylor
Secretary
BRICMICS OS Committee
British Library Map Library
Great Russell Street
London WC1B 3DG
telephone: 0171 323 7703
fax: 0171 323 7780

Mrs Irene Elsom
Secretary
NJUG OS Committee
NJUG
30 Millbank
London SW1P 4RD
telephone: 0171 344 5723
fax: 0171 630 8820

Dr Andrew Tatham
Secretary
Royal Geographical Society OS Education Committee
Royal Geographical Society
1 Kensington Gore
London SW7 2AR
telephone: 0171 589 5466
fax: 0171 584 4447

Miss Annette Cairncross
Secretary
Central Council of Physical Recreation OS Committee
Central Council of Physical
Recreation
Francis House
Francis Street
London SW1P 1DE

telephone: 0171 828 3163
fax: 0171 630 8820

Mr D I Bill
Secretary
Public Services OS Committee
Consultative Committee Liaison
Room N108
Ordnance Survey
Romsey Road
Maybush
Southampton SO16 4GU
telephone: 01703 792545
fax: 01703 792404

Ordnance Survey of Northern Ireland
contact: Mr G Mitchell
NI Geographic Information System
Liaison Committee
Colby House
Stranmillis Court
Belfast
telephone: 01232 255755
fax: 01232 255700
1 meeting per year

PAFEC Ltd
contact: Amanda Ward
PAFEC Ltd
Strelley Hall
Nottingham
NG8 6PE
telephone: 0115 935 7055
fax: 0115 935 7057
1 meeting per year

PD Computing Ltd
contact: Mr T Dawes
IT Manager
Havant Borough Council
Hampshire
PO9 2AX
telephone: 01705 446390
2 meetings per year

Procis Software Ltd
contact: R.Webb
Southern Water Services
Southern House
Lewes Road
Falmer
Brighton
BN1 9PY
telephone: 01273 606766
fax: 01273 675239
4 meetings per year

Property Intelligence plc
contact: c/- Property Intelligence plc
13 - 15 John Adam Street
London
WC2N 6LD
telephone: 0171 839 7684
fax: 0171 839 1060
1 meeting per year

Racal Survey (UK) Ltd
contact: UK Civil Satnav Group
The Royal Institute of Navigation
1 Kensington Gore
London
SW7 2AT
telephone: 0171 589 5021
fax: 0171 823 8671
1 meeting per year

Royal Institute of Navigation
contact: Royal Institute of
Navigation
1 Kensington Gore
London
SW7 2AT
telephone: 0171 589 5021
fax: 0171 823 8671
8 meetings per year

Royal Mail - Address Management Centre
contact: Paul J McCartney
OXFAM
274 Banbury Road
Oxford
OX2 7DZ
telephone: 01865 312252
4 meetings per year

SAS Institute
contact: Helen Hardenberg
SAS Institute
PO Box 105340,
Nevenheimer Landstr 28 - 30
D-69043 Heidelberg
Germany
telephone: +49 6221 4160
fax: +49 6221 474850
1 meeting per year

SIA Limited
contact: Janice Mabert
Secretary
dataMAP User Group
Ebury Gate
23 Lower Belgrave Street
London
SW1W 0NW
telephone: 0171 730 4544
fax: 0171 730 6772
2 meetings per year

Siemens Nixdorf
contact: Andrea Schmitz
c/- Siemens Nixdorf
(Informationssysteme AG)
Carl Wery Strasse 22
0-81730 Munich
Germany
telephone: + 4989 636 44731
fax: + 4989 636 45202
1 meeting per year

Smallworld Systems Ltd
contact: Alan Bourke
Scottish Homes
Thistle House
91 Haymarket Terrace
Edinburgh E12 5YA
telephone: 0131 313 0044
fax: 0131 479 5252
2 meetings per year

SPSS UK Ltd
contact: Dane Phillips
Assess
Manchester Metropolitan University
SITU, Dept. of Social Science
Chatham/Undercroft Buildings
Cavindish Street
Manchester M15 6BR
telephone: 0161 247 6312
4 meetings per year

Star Informatic S.A.
contact: Benoit Fettweis
STAR Informatic SA
PARC Scientifique du Sart-Tilman
Avenue du Pre Aily, 24
B-4031 Angleur-Liege
Belgium
telephone: +32 41 675313
fax: +32 41 671711
6 monthly

Structural Technologies Ltd (STL)
contact: Mr.D.Schindler
Structural Technolgoies Ltd
Woodside
The Slough
Studley
Warwickshire B80 7EN
telephone: 01527 854819
fax: 01527 854819
1 meeting per year

Sun Microsystems Ltd
contact: Sun UK User Group
Owles Hall
Buntingford
Hertfordshire SG9 9PL
telephone: 01763 271894
fax: 01763 273255
4 meetings per year

Survey & Development Services Ltd
contact: Mike Peascod
Drawbase User Group
104 Durley Avenue
Pinner
Middlesex HA5
telephone: 0181 868 0851

Symology Limited
contact:
Millfield Lane
Caddington
Bedfordshire
LU1 4AJ
telephone: 01582 842626
fax: 01582 842600
3 meetings per year

Sysdeco (UK) Ltd
contact: Jerry Fisher
Surrey Heath Borough Council
Surrey Heath House
Knoll Road
Camberley
Surrey GU15 3HD
telephone: 01276 686252
fax: 01276 22277
2 meetings per year

System Options Limited
contact: Jim Pedroza
System Options Ltd
45 Victoria Road
Aldershot
Hampshire
GU11 1SJ
telephone: 01252 334383
fax: 01252 28779
2 meetings per year

Tekla Oy
contact: Risto Raty
Koronakatu 1
02210 Espoo
Finland
telephone: +358 0 8879 432
fax: +358 0 8039 489
1 meeting per year

Tele Atlas
contact: A.Bastiaansen
Tele Atlas User Group
c/- Moutstraat 92
9000 Gent
Belgium
telephone: +32 9 222 5658
fax: +32 9 222 7412
1 meeting per year

Terrafix Ltd
contact: I. Van-Creveld
Falcon House
4th Floor
West Midlands Ambulance Service
Dudley
West Midlands
12 meeting per year

The British Computer Society
contact: Frank Bennion
3 Rowley Close
Loggerheads
Market Drayton
Shropshire
TF9 4DB
telephone: 01630 672067
5 meeting per year

The Data Consultancy
contact: Terry Tysoe
Secretary
MapInfo User Group (MUG UK)
c/- Milton Keynes Borough Council
1 Saxon Gate East
Central Milton Keynes
MK9 3HQ
telephone: 01908 682760
fax: 01908 682319
4 meetings per year

The LGMB
contact: Rosi Somerville
The LGMB
Arndale House
The Arndale Centre
Luton
Bedfordshire
LU1 2TS
telephone: 01582 451166
fax: 01582 412525
1 meeting per year

The NPA Group Ltd
contact: Ren Capes
The NPA Group
1 Fircroft Way
Edenbridge
Kent
TN8 6HS
telephone: 01732 865023
fax: 01732 866521
1 meeting per year

TYDAC Technologies Ltd
contact: James Oliver
Tydac Technologies Ltd
2 Venture Road
Chilworth Research Centre
Southampton
SO1 7NP
telephone: 01703 760824
fax: 01703 760944
1 meeting per year

UNISYS Ltd
contact: Paul Coward
System 9 User Group
GIS Unit
Technology House
Lissadel Street
Salford
M6 6AP
telephone: 0161 957 0012
fax: 0161 737 7700
1 meeting per year

Universal Systems Ltd
contact: Marine Information
Technology
Eekhout straat 2,
P.O. Box 5068
3087 AB
Rotterdam
The Netherlands
telephone: +31 10 428 3385
1 meeting per year

Walker Ladd Surveys
contact: Ian Briffett
Blue Moon Systems
8 Civil Road
Portsmouth
PO1 5JE
telephone: 01243 373504
1 meeting per year

Wallingford Software Limited
contact: Eric Keasberry
WaPUG
Manchester City Council
City Engineer & Surveyor's Dept.
Town Hall
Albert Square
Manchester
ML0 2JT
telephone: 0161 234 4054
2 meetings per year

directory 39:
location directory
UK only

Aberdeen	**ERTEC** 01224 740324
Aberdeen	**Macaulay Land Use Research Institute** 01224 318611
Aberdeen	**Munro Garratt International** 01224 622888
Aberdeen	**Positioning Resources Limited** 01224 581502
Aberdeen	**Racal Survey (UK) Ltd** 01224 249700
Aldermaston	**Colorgraph (UK) Ltd** 01734 819435
Aldershot	**Quorum Information Services** 01252 318884
Aldershot	**System Options Limited** 01252 334383
Alloa	**IME (UK) Ltd** 01259 210210
Alton	**Remote Sensing Applications Consultants** 01420 561377
Ampthill	**Hunting Engineering Ltd** 01525 841000
Andover	**Chroson Ltd** 01264 336339
Arundel	**LASCO Ltd** 01903 882466
Ashford	**KJB Consulting** 01233 756461
Ashstead	**G.L. Consulting Ltd** 01372 272937
Axbridge	**Simmons Survey Partnership Limited** 01934 732122
Bagshot	**GEO-UK Ltd** 01276 473579
Bagshot	**Sun Microsystems Ltd** 01276 451440
Barnet	**John D Leatherdale FRICS** 0181 449 0123
Barton on Sea	**International Map Trade Association** 01425 620532
Basingstoke	**Active Software Ltd** 01256 56629
Basingstoke	**GTX Europe Ltd** 01256 843555
Basingstoke	**Scott Wilson Kirkpatrick** 01256 461161
Bath	**Paul Clasper & Associates Ltd** 01225 444561
Bedford	**COMSULT** 01234 342401
Belfast	**Ordnance Survey of Northern Ireland** 01232 255755
Billingshurst	**Planning & Mapping Ltd** 01403 783314
Birdlip	**Hall & Watts Systems Limited** 01452 864244
Birmingham	**Bull Information Systems Limited** 0121 717 0777
Birmingham	**Dotted Eyes** 0121 445 6150
Birmingham	**Global Surveys Ltd** 0121 421 1414
Birmingham	**ICL Ltd** 0121 456 1111
Bishop's Stortford	**APIC Systems** 01279 466966
Bishop's Stortford	**Flynn & Rothwell** 01279 507346
Blackburn	**CAD - Capture Limited** 01254 583534
Blackburn	**TDS-CAD Graphics Ltd** 01254 676921
Blairgowrie	**GeoMEM Software** 01250 872284
Bo'ness	**Survey & Development Services Ltd** 01506 825121
Borehamwood	**CZ Scientific Instruments Ltd** 0181 953 1688
Borehamwood	**Hunting Aerofilms Limited** 0181 207 0666
Borehamwood	**SDI Ltd** 0181 207 5474
Bracknell	**Bentley Systems UK Ltd** 01344 412233

Bracknell	**MapInfo Ltd** 01344 482888
Bracknell	**Siemens Nixdorf** 01344 850829
Bradford	**EuroDirect Database Marketing Ltd** 01274 737144
Brentford	**AGFA UK** 0181 231 4141
Brighton	**Corbins Consultancy** 01273 553110
Brighton	**European Business Mapping** 01273 702957
Brighton	**IT Southern Ltd** 01273 600444
Bristol	**CATALIST** 0117 923 7113
Bristol	**Dataflow Information Systems** 0117 927 2466
Bristol	**FastCAD GIS Ltd** 0117 942 8195
Bristol	**I.S. Ltd** 0117 925 0553
Bristol	**MPSI Systems Ltd** 0117 927 9653
Bristol	**MVM Consultants plc** 0117 974 4477
Bristol	**Tendron Systems Ltd** 0117 929 4759
Bristol	**Walker Ladd Surveys** 0117 925 1251
Byfleet	**ESR Cartographers Ltd** 01932 348981
Caddington	**Symology Limited** 01582 842626
Camberley	**Data Dictionary Systems Limited** 01276 23519
Camberley	**SHL Vision* Solutions Limited** 01276 677707
Cambridge	**A.Rutherford Ltd** 01223 872646
Cambridge	**AiC Analysts** 01223 300044
Cambridge	**Cambashi Ltd** 01223 460439
Cambridge	**CARTograph Ltd** 01223 67818
Cambridge	**Construction Industry Computing Association** 01223 236336
Cambridge	**DCL Consulting** 01223 314888
Cambridge	**ECM Selection Limited** 01638 742244
Cambridge	**ERDAS (UK) Ltd** 01223 880802
Cambridge	**Geoinformation International** 01223 423020
Cambridge	**Laser-Scan Ltd** 01223 420414
Cambridge	**Panda** 01223 233577
Cambridge	**Smallworld Systems Ltd** 01223 460199
Cambridge	**Sysdeco (UK) Ltd** 01223 420464
Camforth	**Map Data Management Ltd** 015395 67431
Cardiff	**Ove Arup & Partners** 01222 473727
Chatham Maritime	**Southbank Systems PLC** 01634 880141
Chatham Maritime	**University of Greenwich** 0181 331 9800
Chelmsford	**GEC Marconi Research Centre** 01245 473331
Chertsey	**SPSS UK Ltd** 01932 566262
Chester	**EA Technology** 0151 347 2451
Cobham	**Cobham Digital Services Limited** 01932 868133
Colchester	**Aneberie CAD** 01206 331215
Coleraine	**BKS Surveys Ltd** 01265 52311
Cottenham	**Robert Walker Consultants** 01954 251003
Coventry	**Husky Computing Limited** 01203 604040
Crewe	**Sokkia Ltd** 01270 250525
Crowthorne	**Bradly Associates Ltd** 01344 779381
Croydon	**GGP Systems Limited** 0181 656 8562
Croydon	**Infolink Decision Services Limited** 0181 686 7777
Croydon	**Mott MacDonald Ltd** 0181 686 5041
Dagenham	**University of East London** 0181 849 3618
Derby	**Numonics UK (Division of Telmtek Ltd)** 01332 298480

Didcot	**Geosystems** 01235 813913	
Dunfermline	**Mason Land Surveys** 01383 727261	
Dunstable	**Trident Map Services** 01582 867211	
Durham	**NOMIS** 0191 374 2468	
Eastleigh	**Enghouse (UK) Limited** 01703 615228	
Edenbridge	**The NPA Group Ltd** 01732 865023	
Edinburgh	**Conic Systems** 0131 667 2728	
Edinburgh	**Design Computer Aids Limited (DeCAL)** 0131 553 3159	
Edinburgh	**General Register Office for Scotland** 0131 314 4254	
Edinburgh	**GIMMS (GIS) Ltd** 0131 668 3046	
Edinburgh	**Know Edge Ltd** 0131 443 1872	
Edinburgh	**SIAS Limited** 0131 225 7900	
Edinburgh	**University of Edinburgh** 0131 650 2565/2543	
Egham	**Earth Resource Mapping** 01784 430691	
Egham	**GISL Limited** 01956 285077	
Ellesmere Port	**Land Aspects Consultancy Ltd (Parkman Group)** 0151 356 1666	
Enfield	**International Products** 01992 651695	
Epsom	**GEOBASE Consultants Ltd** 01372 811225	
Epsom	**WS Atkins Planning & Management Consultants** 01372 726140	
Exeter	**Quail Map Company** 01392 430277	
Fareham	**Office of Population Censuses and Surveys (OPCS)** 01329 813536	
Fareham	**Spatial Information Services Ltd (SIS)** 01329 662891	
Farnborough	**National Remote Sensing Centre Ltd (NRSC)** 01252 541464	
Farnham	**Earth Observation Sciences Ltd** 01252 721444	
Fenstanton	**DMAP Ltd** 01480 497673	
Fleet	**Cray Systems** 01252 816816	
Glasgow	**P&L Engineering Surveys Ltd** 0141 644 1690	
Godalming	**Dr Stanley Port** 01483 421970	
Godalming	**Empress Software UK** 01483 861990	
Great Yarmouth	**C.A.Design Services Ltd** 01493 440444	
Great Yarmouth	**Gardline Infotech** 01493 442544	
Great Yarmouth	**Svitzer Limited** 01493 440320	
Guildford	**Autodesk Ltd** 01483 300077	
Guildford	**MR Data Graphics** 01483 575312	
Guildford	**Smith System Engineering Ltd** 01483 442000	
Handforth	**Consensus Information Technology Ltd** 01625 537777	
Harlow	**Longman Group Ltd** 01279 623623	
Harpenden	**Concurrent Appointments International** 01582 712976	
Harpenden	**Plowman Craven & Associates Ltd** 01582 765566	
Harrogate	**Dowling Associates Limited** 01943 880332	
Harrogate	**Geoplan (UK) Ltd** 01423 569538	
Harrogate	**Photarc Surveys Ltd** 01423 871629	
Harrogate	**TACTICIAN UK** 01423 560064	
Havant	**Pear Technology Services Ltd** 01705 499689	
Hayes	**Hitachi Home Electronics (Europe) Limited** 0181 849 2092	
Hemel Hempstead	**Hunting Technical Services Ltd** 01442 231800	
Henley-on-Thames	**Advent Imaging Ltd** 01491 411566	
Henley-on-Thames	**PAX Technology** 01491 572282	
Henley-on-Thames	**Scientific Software Limited** 01491 411727	
High Wycombe	**CDD Ltd** 01494 713769	
High Wycombe	**Dataquest Europe Ltd** 01494 422722	

High Wycombe	**Target Market Consultancy**	01494 712371
Hook	**Trimble Navigation Europe Ltd**	01256 760150
Horsham	**MOSS Systems Limited**	01403 259511
Horsham	**Tenet Systems Ltd**	01403 273173
Huntingdon	**Cambridge Computer Consultants (UK) Ltd**	01480 469577
Huntingdon	**Data Base Builders**	01487 813745
Huntingdon	**Geotronics Limited**	01480 433555
Huntingdon	**Institute of Terrestrial Ecology**	014873 381
Huttingdon	**MAPIT Limited**	01487 813745
Hyde	**Optimal Software Ltd**	0161 367 8715
Keele	**Keele University**	01782 583078
Kendal	**Spatial Data Limited**	01539 721070
Kenilworth	**Peter Thorpe Consulting**	0926 52799
Keyworth	**British Geological Survey**	0115 936 3100
Kilmacolm	**BS International Consultants**	01505 873563
Lancaster	**Lancaster University**	01524 593762
Leatherhead	**ERA Technology Ltd**	01372 367028
Leatherhead	**Logica UK Limited**	0171 637 9111
Leeds	**Carl Bro Group**	0113 2620000
Leeds	**GEOSOFT Ltd**	0113 234 4000
Leeds	**GMAP Ltd**	0113 244 6164
Leeds	**TEAMS (Taylor Woodrow Electronics Asset Mapping Survey)**	0113 242 3802
Leicester	**Midlands Regional Research Laboratory**	0116 252 3825
Leicester	**University of Leicester**	01533 523839
Leigh	**Oaklands I.T.**	01306 611590
Letchworth	**Adept Scientific Micro Systems Ltd**	01462 480055
Liverpool	**Survey Supplies Ltd**	0151 931 3161
London	**A.L.Downloading Services**	0181 994 5471
London	**Bartholomew**	0181 307 4065
London	**Birkbeck College London**	0171 631 6485
London	**CACI Limited**	0171 602 6000
London	**CAM - Centre for Analysis & Modelling Limited**	0171 232 1111
London	**Cambridge Market Intelligence**	0171 924 7117
London	**Citywise**	0171 636 5448
London	**CMG Computer Management Group (UK) Ltd**	0171 233 0288
London	**ColourMap Scanning Ltd**	0181 789 0737
London	**Coopers & Lybrand**	0171 213 2841
London	**Council of European Professional Informatics Societies**	0171 637 5607
London	**Dataview Solutions Ltd**	0171 404 0640
London	**DM Management Consultants Ltd (DMMC)**	0171 499 8030
London	**Dolphin Consulting Group**	0171 798 8465
London	**Elstree Computing Ltd**	0181 906 5656
London	**FileNet Ltd**	0181 944 5111
London	**Graphical Data Capture Ltd (GDC)**	0181 349 2151
London	**Greig Fester Limited**	0171 488 2828
London	**Grove Projects Ltd**	0181 846 2459
London	**Kingswood Consulting Limited**	0181 995 2050
London	**KPMG**	0171 311 1000
London	**London Research Centre**	0171 735 4250
London	**LTG Services**	0181 906 5559
London	**Methods Applications Ltd**	0171 240 1121

London	**Nestor International Ltd** 0171 937 4434
London	**PA Consulting Group** 0171 730 9000
London	**PLANTECH Ltd** 0171 922 8825
London	**Property Intelligence plc** 0171 839 7684
London	**QAS Systems Ltd** 0171 498 7777
London	**Raindrop Information Systems Ltd** 0171 734 1091
London	**Recruit Media Ltd** 0171 704 1227
London	**RICS Books** 0171 222 7000
London	**Royal Geographical Society**
	(with The Institute of British Geographers) 0171 589 5466
London	**Royal Institute of Navigation** 0171 589 5021
London	**Royal Town Planning Institute** 0171 636 9107
London	**SAZTEC Philippines Inc** 0171 702 2906
London	**Sector (UK) Limited** 0171 582 9982
London	**SIA Limited** 0171 730 4544
London	**Taylor & Francis Ltd** 0171 400 3500
London	**Trac Consultancy** 0171 639 9825
Long Hanborough	**Ashtech Europe Limited** 01993 883533
Loughborough	**Tangent Technology Design Associates Ltd** 01509 610910
Luton	**The LGMB** 01582 451166
Luton	**University of Luton** 01582 489264
Lympne	**J.C.White Chartered Land Surveyors** 01303 261212
Lyndhurst	**Cartwright Associates** 01703 812472
Manchester	**Fairbairn Services Limited** 0161 976 3536
Manchester	**Genasys II Limited** 0161 232 9444
Manchester	**Manchester Metropolitan University** 0161 247 1581
Manchester	**PD Computing Ltd** 0161 747 7110
Marlow	**SAS Institute** 01628 486933
Matlock	**Derek Hunter & Partners Ltd** 01629 822100
Merton	**CADAC Ltd** 0181 543 3411
Middleton-on-Sea	**Data Collection Ltd** 01243 587390
Milton Keynes	**ITS : Intertrade Scientific Ltd** 01908 676633
Milton Keynes	**Leica UK Ltd** 01908 666663
Milton Keynes	**Midsummer Computing** 01908 668866
N-0506 Oslo	**Xcon Data**
Nantwich	**Graphtec (UK) Ltd** 01270 611234
Near Farnham	**Unistride Sewer Technology** 01420 23456
near Trowbridge	**Action Information Management Ltd** 01225 777288
Newbury	**Land and Satellite Surveys** 01635 49512
Newbury	**Modern Maps** 01635 34251
Newcastle-upon-Tyne	**IMASS Limited** 0191 213 5555
Newcastle-upon-Tyne	**University of Newcastle Upon Tyne** 0191 222 6445
Newhaven	**Business Information Management** 01273 515018
Northampton	**HJM Imaging Systems** 01604 39792
Northampton	**Morgan Collis Group Ltd** 01604 580980
Northampton	**Navstar Systems Ltd** 01604 585588
Norwich	**CCTA - The Government Centre for Information Systems** 01603 704844
Norwich	**Geografix Limited** 01603 788940
Norwich	**HMSO Books** 01603 695911
Nottingham	**CCN Marketing** 0115 941 0888
Nottingham	**Graphite Management Services Ltd** 0115 969 1114

Nottingham	**PAFEC Ltd**	0115 935 7055
Oldham	**Guild of Incorporated Surveyors**	0161 627 2389
Orpington	**Glen Computing Ltd**	01689 875577
Oxford	**McLintock Limited**	01865 749957
Oxford	**NAG Ltd**	01865 511245
Oxford	**NCC Blackwell Ltd**	01865 791100
Oxford	**Organisation Management Systems**	01865 372161
Oxford	**Oxford Institute of Retail Management**	01865 735422
Oxford	**StatSci Europe**	01865 200952
Peterborough	**Photoair**	01733 241850
Peterborough	**Posford Duvivier**	01733 334455
Pewsey	**Geo/SQL (UK)**	01672 562012
Pinner	**Sovereign C.S. Ltd**	0181 866 0713
Pontypridd	**HollyBush Software Limited**	01443 482785
Portsmouth	**Royal Mail - Address Management Centre**	01705 838518
Portsmouth	**Solent Mapping and Charting (SMAC)**	01705 842477
Reading	**Babtie Shaw & Morton Limited**	01734 234780
Reading	**CalComp Limited**	01734 320032
Reading	**Digital Equipment Corporation**	01734 203546
Reading	**GID Ltd**	01491 671964
Reading	**Silicon Graphics Limited**	01734 257500
Reading	**Sir Alexander Gibb & Partners Ltd**	01734 261061
Reading	**The Data Consultancy**	01734 588181
Redhill	**Binnie Black & Veatch**	01737 774155
Redhill	**Evox Facilities Ltd**	01737 764137
Reigate	**L.E.S. (Computer Services) Ltd**	01737 223899
Renfrew	**LOY Surveys Ltd**	0141 885 0800
Richmond	**Spacesense consultants**	0181 940 6290
Romsey	**Effective Solutions (Data Products)**	01794 514233
Romsey	**LiveChart**	01794 518085
Romsey	**Mathshop**	01794 523423
Royal Leamington Spa	**CODEC Facilities Limited**	01926 330112
Royston	**Primagraphics Ltd**	01763 262041
Runcorn	**Pinpoint Digitising Services**	01928 579148
Salford	**Environment & Planning Library**	0161 708 9799
Salford	**Salford University Business Services Ltd**	0161 957 0012
Salisbury	**Cartographical Services (Southampton) Limited**	01794 390321
Sheffield	**CDR Group**	01433 621282
Sheffield	**Cliffe House Associates**	0114 285 0663
Sheffield	**Geographical Association, The**	0114 267 0666
Sheffield	**GISDATA**	0114 272 0185
Sheffield	**Graphics Online Limited**	0114 279 7972
Shoreham by Sea	**Photogrammetric Data Services Ltd**	01273 464883
Shrewsbury	**The Severn Partnership**	01743 874135
Sidcup	**Datatechnology Datech Ltd**	0181 308 1800
Silsoe	**Silsoe College, Cranfield University**	01525 863060
Slough	**The Business Database from Yellow Pages**	01753 583311
Solihull	**GIS Services Ltd**	01564 779656
Southampton	**GeoData Institute**	01703 592719
Southampton	**Ordnance Survey**	01703 792773
Southampton	**TYDAC Technologies Ltd**	01703 760824

St Albans	**Oscar Faber** 0181 784 5784
St Albans	**TerraHunt GeoScience Ltd** 01727 822287
Stalybridge	**Kirstol Ltd** 0161 338 7512
Stevenage	**Computer Aided Development (CADCORP) Ltd** 01438 747996
Stockfield	**CSI** 01661 842741
Stockport	**Informed Solutions Limited** 0161 476 6716
Stoke-on-Trent	**Terrafix Ltd** 01782 577015
Stratford Upon Avon	**WDV (UK)** 01789 297000
Studley	**Structural Technologies Ltd (STL)** 01527 854819
Sudbury	**AP3 Imaging Services Limited** 01787 378242
Swansea	**Longdin & Browning** 01792 202244
Swindon	**Intergraph (UK) Ltd** 01793 619999
Swindon	**Natural Environment Research Council** 01793 411996
Swindon	**NERC** 01793 411683
Swindon	**Procis Software Ltd** 01793 541200
Swindon	**Royal Commission on the Historical Monuments of England** 01793 414727
Swindon	**Sir William Halcrow and Partners** 01793 812479
Swindon	**The British Computer Society** 01743 417417
Swindon	**WRc (Water Research centre)** 01793 511711
Teddington	**The Marketing Information Consultancy (MIC)** 0181 213 5500
Tehran 19688	**Kamyco International**
Thatcham	**GEO-Marketing Systems Ltd (GMSL)** 01635 872382
Thetford	**Anglian Engineering & International Consultancy** 01842 750329
Tipton	**TerraQuest Group Limited** 0121 520 0111
Towcester	**Assist Applications Limited** 01908 543323
Tunbridge Wells	**Laser Technology International Ltd** 01892 863351
Tyne & Wear	**Survey Control Services** 01207 544996
Uxbridge	**UNISYS Ltd** 01895 237137
Wallingford	**H R Wallingford Ltd** 01491 835381
Wallingford	**Institute of Hydrology** 01491 838800
Wallingford	**Wallingford Software Limited** 01491 824777
Wallington	**Cromwell House Technical Services** 0181 647 1686
Warwick	**IBM UK Ltd** 01926 464336
Watford	**ALLM Systems & Marketing** 01923 230150
Watford	**ESRI (UK) Ltd** 01923 210450
Watford	**FastCAD GIS Ltd** 01923 240216
Watlington	**Chiltern Digitising Services** 01491 612581
Welwyn Garden City	**Carl Zeiss Limited** 01707 331144
Wetherby	**Dataman Computer Solutions UK Ltd** 01423 358226
Windsor	**Saztec Europe Limited** 01753 833131
Witney	**Lovell Johns Ltd** 01993 883161
Witney	**Spatial Geographic Services & Applications Ltd** 01865 881753
Woking	**Graphic Data Systems Corporation (GDS)** 01483 725225
Woking	**Mentis Management Consultants Ltd** 01483 776717
Woking	**MVA Systematica** 01483 728051
Worcester	**Computer Graphic Suppliers Association** 01905 613236
Worcester	**The Survey Centre** 01905 21073
Wotton-under-Edge	**Geographic Management Solutions Ltd** 01454 281802
York	**Beacon Dodsworth Limited** 01904 638997
York	**R.W.A. Dallas FRICS** 01904 652408

directory 39a
location directory
outside the UK

Australia	Sydney	**ARC Systems Pty Ltd** +612 290 2400
Austria	9500 Villach	**Progis GmbH** +43 4242 26332
Belgium	B-1780 Wemmel	**Eurosense Technologies N.V.** +32 2 460 7000
	B-4031 Angleur-Liege	**Star Informatic S.A.** +32 41 675313
	B-9000 Gent	**Tele Atlas** +32 9 222 5658
	Brussels	**Summagraphics Europe N.V.** +32 2 721 5033
Bulgaria	Sofia 1113	**CAD R&D Centre Limited** +359 2 705257
Canada	Fredericton	**Universal Systems Ltd** +1506 458 8533
	Hull	**ACDS Graphic System Inc** +1819 770 9631
	Nepean	**Intera Information Technologies** +1613 226 5442
	Victoria	**EPS - Essential Planning Systems Limited** +1604 652 8895
Finland	02210 Espoo	**Tekla Oy** +358 0 803 7722
France	31 526 Ramonville	**SCOT Conseil** +33 613 94600
	94110 Arcueil	**Logitrans** +331 498 51516
	94742 Arcueil	**APIC Systemes** +331 496 99090
	BP68, 94160 St. Mande	**MEGRIN Group** +331 43 988440
Germany	D-10709 Berlin	**regioplanDATA GmbH** +49 30 896704 18
	D-48145 Munster	**Hansa Luftbild** +49 251 23300
Hungary	Budapest	**Geoview Systems Kft** +361 269 2099
	H-1025 Budapest	**Geometria GIS Systems House Ltd** +361 250 0989
Iran	Tehran 19688	**Kamyco International**
Ireland	Cork	**QC Data (Ireland) Limited** +353 21 341700
	Dublin 2	**ERA-Maptec Ltd** +3531 676 6266
	Dublin 2	**Gamma Ltd** +3531 6713066
	Millstreet Town	**Kalidor Europe (ALPS Electric (Ireland) Ltd)** +353 29 21212
Israel	Haifa Bay	**Scan Group Limited** +9724 410339

Netherlands	2611 HB Delft	**Geodelta** +31 1515 8188
	2740 AE Waddinxveen	**Geo-Perfect TWI B.V.** +31 1828 30477
	3011 ED Rotterdam	**AND Mapping B.V.** +31 10 433 3440
	3121 XA Schiedam	**Geo2 Consulting** +31 10 4712372
	3800 BJ Amersfourt	**DMV Consultants BV** +31 33 682300
	5680 AB Best	**European Geographic Technologies BV** +31 4998 93385
	6708 PW Wageninggen	**Geops BV** +31 8370 79636
	6720 AD Bennekom	**Ingecon B.V.** +31 8389 16681
	7500 AA Enschede	**ITC** +31 53 874444
Norway	1370 Asker	**Corena A/S** +47 66 794500
	N-0506 Oslo	**Xcon Data**
Portugal	2765 Estoril	**Magdala Sociedade** +351 1 460 0684
Spain	10.001 Caceres	**Foto Res** +34 27 216455
	17181 Aiguaviva	**Audifilm Girona S.L.** +34 72 242611
Sweden	S-111 20 Stockholm	**Lantmatenet GIS-centrum** +46 8402 1700
Switzerland	CH-4005 Basel	**AM/FM International - European Division** +41 61 6915111
United Arab Emirates	Sharjah	**MAPS geosystems** +9716 356411
USA	Anchorage	**DAT/EM Systems International** +1907 274 3681
	Atlanta	**Byers Engineering Company** +1404 843 1000
	Boulder	**Smartscan Inc** +1303 443 7226
	Clearwater	**Baymont Technologies Inc** +1813 539 1661
	Columbia	**GTCO Corporation** +1410 381 6688
	Frankfort	**PlanGraphics Inc** +1502 223 1501
	Lanham	**EOSAT** +1301 552 0525
	San Dimas	**Magellan Systems Corporation** +1909 394 5000
	Silver Spring	**ALTEK Corporation** +1301 572 2555

directory 40:
supplier contacts

A.L.Downloading Services
John Farrant
telephone: 0181 994 5471
fax: 0181 994 4959
email: sales@aldown.algroup.com

A.Rutherford Ltd
Allan R Rutherford
telephone: 01223 872646
fax: 01223 872646

ACDS Graphic System Inc
Jean-Guy Laplante
telephone: +1819 770 9631
fax: +1819 770 9267

Action Information Management Ltd
Tony Hay/John Page
telephone: 01225 777288
fax: 01225 751616

Active Software Ltd
Paul Smith
telephone: 01256 56629
fax: 01256 56708
email: 100140.472@compuserve.com

Adept Scientific Micro Systems Ltd
Stephen Hawkins
telephone: 01462 480055
fax: 01486 480213
email: atlasgis@adeptscience.co.uk

Advent Imaging Ltd
Clare Bamforth
telephone: 01491 411566
fax: 01491 411577
email: sales@advent.co.uk

AGFA UK
Paresh M Patel
telephone: 0181 231 4141
fax: 0181 231 4957

AiC Analysts
Trevor Jarvis
telephone: 01223 300044
fax: 01223 302005
email: 100067,1364@compuserve

ALLM Systems & Marketing
Alan Pritchard
telephone: 01923 230150
fax: 01923 211148
email: apritchard@cix.compulink.co.uk

ALTEK Corporation
Shawn Richards
telephone: +1301 572 2555
fax: +1301 572 2510

AM/FM International - European Division
Ing.Hans J.Festen
telephone: +41 61 6915111
fax: +41 61 6918189

AND Mapping B.V.
John Heofnagels
telephone: +31 10 433 3440
fax: +31 10 414 0606
email: @andmap.nl

Aneberie CAD
Ralph Massie
telephone: 01206 331215
fax: 01206 330313

Anglian Engineering & International Consultancy
Reuben Hickin
telephone: 01842 750329

AP3 Imaging Services Limited
Peter Wigmore
telephone: 01787 378242
fax: 01787 374017

APIC Systemes
Jean-Pierra Rogala
telephone: +331 496 99090
fax: +331 396 99293
email: info@apic.fr

APIC Systems
Nick Chisnall
telephone: 01279 466966
fax: 01279 466788
email: info@apic.fr

ARC Systems Pty Ltd
Harry Clarsen
telephone: +612 290 2400
fax: +612 261 3472

Ashtech Europe Limited
Barrie Hogarth
telephone: 01993 883533
fax: 01993 883977

Assist Applications Limited
Steve Kurle
telephone: 01908 543323
fax: 01908 543324
email: axis@assistap.demon.co.uk

Audifilm Girona S.L.
Mr Joan Font
telephone: +34 72 242611
fax: +34 72 242311

Autodesk Ltd
Sales Department
telephone: 01483 300077
fax: 01483 304556

Babtie Shaw & Morton Limited
Chris Gower
telephone: 01734 234780
fax: 01734 310268

Bartholomew
Dave Benson
telephone: 0181 307 4065
fax: 0181 307 4813
email: barts_twr@geovax.ed.uk

Baymont Technologies Inc
William Reid
telephone: +1813 539 1661
fax: +1813 539 1749

Beacon Dodsworth Limited
Simon Perry
telephone: 01904 638997
fax: 01904 638999

Bentley Systems UK Ltd
Kevin Twigger
telephone: 01344 412233
fax: 01344 412386
email: kevin.twigger@bentley.nl

Binnie Black & Veatch
Ian G Bush
telephone: 01737 774155
fax: 01737 772767
email: igbush@binnie.demon.co.uk

Birkbeck College London
Professor David Unwin
telephone: 0171 631 6485
fax: 0171 631 6498
email: d.unwin@uk.ac.bbk.geog

BKS Surveys Ltd
Jon McNally
telephone: 01265 52311
fax: 01265 57637

Bradly Associates Ltd
Peter Kelly
telephone: 01344 779381
fax: 01344 773168

British Geological Survey
Dr.Alan Dobinson
telephone: 0115 936 3100
fax: 0115 936 3200
email: k_ald@uk.ac.nkw.va

BS International Consultants
Bob Stirling
telephone: 01505 873563
fax: 01505 873563

Bull Information Systems Limited
Nigel Sheath
telephone: 0121 717 0777
fax: 0121 626 1550
email: nsheath@uk22p.bull.co.uk

Business Information Management
Rob Mahoney
telephone: 01273 515018
fax: 01273 515557
email: 100411.3323@compuserve.com

Byers Engineering Company
Bonnie Owen
telephone: +1404 843 1000
fax: +1404 843 2000
email: 74677.1174@compuserve.com

C.A.Design Services Ltd
Matt J Tuohy
telephone: 01493 440444
fax: 01493 442480

CACI Limited
John Rae
telephone: 0171 602 6000
fax: 0171 603 5862
email: jrae@cacipo.caci.co.uk

CAD - Capture Limited
Sarah Pickering
telephone: 01254 583534
fax: 01254 665528
email: info@cadcap.co.uk

CAD R&D Centre Limited
Plamen Mateev
telephone: +359 2 705257
fax: +359 2 703556

CADAC Ltd
Paul W Bennett
telephone: 0181 543 3411
fax: 0181 543 6844

CalComp Limited
Sales
telephone: 01734 320032
fax: 01734 341215

CAM - Centre for Analysis & Modelling Limited
Gurmukh Singh
telephone: 0171 232 1111
fax: 0171 237 4247

Cambashi Ltd
Mrs Jenny.R.Jacobsberg
telephone: 01223 460439
fax: 01223 461055
email: 100431.3342@compuserve.com

Cambridge Computer Consultants (UK) Ltd
Colin Hookham
telephone: 01480 469577
fax: 01480 466784
email: ccc_uk_ltd@online.rednet.co.uk

Cambridge Market Intelligence
Peter Bomer
telephone: 0171 924 7117
fax: 0171 403 6729

Carl Bro Group
Dr Chris McDermott
telephone: 0113 2620000
fax: 0113 2620737

Carl Zeiss Limited
Mr.E.H.Wickens
telephone: 01707 331144
fax: 01707 373210

CARTograph Ltd
Nigel Payne
telephone: 01223 67818
fax: 01223 464142

Cartographical Services (Southampton) Limited
John B Waterman
telephone: 01794 390321
fax: 01794 390867

Cartwright Associates
Jac Cartwright
telephone: 01703 812472
fax: 01703 812472
email: jac@caas.demon.co.uk

CATALIST
Nigel Lang
telephone: 0117 923 7113
fax: 0117 923 7166

CCN Marketing
Elaine Peters
telephone: 0115 941 0888
fax: 0115 934 4903

CCTA - The Government Centre for Information Systems
Pat Middleton
telephone: 01603 704844
fax: 01603 704817
email: pmiddleton@ccta.gov.uk

CDD Ltd
Derek Pavely
telephone: 01494 713769
fax: 01494 713769

CDR Group
Martin Waters
telephone: 01433 621282
fax: 01433 621292
email: 100436.1330@compuserve.com

Chiltern Digitising Services
Dr.M.J.Lowing
telephone: 01491 612581
fax: 01491 612930
email: 75337.2741@compuserve.com

Chroson Ltd
Mr.N.C.Adnitt
telephone: 01264 336339

Citywise
Tim J.Craine
telephone: 0171 636 5448
fax: 0171 636 5451

Cliffe House Associates
Peter Clegg
telephone: 0114 285 0663
fax: 0114 285 0663

CMG Computer Management Group (UK) Ltd
Andy Downie
telephone: 0171 233 0288
fax: 0171 799 2017

Cobham Digital Services Limited
Roger K Dollimore
telephone: 01932 868133
fax: 01932 867024

CODEC Facilities Limited
Dennis J.Caldwell
telephone: 01926 330112
fax: 01926 316728

Colorgraph (UK) Ltd
Simon Johnson
telephone: 01734 819435
fax: 01734 815197

ColourMap Scanning Ltd
D.J.Brooker
telephone: 0181 789 0737
fax: 0181 780 2663

Computer Aided Development (CADCORP) Ltd
Nicola Radford
telephone: 01438 747996
fax: 01438 747997
email: 100113.2367@compuserve.com

Computer Graphic Suppliers Association
R.Crumpton
telephone: 01905 613236
fax: 01905 29138
email: 100013,427@compuserve.com

COMSULT
P.N.Careless
telephone: 01234 342401
fax: 01234 328609

Concurrent Appointments International
Alan Carnell
telephone: 01582 712976
fax: 01582 764858

Conic Systems
Mrs Kailash Watchman
telephone: 0131 667 2728
fax: 0131 667 2728
email: klatchman@conic.com

Consensus Information Technology Ltd
Janet Sheard
telephone: 01625 537777
fax: 01625 539621

Construction Industry Computing Association
Erik G Winterkorn
telephone: 01223 236336
fax: 01223 236337
email: robhoward@constcom.demon.co.uk

Coopers & Lybrand
Dr Helen Mounsey
telephone: 0171 213 2841
fax: 0171 213 2850
email: helen.mounsey@coopers.colybrand.gold400.gb

Corbins Consultancy
Chris Corbin
telephone: 01273 553110
fax: 01273 389497

Corena A/S
Stuart Hodgson
telephone: +47 66 794500
fax: +47 66 794590

Council of European Professional Informatics Societies
Mrs Peta Walmisley
telephone: 0171 637 5607
fax: 0171 637 5607

Cray Systems
Guy Pullen
telephone: 01252 816816
fax: 01252 812163

Cromwell House Technical Services
Paul L.Asquith
telephone: 0181 647 1686
fax: 0181 773 3110

CSI
Gilbert H Scott
telephone: 01661 842741
fax: 01661 842288

CZ Scientific Instruments Ltd
Nigel Harding
telephone: 0181 953 1688
fax: 0181 953 9456

DAT/EM Systems International
Jim Cucurull
telephone: +1907 274 3681
fax: +1907 272 6413
email: jrogers@datem.com

Data Base Builders
Douglas Cross
telephone: 01487 813745
fax: 01487 813745

Data Collection Ltd
Steve Batchelor
telephone: 01243 587390
fax: 01243 587390

Data Dictionary Systems Limited
David J.L.Gradwell
telephone: 01276 23519
fax: 01276 676670
email: awpl@applelink.apple.com

Dataflow Information Systems
Clare Dorey
telephone: 0117 927 2466
fax: 0117 929 0768

Dataman Computer Solutions UK Ltd
Paul Potts
telephone: 01423 358226
fax: 01423 358262
email: potts@dataman.co.uk

Dataquest Europe Ltd
Ms Petra Gartzen
telephone: 01494 422722
fax: 01494 422742
email: pgartzen@dqeurope.com

Datatechnology Datech Ltd
Alistair Brook
telephone: 0181 308 1800
fax: 0181 308 0802
email: alistair.brook@datech.co.uk

Dataview Solutions Ltd
Geoff Kendall
telephone: 0171 404 0640
fax: 0171 404 0664
email: geoff@dataview.demon.co.uk

DCL Consulting
David Litton
telephone: 01223 314888

Derek Hunter & Partners Ltd
Derek Hunter
telephone: 01629 822100
fax: 01629 822030
email: 100345,3001@compuserve.com

Design Computer Aids Limited (DeCAL)
Lorraine Sinclair
telephone: 0131 553 3159
fax: 0131 553 5121

Digital Equipment Corporation
Malcolm Wicks
telephone: 01734 203546
fax: 01734 204757

DM Management Consultants Ltd (DMMC)
Dr Peter M.Thompkins
telephone: 0171 499 8030
fax: 0181 948 6306

DMAP Ltd
Paul Holroyd
telephone: 01480 497673
fax: 01480 492281

DMV Consultants BV
R.Beck
telephone: +31 33 682300
fax: +31 33 682601

Dolphin Consulting Group
Nic Walker
telephone: 0171 798 8465
fax: 0171 798 8692

Dotted Eyes
Jamie M Justham
telephone: 0121 445 6150
fax: 0121 445 6150

Dowling Associates Limited
W.J.Dowling
telephone: 01943 880332
fax: 01943 880634
email: bdowling@cix.compulink.co.uk

Dr Stanley Port
Stanley Port
telephone: 01483 421970
fax: 01483 861023

EA Technology
Gary Marsden
telephone: 0151 347 2451
fax: 0151 347 2135
email: ggm@eatl.co.uk

Earth Observation Sciences Ltd
Matthew Stuttard
telephone: 01252 721444
fax: 01252 712552
email: matthews@eos.co.uk

Earth Resource Mapping
Brian Talbot
telephone: 01784 430691
fax: 01784 430692
email: brian@ermapper.co.uk

ECM Selection Limited
Michael Gernat
telephone: 01638 742244
fax: 01638 743066
email: postmaster@ecmsel.co.uk

Effective Solutions (Data Products)
Graham Collins
telephone: 01794 514233
fax: 01794 514244

Elstree Computing Ltd
Steve Pittard
telephone: 0181 906 5656
fax: 0181 906 5666

Empress Software UK
Dennis Flavell
telephone: 01483 861990
fax: 01483 860064

Enghouse (UK) Limited
Simon Crowley
telephone: 01703 615228
fax: 01703 615253

Environment & Planning Library
Harry Z A Orenstein
telephone: 0161 708 9799

EOSAT
Annamarie De Carlo
telephone: +1301 552 0525
fax: +1301 794 4243

EPS - Essential Planning Systems Limited
Alison Malis
telephone: +1604 652 8895
fax: +1604 652 8896
email: marketing@eps.bc.ca

ERA Technology Ltd
Peter Tucker
telephone: 01372 367028
fax: 01372 367099

ERA-Maptec Ltd
Paul Kidney
telephone: +3531 676 6266
fax: +3531 661 9785

ERDAS (UK) Ltd
Jonathan Shears
telephone: 01223 880802
fax: 01223 880160
email: jshears@erdas-uk.demon.co.uk

ERTEC
David R Green
telephone: 01224 740324
fax: 01224 740324

ESR Cartographers Ltd
Alan Smith
telephone: 01932 348981
fax: 01932 344882

ESRI (UK) Ltd
Carole Smith
telephone: 01923 210450
fax: 01923 210739
email: csmith@esriuk.com

EuroDirect Database Marketing Ltd
John K Dobson
telephone: 01274 737144
fax: 01274 741126

European Business Mapping
Bruce Mackay
telephone: 01273 702957
fax: 01273 673979

European Geographic Technologies BV
Yiannis Moissidis
telephone: +31 4998 93385
fax: +31 4998 92078

Eurosense Technologies N.V.
Nancy Schryvers
telephone: +32 2 460 7000
fax: +32 2 460 4958

Evox Facilities Ltd
Phil Assender
telephone: 01737 764137
fax: 01737 764137

Fairbairn Services Limited
Mr Kym Soni
telephone: 0161 976 3536
fax: 0161 969 5131
email: ks@fairbairn.co.uk

FastCAD GIS Ltd
Louise Solomon
telephone: 0117 942 8195
fax: 0117 942 8196
email: fcadgis@dircon.co.uk

FastCAD GIS Ltd
Miss Victoria Gregory
telephone: 01923 240216
fax: 01923 228796

FileNet Ltd
Penny Cooper
telephone: 0181 944 5111
fax: 0181 944 5146

Flynn & Rothwell
Linda Elkins
telephone: 01279 507346
fax: 01279 758219
email: 100566.3365@compuserve.com

Foto Res
Jorge Fabricant
telephone: +34 27 216455
fax: +34 27 216455

G.L. Consulting Ltd
Dr Les W Thorpe
telephone: 01372 272937
fax: 01372 279362

Gamma Ltd
Fearoal O'Neill
telephone: +3531 6713066
fax: +3531 6713593
email: gamma@iol.ie

Gardline Infotech
David Pettit
telephone: 01493 442544
fax: 01493 441200

GEC Marconi Research Centre
Mr.A.J.Rye
telephone: 01245 473331
fax: 01245 475244
email: tony.rye@gmrc.gecm.com

Genasys II Limited
John Tarleton
telephone: 0161 232 9444
fax: 0161 232 9453
email: johnt@genasys.co.uk

General Register Office for Scotland
Peter Jamieson
telephone: 0131 314 4254
fax: 0131 314 4344
email: 100431.2507@compuserve.com

GEO-Marketing Systems Ltd (GMSL)
Alan Odham/Phil Durbin
telephone: 01635 872382
fax: 01635 871302

Geo-Perfect TWI B.V.
Geworge J.W.Lavigne
telephone: +31 1828 30477
fax: +31 1828 31280
email: hkersten@knoware.nl

GEO-UK Ltd
Roy Wood
telephone: 01276 473579
fax: 01276 473603

Geo/SQL (UK)
Phil Williams
telephone: 01672 562012
fax: 01672 63001

Geo2 Consulting
Chris W.Nelis
telephone: +31 10 4712372

GEOBASE Consultants Ltd
John R Rowley
telephone: 01372 811225
fax: 01372 811226
email: geobase@cix.compulink.co.uk

GeoData Institute
Chris Hill
telephone: 01703 592719
fax: 01703 592849
email: geodata@soton.ac.uk

Geodelta
Jr. R.J.G.A.Kroon
telephone: +31 1515 8188
fax: +31 1515 8154

Geografix Limited
Laurence Taylor
telephone: 01603 788940
fax: 01603 788964

Geographic Management Solutions Ltd
John Standerline
telephone: 01454 281802
fax: 01454 419417

Geographical Association, The
Graham Ranger
telephone: 0114 267 0666
fax: 0114 267 0688

Geoinformation International
Elizabeth Wijnmaalen
telephone: 01223 423020
fax: 01223 425787

GeoMEM Software
Mr.Marlon.P.Binner
telephone: 01250 872284
fax: 01250 873290
email: sales@geomem.win-uk.net

Geometria GIS Systems House Ltd
Mr Tibor Tenke
telephone: +361 250 0989
fax: +361 250 1231
email: 73501.173@compuserve.com

Geoplan (UK) Ltd
Richard Ives
telephone: 01423 569538
fax: 01423 525545

Geops BV
Ir.G.J.M.Kreuwel
telephone: +31 8370 79636
fax: +31 8370 79704

GEOSOFT Ltd
Chris Inie
telephone: 0113 234 4000
fax: 0113 246 5071
email: sales@geosoft.co.uk

Geosystems
Dr Roger F Templeman
telephone: 01235 813913
fax: 01235 813913

Geotronics Limited
Alan Sharp
telephone: 01480 433555
fax: 01480 432480

Geoview Systems Kft
Istvan Nikl
telephone: +361 269 2099
fax: +361 112 6861
email: nikkel@bp.geoview.hu

GGP Systems Limited
Mrs.A.Maxwell
telephone: 0181 656 8562
fax: 0181 656 8562

GID Ltd
Andrew Greener
telephone: 01491 671964
fax: 01491 671964
email: andy@gid.co.uk

GIMMS (GIS) Ltd
M.A.Ferenth
telephone: 0131 668 3046
fax: 0131 668 2104

GIS Services Ltd
Stewart McAusland
telephone: 01564 779656
fax: 01564 779656

GISDATA
Max Craglia
telephone: 0114 272 0185
fax: 0114 272 2199
email: gisdata@sheffield.ac.uk

GISL Limited
Justin Saunders/James Cutler
telephone: 01956 285077
fax: 01428 707132

Glen Computing Ltd
Derek.G.Prior
telephone: 01689 875577
fax: 01689 828735

Global Surveys Ltd
Alan F.Wright
telephone: 0121 421 1414
fax: 0121 423 1480

GMAP Ltd
Paul Kelley
telephone: 0113 244 6164
fax: 0113 234 3173

Graphic Data Systems Corporation (GDS)
Johanna Afors
telephone: 01483 725225
fax: 01483 725221

Graphical Data Capture Ltd (GDC)
Peter.M.Klein
telephone: 0181 349 2151
fax: 0181 349 4095

Graphics Online Limited
Edwin Guiton
telephone: 0114 279 7972
fax: 0114 275 3708

Graphite Management Services Ltd
G.Adrian & K.Hardy
telephone: 0115 969 1114
fax: 0115 969 1115

Graphtec (UK) Ltd
Peter Mitchell
telephone: 01270 611234
fax: 01270 626733

Greig Fester Limited
Andrew Mitchell
telephone: 0171 488 2828
fax: 0171 265 1234

Grove Projects Ltd
M.C.H.Sumner
telephone: 0181 846 2459
fax: 0181 846 3388

GTCO Corporation
Sales Dept.
telephone: +1410 381 6688
fax: +1410 290 9065

GTX Europe Ltd
Robert Brown
telephone: 01256 843555
fax: 01256 246634

Guild of Incorporated Surveyors
Brian Birchenall
telephone: 0161 627 2389
fax: 0161 627 3336

H R Wallingford Ltd
Dr Peter A Bradbury
telephone: 01491 835381
fax: 01491 826352
email: pab%hydres.uucp@uknet.ac.uk

Hall & Watts Systems Limited
S.McCarthy
telephone: 01452 864244
fax: 01452 864194

Hansa Luftbild
Hans-Dieter Arnold
telephone: +49 251 23300
fax: +49 251 2330112

Hitachi Home Electronics (Europe) Limited
Mark Wilkin
telephone: 0181 849 2092
fax: 0181 569 2763

HJM Imaging Systems
Les Bootes/John Hale
telephone: 01604 39792
fax: 01604 30919

HMSO Books
Lisa Hallett
telephone: 01603 695911
fax: 01603 696784

HollyBush Software Limited
Peter Jolly
telephone: 01443 482785
fax: 01443 482788

Hunting Aerofilms Limited
R.C.A.Cox
telephone: 0181 207 0666
fax: 0181 207 5433

Hunting Engineering Ltd
Robert Brevitt
telephone: 01525 841000
fax: 01525 405861

Hunting Technical Services Ltd
Graham Deane/Mike Whitelegge
telephone: 01442 231800
fax: 01442 219886
email: remote-
sensing.hts@cityscape.co.uk

Husky Computing Limited
Christine Smith
telephone: 01203 604040
fax: 01203 603060

I.S. Ltd
Neil Quarmby/Richard Selby
telephone: 0117 925 0553
fax: 0117 925 0663
email: 100044.745@compuserve.com

IBM UK Ltd
GIS Manager
telephone: 01926 464336
fax: 01926 311345
email: gbibm9t5@ibmmail

ICL Ltd
Alan Roden
telephone: 0121 456 1111
fax: 0121 455 0358

IMASS Limited
Ursula Cooke
telephone: 0191 213 5555
fax: 0191 213 0526
email: Ursula.Cooke@IMASS.co.uk

IME (UK) Ltd
Lisa Currid
telephone: 01259 210210
fax: 01259 217303

Infolink Decision Services Limited
Ian Liddicoat
telephone: 0181 686 7777
fax: 0181 680 8295

Informed Solutions Limited
Dr Mark Ketteman
telephone: 0161 476 6716
fax: 0161 476 6710
email: markk@informed.co.uk

Ingecon B.V.
Maarten van Heest
telephone: +31 8389 16681
fax: +31 8389 18648

Institute of Hydrology
Mr.R.V.Moore
telephone: 01491 838800
fax: 01491 832256

Institute of Terrestrial Ecology
Timothy Moffat
telephone: 014873 381
fax: 014873 467
email: t.moffat@uk.ac.ite

Intera Information Technologies
Robert Dams
telephone: +1613 226 5442
fax: +1613 226 5529

Intergraph (UK) Ltd
Peter Mingins
telephone: 01793 619999
fax: 01793 618508
email:
pmingins@swindon.swindon.ingr.com

International Map Trade Association
Mike Cranidge
telephone: 01425 620532
fax: 01425 620532

International Products
John Lessing
telephone: 01992 651695
fax: 01992 651695
email: 100342.12@compuserve

IT Southern Ltd
Charles Gray
telephone: 01273 600444
fax: 01273 675299
email: ah78@solo.pipex.com

ITC
Hans Melos
telephone: +31 53 874444
fax: +31 53 874400
email: ilwis@itc.nl

ITS : Intertrade Scientific Ltd
Matt Costin
telephone: 01908 676633
fax: 01908 666595

J.C.White Chartered Land Surveyors
M P Heiman
telephone: 01303 261212
fax: 01303 264040

John D Leatherdale FRICS
John Leatherdale
telephone: 0181 449 0123
fax: 0181 449 0123

Kalidor Europe (ALPS Electric (Ireland) Ltd)
Judy Costigan
telephone: +353 29 21212
fax: +353 29 21213
email: jcostigan@alps.ie

Kamyco International
Dr.Ahmadi
email: kamyco5-E

Keele University
Dr Michael F Worboys
telephone: 01782 583078
fax: 01782 713082
email: michael@cs.keele.ac.uk

Kingswood Consulting Limited
Dan Rickman
telephone: 0181 995 2050
fax: 0181 747 8047
email: 100322.205@compuserve.com

Kirstol Ltd
Peter Dickinson
telephone: 0161 338 7512
fax: 0161 338 8097

KJB Consulting
Keith Burleton
telephone: 01233 756461

Know Edge Ltd
Robin A McLaren
telephone: 0131 443 1872
fax: 0131 443 1872
email: 100042.3122@compuserve.com

KPMG
Richard Goodwin
telephone: 0171 311 1000
fax: 0171 311 3311

L.E.S. (Computer Services) Ltd
W.D.Rees
telephone: 01737 223899
fax: 01737 223911

Lancaster University
Dr A.C.Gatrell
telephone: 01524 593762
fax: 01524 847099
email: ha001@lancaster.ac.uk

Land and Satellite Surveys
Richard Andrews
telephone: 01635 49512
fax: 01635 523809

Land Aspects Consultancy Ltd (Parkman Group)
Janet Bridge
telephone: 0151 356 1666
fax: 0151 356 1119
email: jbridge@parkman.co.uk

Lantmatenet GIS-centrum
Ann Grengmark
telephone: +46 8402 1700
fax: +46 8791 7151

LASCO Ltd
D.H.Rollason
telephone: 01903 882466
fax: 01903 882599

Laser Technology International Ltd
Peter Fasey
telephone: 01892 863351
fax: 01892 863431

Laser-Scan Ltd
Jenny Nash
telephone: 01223 420414
fax: 01223 420044
email: annef@lsl.co.uk

Leica UK Ltd
Deborah Saunders
telephone: 01908 666663
fax: 01908 609992

LiveChart
Garry Symes
telephone: 01794 518085
fax: 01794 518086

Logica UK Limited
Stephen Darvill
telephone: 0171 637 9111
fax: 01372 227007

Logitrans
William Gorter
telephone: +331 498 51516
fax: +331 498 51550

London Research Centre
John Hollis
telephone: 0171 735 4250
fax: 0171 627 9606

Longdin & Browning
Nick Eales
telephone: 01792 202244
fax: 01792 203333
email: 100271.2433@compuserve.com

Longman Group Ltd
Nila Patel
telephone: 01279 623623
fax: 01279 431059

Lovell Johns Ltd
Richard Hewish
telephone: 01993 883161
fax: 01993 883096

LOY Surveys Ltd
Jim Loy
telephone: 0141 885 0800
fax: 0141 885 1202

LTG Services
John Callaghan
telephone: 0181 906 5559
fax: 0181 906 5248

Macaulay Land Use Research Institute
Dr Richard J.Aspinall
telephone: 01224 318611
fax: 01224 311556
email: r.aspinall@uk.ac.sari.mluri

Magdala Sociedade
Peter Wallace
telephone: +351 1 460 0684

Magellan Systems Corporation
Jim White
telephone: +1909 394 5000
fax: +1909 394 7050

Manchester Metropolitan University
Ms Beverly Heyworth
telephone: 0161 247 1581
fax: 0161 247 6344
email: b.heyworth@mmu.ac.uk

Map Data Management Ltd
Philip Storey
telephone: 015395 67431
fax: 015395 67837

MapInfo Ltd
Matthew Spencer
telephone: 01344 482888
fax: 01344 482777
email: matthew.spencer@mapinfo.com

MAPIT Limited
Douglas Cross
telephone: 01487 813745
fax: 01487 813745

MAPS geosystems
Alistair MacKenzie
telephone: +9716 356411
fax: +9716 354057

Mason Land Surveys
Robert Owen
telephone: 01383 727261
fax: 01383 739480

Mathshop
Timothy J.Coffey
telephone: 01794 523423

McLintock Limited
Chris Hallos-Johnson
telephone: 01865 749957
fax: 01865 749434

MEGRIN Group
Frangois Salge
telephone: +331 43 988440
fax: +331 43 988443
email: megrin@megrin.ign.fr

Mentis Management Consultants Ltd
Mike Gunner
telephone: 01483 776717
fax: 01483 747337
email: 100646.3576@compuserve.com

Methods Applications Ltd
Peter Rowlins
telephone: 0171 240 1121
fax: 0171 379 8561
email: mal@methods.win-uk.net

Midlands Regional Research Laboratory
Dr.Alan.J.Strachan
telephone: 0116 252 3825
fax: 0116 252 3854
email: ajs@le.ac.uk

Midsummer Computing
Richard Avery
telephone: 01908 668866
fax: 01908 667623

Modern Maps
Tony Vickers
telephone: 01635 34251
fax: 01635 34251

Morgan Collis Group Ltd
Ian Holroyd
telephone: 01604 580980
fax: 01604 589208
email: ian@mcgltd.demon.co.uk

MOSS Systems Limited
Carol Heaton
telephone: 01403 259511
fax: 01403 217746

Mott MacDonald Ltd
Graham Bugler
telephone: 0181 686 5041
fax: 0181 681 5706
email: gsb@mm-croy.mottmac.co.uk

MPSI Systems Ltd
Mr.A.Renshaw
telephone: 0117 927 9653
fax: 0117 929 2056

MR Data Graphics
Pat Jarvis
telephone: 01483 575312
fax: 01483 300167

Munro Garratt International
David Field
telephone: 01224 622888
fax: 01224 622229

MVA Systematica
Andy Heath
telephone: 01483 728051
fax: 01483 755207

MVM Consultants plc
David Rix
telephone: 0117 974 4477
fax: 0117 970 6897

NAG Ltd
Dr Terry Burgess
telephone: 01865 511245
fax: 01865 310139
email: infodesk@nag.co.uk

National Remote Sensing Centre Ltd (NRSC)
Peter Beaumont
telephone: 01252 541464
fax: 01252 375016

Natural Environment Research Council
Mr.G.H.D.Darwell
telephone: 01793 411996
fax: 01793 411959
email: g.darwall@nss.nerc.ac.uk

Navstar Systems Ltd
Clive de la Fuente
telephone: 01604 585588
fax: 01604 585599

NCC Blackwell Ltd
Anne Kitson
telephone: 01865 791100
fax: 01865 798210
email: akitson@cix.compulink.co.uk

NERC
George H D Darwall
telephone: 01793 411683
fax: 01793 411610
email: ghdd@nss.nerc.ac.uk

Nestor International Ltd
Jill Farmer
telephone: 0171 937 4434
fax: 0171 937 4180

NOMIS
Michael Blakemore
telephone: 0191 374 2468
fax: 0191 384 4971
email:
michael.blakemore@uk.ac.durham

Numonics UK (Division of Telmtek Ltd)
Clive Rowe
telephone: 01332 298480
fax: 01332 290667

Oaklands I.T.
Robert Coote
telephone: 01306 611590
fax: 01306 611590

Office of Population Censuses and Surveys (OPCS)
Alan Taylor
telephone: 01329 813536
fax: 01329 813532

Optimal Software Ltd
D Brady
telephone: 0161 367 8715
fax: 0161 367 9328

Ordnance Survey
Digital Sales
telephone: 01703 792773
fax: 01703 792324
email: ordsvy.govt.uk

Ordnance Survey of Northern Ireland
Michael Brand
telephone: 01232 255755
fax: 01232 255700

Organisation Management Systems
Jon J Gibbons
telephone: 01865 372161
fax: 01865 842664

Oscar Faber
Shaun Everett
telephone: 0181 784 5784
fax: 0181 784 5700
email:
marketing@oscarfab.demon.co.uk

Ove Arup & Partners
Simon Power
telephone: 01222 473727
fax: 01222 472277

Oxford Institute of Retail Management
Christopher Talbot
telephone: 01865 735422
fax: 01865 736374
email: oxirm@uk.ac.ox.temcol

P&L Engineering Surveys Ltd
Richard Lennox
telephone: 0141 644 1690
fax: 0141 644 5361

PA Consulting Group
Phil Jeanes
telephone: 0171 730 9000
fax: 0171 333 5050
email: phil.jeanes@pa-consulting.com

PAFEC Ltd
Amanda Ward
telephone: 0115 935 7055
fax: 0115 935 7057
email: appwmm@pafec.co.uk

Panda
Denis.W.Payne
telephone: 01223 233577
fax: 01223 233577

Paul Clasper & Associates Ltd
Paul Clasper
telephone: 01225 444561
fax: 01225 422187
email: clasper@csm.uwe.ac.uk

PAX Technology
David J Bethel
telephone: 01491 572282
fax: 01491 411040

PD Computing Ltd
Mr.N.Hoe
telephone: 0161 747 7110

Pear Technology Services Ltd
John Cowling
telephone: 01705 499689
fax: 01705 499689
email: 100554.2753@compuserve.com

Peter Thorpe Consulting
Peter Thorpe
telephone: 0926 52799
fax: 0926 52799

Photarc Surveys Ltd
Rory M.Stanbridge
telephone: 01423 871629
fax: 01423 871639

Photoair
Richard Young
telephone: 01733 241850
fax: 01733 242964

Photogrammetric Data Services Ltd
Robert Finch
telephone: 01273 464883
fax: 01273 454238

Pinpoint Digitising Services
Suzanne Soper
telephone: 01928 579148
fax: 01928 579192

PlanGraphics Inc
Dennis Kunkle
telephone: +1502 223 1501
fax: +1502 223 1235
email: plang@ipx.netcom.com

Planning & Mapping Ltd
Tony Leeds
telephone: 01403 783314
fax: 01403 784596

PLANTECH Ltd
Sylvie Temperley
telephone: 0171 922 8825
fax: 0171 928 8066

Plowman Craven & Associates Ltd
Mark Phillips
telephone: 01582 765566
fax: 01582 765370
email: mark@plowcrav.demon.co.uk

Posford Duvivier
Tim Jeffries-Harris
telephone: 01733 334455
fax: 01733 262243

Positioning Resources Limited
Judith Collier
telephone: 01224 581502
fax: 01224 574354

Primagraphics Ltd
Dr Glynn Wright
telephone: 01763 262041
fax: 01763 262551

Procis Software Ltd
Pammi Panesar
telephone: 01793 541200
fax: 01793 541025

Progis GmbH
Di Walter H Mayer
telephone: +43 4242 26332
fax: +43 4242 263327

Property Intelligence plc
Michael Nicholson
telephone: 0171 839 7684
fax: 0171 839 1060

QAS Systems Ltd
Catherine Meader
telephone: 0171 498 7777
fax: 0171 498 0303

QC Data (Ireland) Limited
Brendan Walshe
telephone: +353 21 341700
fax: +353 21 343645

Quail Map Company
John Yonge
telephone: 01392 430277
fax: 01392 430277

Quorum Information Services
Malcolm Orchard
telephone: 01252 318884
fax: 01252 313120

R.W.A. Dalls FRICS
R.W.A.Dallas
telephone: 01904 652408
fax: 01904 652408

Racal Survey (UK) Ltd
D.Inglis
telephone: 01224 249700
fax: 01224 249446

Raindrop Information Systems Ltd
Catherine Flannigan
telephone: 0171 734 1091
fax: 0171 734 1095

Recruit Media Ltd
Melissa Coxon
telephone: 0171 704 1227
fax: 0171 704 1370
email: 1005363201

regioplanDATA GmbH
Georg Egger
telephone: +49 30 896704 18
fax: +49 30 896704 10

Remote Sensing Applications Consultants
Peter Fletcher
telephone: 01420 561377
fax: 01420 561388
email: consultants@rsac.demon.co.uk

RICS Books
Diane Williams
telephone: 0171 222 7000
fax: 0171 222 9430

Robert Walker Consultants
Rob Walker
telephone: 01954 251003
fax: 01954 251003

Royal Commission on the Historical Monuments of England
Neil Lang
telephone: 01793 414727
fax: 01793 414770

Royal Geographical Society (with The Institute of British Geographers)
Dr A F Tatham
telephone: 0171 589 5466
fax: 0171 584 4447
email: info@rgs.org

Royal Institute of Navigation
Group Capt D.W.Broughton
telephone: 0171 589 5021
fax: 0171 823 8671

Royal Mail – Address Management Centre
Bill Haken
telephone: 01705 838518
fax: 01705 838518

Royal Town Planning Institute
Michael Napier
telephone: 0171 636 9107
fax: 0171 323 1582

Salford University Business Services Ltd
Paul Coward
telephone: 0161 957 0012
fax: 0161 737 7700

SAS Institute
Alastair Sim
telephone: 01628 486933
fax: 01628 483203

Saztec Europe Limited
Michael McLellan
telephone: 01753 833131
fax: 01753 832454

SAZTEC Philippines Inc
Conrad Lealand
telephone: 0171 702 2906
fax: 0171 702 4507
email: compuserve 100535,2213

Scan Group Limited
Samuel Levin
telephone: +9724 410339
fax: +9724 413924

Scientific Software Limited
David Finlay
telephone: 01491 411727
fax: 01491 411627

SCOT Conseil
S Gobin
telephone: +33 613 94600
fax: +33 613 94610

Scott Wilson Kirkpatrick
Richard Metcalfe
telephone: 01256 461161
fax: 01256 460582
email: 1005613337

SDI Ltd
Steve Morris
telephone: 0181 207 5474
fax: 0181 207 2755

Sector (UK) Limited
Stephen Price
telephone: 0171 582 9982
fax: 0171 587 1908

SHL Vision* Solutions Limited
Kevin Challen
telephone: 01276 677707
fax: 01276 676567
email: info@vision.solns.co.uk

SIA Limited
Janice Mabert
telephone: 0171 730 4544
fax: 0171 730 6762

SIAS Limited
Sarah E.Muirhead
telephone: 0131 225 7900
fax: 0131 225 9229
email: lucy@sias.demon.co.uk

Siemens Nixdorf
Miss C.A.J.Twentyman
telephone: 01344 850829
fax: 01344 850943

Silicon Graphics Limited
Andrew Cresci
telephone: 01734 257500
fax: 01734 257569
email: cresci@reading.sgi.com

Silsoe College, Cranfield University
Chris Bird
telephone: 01525 863060
fax: 01525 863099
email: c.bird@cranfield.ac.uk

Simmons Survey Partnership Limited
Grel Simmons
telephone: 01934 732122
fax: 01934 732938

Sir Alexander Gibb & Partners Ltd
Dr R S Steedman
telephone: 01734 261061
fax: 01734 491054

Sir William Halcrow and Partners
Robert Deakin
telephone: 01793 812479
fax: 01793 812089

Smallworld Systems Ltd
Joy Haigh
telephone: 01223 460199
fax: 01223 460210
email: smallworld.@smallworld.co.uk

Smartscan Inc
Rebecca J Culp
telephone: +1303 443 7226
fax: +1303 443 2997

Smith System Engineering Ltd
Mike O'Boyle
telephone: 01483 442000
fax: 01483 442144
email: mwoboyle@smithsys.co.uk

Sokkia Ltd
Mark Harper
telephone: 01270 250525
fax: 01270 250533
email: cix@nobbey

Solent Mapping and Charting (SMAC)
Dominic Fontana
telephone: 01705 842477
fax: 01705 842512

Southbank Systems PLC
Christopher Megan
telephone: 01634 880141
fax: 01634 880383

Sovereign C.S. Ltd
Roger Werry
telephone: 0181 866 0713
fax: 0181 429 0959

Spacesense consultants
Dr Adrian Lloyd-Lawrence
telephone: 0181 940 6290
fax: 0181 332 2786

Spatial Data Limited
Mr.N.A.Richardson
telephone: 01539 721070

Spatial Geographic Services & Applications Ltd
Ron Linton
telephone: 01865 881753
fax: 01865 881753

Spatial Information Services Ltd (SIS)
Peter D.Lever
telephone: 01329 662891
fax: 01329 668440

SPSS UK Ltd
John Davies
telephone: 01932 566262
fax: 01932 567020

Star Informatic S.A.
Manuel Pallage
telephone: +32 41 675313
fax: +32 41 671711

StatSci Europe
Matthew Eagle
telephone: 01865 200952
fax: 01865 200953
email: sales@statsci.co.uk

Structural Technologies Ltd (STL)
David Schindler
telephone: 01527 854819
fax: 01527 854819

Summagraphics Europe N.V.
Carina Govaert
telephone: +32 2 721 5033
fax: +32 2 721 5289

Sun Microsystems Ltd
Colin Taylor
telephone: 01276 451440
fax: 01276 472114
email: colint@sun.co.uk

Survey Control Services
C.Mills
telephone: 01207 544996

Survey & Development Services Ltd
Elspeth Rodger
telephone: 01506 825121
fax: 01506 822629

Survey Supplies Ltd
Bob Wells
telephone: 0151 931 3161
fax: 0151 931 2838

Svitzer Limited
Bob Blow
telephone: 01493 440320
fax: 01493 440319

Symology Limited
Edgar Blazier
telephone: 01582 842626
fax: 01582 842600

Sysdeco (UK) Ltd
Brian Dixon
telephone: 01223 420464
fax: 01223 420324

System Options Limited
Jim Pedroza
telephone: 01252 334383
fax: 01252 28779

TACTICIAN UK
Richard Ives
telephone: 01423 560064
fax: 01423 525545

Tangent Technology Design Associates Ltd
Carl Billson
telephone: 01509 610910
fax: 01509 610403

Target Market Consultancy
Peter Sleight
telephone: 01494 712371
fax: 01494 714203
email: compuserve 100130 1723

Taylor & Francis Ltd
Richard Steele
telephone: 0171 400 3500
fax: 0171 831 2035
email: rsteele@tandf.co.uk

TDS-CAD Graphics Ltd
Karen Bamford
telephone: 01254 676921
fax: 01254 581574

TEAMS (Taylor Woodrow Electronics Asset Mapping Survey)
Keith Ricardo
telephone: 0113 242 3802
fax: 0113 242 5172

Tekla Oy
Risto Sajaniemi
telephone: +358 0 803 7722
fax: +358 0 803 9489
email: rs@tekla.fi

Tele Atlas
Ad Bastiaansen
telephone: +32 9 222 5658
fax: +32 9 222 7412

Tendron Systems Ltd
R.A.Brown
telephone: 0117 929 4759
fax: 0117 922 1320

Tenet Systems Ltd
Dr Sharon Cooper
telephone: 01403 273173
fax: 01403 273123

Terrafix Ltd
Mr.J.B.Rosson
telephone: 01782 577015
fax: 01782 835667

TerraHunt GeoScience Ltd
Derek Morris
telephone: 01727 822287
fax: 01727 826570

TerraQuest Group Limited
Alan Ross
telephone: 0121 520 0111
fax: 0121 520 5800

The British Computer Society
Andrew Wilkes
telephone: 01743 417417
fax: 01793 480270

The Business Database from Yellow Pages
Simon Green
telephone: 01753 583311
fax: 01753 594001

The Data Consultancy
Margaret .H.Smee
telephone: 01734 588181
fax: 01734 597637

The LGMB
A.K.Black
telephone: 01582 451166
fax: 01582 412524

The Marketing Information Consultancy (MIC)
Richard Bandell
telephone: 0181 213 5500
fax: 0181 213 5599
email: mic@hyperlink.com

The NPA Group Ltd
Ren Capes
telephone: 01732 865023
fax: 01732 866521
email: info@npagroup.co.uk

The Severn Partnership
Nigel Atkinson
telephone: 01743 874135
fax: 01743 874716

The Survey Centre
R.M.Whitfield
telephone: 01905 21073
fax: 01905 29085

Trac Consultancy
David Griffiths
telephone: 0171 639 9825
fax: 0171 639 9825

Trident Map Services
Paul Shingfield
telephone: 01582 867211
fax: 01582 867689

Trimble Navigation Europe Ltd
Helen Knight
telephone: 01256 760150
fax: 01256 760148

TYDAC Technologies Ltd
James Oliver
telephone: 01703 760824
fax: 01703 760944
email: sales@tydac.demon.co.uk

Unistride Sewer Technology
Paul Marvin
telephone: 01420 23456
fax: 01420 22712

UNISYS Ltd
Edwin Ecob
telephone: 01895 237137
fax: 01895 862984
email: ecobe@uxbpo1.gb.unisys.com

Universal Systems Ltd
Rick Nyarady
telephone: +1506 458 8533
fax: +1506 459 3849
email: nyarady@unb.ca

University of East London
Andrew Larner
telephone: 0181 849 3618
fax: 0181 849 3514

University of Edinburgh
Bruce M.Gittings
telephone: 0131 650 2565/2543
fax: 0131 650 2524
email: gisadmin@geovax.ed.ac.uk

University of Greenwich
Dr Gesche Schmid
telephone: 0181 331 9800
fax: 0181 331 9805

University of Leicester
Dr.Peter.Fisher
telephone: 01533 523839
fax: 01533 523854

email: pffi@le.ac.uk

University of Luton
Ron Beard
telephone: 01582 489264
fax: 01582 489212
email: rbeard@vax2.luton.ac.uk

University of Newcastle Upon Tyne
Dr David Parker
telephone: 0191 222 6445
fax: 0191 222 8691
email: david.parker@ncl.ac.uk

Walker Ladd Surveys
Phil Thompson
telephone: 0117 925 1251
fax: 0117 925 7500

Wallingford Software Limited
Adrian Turner
telephone: 01491 824777
fax: 01491 826392

WDV (UK)
John Lees
telephone: 01789 297000
fax: 01789 298056

WRc (Water Research centre)
John Cima
telephone: 01793 511711
fax: 01793 511712

WS Atkins Planning & Management Consultants
David Eastwood
telephone: 01372 726140
fax: 01372 740055
email: deastwood@wsatkins.co.uk

Xcon Data
Jon Apneseth

miscellaneous reference 1996

GIS organisations in Europe

directory of EUROGI representatives/members

GIS standards

GIS dictionary

GIS courses available

GIS course contacts

calendar worldwide

AGI papers

AGI papers by subject

AGI awards 1990-1995

AGI council

AGI committees

list of AGI members

GIS organisations in europe

EC Organisations

CORINE programme,
Commission of European Communities
DG X1-T174 00/C88
rue de la Loi 200
B-1049 BRUSSELS
Belgium
telephone: +32 22 35 11 11
fax: +32 22 35 01 44

EUROSTAT
Environmental Statistics
Commission of European Communities
Batiment 06/632
L-2920 LUXEMBOURG
telephone: +352 43 01 72 76
fax: +352 43 01 30 15
Multi-disciplinary European Bodies

AM/FM (automated mapping facilities management)   Secretariat
PO Box 6
CH-4005 BASEL
Switzerland
telephone: +41 61 691 51 11
fax: +41 61 691 81 89

Multi-disciplinary European Bodies

EGIS (European conference on Geographic Information System)
University of Utrecht
Postbus 80.115
NL-3508 UTRECHT
The Netherlands
telephone: +31 30 53 42 61
fax: +31 30 52 36 99

EUROGI (European Umbrella Organisation for Geographic Information)
PO Box 508
3800 AM - Amersfoort
The Netherlands
telephone: +31 33 46 04 150
fax: +31 33 46 56 457

UDMS (Urban Data Management Society)
Delft University of Technology
Faculty of Geodesy
Thijsseweg 11
NL-2629 DELFT
The Netherlands
telephone: +31 15 78 45 48
fax: +31 15 78 23 48

National GIS umbrella organisations or advisory and co-odinating bodies

AESIGY (Asociacion Espanola de Sistems de Informacion Geografica y Territorial)
c/- Padilla 66 3
ES-28006 MADRID
Spain
telephone: +341 402 93 91
fax:

AFIGEO
136 bis, rue de Grenelle
75007 PARIS
France
telephone: +33 1 43 98 83 12
fax: +33 1 43 98 85 66

AGI (The Association for Geographic Information)
12 Great George Street
Parliament Square
London
SW1P 3AD
telephone: +44 171 334 3746
fax: +44 171 334 3791

ProGIS
National Land Survey of Finland
PO Box 84
FIN-00521 HELSINKI
Finland
telephone: +358 0 154 5002
fax: +358 0 154 5005

ULI (Utvecklingsradet for Landskaps Information)
Lantmateriverket
801 82 GAVLE
Sweden
telephone: +46 26 15 37 80
fax: +46 26 68 75 94

Raad voor Vastgoedinformatie
Koningin Wilhelminalaan 41
AMERSFOORT
The Netherlands
telephone: +33 33 60 41 00
fax: +31 33 65 64 57

National Mapping Agencies

Administration du Cadastre et de la Topographie
BP 1761
L-1017 LUXEMBOURG
telephone: +352 449 01 312
fax: +352 449 01 333

Bundesamt fur Eich und Vermessungswesen
Schiffamtsgasse 1-3
A-1025 WIEN
Austria
telephone: +43 1 211 76 3602
fax: +43 1 216 10 62

Czech Office for Surveying, Mapping and Cadastre
Cesky Uread Zememericky Kataastralni
CZ-11121 PRAHA 1
HYBERNASKA 2
telephone: +42 2 242 24 811
fax: +42 2 242 17 383

Department of Land and Surveys
29 Michalakopoulou Street
CY-NICOSA
Cyprus
telephone: +357 230 22 10
fax: +357 244 60 56

Department of Lands & Mapping
Ministry of Agriculture
H-1860 BUDAPEST 55 pf.1
Hungary
telephone: +36 1 131 37 36
fax: +36 1 153 05 18

Department of Surveyor General of Poland
Ministry of Physical Planning and Construction
ul. Wspolna 2
PL-00926 WARSAW
Poland
telephone: +48 2 628 73 64
fax: +48 2 628 72 37

Hellenic Military Geographical Service
Pedion Areos
GR-11362 ATHENS
Greece
telephone: +30 1 994 28 11
fax: +30 1 881 73 76

Iceland Geodetic Survey
Laugavegi 178
PO Box 5060
IS-125 REYKJAVIK
Iceland
telephone: +354 1 68 16 11
fax: +354 1 68 06 14

Institut de Geodesie, Photogrammetrie, Cartographie et Amenagement du Territoire
B-dul Expozitiei 1a
RO-BUCURESTI
Rumania
telephone: +40 1 617 75 64
fax: +40 1 617 75 63

Institut Geographique National
Abbaye de la Cambre 13
B-1050 BRUSSELS
Belgium
telephone: +32 2 648 64 80
fax: +32 2 646 25 42

Institut Geographique National
136 bis
rue de Grenelle
F-75700 PARIS
France
telephone: +33 1 439 882 00
fax: +33 1 455 507 85

Instituto Geografico e Cadastral
Praca Da Estrela
P-1200 LISBON
Portugal
telephone: +351 1 664 258
fax: +351 1 397 02 48

Instituto Geografico Militare Italiano
Via C.Battisti
10-12I-50100 FIRENZE
Italy
telephone: +39 55 214 054
fax: +39 55 282 172

Instituto Geografico Nacional
General Ibanez de Ibero 3
E-28071 MADRID
Spain
telephone: +34 1 533 31 21
fax: +34 1 554 67 43

Kort-og Matrikelstyrelsen
Rentemestervej 8
DK-2400 COPENHAGEN NV
Denmark
telephone: +45 35 87 50 50
fax: +45 35 87 50 59

Ministry of National Defence
General Command of Mapping
TR-06100 CEBECIANKARA
Turkey
telephone: +90 312 319 25 15
fax: +90 312 320 14 94

Ministry of Civil Engineering and Environmental Protection
Department of Cadastre and Surveying
Gruska 20
HR-41000 ZAGREB
Croatia
telephone: +385 41 519 305
fax: +385 41 533 419

National Land Board
Mustamae tee 51
PO Box 1635
EE-0006 TALINN
Estonia
telephone: +7 0142 58 22 49
fax: +7 0142 52 84 01

National Land Survey of Finland
Box 84
Opastinsilta 12c
SF-00521 HELSINKI
Finland
telephone: +358 0 154 5000
fax: +358 0 154 5005

National Land Survey of Sweden
S-801 82 GAVLE
Sweden
telephone: +46 26 153 323
fax: +46 26 687 594

Office Federal de Topographie
Seftgenstrasse 264
CH-3084 WABERN
Switzerland
telephone: +41 31 963 21 11
fax: +41 31 963 24 59

Ordnance Survey
Romsey Road
SOUTHAMPTON
SO9 4DH
telephone: +44 1703 792 000
fax: +44 1703 792 660

Ordnance Survey
Phoenix Park
DUBLIN 8
Ireland
telephone: +353 1 8206 100
fax: +353 1 8204 156

Ordnance Survey of Northern Ireland
Colby House
Stranmillis Court
BELFAST
BT9 5BJ
Northern Ireland
telephone: +44 232 661 244
fax: +44 232 683 211

Republica Geodetska Uprava Slovenije
Kristanova 1SI-61000
Ljubjana
Slovenia
telephone: +386 61 31 68 53
fax: +386 61 13 22 021

Staatsminsterium der Finanzen
Abt. V11 (Vermessungsverwaltung)
Odeonsplatz 4
D-80539 MUNCHEN
Germany
telephone: +49 89 2306 2401
fax: +49 89 2306 2807

State Department of Surveying and Mapping of Lithuania
Jaksto, 9
LT-2600 VILNIUS
Lithuania
telephone: +7012 62 78 12
fax: +7012 62 78 18

State Land Service
11 novembra krastm
31LV-1484 RIGA
Latvia
telephone: +371 2 228 459
fax: +371 2 212 320

Statens Kartverk
N-3500 HONEFOSS
Norway
telephone: +47 32 11 81 00
fax: +47 32 11 81 01

Topografische Dienst
Bendienplein 5
PO Box 115
NL-7800 AC EMMEN
The Netherlands
telephone: +31 5910 96 200
fax: +31 5910 96 296

Urad Geodezie Kartografie a Katastra Slovensky Republiky
Hlobka 2
81323 BRATISLAVA
Slovak Republic
telephone: +42 7 49 75 55
fax: +42 7 49 75 73

Pan-European Mono Disciplinary Bodies

BRGM (Bureau de Recherches Geologiques et Minieres)
90 Avenue Lambeau
B-1200 BRUSSELS
Belgium
telephone: +32 2 734 32 38
fax: +32 2 734 32 38

CERCO (Comite European des Responsables de la Cartographic Officielle)
Secretariat
Avenue de la Couronne 172
Bte 26
B-1050 BRUSSELS
Belgium
telephone: +32 2 649 84 85
fax: +32 2 647 74 90

DWIG
Directorate General of Military Survey
Elmwood Avenue
Feltham
Middlessex
TW13 7AE
telephone: +44 181 890 36220
fax: +44 181 890 36220 x 4148

EARSEL (European Association of Remote Sensing Laboratories)
Secretariat
Instituto di Gasdinamica
University of Naples
P.le Tecchio 80
I-80125 NAPLES
Italy
telephone: +39 81 768 21 59
fax: +39 81 768 21 60

ESF (European Science Foundation)
1 quai Lezay-Marnesia
F-67000 STRASBOURG CEDEX
France
telephone: +33 88 76 71 00
fax: +33 88 37 05 32

Eureau (Water)
Syndicat des Eaux de Sud (SES)
L-8388 KOERICH
Luxembourg
telephone: +352 39 91 96

IGU (International Gas Union)
Grutlistrasse 44
Postfach 658
CH-8027 ZURICH
Switzerland
telephone: +41 1 81 78 16

OEEPE (Organisation Europeenne d'Etudes Photogrammetriques Experimentales)
Secretariat, International Institute for Aerospace
Survey and Earth Sciences
Dept of Geoinformatics
PO Box 67500
AA ENSCHEDE
The Netherlands
telephone: +31 53 874 339
fax: +31 53 874 335

TELECOM
Office de Liaison
BP1283
CH-3001 BERN
Switzerland

UNIPEDE (Electricity)
c/- Director
Schleswag AG
Kieler Strasse 19
D-24768 RENDSBERG
Germany
telephone: +49 43 31 201 2230
fax: +49 43 31 201 2166

URSA
18 Pythagora str
GR-155 62 HOLARGOS
Greece
telephone: +30 1 652 66 48
fax: +30 1 652 17 25

directory of EUROGI
representatives/members

AESIG
Mr J Tejero de la Cuesta
C/Jose Abascal, 44-4 o D
E28003
Spain
telephone: +34 14417799
fax: +34 14424889

AFIGEO
Mr.J.C.Lummaux
136 bis, Rue de Grenelle
75700 Paris 07 SP
France
telephone: +33 143988312
fax: +33 143988566

AGI
Mr. S. Leslie
12 Great George Street
London SW1P 3AD
United Kingdom
telephone: +44 71 334 3746
fax: +44 71 334 3791
email: agi@geo.ed.ac.uk

AM/FM Europe
Mr.G.Liesenfelt
136 bis, Rue de Grenelle
75700 Paris 07 SP
France
telephone: +33 143988312
fax: +33 143988566

AM/FM Italy
Instituto di Fisiologia Clinica
Dr.A.Fernandez Perez de Talens
v.Paolo Savi, 8
56100 Pisa
Italy
telephone: +39 50562721
fax: +39 50553461
email: antonio@po.1fc.pl.cnr.it

CNIG
Mr.Rui Gonsalves Henrique
R.Braamcamp, 82 1o Dto
1200 Lisboa
Portugal
telephone: +351 13860011
fax: +351 13862877
email: nuno@helios.cnig.pt

DDGI
IKS Ingenieurgesellshaft
Kramer-Schuller MBH
Dr.Ing.K.Schuller
Magdeburgerstrasse 37
47800 Krefeld
Germany
telephone: +49 2151478941
fax: +49 2151477580

EGIS
University of Utrecht
Faculty of Geographical Sciences
Prof.P.A.Burrough
P.O. Box 80.115
3508 TC Utrecht
The Netherlands
telephone: +31 30534261
fax: +31 30523699

GISIG
Mr.G.Saio
Via dell'Acciaio
16152 Genova
Italy
telephone: +39 106514000
fax: +39 106508782
email: gisig@itd.ge.cnr.it

GIVE
Ing.H.J.Festen
P.O. Box 3057
60 DB Soest
The Netherlands
telephone: +31 215520849
fax: +31 215523488

GTIM-SIG
Mr.A.Majerus
54 Ave. Gaston Diderich
B.P. 1761
L-1017 Luxembourg
Luxembourg
telephone: +352 44901274
fax: +352 44901333

IRLOGI
c/- IEI
Dr.R.C.Cox
22 Clyde Road
Dublin 4
Eire
telephone: +353 16684341
fax: +353 16685208

NDC
Nat.Hellinic Research Foundation
Prof.N.D.Polydorides
48, Vas. Constantinou Ave
116 35 Athens
Greece
telephone: +30 17242172
fax: +30 17246824
email: ekt@apollon.servicenet.ariadne-t.gr

ProGIS
National Land Survey of Finland
Mr.Anttii Rainio
P.O. Box 84
FIN-00521 Helsinki
Finland
telephone: +35 801 545 446
fax: +34 801 545 454
email: p<ORG1>rogis.@mmh.fi
EUROGI Member

RAVI
secretaiat
Mr.B.C.Kok
P.O. Box 508
3800AM Amersfoort
The Netherlands
telephone: +31 336 04100
fax: +31 335 65457
email: ravi@euronet.nl

SOGI
AKM
Mr.H.Lindenmann
P.O. Box 6
4005 Basel
Switzerland
telephone: +41 616 915111
fax: +41 616 918189

UDMS
Delft University of Technology
Faculty of Geodesy
Ms.E.M.Fendel
P.O. Box 5030
2600 GA Delf
The Netherlands
telephone: +31 15784548
fax: +31 15782745

ULI
Mr.F.Sundberg
c/o Lantmateriverket
S-801 82 Gavle
Sweden
telephone: +46 26153000
fax: +46 26687594
email: christwal@imv.im.se

GIS standards

ANSI X3.135-1986	Database Language - SQL (SQL-86)
ANSI X3.135-1989	Database Language - SQL (SQL-89)
ANSI X3.135-1992	Database Language - SQL (SQL-92)
AQAP-13	NATO Software Quality Control System Requirements
BS 4730	Specification for UK 7-bit coded character set
BS 5750	Quality Systems
	Part 1 : 1979 Specification for design, manufacture and installation
	Part 2 : 1979 Specification for manufacture and installation
	Part 3 : 1979 Specification for final specification and test
	Part 4 : 1981 Guide to the use of BS 5750 Part 1
	Part 5 : 1981 Guide to the use of BS 5750 Part 2
	Part 6 : 1981 Guide to the use of BS 5750 Part 3
	Part 7 :
	Part 8 : Specification for Service providers
BS 6690	Specification for a data descriptive file for information interchange (= ISO 8211 : 1985)
BS 6692	Coded character sets for text communication (= ISO 6937)
	BS 6692 : Part 1 : 1986 General introduction
	BS 6692 : Part 2 : 1990 Specification for Latin alphabetic and non-alphabetic graphic characters.
BS 6856	Specification for code extension techniques for United Kingdom 7-bit and 8-bit coded character sets (= ISO 2022)
BS 6945	Specification for Computer Graphics: metafile for storage and transfer of picture description information (CGM)
BS 6964	Specification for database language, SQL (=ISO 9075 : 1987)
BS 7567	Electronic transfer of geographic information (NTF)
	Part 1. Specification for NTF structures
	Part 2. Specification for implementing plain NTF
	Part 3. Specification for implementing NTF using BS 6690
BS 7666	Spatial datasets for geographic referencing
	Part 1. Street gazetteer
	Part 2. Land and Property Gazetteer
	Part 3. Address specification
DIGEST	Digital Geographic Information Exchange Standards
IEEE POSIX P1003	Portable Operating System for Computer Environments
ISO 10021	Text Communication - Message-Oriented Text Interchange Systems (MOTIS)
ISO 10027	Information Resource Dictionary System (IRDS) framework
ISO 10303	Industrial automation systems and integration - Product data representation and exchange
ISO 10746	A Basic Model for Open Distributed Processing (RM-ODP)
ISO 12382	Permuted index of the vocabulary of information processing
ISO 2382	Data Processing - Vocabulary
ISO 646	Information technology - ISO 7-bit coded character set for information interchange

ISO 6709	Standard representation of latitude, longitude and altitude for geographical point locations.
ISO 7498	Open Systems Interconnection - Basic Reference Model
ISO 7942	Computer graphics - Graphical Kernal System (GKS) functional specification
ISO 8072	Open Systems Interconnection : Transport Service Definition
ISO 8073	Information Processing Systems - OSI - Connection oriented transport protocol specification
	Addendum 1 : Network Connection Management Subprotocol
ISO 8211	Information processing - Specification for a data descriptive file for information interchange
ISO 8613	Open(Office) Document Architecture
	Part 1 : General Introduction
	Part 2 : Document Structures
	Part 3 :
	Part 4 : Document Profile
	Part 5 : Office Document Interchange Format (ODIF)
	Part 6 : Character Content Architecture
	Part 7 : Raster Graphics Content Architecture
	Part 8 : Geometric Graphics Content Architecture
ISO 8632	Computer graphics - Metafile for the storage and transfer of picture description information
	Part 1 : 1987 : Functional specification
	Part 2 : 1987 : Character encoding
	Part 3 :
	Part 4 :
ISO 8802	Information Processing Systems - Local Area Networks
ISO 8805	Computer Graphics - GKS for three dimensions (GKS-3D) functional description
ISO 8824	OSI - Specification of Abstract Syntax Notation One (ASN.1)
ISO 8825	OSI - Specification of basic encoding rules for Abstract Syntax Notation One (ANS.1)
ISO 8859	8 - bit single-byte coded graphic character sets
	Part 1 : 1987 Latin alphabet no 1
	Part 2 : 1987 Latin alphabet no 2
	Part 3 : 1988 Latin alphabet no 3
	Part 4 : 1988 Latin alphabet no 4
	Part 5 :
	Part 6 : 1987 Latin/Arabic alphabet
	Part 7 : 1987 Latin/Greek alphabet
	Part 8 : 1988 Latin/Hebrew alphabet
	Part 9 :
ISO 9000-3	Guidelines for the Application of ISO 9001 to the Development, Supply and Management of Software
ISO 9001	Quality Systems-Model for Quality Assurance in Design/Development, Production, Installation and Servicing
ISO 9075	Database Language - SQL
ISO 9592	Computer graphics - Programmer's Hierarchical Interactive Graphics System (PHIGS)
ISO 9735	Electronic data interchange for administration, commerce, and transport (EDIFACT) - application level syntax rules
NJUG 13	Quality Control Procedure for Large Scale Ordnance Survey Maps Digitised to OS 1988

GIS dictionary

introduction

The subject of Geographic Information has seen rapid development in recent years. It has introduced many new terms while at the same time producing new meaning to some existing terms in a particular context. Consequently, many publications include a glossary of terms for their own particular area of interest. The Standards Committee of the Association for Geographic Information published the first version of its GIS Dictionary in 1991. This dictionary attempts to produce a common set of definitions for general usage.

The new edition reflects the changes in the field of GIS that have occurred since publication of the first edition. In this time, GIS has become more a part of mainstream Information Technology (IT). This is reflected by the inclusion of more standard IT terms, especially those relating to databases and graphics, together with less emphasis on cartography.

The need to standardise terminology has been reflected in the work of both the International Standards Organisation (ISO) and the European Standards Committee (CEN), both of whom are producing sets of standard terminology. This internationalisation of GIS is reflected not only in the adoption of standard definitions where appropriate, but also in the removal from the Dictionary of terms that were totally specific to the UK.

Where definitions are taken from a definitive source, that source is identified in bold as a suffix. Full references are given to these sources in the list of abbreviations. Where no source is identified, then either the definition came from a non-definitive source, or it is a hybrid.

Comments received on the first edition have been taken into account in the preparation of the current edition. Any comments, errors or omissions should be notified to AGI, and will be taken into consideration in the preparation of the next edition.

Robert Walker
Editor

sources (and abbreviations):

AC	American Cartographer vol 15 No 1, Jan 1988.
EXPRESS	ISO CD 10303-11, Product Data Representation and Exchange - Part 11: The EXPRESS Language Reference Manual, 1991.
ISO 2382-1	Information Technology — Vocabulary – Part 1: Fundamental Terms, 1992.
ISO 2382-13	Information Technology – Vocabulary – Part 13: Computer Graphics, 1992.
ISO 2382-24	Information Technology – Vocabulary – Part 24: Computer Integrated Manufacture, 1992.
OS1	Working Party to produce National Standards for the transfer of digital data. Glossary of Terms, January 1987.
OS2	Ordnance Survey. Sample object data specification. March 1993.

dictionary

Term	Definition
2.5D	A system in which the third dimension is constrained to a very simple relationship with the other two dimensions (e.g. where Z is a single valued function of X and Y).
Absolute	One of the coordinates identifying the
Coordinate	position of an addressable point with respect to the origin of a specified coordinate system. **(ISO 2382-13)**
Abstraction	A way of viewing a real world object. For example, a road may be a centre-line in one application and an area bounded by kerblines in another.
Accuracy	The closeness of results of observations, computations or estimates to the true values or the values accepted as being true. Accuracy relates to the exactness of the result, and is distinguished from precision which relates to the exactness of the operation by which the result is obtained.
Address	A means of referencing an object for the purposes of unique identification and location.
Adjacency	The sharing of a common side by two areas (polygons).
Aggregation	The grouping together of a selected set of like entities to form one entity. For example, grouping sets of adjacent areal units to form larger units, often as part of a spatial unit hierarchy (e.g. wards grouped into districts). Any attribute data is also grouped or is summarised to give statistics for the new spatial unit.
Algorithm	A finite, ordered set of well-defined rules for the solution of a problem. **(ISO 2382-1)**
Aliasing	Unwanted visual effects caused by insufficient sampling resolution or inadequate filtering to completely define the object, most commonly seen as a jagged edge along the object's boundary, or along a line. **(ISO 2382-13)**
Alphanumeric	Display of information in character format.
Analogue	In the context of remote sensing and mapping, the term refers to information in graphical or pictorial form as opposed to digital form. Generally, analogue refers to a quantity which is continuously variable, rather than one which varies only in discrete steps.
Annotation	The alphanumeric text or labels on a map, such as street or place name.
ANSI (American National Standards Institute)	The US national standards body.
Application	A process that uses data from a system.
Applications package	A set of specialised programs and associated documentation, usually supplied by an outside agency or software house, to carry out a particular application (such as maintenance scheduling).

Arc	A line described by an ordered sequence of points and connections between them defined by a mathematical function.
Archive	Accessible store for historical records and data.
Area	A bounded continuous two dimensional object which may or may not include its boundary. Usually defined in terms of an external polygon or in terms of a set of grid cells. **(AC)**
ASCII	American Standard Code for Information Interchange. A standard binary coding system used to represent characters within a computer.
AM/FM (Automated Mapping/ Facilities Management)	North American term for digital records, and applications thereof, particularly in the utilities.
Associated Data	Alphanumeric information associated with a specific spatially referenced object.
Attribute	A trait, quality or property that is a characteristic of an entity. **(EXPRESS)**
Attribute Class	A specified group of attributes. For example, those describing measure, serviceability, structure, or composition. **(AC)**
Attribute Code	An alphanumeric identifier for an attribute. For example, 30-road type, Hg-concentration of mercury. **(OS1)**
Attribute Value	A specific quality or quantity assigned to an attribute. For example, for the attribute 'type' the value 'steel'.
Automated Cartography	The process of producing maps with the aid of computer driven devices such as plotters and graphical displays. The term does not imply any information processing, beyond that required to support map production.
Automated Digitising	Conversion of a map to digital form using a method which involves little or no operator intervention during the digitising stage, for example scanning. **(OS1)**
Automated Feature Recognition	The identification of map-based features using computer software incorporating pattern recognition techniques.
Automated Mapping	See Automated Cartography.
Base Map	A set of topographic data displayed in map form, providing a frame of reference for users' data.
Basic Spatial Unit (BSU)	Fundamental areal unit having homogeneous properties in the context of any required theme such as administrative responsibility or ownership.
Benchmark	Subject to context: (i) A standard test devised to enable comparisons to be made between computer systems. (ii) Surveying term for a mark whose height relative to a datum is known.
Blind digitising	A method of manual digitising where the operator has no immediate graphic feedback to register progress. **(OS1)**
BLOB (Binary large object)	An area within a raster dataset that can be considered as a contiguous feature.

BLPU (Basic land and property unit)	The physical extent of a contiguous area of land under uniform property rights.
Buffer	A corridor of a specified width around a point, line or area.
Byte	A unit of computer storage of binary data usually comprising 8 bits, equivalent to a character. Hence Megabyte, one million bytes, and Gigabyte, one thousand million bytes.
CAD (Computer-Aided Design)	The design activities, including drafting and illustrating, in which information processing systems are used to carry out functions such as designing or improving a part or a product. **(ISO 2382-24)**
Cadastral survey	A survey based on the precise measurement and marking of land parcel boundaries.
Cadastre	The public register of the quality, value and ownership of the land of a country.
Cartography	The organisation and communication of geographically related information in either graphic or digital form. It can include all stages from data acquisition to presentation and use.
CCITT (International Telegraph and Telephone Consultative Committee)	An international body primarily addressing telecommunications standards.
Cell	The basic element of spatial information in the raster (grid square) description of spatial entities. **(OS1)**
CEN (Comite Europeen de Normalisation)	The regional standards group for Europe. It is not a recognized standards development organization, and so cannot contribute directly to ISO. It functions broadly as a European equivalent of ISO and its key goal is to harmonize standards produced by the standards bodies of its member countries. Membership is open to EC and EFTA countries.
CENELEC (Comite Europeen de Normalization Electronique)	This is the European equivalent of the IEC.
Centre Line	A line digitised along the centre of a linear feature. **(OS2)**
Centroid	The position of the centre of gravity of an entity, often used to reference polygons.
CERCO (Comite Europeen des Responsables de la Cartographie Officielle)	The European Committee of Representatives of Official Cartography, under the auspices of the Council of Europe.
CGM (Computer Graphics Metafile)	A standard (ISO 8632) file format specification for the storage and transfer of picture description information. **(ISO 2382-13)**
Chain	A directed sequence of non-intersecting line segments and/or arcs with nodes at each end. **(AC)**
Chainage	Distance measured along a link from a node.
Check Plot	A graphic output used to verify either the content or positional accuracy of digital data by direct superimposition on the graphic original used to create the digital record. **(AC)**
Choropleth	A thematic map shaded by value of a parameter.

Clipping	The action of truncating data or an image by removing all the display elements that lie outside a boundary. **(ISO 2383-13)**
Compaction	See data compression.
Complementary BSUs	A family of BSUs designed around agreed definitions, the larger areas being the sum of a number of smaller areas etc.
Conceptual Model	A model that defines the types of entities or objects which are of immediate interest and the relationships between them.
Connectivity	How the links and nodes in a network or polygons are joined to each other, and the degree to which it is applied (e.g. simply, multiply).
Contiguous	Literally adjacent, touching. In the context of digital mapping, it implies a connected polygonal entity.
Continuous Mapping	A system of mapping in which the total extent of the mapped area is represented as a whole, without any spatial sub-divisions being apparent.
Contour	A set of points representing the same value of a given attribute and forming a line that may serve as a bounding of an area. **(ISO 2382-13)**
Control (mapping)	A system of points with established horizontal and vertical positions which are used as fixed references in positioning and relating map features. **(AC)**
Coordinated Point	A point defined by a set of coordinates relative to a well-defined coordinate system.
Coordinates	Pairs of numbers expressing horizontal distances along orthogonal axes, or triplets of numbers measuring horizontal and vertical distances. **(AC)**
Currency	The level to which data is kept up to date.
Cursor	(i) A pointer appearing on a screen, under the control of a mouse or other pointing device. (ii) A hand-held device on a digitising table or tablet used for picking menus or accurately digitising graphic objects.
Data	A collection of facts, concepts or instructions in a formalized manner suitable for communication or processing by human beings or by automatic means.
Databank	A set of data related to a given subject and organised in such a way that it can be consulted by users. **(ISO 2382-1)**
Database	A collection of data organised according to a conceptual structure describing the characteristics of the data and the relationships among their corresponding entities, supporting applications areas. **(ISO 2382-1)**
Database Management System (DBMS)	A collection of software for organising the information in a database. Typically a DBMS contains routines for data input, verification, storage, retrieval and combination.

Data capture	The encoding of data. In the context of digital mapping this includes digitising, direct recording by electronic survey instruments, and the encoding of text and attributes. **(OS1)**
Data classification	A description of the classes into which data is analyzed or divided. **(OS2)**
Data compression	Methods of encoding which reduce the overall data volume. (See Run length encoding)
Data conversion	The transformation of data from paper records or digital form into a form suitable for loading into a GIS.
Data format	A specification that defines the order in which data is stored or a description of the way data is held in a file or record. **(OS2)**
Data Item	A sequence of related characters which can be defined as the smallest logical unit of data that can be independently and meaningfully processed. For example, a single coordinate value. **(OS1)**
Data model	An abstraction of the real world which incorporates only those properties thought to be relevant to the application at hand. The data model would normally define specific groups of entities, and their attributes and the relationships between these entities. A data model is independent of a computer system and its associated data structures. A map is one example of an analogue data model. **(OS1)**
Data quality	Indications of the degree to which data satisfies stated or implied needs. This includes information about lineage, completeness, currency, logical consistency and accuracy of the data.
Dataset	An organised collection of data with a common theme.
Data structure	The logical arrangement of data as used by a system for data management; a representation of a data model in computer form. **(OS1)**
Data transfer	The movement of data from one system to another.
Datum	The fixed starting point of scale or a coordinate system.
Derived map	A map produced from other maps rather than from an original survey.
Device coordinate	A coordinate specified by a device-dependent coordinate system. **(ISO 2382-13)**
Device space	The space defined by the complete set of addressable points of a display device. **(ISO 2382-13)**
Dialogue box	A pop-up window into which data may be entered. **(ISO 2381-13)**
DIGEST	The Digital Geographic Information Working Group (DGIWG) Exchange Standard. A NATO standard for the exchange of geographic data in digital form between defence agencies.
Digital chart of the world	A vector data set based on 1:1,000,000 scale air navigation charts, produced by the US Defence Mapping Agency.
Digital Elevation Model (DEM)	Synonymous with Digital terrain model (DTM).
Digital Map Data	The digital data required to represent a map. **(OS1)**

Digital Mapping	The process of storing and displaying map data in computer form.
Digital Records	Electronically stored representation of user data aligned with and stored as an overlay to a digital map base, together with associated information.
Digital Terrain Model (DTM)	A digital representation of relief (ground surface). Usually a set of elevation values in correspondence with grid cells.
Digitising	The conversion of analogue maps and other sources to a computer readable form. **(OS1)** (This may be point digitising, where points are only recorded when a button is pressed on the puck, or stream digitising where points are recorded automatically at pre-set intervals of either distance or time.)
Digitising table	An electronic draughting table capable of recording the (x,y) coordinates of a point on a table in computer readable form.
Directed link	A link between two nodes with one direction specified. **(AC)**
Distributed data	Data held on several computers and accessible to users at other computers via communications networks.
Distributed system	A complex computer system where the workload is spread between two or more computers linked together by a communications network.
Domesday 2000	A project which aims to create a national computerised archive of property and land data in Britain.
Dpi (Dots per inch)	A unit of measurement for the resolution of a scanning or printing device.
Dragging	Relocating display elements on a screen with a pointing device. This can be done by pressing and holding a pushbutton while moving the pointer on the screen. **(ISO 2382-13)**
DX 90	A format for the supply of digital hydrographic data, developed by the International Hydrographic Organization (IHO). DX 90, together with the IHO's feature coding scheme (Object Catalogue) and a number of digitizing conventions comprise the IHO Transfer Standard for Digital Hydrographic Data.
DXF (Digital Exchange Format)	A format for transferring drawings between Computer Aided Design systems, widely used as a de facto standard in the engineering and construction industries.
Echo	The immediate notification of the current values provided by an input unit to the user at the display console. **(ISO 2382-13)**
Edge	A line between two nodes, bounding one or more faces.
Edge Matching	The process of ensuring that data along the adjacent edges of map sheets or some other unit of storage, matches in both positional and attribute terms. **(OS1)**
EDI (Electronic Data Interchange)	The interchange of processable data between computers electronically.
EDIFACT (Electronic Data Interchange For Administration, Commerce and Transport)	A set of syntax rules for the preparation of messages to be interchanged.

EDIGeO	A data transfer format strongly based on DIGEST, adopted by AFNOR as a French experimental standard Z-13-150.
Edit	The process of adding, deleting and changing data.
Emulator	A software package or program which imitates the functions of a hardware device or other software.
Encoding	The assignment of a unique code to each unit of information, such as encoding of English using the ASCII character set.
Entity	The general term for a real world object or digital phenomenon.
Entity class	A specified group of entities. For example, hydrography, transportation. **(AC)**
Entity relationship model	A logical way of describing entities and their relationships.
Enumeration district	The basic area unit, containing approximately 150 households, used by the UK Office of Population Censuses and Surveys, for the planning and carrying out of population counts and surveys.
EUROGI	European Umbrella Organisation for Geographic Information. An initiative to establish a single organisation covering the GIS community in Europe, made up from representatives of corporate organisations (associations).
EUROSTAT	The European Community statistical agency.
Expert system	A system that provides for solving problems in a particular application area by drawing inferences from a knowledge base acquired by human expertise. **(ISO 2382-1)**
Face	A surface bounded by a closed sequence of edges. Faces are contiguous and fill the spatial extent of the dataset and do not overlap.
Feature	A point or line used to represent one or more real-world objects. **(OS2)**
Feature code	An alphanumeric code which describes and/or classifies geographic features. **(OS1)**
Feature serial number	A unique code to identify an individual feature.
Fenceline	Boundary of a parcel of land physically represented on the ground by a surveyable object such as a wall or a fence.
Field	A set of one or more characters comprising a unit of information.
File	A named set of records stored or processed as a unit. **(ISO 2382-1)**
Format	A language construct that specifies the representation, in character form, of objects on a file.
Generalisation	Simplification of map information, so that information remains clear and uncluttered when map-scale is reduced. Usually involves a reduction in detail, a resampling to larger spacing, or a reduction in the number of points in a line. **(OS2)**
Geocode	A code which represents the spatial characteristics of an entity. For example, a coordinate point or a postcode. **(OS1)**
Geodesy	The science of measuring the Earth.

Geodetic datum	The definition of a particular spheroid and its position and orientation relative to the geoid.
Geographic information	Information about objects or phenomena that are associated with a location relative to the surface of the Earth. A special case of spatial information.
Geographic Information System (GIS)	A system for capturing, storing, checking, integrating, manipulating, analyzing and displaying data which are spatially referenced to the Earth. **(OS1)**
Geoid	An imaginary shape for the Earth defined by mean sea level and its imagined continuation under the continents at the same level of gravitational potential.
Geometry	The shape of the represented entity or entities, in terms of its stored coordinates. **(OS2)**
GKS (Graphics Kernal System)	A standard specification (ISO 7942) for a set of functions for computer graphics programming, and a functional interface between an applications program and the graphical input and output devices. **(ISO 2382-13)**
GPS (Global Positioning System)	A constellation of US satellites which enables users with appropriate receivers to fix their position on or above the surface of the Earth to varying degrees of accuracy depending on the receivers and techniques used.
Graphic primitive	A basic graphic element that can be used to construct a display image (e.g. point, line segment). **(ISO 2382-13)**
Graphical User Interface	A user interface which makes use of graphical objects such as icons, for selecting options, and usually has a windowing capability, enabling multiple window displays on the same screen.
Graphics tablet	A special flat surface with a mechanism for indicating positions thereon, normally used as a locator. **(ISO 2382-13)**
Graticule	The depiction of the lines of latitude and longitude on a map. The lines will not be orthogonal or even, in general, straight.
Grey scale	A range of intensities between black and white. **(ISO 2382-13)**
Grid	An orthogonal set of lines depicting a plane coordinate system.
Grid cell	A two dimensional object that represents an element of a regular or nearly regular tessellation of a surface. **(AC)**
Grid reference	The position of a point on a map expressed in terms of grid coordinates.
Grid squares	A regular array of square cells referenced to a grid, used as a basis for holding spatially referenced information.
Hard copy	A print or plot of output data on paper or some other tangible medium. **(OS1)**
Hardware	All or part of the physical components of an information processing system. **(ISO 2382-1)**
Hidden line	A line or segment of a line which can be masked in a view of a three-dimensional object. **(ISO 2382-13)**

Highlighting	Emphasizing a display element by modifying its visual attributes. **(ISO 2382-13)**
Hotspot	The x, y position that corresponds to the coordinates reported for a pointer. **(ISO 2382-13)**
Icon	A graphic symbol, displayed on a screen, that the user can point to with a device such as a mouse, in order to select a particular function or software application. **(ISO 2382-13)**
IEC (International Electrotechnical Committee)	This has the same status as ISO, but focuses on electrical and electrotechnical issues, especially electricity measurement, testing, use and safety.
IEEE (The Institute of Electrical and Electronics Engineering Inc.)	A major international professional body and an accredited standards setting organization.
IGES	International Graphics Exchange System. An ANSI standard for the exchange in digital form of CAD drawings.
Image	A raster representation of a graphic product (scanned map, photograph, drawing etc) or a remotely sensed surface that consists of one or more spectral bands.
Image processing	The use of a data processing system to create, scan, analyze, enhance, interpret or display images. **(ISO 2382-1)**
Information	Intelligence resulting from the assembly, analysis or summary of data into a meaningful form.
Integration	The combining of data of different types from different sources and systems to provide new information.
Interactive digitising	A method of digitising in which dialogue (interaction) takes place between the operator and the computer. **(OS1)**
Interface	The junction or linking together of two or more systems.
Interior area	An area not including its boundary. **(AC)**
Interoperability	The capability to communicate, execute programs, or transfer data among various functional units (items of hardware or software or both, capable of accomplishing a specified purpose) in a manner that requires the user to have little or no knowledge of the unique characteristics of those units. **(ISO 2382-1)**
Invisible line	A line which is sometimes used to make a logical connection between two parts of a feature or between two different features but which is not displayed. **(OS1)**
IRDS (Information Resource Dictionary System)	An international (IS) standard which includes data dictionaries and software development aspects of building database management systems.
Island	An area delimited by a contour and surrounded by a fill pattern. **(ISO 2382-13)**
ISO (International Organization for Standardization)	The main de jure international standards setting body.
Isoline	A line joining points of equal value of a parameter.

Land Information System (LIS)

A system for capturing, storing, checking, integrating, manipulating, analyzing and displaying data about land and its use, ownership, development, etc. **(OS1)**

Land parcel

An area of land, usually with some implication for land ownership or land use. **(OS1)**

Land terrier

A document system comprising a set of marked maps and ledgers containing textual information to record land and property.

Latitude

The angular distance between the normal at a point on an ellipsoid and the plane of the equator of the ellipsoid.

Layer

A usable subdivision of a dataset, generally containing objects of certain classes. See also, level.

Level

A term synonymous with layer.

Line

A series of connected coordinated points forming a simple feature with homogeneous attribution.

Line segment

A line described by two sets of coordinates and the shortest connection between them.

Lineage/Origin

The ancestry of a dataset describing its origin and the processes by which it was derived from that origin. Lineage is synonymous with provenance, but is more than just the original source or author.

Link

A line without logical intermediate intersections.

Link and node structure

A data structure in which links and nodes are stored with cross-referencing. **(OS1)**stored with cross-referencing. (OS1)

Locational reference

The means by which information can be related to a specific spatial position or location.

Locator

An input unit, such as a graphics tablet or any pointing device, that provides data to generate coordinates of a position. **(ISO 2382-13)**

Longitude

The angular distance between a meridian plane through a point on an ellipsoid, and an arbitrarily defined meridian (usually the Greenwich meridian).

Manual digitising

A method of digitising by an operator moving a cursor over a map on a digitising table. **(OS1)**

Map

A graphic representation of features of the Earth's surface or other geographically distributed phenomena. Examples are topographic maps, road maps, weather maps.

Map projection

A systematic portrayal of geographically distributed features from the (curved) surface of the Earth onto a plane.

MEGRIN

Multi-purpose European Ground-Related Information Network. A consortium of European countries developing joint topographic data such as national and administrative boundaries.

Menu

A list of options displayed by a data processing system, from which the user can select an action to be initiated. **(ISO 2382-1)**

Menu bar

An area along one edge of a window used to display names or icons for menus. **(ISO 2382-13)**

Mereing	The definition of a boundary in relation to topographic features on the ground at the time of survey (eg "one metre from the road edge"). If at a later date the ground detail changes or disappears, then that section is not re-mered, but the mereing changes to "defaced".
Meta data	Information about data. Examples are data quality information or feature classification information. **(OS2)**
Meta model	A model that defines the components (concepts) needed to define conceptual models (application models). The Meta Model also defines the relationships between the components.
Model	A formal description of the real world or part of it. **(EXPRESS)**
Mouse	A device used for pointing and selecting areas of a VDU or graphics screen, moved over another surface.
Multi-media	A combination of a variety of user interfaces and communication elements such as still and moving pictures, sound, graphics and text.
National Grid	The metric grid on a Transverse Mercator Projection used by the Ordnance Survey to provide an unambiguous spatial reference in Great Britain for any place or entity whatever the map scale. **(OS1)**
Network	An arrangement of nodes and interconnecting links.
Node	The start or end of a link or line. A node may be shared by several lines. **(OS1)**
NTF (National Transfer Format)	A British Standard (BS 7567) for the transfer of geographic data, administered by AGI.
Object	(i) A discrete element of the real world, instances of which occur in a dataset. (ii) A collection of entities which form a higher level entity. For example, an object could be built from a collection of links to form a polygon. **(OS2)**
Object Class	A logical classification of objects (e.g. buildings, roads). Classes may contain sub-classes, or may be sub-classes of other classes (e.g. motorways are a sub-class of roads, which are a sub-class of transport systems).
Object-Oriented Programming	The writing of computer programs using object-oriented techniques and languages. These employ a data-centred approach to programming, based on objects and object classes. Data is 'encapsulated' with operations, and commands are executed using message passing.
OEEPE	Organisation Europeene d'Etudes en Photogrammetrie Experimentale. A voluntary group of individuals and organisations that carry out research into photogrammetry and related subjects.
Open System	An information processing system that complies with the requirements of open systems interconnection standards in communication with other such systems.
Origin	The reference point (0,0) from which coordinates are measured. **(OS2)**

OSI	Open Systems Interconnection. This defines the accepted international standards (IS 7498-1984) by which open systems should communicate with each other. It takes the form of a seven-layer model of a network architecture, with each layer performing a different function.
OSTF	Ordnance Survey Transfer Format. A UK data transfer format previously used by Ordnance Survey for the supply of digital data to customers.
Overlay	A set of graphical data that can be superimposed on another set of graphical data through registration to a common coordinate system.
Overshoot	The projection of a line feature beyond the true point of intersection with another line feature. **(OS2)**
Panning	Progressively translating the display elements to give the visual impression of lateral movement of the image. **(ISO 2382-13)**
Parcel	See Land Parcel
Pecked Line	A line drawn as a series of dashes. **(OS1)**
PHIGS (Programmer Hierarchical Interactive Graphics System)	A standard (ISO 9592) set of graphics support functions to control the definition, modification, storage, and display of hierarchical graphics data. **(ISO 2382-13)**
Photogrammetry	The measurement of photographic images for the purpose of extracting useful information, particularly for the creation of accurate maps.
Pixel	The smallest element of a display surface that can be independently assigned attributes such as colour and intensity. **(ISO 2382-13)**
Point	A zero-dimensional abstraction of an object, with location specified by a set of coordinates.
Pointer	A symbol displayed on a screen, that a user can move with a pointing device such as a mouse, to select items. **(ISO 2382-13)**
Polygon	An area bounded by a closed line. **(OS2)**
Polyline	A line made up of a sequence of line segments.
Polymorphism	The ability to use the same consistent methods in different situations.
Pop-up Window	A window that appears rapidly on the display surface in response to some action. **(ISO 2382-13)**
Portability	The capability of a program to be executed on various types of data processing systems without converting the program to a different language and with little or no modification. **(ISO 2382-1)**.
Positional Accuracy	The degree to which a position is measured or depicted, relative to its correct value established by a more accurate process.
Postcode	A coding system for referencing all properties in the UK which have a postal address. The country is divided into 120 Areas, each Area is divided into Districts, each District into Sectors, each Sector into Units. A unit postcode applies to a group of adjacent addresses (approximately 15 in number) of neighbouring properties and does not define an area.
Precision	The exactness with which a value is expressed, whether the value be right or wrong. **(OS1)**

Pre-development Mapping	A concept in which one party prepares a definitive digital map of a proposed development which is then used by all interested parties for detailed planning and recording purposes until the as-built has been surveyed.
Premcode	A set of alphanumeric characters which, when added to the Unit Postcode, serves to identify a specific premise or address in the UK.
Program	A logically connected set of instructions which tell a computer to perform a sequence of tasks.
Projection	See Map Projection.
Property	Land, building or estate which is owned and to which legal title exists.
Protocol	The method by which components of a system communicate with each other.
Puck	A pointing device that must be positioned manually on the pad of a graphics tablet in order to register input points when tracing images. **(ISO 2382-13)**
Pull-down Menu	A menu that appears below the menu bar when the user selects a name or icon from the menu bar. **(ISO 2382-13)**
Quadtree	The expression of a two-dimensional object as a tree structure of quadrants which are formed by recursively subdividing each nonhomogeneous quadrant until all quadrants are homogeneous with respect to a selected property, or until a predetermined cutoff depth is reached. **(ISO 2382-13)**
Raster Data	Spatial Data expressed as a matrix of cells or 'pixels', with spatial position implicit in the ordering of the pixels. **(OS2)**
Raster Scan	A technique for generating or recording the elements of a display image by means of a line-by-line sweep across the entire display space; for example, the generation of a picture of a television screen. **(ISO 2382-13)**
Raster to Vector Conversion	The process of converting an image made up of cells into one described by lines and polygons.
Ray Tracing	A technique for determining, by tracing imaginary rays of light from the viewer's eye to the objects in a scene, the parts of the scene that should be displayed in the resulting image at any given point in time. **(ISO 2382-13)**
RDBMS (Relational Database Management System)	A database management system that supports the relational model.
Real-time System	A system that is able to receive continuously changing data from external sources and to process that data sufficiently rapidly to be capable of influencing the sources of data (e.g. monitoring reservoir levels).
Record	A set of related data fields grouped for processing.
Relative Accuracy	The measure of the internal consistency of the positional measurements in a dataset. For many local area purposes, for example records of utility plant, relative accuracy is more important than absolute accuracy. In this case, accurate measurement of offsets from fixed points is required rather than knowledge of the true position in space.

Relative Coordinate One of the coordinates identifying the position of an addressable point with respect to another addressable point. **(ISO 2382-13)**

Remote Sensing The technique of obtaining data about the environment and the surface of the Earth from a distance, for example, from aircraft or satellites. **(OS1)**

Rendering The conversion of the geometry, colouring, texturing, lighting and other characteristics of a scene into a display image. **(ISO 2382-13)**

Repeatability The ability of a device to perform the same action consistently or to provide the same data given identical conditions. Given identical inputs, the limits within which the output will fall with a given statistical confidence. **(OS1)**

Resolution A measure of the ability to detect quantities. High resolution implies a high degree of discrimination but has no implication as to accuracy. **(OS1)**

Ring A sequence of non-intersecting chains, strings, links, or arcs with closure. (It represents a closed boundary, but not the interior area inside the closed boundary). **(AC)**

Rubber Banding The result of moving a point or an object in a manner that preserves interconnectivity with other points or objects through stretching, shrinking or reorienting their interconnecting lines. **(ISO 2382-13)**

Run Length Encoding (RLE) The process of encoding a digital data stream which defines that stream in terms of the number of successive digital data elements that have the same value. **(ISO 2382-13)**

Scale The ratio or fraction between the distance on a map, chart or photograph and the corresponding distance in the real world.

Scanning A method of data capture whereby an image or map is converted into digital raster form by systematic line-by-line sampling.

Schema A collection of items forming part or all of a model. **(EXPRESS)**

SDTS Spatial Data Transfer Standard. The US Federal Information Processing Standard for the transfer of spatial data (FIPS Publication 173).

Seed A point within an area that can be used to carry the attributes of the whole area, e.g. ownership, address, land use type. **(OS1)**

Semi-Automatic Digitising A method of digitising from a map in which the majority of the line following is controlled by a machine, but which requires an operator to be on hand constantly to assist the machine to identify features and resolve anomalies. **(OS1)**

SIF Intergraph's Standard Interchange Format. Proprietary format primarily used for the transfer of CAD drawings.

Sliver Polygon A small area formed when two polygons which have been overlaid do not abut exactly, but overlap along one or more edges.

Snapping An operation or process whereby the computer will pick a nearby point or closest point on a nearby line.

Software	All or part of the programs, procedures, rules and their associated documentation of an information processing system. **(ISO 2382-1)**
Software Engineering	The systematic application of scientific and technological knowledge, methods and experience to the design, implementation, testing and documentation of software to optimize its production, support and quality. **(ISO 2382-1)**
Software Package	A fully documented program or set of programs, designed to perform a particular task (e.g. a word processor).
Spaghetti	Vector data composed of line segments which are not topologically structured or organised into objects and which may not even be geometrically tidy.
Spatial Analysis	Analytical techniques associated with the study of locations of geographic phenomena together with their spatial dimensions.
Spatial Information	Information which includes a reference to a two or three dimensional position in space as one of its attributes.
Spatial Reference	Co-ordinate, textual description or codified name by which information can be related to a specific position or location on the Earth's surface.
Spline	A smooth curve fitted mathematically to a sequence of points.
SQL (Structured Query Language)	An ISO standard (IS 9075) interface to relational database products. It is used to define and access databases and to manipulate the data stored in them.
STEP	A transfer format for graphics data being developed by ISO (TC184/ SC4) to replace IGES.
String	A sequence of line segments or text items. (A string does not have nodes, node identifiers, or left and right identifiers and may intersect itself or other strings) **(AC)**
Surveying	The measurement and recording of geographically distributed information. Particular types of surveying are topographic, cadastral and geological.
Symbol	A graphic representation of a concept that has meaning in a specific context. **(ISO 2383-1)**
Systems Development Methodology	An integrated set of techniques and methods for effective and efficient planning, analysis, design, construction, implementation and support of computer systems.
Tablet	See graphics tablet.
Tessellation	The subdivision of a 2-dimensional plane (or 3-dimensional volume) into disjoint congruent polygonal tiles (polyhedral blocks). **(OS1)**
Tesseral	A gridded representation of the plane surface into disjoint polygons. These polygons are normally either square (raster), triangular (TIN), or hexagonal. These models can be built into hierarchical structures, and have a range of algorithms available to navigate through them.
Thematic Map	A map depicting one or more specific themes. Examples are land classification, population density, rainfall etc.

Thiesen Polygon	A polygon bounding the region closer to a point than to any adjacent point.
TIGER	Topologically Integrated Geocoding and Referencing. A data format developed by the US Bureau of Census for the 1990 US census.
Tile	A logical rectangular set of data used to subdivide digital map data into manageable units. **(OS2)**
Tint	A stipple dot pattern used to create subdued colour infill within a defined area. **(OS2)**
Topographical Database	A database in which data relating to the physical features and boundaries on the Earth's surface is held. **(OS1)**
Topologically Structured Data	Data structured such that relations and characteristics referred to as topology can be expressed, including concepts such as connectivity, adjacency and containment.
Topology	The relative location of geographic phenomena independent of their exact position. In digital data, topological relationships such as connectivity and relative position are usually expressed as relationships between nodes, links and polygons. **(OS2)**
Transfer Format	The format used to transfer data between computer systems. In general usage this can refer not only to the organisation of data, but also to the associated information, such as attribute codes which are required in order to successfully complete the transfer. **(OS1)**
Transfer Medium	The physical medium on which digital data is transferred from one computer system to another. For example, magnetic tape. **(OS1)**
Transformation	A computational process of converting a position from one coordinate system to another. **(AC)**
Triangulated Irregular Network (TIN)	A form of the tesseral model based on triangles. The vertices of the triangles form irregularly spaced nodes. Unlike the grid, the TIN allows dense information in complex areas, and sparse information in simpler or more homogeneous areas.
Unique Property Reference Number (UPRN)	A short-coded address or number which uniquely identifies a parcel of land.
Undershoot	A line feature which is short of its true intersection with another line feature. **(OS2)**
Universe of Discourse	That aspect of the real world under consideration.
Update	The process of adding to and revising existing information to take account of change. **(OS1)**
Vector Data	Positional data in the form of coordinates of the ends of line segments, points, text position etc. **(OS1)**
Window	A part of a display image with defined boundaries in which information is displayed. **(ISO 2382-13)**
Wireframe Representation	A representation of an object, composed entirely of lines as though constructed of wire. (The lines may represent edges or surface contours in the display including those that may be hidden in the view of a real object) **(ISO 2382-13)**

Workstation	An individual interactive computer with at least a processor, screen and keyboard, that is more powerful than a PC and typically operates under a multitasking operating system.
WYSIWYG (What-you-see-is-what-you-get)	A capability to display information on a screen, exactly as it will be printed or plotted on an output device.
Zooming	Progressively scaling the entire display image to give the visual impression of movement of display elements toward or away from the observer. **(ISO 2382-13)**

GIS courses available

	BSc	Diploma	MA	MPhil	MSc	Phd	post graduate diploma	short course	other qualification
Birkbeck College - University of London	–	–	–	–	–	–	–	◆	–
Cambridge University	–	–	–	◆	–	–	–	◆	–
Chichester College of Arts, Science and Technology	–	◆	–	–	–	–	–	–	–
Coventry University	◆	–	–	–	–	–	–	–	–
Cranfield University	◆	–	–	–	◆	–	◆	–	–
East Anglia University	–	–	–	–	◆	–	–	–	–
Glasgow University	–	–	–	◆	◆	–	–	–	–
Keele University	–	◆	–	–	◆	–	–	–	–
Kingston College	◆	–	–	–	–	–	–	–	–
Kingston University	–	–	–	–	–	–	–	◆	◆
Leicester University	–	◆	–	–	◆	◆	–	–	–
Liverpool University	–	–	–	–	–	–	–	◆	–
London School of Economics	–	–	–	–	–	–	–	◆	–
Manchester Metropolitan University	–	–	–	–	◆	–	◆	–	–
Portsmouth University	–	–	–	–	◆	–	◆	–	–
Sheffield University	–	–	–	–	–	–	–	◆	–
Stirling University	–	◆	–	–	◆	–	–	–	–
Strathclyde University	–	◆	–	◆	–	–	–	–	–
University College London	–	–	–	–	◆	–	–	–	–
University College of Swansea	◆	–	–	◆	–	◆	–	–	–
University of Aberdeen	◆	–	–	–	◆	–	–	–	–
University of East London	◆	–	–	–	–	–	–	–	–
University of Edinburgh	–	◆	–	–	◆	–	–	–	–
University of Glamorgan	–	–	–	–	–	–	–	◆	–
University of Greenwich	◆	–	–	–	◆	–	–	–	–
University of Hertfordshire	–	–	–	–	–	–	–	◆	–
University of Huddersfield	–	–	–	–	◆	–	◆	–	–
University of Leeds	–	◆	◆	◆	◆	◆	–	–	–
University of Luton	◆	–	–	–	–	–	–	–	◆
University of Newcastle upon Tyne	◆	–	–	–	◆	–	–	◆	–
University of Nottingham	–	–	–	–	◆	–	–	–	–
University of Reading	◆	–	–	–	–	–	–	–	–
University of Salford	–	–	–	–	◆	–	◆	–	–
University of Sussex	◆	–	–	–	–	–	–	–	–
University of York	–	–	–	–	◆	–	–	–	–
Wales University	–	◆	–	–	◆	–	–	–	–

GIS course contacts

Birkbeck College - University of London
Dr Jonathan Raper *telephone:* 0171 631 6577
Dr John Shepherd *telephone:* 0171 631 6483

Cambridge University
A D Cliff *telephone:* 01223 334981
Rebecca Simons *telephone:* 01223-332722

Chichester College of Arts, Science and Technology
Peter Thaiwates *telephone:* 011243 786321 X2239

Coventry University
Dr Serwan Baban *telephone:* 01203 838973

Cranfield University
Student recruitment executive
telephone: 01525 860428

East Anglia University
Dr Andrew Lovett *telephone:* 01603 593126

Glasgow University
David Forrest *telephone:* 0141 339 8855
Dr Alastair Morrison *telephone:* 0141 339 8855 x 5291

Keele University
Dr Mike Worboys *telephone:* 01782 713082

Kingston College
Mr Alan Wood

Kingston University
Dr Alun Jones *telephone:* 0181 547 7510
GIS Short Course Administrator
telephone: 0181 547 2000 x 2508

Leicester University
Peter Fisher *telephone:* 01162523839

Liverpool University

London School of Economics
Dr Elsa Joao *telephone:* 0171 955 7585

Manchester Metropolitan University
Beverley Heyworth *telephone:* 0161 247 1581

Portsmouth University
Dr Peter Collier *telephone:* 01705 842473

Sheffield University
Mr S M Wise *telephone:* 01142824749

Stirling University
D A Davidson *telephone:* 01786 467840

Strathclyde University
Dr Gareth Jones *telephone:* 0141 552 4400 x 3794

University College London
Prof Ian Harley *telephone:* 0171 380 7225

University College of Swansea
Martin G Coulson *telephone:* 01792 295546

University of Aberdeen
Dr David R Green *telephone:* 01224 272324

University of East London
Dr Brian Whiting *telephone:* 0181 590 7722

University of Edinburgh
Bruce Gittings *telephone:* 0131 650 2543

University of Glamorgan
Professor Chris Jones *telephone:* 011443 480722

University of Greenwich
telephone: 0181 316 9800

University of Hertfordshire
Science Training Centre *telephone:* 01707 284590

University of Huddersfield

University of Leeds
Dr S Carver *telephone:* 01132333318
Val Marrison *telephone:* 01132333321

University of Luton
Ron Beard *telephone:* 01582 362028

University of Newcastle upon Tyne
Dr David Parker *telephone:* 0191 222 6443

University of Nottingham
Dr R H Haines-Young *telephone:* 01159484848 x3381

University of Reading
Mr R Parry *telephone:*

University of Salford
James Petch *telephone:* 0161 745 5261

University of Sussex
Dr Tom Browne *telephone:* 01273 606755

University of York
Dr Chris Elliot *telephone:* 01904 433806

Wales University
Dr Gary Higgs *telephone:* 01222 874000 x5272

calendar worldwide

Acknowledgement: this calendar has been reproduced from GIS Europe with kind permission of the publishers, GeoInformation International, 307 Cambridge Science Park, Cambridge CB4 4ZD, UK

The first part of the calendar comprises GIS-specific categories: cartography, remote sensing, user groups and so on. The second part is devoted to events that address market sectors, for instance the environment, transport, etc.

AM/FM

24–27 March 96. Seattle, Washington, USA. **AM/FM International Annual Conference XIX.** Contact: Elizabeth A. Clark, AM/FM International, 14456 East Evans Avenue, Aurora, CO 80014, USA. Tel: +1 303 337 0513; *Fax:* +1 303 337 1001

CARTOGRAPHY

9–10 February 96. Derby, UK. **International Map Trade Association Third Annual European Conference and Trade Show.** Contact: Mike Cranidge, Executive Director (Europe), 5 Spinacre, Becton Lane, Barton on Sea, Hants BH25 7DF. Tel/*Fax:* 01425 620532

12–15 September 96. Reading, UK. **Annual Technical Symposium of the British Cartography Society.** Contact: David Fairbairn, Department of Surveying, University of Newcastle, Newcastle-upon-Tyne NE1 7RU, UK. Tel: +44 191 222 6353

COMPUTING/COMPUTER GRAPHICS

28–30 November 95. London, UK. **1995 Computer Graphics Expo 95.** Contact: Debbie Brown, Digital Media International, 10 Barley Mow Passage, London W4 4PH, UK. Tel: +44 181 995 3632; *Fax:* +44 181 995 3633; *Email:* digmedia@atlas.co.uk

GENERAL GIS

21–23 November 95 Birmingham. **AGI 95.** For conference details contact: Westrade Fairs, 28 Church Street, Rickmansworth, Herts WD3 1DD. Tel: 01923 778311; *Fax:* 01923 776820. For exhibition details contact: Christine Prentice, Excell Exhibitions, 2 Little Grange, Portley Wood Road, Caterham, Surrey CR3 0BQ. Tel: 01883 343139; *Fax:* 01883 330900

13–15 February 96. Wiesbaden, Germany. **GIS 96.** Contact: Sabine Jossé, Institute for International Research, Lyoner Strasse 15, D–60528, Frankfurt am Main, Germany. Tel: +49 69 664 43490; *Fax:* +49 69 664 43240

19–21 February 96. Gauteng, South Africa. **Earth Data Information Systems Conference.** Contact: Rob Truter, EDIS 96, PO Box 69, Newlands 7725, South Africa. Tel: +27 21 685 4070

6–8 March 96. Milan, Italy. **New Developments in GIS.** Contact: James B. Johnston, National Biological Service, 700 Cajundome Boulevard, Lafayette, Louisiana 70506, USA. Tel: +1 318 266 8556; *Fax:* +1 318 266 8616; *Email:* johnstonj@nwrc.gov

27–29 March 96. Barcelona, Spain. **Joint European Conference and Exhibition on Geographic Information.** Contact: JEC Conference on Geographic Information, AKM Congress Service, Clarastrasse 57, PO Box, CH–4005 Basel, Switzerland. Tel: +41 61 691 51 11; *Fax:* +41 61 691 81 89

10–12 April 96. Canterbury, UK. **GISRUK 96.** Contact: Zarine Kemp, Computing Laboratory, University of Kent at Canterbury, Canterbury CT2 7NF. Tel: 01227 827698; *Fax:* 01227 762811; E-mail: Z.Kemp@ukc.ac.uk

9–13 April 96. Charlotte, North Carolina, USA. **92nd Annual AAG Meeting.** Contact: Association of American Geographers, 1710 Sixteenth Street NW, Washington DC 20009–3198, USA. Tel: +1 202 234 1450; *Fax:* +1 202 234 2744

15–17 April 96. Manama, Bahrain. **Geo 96: 2nd Middle East Geosciences Conference and Exhibition.** Contact: Will Martin, Overseas Exhibition Services Ltd, 11 Manchester Square, London W1M 5AB, UK. Tel: +44 171 486 1951; *Fax:* +44 171 486 8773

26–28 April 96. West Palm Beach, Florida, USA. **GeoInformatics 96: GIS/Remote Sensing Research, Development and Applications.** Contact: Bin Li, Tel: +1 305 284 4087; *Fax:* +1 305 284 5430; *Email:* cpgis96@umgis.merrick.miami.edu

21–23 May 96. Fort Collins, Colorado, USA. **2nd International Symposium on Spatial Accuracy Assessment in Natural Resources and Environmental Sciences.** Contact: H. Todd Mowrer, Chair, Spatial Accuracy Symposium, Rocky Mountain Forest and Range Experiment Station, 240 W. Prospect, Fort Collins, CO 80526–2098, USA.

23–25 May 96. Birmingham, UK **GIS 96.** Contact: Philippa Awford, Blenheim Group plc, 630 Chiswick High Road, London W4 5BG. Tel: 0181 742 2828; *Fax:* 0181 747 3856; E-mail: mhs:blenheim@blen-uk

4–7 June 1996. Baltimore, Maryland, USA. **9th Annual Towson State University GIS Conference (TSUGIS 96).** Contact: Jay Morgan, Department of Geography and Environmental Planning, Towson State University, Baltimore, Maryland 21204–7097, USA. Tel: +1 410 830 2964; *Fax:* +1 410 830 3888; *Email:* e7g4mor@toe.towson.edu

4–14 August 96. Beijing, China. **30th International Geological Congress.** Contact: 30th International Geological Congress, PO Box 823, Beijing 100037, China

16–22 November 96. Denver, Colorado, USA. **GIS/LIS 96.** Contact: GIS/LIS 96, 5410 Grosvenor Lane, Suite 100, Bethesda, MD 20814–2122, USA. Tel: +1 301 493 0200; *Fax:* +1 301 493 8245

PHOTOGRAMMETRY & REMOTE SENSING

19 December 95. Chatham Maritime, UK. **Remote Sensing and GIS for Natural Resources Management One-Day Technical Meeting and Workshop.** Contact: Clare Power, School of Earth Sciences, University of Greenwich, Medway Towns Campus, (Pembroke Area) Chatham Maritime, Kent ME4 4AW, UK. Tel: +44 181 331 9803; *Fax:* +44 181 331 9805; *Email:* C.H. Power@greenwich.ac.uk

27–29 February 96. Las Vegas, Nevada, USA. **11th Thematic Conference on Geologic Remote Sensing.** Contact: ERIM, PO Box 134001, Ann Arbor, MI 48113–4001, USA. Tel: +1 313 994 1200 ext. 3453; *Fax:* +1 313 994 5123

25–29 March 96. Canberra, Australia. **8th Australian Remote Sensing Conference.** Tel: +61 6 257 3299; *Fax:* +61 6 257 3256

20–26 April 96. Baltimore, Maryland, USA. **1996 ASPRS/ACSM Annual Convention.** Contact: Denise Cranwell, ASPRS/ACSM 96, 5410 Grosvenor Lane, Suite 100, Bethesda, MD 20814, USA. Tel: +1 301 493 0200; *Fax:* +1 301 493 8245

24–27 June 96. San Francisco, California, USA. **2nd International Airborne Remote Sensing Conference and Exhibition.** Contact: ERIM/Airborne Conference, PO Box 134001, Ann Arbor, MI 48113–4001, USA. Tel: +1 313 994 1200 ext. 3234; *Fax:* +1 313 994 5123; *Email:* wallman@erim.org

9–19 July 96. Vienna, Austria. **18th International Congress of Photogrammetry and Remote Sensing.** Contact: ISPRS Congress Director, Vienna University of Technology, Gusshausstrasse 27–29, A–1040 Vienna, Austria. *Fax:* +43 1 505 6268

SURVEYING

15–19 April 96. Buenos Aires, Argentina. **63rd FIG Permanent Committee Meeting and International Symposium.** *Fax:* +54 1 393 1750

20–26 April 96. Baltimore, Maryland, USA. **1996 ASPRS/ACSM Annual Convention.** Contact: Denise Cranwell, ASPRS/ACSM 96, 5410 Grosvenor Lane, Suite 100, Bethesda, MD 20814, USA. Tel: +1 301 493 0200; *Fax:* +1 301 493 8245

MARKET SECTORS

COMMERCIAL

1 December 96. Oxford, UK. **Introduction to Geodemographics.** Contact: Peter Sleight, Target Market Consultancy, 'Woodlands', Woodlands Close, Holmer Green, High Wycombe, Buckinghamshire HP15 6QG. Tel: +44 1494 712371; *Fax:* +44 1494 714203

15–18 January 96. Tampa Bay, Florida, USA. **The 1996 DA/DSM DistribuTECH Conference.** Contact: James D. Black. Tel: +1 303 604 2566; *Fax:* +1 303 666 9359; *Email:* jdblack@csn.net

24–27 September 96. Singapore. **Offshore South East Asia 96.** Contact: Bob Goh, Singapore Exhibition Services Pte Ltd, 2 Handy Road, 15-09 Cathay Building, Singapore 0922. Tel: +65 338 4747; *Fax:* +65 339 5651

ENVIRONMENT

19–20 December 95. Dundee. **Monitoring Coastal Processes: applications and research.** Contact: A. P. Cracknell, Dundee Centre for Coastal Zones Research, Department of Applied Physics & Electronic and Mechanical Engineering, University of Dundee, Dundee DD1 4HN. Tel: 01382 344549; *Fax:* 01382 202830; E-mail: apc@ewing.dundee.ac.uk

16–19 April 96. Vienna, Austria. **International Conference on GIS in Hydrology and Water Resource Management.** Contact: HydroGIS 96, c/o Austropa-Interconvention, PO Box 30, A–1043 Vienna, Austria. Tel: +43 1 588 00 110; *Fax:* +43 1 586 71 27

21–23 May 96. Fort Collins, Colorado, USA. **2nd International Symposium on Spatial Accuracy Assessment in Natural Resources and Environmental Sciences.** Contact: H. Todd Mowrer, Chair, Spatial Accuracy Symposium, Rocky Mountain Forest and Range Experiment Station, 240 W. Prospect, Fort Collins, CO 80526–2098, USA.

TRANSPORT

15–18 April 96. Houston, Texas, USA. **IVHS AMERICA 6th Annual Meeting.** Contact: Bonnie Jessup, IVHS AMERICA, 1776 Massachusetts Ave NW, Suite 510, Washington, DC 20036–1993, USA. Tel: +1 202 484 2896

AGI papers

Listed are all the papers presented at AGI conferences since 1989 and published in the conference proceedings. Copies of individual papers are available at £5.00 or £10 for five, from the AGI Secretariat, 12 Great George Street, London, SWA1P 3AD. Copies of the full AGI 1993 proceedings are available at £50 and the AGI 1994 proceedings at £65 (earlier years out of print).

Abbott.G., & Dickason.P.
Digital Chart of the World - A Global Perspective
AGI92 Conference Proceedings AGI89. Paper 2.28.
Birmingham. 1992

Abbott.T.
GIS and Interactive multi-media in retail
AGI91 Conference Proceedings. Paper 1.18.
Birmingham. 1991

Adnitt.N.
The Relational Database, Ideal for a Corporate GIS
AGI92 Conference Proceedings. Paper 2.3. Birmingham.
1992

Adnitt.N.
Data Capture - The Technology, The Issues
AGI91 Conference Proceedings. Paper 2.2. Birmingham.
1991

Adnitt.N.
Of Course we could always scan it
AGI93 Conference Proceedings. Paper 2.14.
Birmingham. 1993

Adnitt.N.C.
Number 1 in the Charts , Yorkshire Water GIS, currently available on CD
AGI94 Conference Proceedings. Paper 7.2. Birmingham.
1994

Adrian.G. & Hardy.K.
Facilities & Estates Management with GIS
AGI93 Conference Proceedings. Paper 2.21.
Birmingham. 1993

Afors.J., Chiu.M., & Martin.J.
Designing for corporate information
AGI93 Conference Proceedings. Paper 2.7. Birmingham.
1993

Albaredes.G.
GIS Usability: The Challenge of the 1990's
AGI92 Conference Proceedings. Paper 2.32.
Birmingham. 1992

Albaredes.G.
Data Sharing - a hurdle for GIS
AGI93 Conference Proceedings. Paper 2.12.
Birmingham. 1993

Albaredes.G.
Corporate GIS and access to information
AGI94 Conference Proceedings. Paper 9.2. Birmingham.
1994

Alla.P. & Trow.S.W.
Implementation of a GIS - A way to optimising utilities operations
AGI90 Conference Proceedings. Paper 3.2. Brighton.
1990

Allen.M., Schumacher.C., & Kutz.A.
The Use of GIS in the Environmental Assessment of the WW2 Mustard Gas Factory at Ergethan in the former East Germany
AGI94 Conference Proceedings. Paper 23.1.
Birmingham. 1994

Allen.P.M., & Jackson.I.,
Digital Geological Maps : The Extra Dimension
AGI94 Conference Proceedings. Paper 5.1. Birmingham.
1994

Allinson.J.
The Breaking of the third wave: The Demise of GIS
AGI94 Conference Proceedings. Paper 22.1.
Birmingham. 1994

Andrews.D.E., Hartley.W.S. & Walter.J.
GIS and ROMANSE - Integrating Traffic Systems
AGI92 Conference Proceedings. Paper 2.27.
Birmingham. 1992

Annand.K.
BPR: A Golden Opportunity for GIS
AGI94 Conference Proceedings. Paper 19.3.
Birmingham. 1994

Annand.K., & Hookham.C.
A GIS Strategy for guarding the water environment
AGI93 Conference Proceedings. Paper 2.2. Birmingham.
1993

Antenucci.J.C.
20 - 20 Hindsight: Or if I had to do it again
AGI90 Conference Proceedings. Paper 2.1. Brighton.
1990

Antenucci.J.C.
Redefinition of GIS for the 90's: Strategic Trends
AGI93 Conference Proceedings. Paper 2.10.
Birmingham. 1993

Archibald.I. & Jeffries-Harris.T.
A Vectorised Spaghetti-Hooped Boundary 10000 kms long
AGI91 Conference Proceedings. Paper 2.18.
Birmingham. 1991

Arnott.D. & Keddie.A.R.
Data Capture - The Standards and Procedures Utilised within Northumbrian Water Group
AGI92 Conference Proceedings. Paper 2.12.
Birmingham. 1992

Asabere.R.K. Durucan.S. & Owen.D.B.
An Expert System for Application of Geographical Information Systems in Mineral Resources Management
AGI92 Conference Proceedings. Paper 1.14.
Birmingham. 1992

Ashton.C.H.
Desktop Mapping the Future view of the Enterprise GIS?
AGI92 Conference Proceedings. Paper 2.23.
Birmingham. 1992

Ashton.G.
Partners in Progress - The Open cast Experience
AGI91 Conference Proceedings. Paper 1.9. Birmingham.
1991

Aspinall.R.J.
A Landscape Ecological Approach to Mapping Distribution and Abundance of Birds with GIS
AGI91 Conference Proceedings. Paper 2.19.
Birmingham. 1991

Atkinson.R.
GIS: A catalyst for positive change
AGI93 Conference Proceedings. Paper 1.25.
Birmingham. 1993

Aybet.J.
Spatial data modelling for Geographic Information System design
AGI93 Conference Proceedings. Paper 3.1. Birmingham.
1993

Aybet.J., Martin.J., & Taskis.D.
Protecting investment by capturing asset data from existing systems
AGI91 Conference Proceedings. Paper 3.2. Birmingham.
1991

Aybet.J., & Walpole.R.
An Integrated GIS Approach to the Use of Earth Observation Data - Sugar Beet prediction and management system
AGI94 Conference Proceedings. Paper 15.1.
Birmingham. 1994

Azizi.A. Clark.M.J. & Davenport.J.
Air Photo or Video Inputs to Vector or Raster GIS
AGI91 Conference Proceedings. Paper 1.20.
Birmingham. 1991

Bainbridge.P., Hinton.P. & Curry.D.
GIS: Unlocking the Benefits by Implementing Applications
AGI93 Conference Proceedings. Paper 1.6. Birmingham.
1993

Barker.P.
Geographic Information Systems in BT : BT Uses GIS 40 different ways to manage its business
AGI94 Conference Proceedings. Paper 10.3.
Birmingham. 1994

Barr.R.
Micro-analysis in GIS
AGI91 Conference Proceedings. Paper 3.27.
Birmingham. 1991

Barr.R. & Campbell.K.
The Postal Address as a fundamental Geo-Reference
AGI89 Conference Proceedings. Paper D.2.
Birmingham. 1989

Batty.P.
An Introduction to GIS Database Issues
AGI92 Conference Proceedings. Paper 2.1. Birmingham.
1992

Bibby.P. & Shepherd.J.
GIS: Strategic Choice or Laundry Lists?
AGI92 Conference Proceedings. Paper 2.31.
Birmingham. 1992

Bingham.P. & Thorneycroft.R.
GIS in a Small Shire County
AGI90 Conference Proceedings. Paper 5.2. Brighton.
1990

Bird.A.C.
GIS Based Data on Land cover Change in the National Parks of England and Wales
AGI91 Conference Proceedings. Paper 2.12.
Birmingham. 1991

Blakemore.M.
UK Education and Research in GIS. A Selective Review
AGI91 Conference Proceedings. Paper 1.28.
Birmingham. 1991

Blakemore.M.
Sharing Data - Whose why how?
AGI90 Conference Proceedings. Paper 4.4. Brighton.
1990

Blakemore.M.
Management issues in a Nationally Networked Geographic Information System
AGI89 Conference Proceedings. Paper 6.2. Birmingham.
1989

Booth.S.
GIS - are we missing the point?
AGI90 Conference Proceedings. Paper 1.4. Brighton.
1990

Bosworth.M.
The Proposed National Street Gazetteer to a Standard Specification
AGI91 Conference Proceedings. Paper 3.15.
Birmingham. 1991

Bourke.A.
How we Integrated GIS with Existing Central RDBMS over a wide area network
AGI92 Conference Proceedings. Paper 2.16.
Birmingham. 1992

Boxer.A.
IS Government different?
AGI93 Conference Proceedings. Paper 1.27.
Birmingham. 1993

Boyd.G., & Schaap.D.
Integrating Marine Geographical Information
AGI94 Conference Proceedings. Paper 13.1.
Birmingham. 1994

Brand.M.J.D. & Gray.S.
Experience with Structured Data -Millstone or Milestone?
AGI92 Conference Proceedings. Paper 1.24.
Birmingham. 1992

Brand.M.J.D. & Gray.S.
From Concept to reality
AGI89 Conference Proceedings. Paper 5.2. Birmingham.
1989

Brand.M.J.D., & Mitchell.G.
CORINE Land Cover (Ireland) - The Northern Perspective
AGI93 Conference Proceedings. Paper 3.13.
Birmingham. 1993

Brewer.A. & Shepherd.J.
Post codes as a Geographic Database: Added Value and Added Analysis
AGI92 Conference Proceedings. Paper 1.2. Birmingham.
1992

Bridger.I. & Fukushima.Y.
Towards Intelligent Data
AGI89 Conference Proceedings. Paper B.2. Birmingham.
1989

Bridger.I. & Land.N.
Structured Data - An Ordnance Survey Perspective
AGI92 Conference Proceedings. Paper 1.23.
Birmingham. 1992

Brown.N.
Mapping of Land Cover from space and its use in a Geographical Information System
AGI92 Conference Proceedings. Paper 2.26.
Birmingham. 1992

Brown.N., Gerrard.F., & Parr.T.
Defining English Chalk and Limestone Grassland areas using GIS
AGI93 Conference Proceedings. Paper 2.26.
Birmingham. 1993

Brown.P.J.B., Batey.P.W.J., Hirschfield.A., & Marsden.J.
Poisson Chi Square Mapping, GIS & Geodemographic Analysis: The Spatial & Aspatial Analysis of Relatively Rare Conditions
AGI90 Conference Proceeding. Paper A.3. Brighton.
1990

Brown.R., Slater.J., & Askew.D.
Environmental Monitoring of Protected Landscapes - monitoring environmentally sensitive areas using satellite imagery, within GIS
AGI94 Conference Proceedings. Paper 15.2.
Birmingham. 1994

Browne.T.
Teaching through Training?
AGI93 Conference Proceedings. Paper 2.17.
Birmingham. 1993

Buchanan.H.
A Young Professional's View of the GIS Industry
AGI91 Conference Proceedings. Paper 3.19.
Birmingham. 1991

Bundock.M.S. & Theriault.D.
Integration of Case Technology into the GIS Environment
AGI92 Conference Proceedings. Paper 2.25.
Birmingham. 1992

Bundred.P., Hirschfield.A., & Marsden.J.
GIS in the Planning of Health Services in a District Health Authority
AGI93 Conference ProceedingsAGI89. Paper 1.1.
Birmingham. 1993

Burkmar.R.J. Petch.J.R. Basden.A. & Yip.J.
Integrating GIS and Knowledge-based systems for Landscape Management
AGI92 Conference Proceedings. Paper 1.28.
Birmingham. 1992

Burrough.P.A.
Accuracy issues for future GIS
AGI94 Conference Proceedings. Paper 25.1.
Birmingham. 1994

Buxton.R. & Ball.C.E.
Geographic Information and the Expert System in Local Authority Development Control
AGI89 Conference Proceedings. Paper A.1. Birmingham.
1989

Callaghan.J.G.
GIS - Increasing business efficiency and profit
AGI89 Conference Proceedings. Paper 1.4. Birmingham.
1989

Calvert.C., & Holland.D.
GPS and Ordnance Survey map data: Issues and solutions
AGI94 Conference Proceedings. Paper 26.1.
Birmingham. 1994

Campbell.H. & Masser.I.
The Impact of GIS on Local Government in Great Britain
AGI91 Conference Proceedings. Paper 2.5. Birmingham.
1991

Campbell.H. & Masser.I.
Implementing GIS : The Organisational dimension
AGI93 Conference Proceedings. Paper 1.28.
Birmingham. 1993

Cane.S.
Adding Corporate Value to Data - a sugGIStion
AGI89 Conference Proceedings. Paper 2.1. Birmingham.
1989

Capell.P.J. & Singh.G.
London Area Transport Survey; Automated Address Coding
AGI93 Conference Proceedings. Paper 1.7. Birmingham.
1993

Carig.W.J.
Data to the People : North American efforts to empower communities with data and information
AGI94 Conference Proceedings. Paper 1.1. Birmingham. 1994

Cassettari.S.
Introducing GIS into the National Curriculum for Geography
AGI91 Conference Proceedings. Paper 3.4. Birmingham. 1991

Cassettari.S.
Using Aerial Photography in your GIS
AGI90 Conference Proceedings. Paper C.3. Brighton. 1990

Cassettari.S.
Using Animation to improve visualisation in GIS
AGI93 Conference Proceedings. Paper 3.5. Birmingham. 1993

Chapman.D. Dowman.I. & Muller.J.P.
Digital photogrammetry - interfaces with GIS
AGI91 Conference Proceedings. Paper 2.24. Birmingham. 1991

Chell.M.
Patterns of energy use and air pollution
AGI93 Conference Proceedings. Paper 3.18. Birmingham. 1993

Clark.M.J.
GIS Awareness: The Technical and Educational Challenge
AGI91 Conference Proceedings. Paper 3.9. Birmingham. 1991

Clark.M.J. Ball.J.H. & Gatward.J.
Costing the Impact of Sea-level rise: A GIS solution for the South Coast of England
AGI92 Conference Proceedings. Paper 3.12. Birmingham. 1992

Clark.M.J., Gurnell.A.M., Candish.C., & Mills.D.,
Flood Defence Assessment through GIS
AGI90 Conference Proceedings. Paper 5.1. Brighton. 1990

Clarke.K.E., & Norris.M.T.
Data and information highways
AGI94 Conference Proceedings. Paper 1.2. Birmingham. 1994

Clegg.P.
GIS Implementation - The Sheffield Experience
AGI92 Conference Proceedings. Paper 3.9. Birmingham. 1992

Clegg.P.
Basic Issues in Data Capture
AGI93 Conference Proceedings. Paper 1.18. Birmingham. 1993

Clegg.P. & Keddie.A.R.
The Chorley Report 4 Years on - A User Perception
AGI91 Conference Proceedings. Paper 1.1. Birmingham. 1991

Cooke.I.C.
Field Data Capture with Pen Computers
AGI93 Conference Proceedings. Paper 2.13. Birmingham. 1993

Coote.A.M.
Managing a large Spatial Archive
AGI89 Conference Proceedings. Paper B.3. Birmingham. 1989

Coote.A.M. & Rackham.L.J.
Handling Update: Issues in Spatial Data Maintenance
AGI92 Conference Proceedings. Paper 2.30. Birmingham. 1992

Corbin.C.E.H.
GIS: A Catalyst for Data Integration
AGI92 Conference Proceedings. Paper 1.9. Birmingham. 1992

Cory.M.J.
Portable Data Capture
AGI91 Conference Proceedings. Paper 3.23. Birmingham. 1991

Coulson.M.G.
Methodologies for Assessing cost and value of GIS
AGI90 Conference Proceedings. Paper 6.1. Brighton. 1990

Cross.A. & Openshaw.S
Crime Pattern Analysis: The development of ARC/ CRIME
AGI91 Conference Proceedings. Paper 3.28. Birmingham. 1991

Cross.D.
How to populate your GIS Database, on budget and on time
AGI93 Conference Proceedings. Paper 1.16. Birmingham. 1993

Cross.D.
Has the GIS Workstation been consigned to Jurassic Park? A review of "Desktop GIS" - What it is possible to achieve with an unmodified PC
AGI94 Conference Proceedings. Paper 22.3. Birmingham. 1994

Crowder.J.
GIS - Engineering or Science?
AGI94 Conference Proceedings. Paper 6.1. Birmingham. 1994

Dale.P.
GIS in property management - strengths, weaknesses, opportunities and threats
AGI94 Conference Proceedings. Paper 11.4. Birmingham. 1994

Dale.P.F.
Domesday 2000: The Professional as a Politician
AGI91 Conference Proceedings. paper 3.21. Birmingham. 1991

Daniels.D.J.
High Performance Ground Penetrating Radar System
AGI91 Conference Proceedings. Paper 1.22. Birmingham. 1991

Davies.A., & Pittard.S.
Photogrammetry : The first step to better process industry management
AGI94 Conference Proceedings. Paper 21.2. Birmingham. 1994

Davies.J.
Raster - The Forgotten Frontier
AGI91 Conference Proceedings. Paper 2.3. Birmingham. 1991

Deakin.R. & Diment.R.P.
Strategic Planning and Decision Support: The Use of GIS in the Coastal Environment
AGI94 Conference Proceedings. Paper 13.3. Birmingham. 1994

Devereux.B. & Mayo.T.
Task Oriented Tools for Cartographic Data Capture
AGI92 Conference Proceedings. Paper 2.14. Birmingham. 1992

Devereux.B. & Mayo.T.
In the Foreground: Intelligent Techniques for Cartographic Data Capture
AGI90 Conference Proceedings. Paper 6.2. Brighton. 1990

Dhillon.P.
Towards an Integrated Future
AGI92 Conference Proceedings. Paper 2.17. Birmingham. 1992

Dirdal.P.
A Vendor's Perspective of the Chorley report four Years On
AGI91 Conference Proceedings. Paper 1.2. Birmingham. 1991

Dixon.B.
How the vendor can help
AGI91 Conference Proceedings. Paper 1.13. Birmingham. 1991

Dixon.B.
Transferring Data into Information - Tools and Technologies
AGI92 Conference Proceedings. Paper 2.22. Birmingham. 1992

Dixon.P., Smallwood.J., & Dixon.M.
Development of a mobile GIS: Field data capture using a pen based note pad computer system
AGI93 Conference Proceedings. Paper 2.16. Birmingham. 1993

Dodson.A.H. & Basker.G.A.
GPS: GIS Problems Solved?
AGI92 Conference Proceedings. Paper 1.12. Birmingham. 1992

Doig.J., Atkinson.R. & Taylor.M.
Turning Public Registers into Public Information
AGI92 Conference Proceedings. Paper 1.20. Birmingham. 1992

Dowers.S.
SQL - The Way Forward
AGI91 Conference Proceedings. Paper 3.13. Birmingham. 1991

Drake.P.
The use of different Data Conversion options for a GIS application in the Water Industry
AGI89 Conference Proceedings. Paper 4.2. Birmingham. 1989

Dugmore.K.
1991 Census: Outputs and Opportunities
AGI91 Conference Proceedings. Paper 1.24. Birmingham. 1991

Dunlea.M.
The use of G.I.S. in Mobilising Irish Emergency Services - the C.A.M.P. Experience
AGI91 Conference Proceedings. Paper 2.10. Birmingham. 1991

Dunn.R.
Working towards a National Land Use Stock System
AGI94 Conference Proceedings. Paper 8.1. Birmingham. 1994

Dunn.R. & Harrison.A.
A Feasibility Study for a National Land Use Stock Survey
AGI92 Conference Proceedings. Paper 1.13. Birmingham. 1992

Dunn.R. & Harrison.A.
Assessing User Requirements for GIS - the critical stage in implementation: how prototyping can help you get it right
AGI91 Conference Proceedings. Paper 1.12. Birmingham. 1991

Dunn.R., Harrison.A.R., Brown.L., & Turton.P.J.
The Use of Geographic Information Systems in the Analysis of Countryside Data
AGI90 Conference Proceedings
Paper 4.3. Brighton. 1990.

Dunn.R., Whitelegge.M., Barr.M., Swanwick.C., & Warnock.S.
The use of GIS in landscape classification : A case study of Southwest England
AGI93 Conference Proceedings. Paper 2.28. Birmingham. 1993

Dyson.J.
Data Integration via the Gazetteer
AGI92 Conference Proceedings. Paper 1.1. Birmingham. 1992

Easterfield.M.E., Newell.R.G., & Theriault.D.G.
Management of Time and Space in GIS
AGI90 Conference Proceedings. Paper A.2. Brighton. 1990

Elliott.L., McCallum.D., & Pretty.S.
Address-Point : Fusion of OS Mapping Precision and Royal Mail Postal Addresses
AGI93 Conference Proceedings. Paper 1.9. Birmingham. 1993

Ellis.S.
How not to use the 1991 Census
AGI93 Conference Proceedings. Paper 3.20. Birmingham. 1993

Ellis.S., & Gower.C.
A Generic data model for GIS Local Government
AGI93 Conference Proceedings. Paper 2.5. Birmingham.
1993

Ellis.S.P.
GIS 2000BC - AD2000
AGI94 Conference Proceedings. Paper 12.2.
Birmingham. 1994

Evans.C.M., & Carty.A.
A Survey of Geographic Information Systems in Government
AGI94 Conference Proceedings. Paper 19.4.
Birmingham. 1994

Farrell.B.
Corporate GIS in a Scottish Regional Council
AGI92 Conference Proceedings. Paper 3.11.
Birmingham. 1992

Flowerdew.R. Green.M. & Lucas.S.
Analysing Local House Price Variations with GIS
AGI91 Conference Proceedings. Paper 3.25.
Birmingham. 1991

Forrest.D., & Pearson.W.A.,
Somewhere Over the Rainbow
AGI94 Conference Proceedings. Paper 5.3. Birmingham.
1994

Freeman.D.
Development of an Educational GIS for the National Curriculum
AGI91 Conference Proceedings. Paper 3.6. Birmingham.
1991

Gadd.S.
GIS - Forging a Key Role in Data Integration
AGI92 Conference Proceedings. Paper 1.15.
Birmingham. 1992

Gadd.S.
Training for Effective Implementation. Are we getting it right?
AGI93 Conference Proceedings. Paper 2.20.
Birmingham. 1993

Gaffney.V. & Stancic.Z.
Illyrians, Greeks, Ancient Battles and GIS on the Island of Hvar, Daimatia
AGI91 Conference Proceedings. Paper 2.17.
Birmingham. 1991

Gallagher.S.
The Personal GIS Revolution
AGI90 Conference Proceedings. Paper C.4. Brighton.
1990

Gannon.P.J.
Managing Infrastructure Impacts
AGI94 Conference Proceedings. Paper 23.3.
Birmingham. 1994

Gartzen.P.
GIS in the eye of a storm
AGI93 Conference Proceedings. Paper 2.9. Birmingham.
1993

Giddings.T.
Ingredients for Successful GIS Implementation : A Hong Kong Case Study
AGI91 Conference Proceedings. Paper 1.10.
Birmingham. 1991

Gittings.B.M. Dowers.S. Sloan.T.M. Healey.R.G. & Waugh.T.C.
Turbo-Charging your GIS: Dealing with Performance Issues
AGI92 Conference Proceedings. Paper 2.6. Birmingham.
1992

Gittings.B. & Mounsey.H.
GIS and LIS Training in Britain: The present situation
AGI89 Conference Proceedings. Paper 4.4. Birmingham.
1989

Glickman.V.
Common Responses to Common Challenges Towards a Charter for the Marketing of Geographic Information
AGI92 Conference Proceedings. Plenary. Birmingham.
1992

Godlement.H.J.R.
Managing the Geographic Information
AGI94 Conference Proceedings. Paper 19.2.
Birmingham. 1994

Gordon.M., & Gennery.D.
Data as a Corporate Resource
AGI93 Conference Proceedings. Paper 2.8. Birmingham.
1993

Gower.C.J.
GIS Implementation Standards
AGI91 Conference Proceedings. Paper 3.12.
Birmingham. 1991

Gradwell.D.J.L.
Can SQL Handle Geographic Data?
AGI90 Conference Proceedings. Paper D.3. Brighton.
1990

Gray.T.
Advanced Methods of Capturing and Using GIS Data in the Field
AGI91 Conference Proceedings. Paper 2.25.
Birmingham. 1991

Green.D.R. & Kemp.A.
A Geographical Information System (GIS) may be the answer to every pipeline manager's dream
AGI93 Conference Proceedings. Paper 1.26.
Birmingham. 1993

Green.D.R. & McEwen.L.J.
Turning Data into Information: Assessing GIS User Interfaces
AGI92 Conference Proceedings. Paper 1.27.
Birmingham. 1992

Green.D.R. & McEwen.L.J.
GIS as a component of Information Technology courses in Higher Education: Meeting the requirements of employers
AGI89 Conference Proceedings. Paper C.1. Birmingham.
1989

Green.D.R., & Morton.D.
Acquiring Environmental Remotely Sensed Data from Model Aircraft for Input to Geographic Information Systems
AGI94 Conference Proceedings. Paper 15.3. Birmingham. 1994

Green.P.
GIS for Market Analysis
AGI91 Conference Proceedings. Paper 1.17. Birmingham. 1991

Grimshaw.D.J.
The Transformation of Customer Databases
AGI92 Conference Proceedings. Paper 1.7. Birmingham. 1992

Grimshaw.D.J.
Towards an appropriate GIS Strategy
AGI93 Conference Proceedings. Paper 2.3. Birmingham. 1993

Grimshaw.D.J.
Broadening Your horizons: Shifting your thinking
AGI94 Conference Proceedings. Paper 9.1. Birmingham. 1994

Gross.H.
Very Clever....but does it really mean anything?
AGI90 Conference Proceedings. Paper 5.3. Brighton. 1990

Gugan.D.J.
Photogrammetry and GIS
AGI91 Conference Proceedings. Paper 2.22. Birmingham. 1991

Gugan.D.J. & Gliddon.D.J.
User Requirements for an Integrated GIS
AGI91 Conference Proceedings. Paper 1.15. Birmingham. 1991

Haines-Young.R.
GIS for Environmental Planning: Problems and Prospects
AGI89 Conference Proceedings. Paper 6.1. Birmingham. 1989

Haines-Young.R.,
The Tradable Information Initiative: A Review
AGI91 Conference Proceedings. Paper 1.26. Birmingham. 1991

Haines-Young.R.H., Bunce.R.G.H., & Parr.T.W.
Countryside Information System
AGI93 Conference Proceedings. Paper 1.23. Birmingham. 1993

Hall.J., Ullyett.J., Hornung.M., & Hancock.S.
Using GIS for mapping the sensitivity of fresh waters to acidification
AGI93 Conference Proceedings. Paper 3.16. Birmingham. 1993

Hallett.S.H., Keay.C.A., Jarvis.M.G., & Jones.R.J.A.
INSURE: Subsidence Risk Assessment from Soil and Climate Data
AGI94 Conference Proceedings. Paper 16.2. Birmingham. 1994

Halls.P.J.
Video for Capture of Paper Material - Worthless for Some but Useful for Others?
AGI94 Conference Proceedings. Paper 7.3. Birmingham. 1994

Halls.P.J. & Tealby.J.M.
GIS - The Web if coordination enabling both research and management
AGI93 Conference Proceedings. Paper 3.11. Birmingham. 1993

Hardy.G.A.K.
From GIS to Graphical MIS
AGI94 Conference Proceedings. Paper 19.1. Birmingham. 1994

Harrison.A. Dunn.R. & Turton.P.
Environmental GIS: Technology, Data and Policy
AGI91 Conference Proceedings. Paper 2.14. Birmingham. 1991

Hartley.J.C. Homer.I.R. Trow.S. Hinton.P.H. & Everett.C.
Inter Utility Exchange of Electronic Map Based Records in the North East of England
AGI92 Conference Proceedings. Paper 1.19. Birmingham. 1992

Havercroft.M., & Fox.D.
The creation of processed satellite imagery products compatible with Ordnance Survey digital mapping
AGI93 Conference Proceedings. Paper 3.9. Birmingham. 1993

Hawker.L. & Goodwin.R.
The Computerised Street Works Register - Streets ahead?
AGI91 Conference Proceedings. Paper 1.23. Birmingham. 1991

Haywood.P.
AGI and NTF (National Transfer Format): The Way Forward
AGI90 Conference Proceedings. Paper D.1. Brighton. 1990

Haywood.P.
Structured Topographic Data - The Key to GIS
AGI89 Conference Proceedings. Paper B.1. Birmingham. 1989

Heard.M., Higgins.C., & Mather.P.,
Prototype Expert Systems for the Geometric Correction of Remotely-sensed Images
AGI89 Conference Proceedings. Paper A.2. Birmingham. 1989

Hearnshaw.H.M. & Medyckyj-Scott.D.
How Usable is your GIS?
AGI91 Conference Proceedings. Paper 2.26. Birmingham. 1991

Hendley.D.J.
Parallel Computing in GIS
AGI90 Conference Proceedings. Paper B.1. Brighton. 1990

Heywood.D.I. & Petch.J.R.
GIS: A Toybox Approach
AGI91 Conference Proceedings. Paper 3.5. Birmingham.
1991

Hinton.M.A., & Wheeler.K.
Data Management for the Single Financial Community
AGI92 Conference Proceedings. Paper 2.8. Birmingham.
1992

**Hirschfield.A. Brown.P.J.B. Marsden.J. &
Bundred.P.**
*A GIS-linked Database for Analysing the Deployment of
Community-based Health and Social Services on the
Wirral*
AGI91 Conference Proceedings. Paper 2.11.
Birmingham. 1991

Hobbs.K.F.
The Visualisation of Contour Data Via Profiles
AGI94 Conference Proceedings. Paper 5.2. Birmingham.
1994

Hobson.S.A.
Bridging Islands to Increase Return on Investment
AGI92 Conference Proceedings. Paper 1.4. Birmingham.
1992

Homer.A.R. & Watson.M.
Integrated Data Saves Money!!
AGI92 Conference Proceedings. Paper 3.10.
Birmingham. 1992

Hookham.C.
*The Need for Public-Sector Policies for Information
Availability and Pricing*
AGI94 Conference Proceedings. Paper 8.2. Birmingham.
1994

Hooper.B.D.
*Operational Management of Geographical Information
in South West Water*
AGI89 Conference Proceedings. Paper 5.1. Birmingham.
1989

Hooper.B.D.
Five Years on and Meeting new challenges
AGI94 Conference Proceedings. Paper 10.1.
Birmingham. 1994

Hornby.R.
Successful Application of GIS at Nationwide Anglia
AGI90 Conference Proceedings. Paper 1.2. Brighton.
1990

Hosken.E. & Parker.D.
Civil Engineers need Geographic Information
AGI90 Conference Proceedings. Paper 4.1. Brighton.
1990

Howes.D.A., Green.M., & Kurtz.T.
Issues in the assessment of rural deprivation in England
AGI93 Conference Proceedings. Paper 3.21.
Birmingham. 1993

Ives.M.
Reaping the Benefits
AGI92 Conference Proceedings. Paper 1.6. Birmingham.
1992

Ives.M.J., & Hartley.J.C.
*Management of the Implementation of a Digital Records
Systems*
AGI90 Conference Proceedings. Paper 3.3. Brighton.
1990

Jackson.G.H., & Platt.P.
*Machine transfer of Circuit Information from film to
computer*
AGI90 Conference Proceedings. Paper 6.3. Brighton.
1990

Jay.T.
*The Integration of Network and Polygon Topologies
within a Single Geographic Information System*
AGI92 Conference Proceedings. Paper 3.1. Birmingham.
1992

Jeanes.P.J., & Brand.M.J.D.
Wide Area Distribution of Geographic Information
AGI90 Conference Proceedings. Paper 4.2. Brighton.
1990

Jenkins.J.C. & Yardy.A.S.
GIS : A strategic review for Northamptonshire
AGI93 Conference Proceedings. Paper 2.24.
Birmingham. 1993

Jones.A.C.
GIS for Risk Assessment?
AGI94 Conference Proceedings. Paper 4.4. Birmingham.
1994

Jones.R.J.A., Bradley.R.I., & Siddons.P.A.
*A Land Information System for environmental risk
assessment*
AGI93 Conference Proceedings. Paper 3.15.
Birmingham. 1993

Kawalek.J.P.
*Information Systems and the use of multi-media in the
water industries*
AGI93 Conference Committee. Paper 3.7. Birmingham.
1993

Keddie.J.G., & Docherty.E.
*GIS and the 1991 Census as a cohesive force in Local
Government*
AGI93 Conference Proceedings. Paper 3.23.
Birmingham. 1993

Kemp.K.K.
*GIS Education Around the World: Year Three of the
NVCGIA Core Curriculum Project*
AGI91 Conference Proceedings. Paper 3.8. Birmingham.
1991

Kennie.T. & Mather.P.
Do GIS Specialists need the Professional Institutions?
AGI91 Conference Proceedings. Paper 3.20.
Birmingham. 1991

Kjenstad.K., & Hvashovd.J.
*Which concepts should form the basis for a geographic
information management system?*
AGI93 Conference Proceedings. Paper 3.3. Birmingham.
1993

Knott.J. & Shiers.D.
Mobile Maps - Sending vector map based maps across radio networks for use in unplanned locations by GIS Users
AGI93 Conference Proceedings. Paper 1.24. Birmingham. 1993

Knox.B.
Maps for the Past - Plans for the future
AGI91 Conference Proceedings. Paper 2.16. Birmingham. 1991

Kojakovic.M, Kerekovic.D., Barakat.S., & Halls.P.
GIS in Planning Post-War reconstruction
AGI94 Conference Proceedings. Paper 23.2. Birmingham. 1994

Land.N.
The Classification of Spatial Data - A proposal for a National Standard
AGI91 Conference Proceedings. Paper 3.16. Birmingham. 1991

Lang.N.
Why reinvent the Wheel? Solutions using off-the-shelf GIS
AGI94 Conference Proceedings. Paper 2.2. Birmingham. 1994

Langford.M., & Fisher.P.
Measuring the accuracy of cross areal population estimates
AGI93 Conference Proceedings. Paper 3.22. Birmingham. 1993

Langford.M. & Strachan.A.J.
Getting Started in GIS: Cost Effective Solutions and Academic Support
AGI91 Conference Proceedings. Paper 3.7. Birmingham. 1991

Law.H.C., & Ekbolm.P.
Application of GIS: Evaluation of safe cities programme - from knowledge to data representation and transformation
AGI94 Conference Proceedings. Paper 17.2. Birmingham. 1994

Leeson.R.
The Computerised Street Works Register Service - on the road to delivery
AGI94 Conference Proceedings. Paper 14.1. Birmingham. 1994

Leggett.D. & Dowie.P.J.
Avoiding deep water - getting the job done with GIS?
AGI93 Conference Proceedings. Paper 1.2. Birmingham. 1993

Leith.M.
The Geographic Information Market place of the Future
AGI92 Conference Proceedings. Paper 1.30. Birmingham. 1992

Levett.G.R., & Rowley.J.
Don't let GIS be the tail that wags the dog!
AGI93 Conference Proceedings. Paper 2.1. Birmingham. 1993

Linsey.T., & Woods.A.
GIS Education - The Kingston Way
AGI94 Conference Proceedings. Paper 6.3. Birmingham. 1994

Litton.D.
GIS in action in Local Government - The benefits of partnerships
AGI93 Conference Proceedings. Paper 1.3. Birmingham. 1993

Lloyd.J.
Is Data Modelling Enough?
AGI92 Conference Proceedings. Paper 3.2. Birmingham. 1992

Logan.I.T.
Professionalism in GIS: The Role of the Royal Institution of Chartered Surveyors
AGI91 Conference Proceedings. Paper 3.18. Birmingham. 1991

Longley.P. Batty.M. Shepherd.J. & Sadler.G.
On the use of Digitized Boundaries of Urban Areas in the Spatial Analysis of Settlements: The Case of South East England
AGI91 Conference Proceedings. Paper 3.26. Birmingham. 1991

Longstaff.J.J. Denn.J.M. Dunne.S.E. & Massey.P.C.
GIS Design and Database Integration Using Object Orientated Techniques
AGI92 Conference Proceedings. Paper 3.4. Birmingham. 1992

Lukes.D. & Vann.P.
Practical Implementation issues in a multi-function organisation
AGI93 Conference Proceedings. Paper 1.22. Birmingham. 1993

Maguire.D.J.
GIS Employment : Matching supply and Demand
AGI93 Conference Proceedings. Paper 2.19. Birmingham. 1993

Maguire.D.J.
Spatial Data Management : Multi-User Access to Continuous Spatial Databases
AGI94 Conference Proceedings. Paper 3.1. Birmingham. 1994

Maguire.D.J., Kimber.B., & Laming.R.
A Corporate Map Server!
AGI92 Conference Proceedings. Paper 2.15. Birmingham. 1992

Mahoney.R.P.
The Organisational Implications of Corporate Data Management for GIS
AGI92 Conference Proceedings. Paper 3.5. Birmingham. 1992

Mahoney.R.P.
Does it work and does it fit?
AGI90 Conference Proceedings. Paper 3.1. Brighton. 1990

Mahoney.R.P.
GIS Applications for Corporate Data
AGI89 Conference Proceedings. Paper 1.2. Birmingham.
1989

Mahoney.R.P. & McLaren.R.A.
*Best practice guidelines for GIS Implementation in GB
Local Government*
AGI93 Conference Proceedings. Paper 1.20.
Birmingham. 1993

Manthorpe.J.,
*Developments in the Land Register for England and
Wales and the future national land information system.*
AGI93 Conference. Opening Plenary. Birmingham. 1993

Manthorpe.J. & Beardsall.T
Land Information and the Land Register
AGI93 Conference Proceedings. Paper 1.8. Birmingham.
1993

Markham.R. & Rix.D.
*Local Land Charges Corporate Application or
Corporate Report?*
AGI91 Conference Proceedings. Paper 2.8. Birmingham.
1991

Mason.R.W.
Data Conversion - Raster and Vector Formats
AGI89 Conference Proceedings. Paper 4.1. Birmingham.
1989

Masser.I.
*GIS in Britain: The Regional Research Laboratory
Initiative*
AGI89 Conference Proceedings. Paper D.1.
Birmingham. 1989

Masser.I., & Blakemore.M.
*The Achievements and Impact of the Regional Research
Laboratory Initiative*
AGI92 Conference Proceedings. Paper 1.32.
Birmingham. 1992

Masser.I., & Campbell.H.
*The take-up of GIS in Local Government: The LGMB/
University of Sheffield Project*
AGI94 Conference Proceedings. Paper 14.2.
Birmingham. 1994

Matthews.S.A.
*GIS Methodology and Spatial Statistics: The potential
for Epidemiological Study*
AGI89 Conference Proceedings. Paper 6.4. Birmingham.
1989

McAdam.D.
Mustang 2 :The Silk Road
AGI94 Conference Proceedings. Paper 4.1. Birmingham.
1994

McAllister.C.S.
*A Scottish Approach - The Scottish Geographic
Information Systems Forum*
AGI91 Conference Proceedings. Paper 1.3. Birmingham.
1991

McAusland.S. & Summerside.A.
First Steps to Success
AGI93 Conference Proceedings. Paper 1.17.
Birmingham. 1993

McDonald.A.J.W.
Remote mapping of UK geology
AGI94 Conference Proceedings. Paper 15.4.
Birmingham. 1994

McLaren.R.A.
*Casual User Interfaces for GIS Through the Use of
Hypertext Technology*
AGI91 Conference Proceedings. Paper 1.5. Birmingham.
1991

McLaren.R.A. & Healey.R.
Corporate Harmony - A Review of GIS Integration Tools
AGI92 Conference Proceedings. Paper 1.17.
Birmingham. 1992

McMillan.R. & Williams.D.S.
Reaping the Benefits of a GIS System
AGI93 Conference Proceedings. Paper 1.7a.
Birmingham. 1993

Meaden.G.J.
*Fishing for a marine GIS or how the world's first marine
fisheries GIS might be implemented*
AGI93 Conference Proceedings. Paper 3.17.
Birmingham. 1993

Medyckyj-Scott.D.
*User and organisational acceptance of Geographical
Information Systems: The route to failure or success*
AGI89 Conference Proceedings. Paper 4.3. Birmingham.
1989

Miller.D.R., Aspinall.R.J., Finch.P., & Fon.T.C.
*A model of DEM and orthophotography quality using
aerial photography*
AGI94 Conference Proceedings. Paper 21.1.
Birmingham. 1994

Miller.D.R. Aspinall.R.J. Morrice.J.G. & Ferrier.R.
Data Models for Environmental Applications
AGI92 Conference Proceedings. Paper 3.3. Birmingham.
1992

Miller.D.R. Gauld.J.H. Bell.J.S. & Towers.W.
Land Cover Changes in the Cairngorms
AGI91 Conference Proceedings. Paper 2.13.
Birmingham. 1991

**Miller.D.R., Morrice.J.G., Horne.P.L., &
Aspinall.R.J.**
Analysis of Landscape views for visual impact
AGI93 Conference Proceedings. Paper 2.27.
Birmingham. 1993

Miller.P., & Oxley.J.,
Archaeology and Planning: GIS to the Rescue?
AGI94 Conference Proceedings. Paper 4.2. Birmingham.
1994

Mills.D.
The Role of GIS in Safeguarding the Water Environment
AGI92 Conference Proceedings. Paper 3.16.
Birmingham. 1992

Mitchell.I.
Corporate GIS - Do the Real Users get the systems they want?
AGI94 Conference Proceedings. Paper 22.2. Birmingham. 1994

Monckton.C.
Meta-Information and Future GIS
AGI93 Conference Proceedings. Paper 2.6. Birmingham. 1993

Moore.A.
GIS Implementation in Local Government - the central approach
AGI94 Conference Proceedings. Paper 20.1. Birmingham. 1994

Morgan.R., & McKay.I.
Corporate Mapping Libraries a Local Authority experience
AGI94 Conference Proceedings. Paper 20.3. Birmingham. 1994

Morris.C.
Lifestyle Data in GIS - Treating Consumers as Individuals
AGI92 Conference Proceedings. Paper 1.11. Birmingham. 1992

Mounsey.H.
From Research to Reality - the Diffusion of Innovation
AGI90 Conference Proceedings. Paper B.2. Brighton. 1990

Mullins.S.
Business Strategy vs Data Management - No Contest!
AGI92 Conference Proceedings. Paper 1.10. Birmingham. 1992

Murray.D., & Dixon.P.,
GIS in Archaeological Survey
AGI94 Conference Proceedings. Paper 4.3. Birmingham. 1994

Murray.K.J., & Watts.P.A.G.
Ordnance Survey national height models - mapping the third dimension
AGI93 Conference Committee. Paper 3.6. Birmingham. 1993

Nagioff.O. & Kinch-James.D.
Use of Geographic Information Systems in Telecommunications Location Planning
AGI93 Conference Proceedings. Paper 1.13. Birmingham. 1993

Nathanail.P., & Nathanail.J.
Spatial Risk Assessment for Contaminated Land
AGI94 Conference Proceedings. Paper 11.3. Birmingham. 1994

Nathanail.P. & Rosenbaum.E.I.M.
Spatial Interpolation in GIS for environmental studies
AGI91 Conference Proceedings. Paper 2.20. Birmingham. 1991

Newell.D.
Strategies for managing large AM/FM/GIS systems
AGI94 Conference Proceedings. Paper 3.2. Birmingham. 1994

Newell.R.G., & Batty.P.M.
GIS Databases are different
AGI93 Conference Proceedings. Paper 3.2. Birmingham. 1993

Newman.B.
Developing a Major GIS Project
AGI92 Conference Proceedings. Paper 3.8. Birmingham. 1992

Newman.B.
Setting up a major GIS Implementation
AGI91 Conference Proceedings. Paper 2.4. Birmingham. 1991

Openshaw.S. Cross.A. & Charlton.M.
Using a super computer to improve GIS analysis
AGI90 Conference Proceedings. Paper C.1. Brighton. 1990

Openshaw.S. , Cross.A., Charlton.M., Brunsden.C., & Lillie.J.
Lessons learnt from a Post Mortem of a failed GIS
AGI90 Conference Proceedings. Paper 2.3. Brighton. 1990

Ovadia.D.C., & Loudon.T.V.
GIS in a geological survey's migration strategy
AGI93. Paper 3.12. Birmingham. 1993

Palmer.B.D.
GIS or GIM?
AGI91 Conference Proceedings. Paper 2.7. Birmingham. 1991

Parker.D. Taylor.G.
The 'S' in GIS
AGI91 Conference Proceedings. Paper 1.16. Birmingham. 1991

Parker.D. Taylor.G. & Dauncey.D.
The Structured Concept: Realising the World
AGI92 Conference Proceedings. Paper 1.25. Birmingham. 1992

Parsons.E.
The Development of a Multimedia Hypermap
AGI92 Conference Proceedings. Paper 2.24. Birmingham. 1992

Parsons.E
Virtual Worlds Technology: The Ultimate GIS Visualisation Tool?
AGI93 Conference Proceedings. Paper 3.24. Birmingham. 1993

Paschoud.J. & Bell.A.
Quality GIS Applications - Easier said than done?
AGI93 Conference Proceedings. Paper 1.4. Birmingham. 1993

Peacock.D. & Rutherford.I.
Concepts into reality: An Account of a GIS Implementation in the South Western Electricity Board
AGI89 Conference Proceedings. Paper 2.3. Birmingham. 1989

Pearce.N.
Towards a Formula for Success
AGI90 Conference Proceedings. Paper 2.2. Brighton.
1990

Petrie.G.
Photogrammetric input to an environmental GIS
AGI94 Conference Proceedings. Paper 21.3.
Birmingham. 1994

Pitticas.N.
Land Value Information Unit. The Scottish NLIS?
AGI92 Conference Proceedings. Paper 3.13.
Birmingham. 1992

Platt.P.G.
*Developments in Scanning, Application & Compression
of Colour Raster Data*
AGI91 Conference Proceedings. Paper 1.21.
Birmingham. 1991

Porter.L., & McNamee.C.
*GIS and the review of parliamentary constituencies in
Northern Ireland*
AGI94 Conference Proceedings. Paper 16.3.
Birmingham. 1994

Power.S., & Lay.M.
*TITAN CD: New Horizons in GIS for Electronic
Publishing of travel information*
AGI94 Conference Proceedings. Paper 18.2.
Birmingham. 1994

Pritchard.M. & Benson.L.
*Geographical Information Systems and the NHS Internal
Market*
AGI92 Conference Proceedings. Paper 3.6. Birmingham.
1992

Pugh.D. Black.T. & Mounsey.H.
*A National Land and Property Gazetteer? Opportunities
for development*
AGI91 Conference Proceedings. Paper 1.25.
Birmingham. 1991

Pugh.D. & Cushnie.J.
The Land and Property Gazetteer
AGI92 Conference Proceedings. Paper 2.20.
Birmingham. 1992

Purvis.J.
GIS - Moving Outside the Square
AGI89 Conference Proceedings. Paper 1.1. Birmingham.
1989

Rackham.L.J., Coote.A.M., & Gower.R.J.
An operational requirement for spatial data
AGI93 Conference Proceedings. Paper 3.4. Birmingham.
1993

**Ralphs.M. Wyatt.P. Sabel.C. Fovargue.A.
Larner.A. & Lopez.X.**
*Towards a National Land Information System for Britain
- a Coordinated Research Strategy*
AGI92 Conference Proceedings. Paper 1.31.
Birmingham. 1992

Raper.J. & Bundock.M.
*Extending GIS user interface concepts in the UGIX
project*
AGI91 Conference Proceedings. Paper 1.7. Birmingham.
1991

Raper.J.F., & Bundock.M.S.
GIS User Interfaces: A Window on the Future
AGI90 Conference Proceedings. Paper B.3. Brighton.
1990

Raper.J., & Green.N.
*GIST: A new approach to a Geographical Information
System Tutor*
AGI89 Conference Proceedings. Paper C.4. Birmingham.
1989

Raper.J., & McCarthy.T.
Virtually GIS: The New Media Arrive
AGI94 Conference Proceedings. Paper 18.1.
Birmingham. 1994

Raper.J., McCarthy.T., & Livingstone.D.
Interfacing GIS with Virtual Reality Technology
AGI93 Conference Proceedings. Paper 3.25.
Birmingham. 1993

Rasmussen.K. & Robson.P.
A Data Architecture for Geoscientific Data
AGI92 Conference Proceedings. Paper 2.18.
Birmingham. 1992

Rayner.R.S.
*Geographical Information Systems for the Marine
Environment The Professional as a Consultant*
AGI91 Conference Proceedings. Paper 3.22.
Birmingham. 1991

Reid.E.
Machiavelli and Data
AGI92 Conference Proceedings. Paper 2.4. Birmingham.
1992

Reid.J.A.
Getting Started in GIS - The Cornish Experience
AGI91 Conference Proceedings. Paper 2.1. Birmingham.
1991

Rhind.D.
*Policy on the Supply and Availability of Ordnance
Survey Information over the next five years*
AGI92 Conference Proceedings. Paper 1.22.
Birmingham. 1992

Rickman.D.
*PLEIADES - Evaluating the Use of GIS for a Planned
Traffic Corridor Between London and Paris*
AGI92 Conference Proceedings. Paper 2.10.
Birmingham. 1992

Rickman.D.
Open Systems GIS - Do We Need Them?
AGI91 Conference Proceedings. Paper 3.14.
Birmingham. 1991

Rickman.D.
*Pleiades - Evaluating the use of GIS for a planning
traffic corridor between London and Paris*
AGI93 Conference Proceedings. Paper 2.30.
Birmingham. 1993

Rietman.J.
The Utility base map of Rotterdam
AGI89 Conference Proceedings. Paper 5.3. Birmingham.
1989

Rix.D., & Markham.R.
GIS Certification, Ethics and Professionalism
AGI94 Conference Proceedings. Paper 6.2. Birmingham.
1994

Robbins.R.
*The Automobile Association - European Relationships
and Activities*
AGI93 Conference Proceedings. Paper 1.10.
Birmingham. 1993

Roberts.G.V. Park.G.W.A. & Cottle.E.J.T.
Manweb's Implementation of a Mapless GIS
AGI90 Conference Proceedings. Paper C.2. Brighton.
1990

Roberts.J.
The Hertsmere Experience
AGI89 Conference Proceedings. Paper 3.2. Birmingham.
1989

Robinson.G.J.
*Application of Expert Systems to Topographic Map
Generalisation*
AGI89 Conference Proceedings. Paper A.3. Birmingham.
1989

Robinson.G.J. Pearson.E.J. & Settle.J.J.
The use of Digital Elevation Models in GIS Applications
AGI89 Conference Proceedings. Paper D.3.
Birmingham. 1989

Robinson.K.W., & Gill.S.J.
Property Management and O.S. Structured Data
AGI90 Conference Proceedings. Paper 1.3. Brighton.
1990

Rowland.J., & Lilley.C.
The wide area connection
AGI94 Conference Proceedings. Paper 7.1. Birmingham.
1994

Rowley.J.
Summary of Standards Papers
AGI94 Conference Proceedings. Paper 24.1.
Birmingham. 1994

Rowley.J.R.
A Strategy for Geographic Information Standards
AGI92 Conference Proceedings. Paper 2.21.
Birmingham. 1992

Rowley.J.R.
Geographic Information - The Standards Viewpoint
AGI91 Conference Proceedings. Paper 3.10.
Birmingham. 1991

Rowley.J.R.
*Standards - A key building block of Geographic
Information Strategy?*
AGI93 Conference Proceedings. Paper 1.29.
Birmingham. 1993

Salge.F.,
Is the rest of Europe any different?
AGI93 Conference. opening plenary. Birmingham. 1993

Shapiro.J.
GIS Data Conversion for Complex Documents
AGI92 Conference Proceedings. Paper 2.11.
Birmingham. 1992

Sharman.D. & Haywood.P.
The National Transfer Format in 1991
AGI91 Conference Proceedings. Paper 3.17.
Birmingham. 1991

Sheath.N.
Geography - A Corporate Resource
AGI89 Conference Proceedings. Paper 2.2. Birmingham.
1989

Sheehan.R.
*The UK Digital Marine Atlas Project and Issues in Data
Capture*
AGI93 Conference Proceedings. Paper 3.14.
Birmingham. 1993

Sheehan.R.
*Coastal Zone Management is okay but it's not as good as
the real thing!*
AGI94 Conference Proceedings. Paper 13.2.
Birmingham. 1994

Shepherd.I.D.H.
Meeting the design challenge of Multi-Sensory GIS
AGI94 Conference Proceedings. Paper 12.1.
Birmingham. 1994

Shore.D.G.
*Resourcing a Corporate GIS Implementation - A
Strategic Viewpoint*
AGI92 Conference Proceedings. Paper 1.8. Birmingham.
1992

Short.M.
*Time is Money - Vehicle Routing and Scheduling can
save both*
AGI93 Conference Proceedings. Paper 1.15.
Birmingham. 1993

Sian.A.J. & Xavier.R.L.
*Integrating Data Derived from within the British Local
Government Institutional Framework*
AGI92 Conference Proceedings. Paper 1.18.
Birmingham. 1992

Sinclair.N.
Selecting Locational Referencing Methods for GIS
AGI90 Conference Proceedings. Paper A.1. Brighton.
1990

Smallwood.J.,Dixon.P., & Orange.G.,
GIS - Satisfying Business Needs?
AGI94 Conference Proceedings. Paper 2.1. Birmingham.
1994

Smith.A.B.
Small Scale Data - A Large Scale Problem
AGI89 Conference Proceedings. Paper B.4. Birmingham.
1989

Smith.P.J.
Instant Access to Land Registry Data - Now a Reality
AGI94 Conference Proceedings. Paper 11.2.
Birmingham. 1994

Smith.W.,
Here now, where next
AGI91. Keynote Paper. Birmingham. 1991

Snoxell.J.
Portable GIS in the Water Industry - Dealing with HAUC
AGI92 Conference Proceedings. Paper 1.21.
Birmingham. 1992

Snoxell.J.D.
GIS - Satisfying business needs
AGI94 Conference Proceedings. Paper 10.2.
Birmingham. 1994

Sowton.M.
European Standards - Where are they going?
AGI92 Conference Proceedings. Paper 2.9. Birmingham.
1992

State.R
A practical application of data capture techniques
AGI93 Conference Proceedings. Paper 1.14.
Birmingham. 1993

Stringer.P. & Haslett.J.
*Exploratory, Interactive analysis of spatial data: an
illustration in the area of health inequalities*
AGI91 Conference Proceedings. Paper 3.29.
Birmingham. 1991

Stuart.N.
Spatial Analysis in GIS: Value Added or Vapour?
AGI92 Conference Proceedings. Paper 1.26.
Birmingham. 1992

**Stuttard.M.J., Hayball.J.B., Narciso.G., Oroda.A.
& Suppo.M.**
*Use of a GIS to Assist Hydrological Modelling of Lake
Basins in the Kenyan Rift Valley*
AGI94 Conference Proceedings. Paper 13.4.
Birmingham. 1994

Swainston.R.
*Integration of data within GIS - A case study in the
Metropolitan Borough of Dudley, West Midlands*
AGI93 Conference Proceedings. Paper 1.21.
Birmingham. 1993

Taylor.N.
The Third Dimension - Who Needs it?
AGI92 Conference Proceedings. Paper 1.3. Birmingham.
1992

Thomas.B.D.
*Off-line Calibration and Validation Facilities at the
Earth Observation Data Centre*
AGI91 Conference Proceedings. Paper 1.6. Birmingham.
1991

Thomas.J.
Picking the Perfect Peripheral
AGI93 Conference Proceedings. Paper 1.19.
Birmingham. 1993

Timms.T.
Relationships and Behaviour in GIS
AGI92 Conference Proceedings. Paper 2.2. Birmingham.
1992

Todd.P.
Cost-benefit - The Solution
AGI89 Conference Proceedings. Paper 3.4. Birmingham.
1989

Todd.P., Bundred.P., & Brown.P.
*The Demography of Demand for Oncology Services: A
Health Care Planning GIS Application*
AGI94 Conference Proceedings. Paper 17.1.
Birmingham. 1994

Trevor.I., & Russell.D.J.
Data Capture - in weeks not years
AGI93 Conference Proceedings. Paper 2.15.
Birmingham. 1993

Tyler.J.
*British Rail and its passengers: How Geographic
Information could bring them closer together*
AGI89 Conference Proceedings. Paper 1.3. Birmingham.
1989

Unwin.D. & Dale.P.
An Educationalist's view of GIS
AGI89 Conference Proceedings. Paper C.3. Birmingham.
1989

**Unwin.D.J., Dykes.J.A., Fisher.P.F., Stynes.K.,
& Wood.J.D.**
WYSIWYG? Visualisation in the Spatial Sciences
AGI94 Conference Proceedings. Paper 5.4. Birmingham.
1994

Vaughan.D.
Project-based Environmental GIS
AGI91 Conference Proceedings. Paper 2.21.
Birmingham. 1991

Vicars.D.
Modelling Generalisation
AGI94 Conference Proceedings. Paper 12.3.
Birmingham. 1994

Vincent.S.
Is GIS putting the cart before the horse?
AGI93 Conference Proceedings. Paper 2.4. Birmingham.
1993

Vipond.D.L. & Harvey.R.A.
Gluing GIS into the System Integration environment
AGI93 Conference Proceedings. Paper 2.22.
Birmingham. 1993

Wachowicz.M., & Broadgate.M.L.
*A Significant Challenge : Prediction of Environmental
Changes using a temporal GIS*
AGI93 Conference Committee. Paper 2.25. Birmingham.
1993

Walker.J.C. & Hampson.E.
When the wind blows the GIS will be used
AGI93 Conference Proceedings. Paper 1.12.
Birmingham. 1993

Walker.J.C., & Hampson.E.,
Integrating GIS with the Network using ISDN
AGI94 Conference Proceedings. Paper 7.4. Birmingham.
1994

Walker.P.
The Digital Geographic Information Working Group
AGI90 Conference Proceedings. Paper D.2. Brighton.
1990

Walker.R.S.
Standards for GIS Data
AGI91 Conference Proceedings. Paper 3.11.
Birmingham. 1991

Walker.R.S.
Data quality standards
AGI94 Conference ProceedingsAGI89. Paper 24.2.
Birmingham. 1994

Waller.R.
GIS - A Cost Benefit study
AGI93 Conference Proceedings. Paper 1.5. Birmingham.
1993

Wallwork.P. & Goodwin.R.
*The Computerised Street and Road Works Register -
Sharing Information*
AGI92 Conference Proceedings. Paper 2.19.
Birmingham. 1992

Waters.R.
Air Photos to GIS, do we need the map?
AGI91 Conference Proceedings. Paper 2.23.
Birmingham. 1991

Waters.R. & Ternouth.P.
Kill the 'G' in GIS!
AGI92 Conference Proceedings. Paper 2.5. Birmingham.
1992

Watts.C.D.
Implementation of a 4-D GIS at a Utility Company
AGI93 Conference Proceedings. Paper 3.10.
Birmingham. 1993

Weatherill.D. & Buchanan.D.
*Effective Transition: The Management Skills and
Processes*
AGI89 Conference Proceedings. Paper 3.1. Birmingham.
1989

Weatherill.D. & Keddie.J.
Building a Financial Case for GIS
AGI92 Conference Proceedings. Paper 1.5. Birmingham.
1992

Webb.G. & Todd.P.
*Identifying GIS Benefits in the Regional Electricity
Companies*
AGI91 Conference Proceedings. Paper 1.11.
Birmingham. 1991

Webb.H.
The Digitising Table - Is there a viable alternative?
AGI91 Conference Proceedings. Paper 3.1. Birmingham.
1991

Webb.R.A., & Elkins.P.G.
*Guidelines for the Implementation and Management of a
GIS*
AGI89 Conference Proceedings. Paper 3.3. Birmingham.
1989

White.A.
Managing the Cultural Change Minimising the Risks!!
AGI92 Conference Proceedings. Paper 3.7. Birmingham.
1992

White.A.
*Bringing together the right team to turn GIS pilots into
Production Systems*
AGI91 Conference Proceedings. Paper 1.8. Birmingham.
1991

Wilkinson.C.R.
*Store Performance - Evaluating the Ingredients of a
successful site*
AGI90 Conference Proceedings. Paper 1.1. Brighton.
1990

Willis.J.
The need for skilled staff
AGI89 Conference Proceedings. Paper C.2. Birmingham.
1989

Windsor.P. & Aybet.J.
*What Object-Orientation Offers to Geographical
Information Systems*
AGI91 Conference Proceedings. Paper 2.27.
Birmingham. 1991

Winter.B.
Using GIS - The Next Generation
AGI91 Conference Proceedings. Paper 1.19.
Birmingham. 1991

Winter.P.
Selling a Corporate GIS
AGI91 Conference Proceedings. Paper 2.6. Birmingham.
1991

Winterkorn.E.
*Local Authority GIS Usage in England, Scotland and
Wales*
AGI94 Conference Proceedings. Paper 14.3.
Birmingham. 1994

Wise.S. & Haining.R.
*The Role of Spatial Analysis in Geographical
Information Systems*
AGI91 Conference Proceedings. Paper 3.24.
Birmingham. 1991

Wise.S. Haining.R. & Blake.M.
*Analysing intra-urban variations in colorectal cancer -
the role of GIS*
AGI92 Conference Proceedings. Paper 3.14.
Birmingham. 1992

Wiseman.P.G.
Suffolk County Council's corporate GIS pilot study
AGI89 Conference Proceedings. Paper 6.5. Birmingham.
1989

Witten.A.P., Peacock.B., & Jacobs.R.
The Role of GIS in managing corporate data within Northeast Water's System architecture
AGI93 Conference Proceedings. Paper 2.23.
Birmingham. 1993

Wood.A.J.
Geographic Information Systems: An Executive Decision Perspective
AGI92 Conference Proceedings. Paper 1.16.
Birmingham. 1992

Wood.T.
Using GIS to Develop the European Market for Geographic Information: The OMEGA CD
AGI94 Conference Proceedings. Paper 16.1.
Birmingham. 1994

Wood.T.R.
Experience of an operational GIS in the Water Industry
AGI89 Conference Proceedings. Paper 6.3. Birmingham.
1989

Woodcock.M.E. & Clennell.R.
Can I Use a Digitising Bureau?
AGI91 Conference Proceedings. Paper 3.3. Birmingham.
1991

Woodhouse.T. Todd.P. & Hilder.D.
Possibilities for Force-Wide Geographic Information Systems in West Yorkshire Police
AGI91 Conference Proceedings. Paper 2.9. Birmingham.
1991

Woodsford.P.
The AGI Today
AGI91 Conference Proceedings. Keynote Paper.
Birmingham. 1991

Woodsford.P.A.
Using Geographic Data - Practicalities and Politics
AGI93 Conference Proceedings. Paper 2.29.
Birmingham. 1993

Woodsford.P.A. & Meader.D.
The State of the Art in Raster-to-Vector Conversion
AGI92 Conference Proceedings. Paper 2.13.
Birmingham. 1992

Worboys.M.
Putting Time into GIS
AGI93 Conference Proceedings. Paper 3.8. Birmingham.
1993

Worrall.L.
Developing a Corporate Information Systems Strategy
AGI92 Conference Proceedings. Paper 3.17.
Birmingham. 1992

Worrall.L., & Bond.D.
Geographical information systems, spatial analysis and Public Policy
AGI94 Conference Proceedings. Paper 14.4.
Birmingham. 1994

Wren.G., & MacKenzie.R.,
Deadline Oriented GIS - Task Specific
AGI94 Conference Proceedings. Paper 20.2.
Birmingham. 1994

Wyatt.P.
Using a Geographical Information System for Property Valuation
AGI94 Conference Proceedings. Paper 11.1.
Birmingham. 1994

Young.G.M. & Duley.C.
Spatial Marketing Information and its Analysis: A Pan European Perspective
AGI92 Conference Proceedings. Paper 2.29.
Birmingham. 1992

AGI papers by subject

Address management

Brewer.A. & Shepherd.J.	Post codes as a Geographic Database: Added Value and Added Analysis
Capell.P.J. & Singh.G.	London Area Transport Survey; Automated Address Coding
Dunlea.M.	The use of G.I.S. in Mobilising Irish Emergency Services - the C.A.M.P. Experience
Elliott.L., McCallum.D., & Pretty.S.	Address-Point : Fusion of OS Mapping Precision and Royal Mail Postal Addresses
Sinclair.N.	Selecting Locational Referencing Methods for GIS

Benefits

Alla.P. & Trow.S.W.	Implementation of a GIS - A way to optimising utilities operations
Bainbridge.P., Hinton.P. & Curry.D.	GIS: Unlocking the Benefits by Implementing Applications
Callaghan.J.G.	GIS - Increasing business efficiency and profit
Cane.S.	Adding Corporate Value to Data - a sugGIStion
Capell.P.J. & Singh.G.	London Area Transport Survey; Automated Address Coding
Corbin.C.E.H.	GIS: A Catalyst for Data Integration
Coulson.M.G.	Methodologies for Assessing cost and value of GIS
Cross.D.	How to populate your GIS Database, on budget and on time
Drake.P.	The use of different Data Conversion options for a GIS application in the Water Industry
Gadd.S.	GIS - Forging a Key Role in Data Integration
Grimshaw.D.J.	The Transformation of Customer Databases
Gross.H.	Very Clever....but does it really mean anything?
Ives.M.	Reaping the Benefits
Kennie.T. & Mather.P.	Do GIS Specialists need the Professional Institutions?
Knott.J. & Shiers.D.	Mobile Maps - Sending vector map based maps across radio networks for use in unplanned locations by GIS Users
Maguire.D.J., Kimber.B., & Laming.R.	A Corporate Map Server!
Markham.R. & Rix.D.	Local Land Charges Corporate Application or Corporate Report?
McMillan.R. & Williams.D.S.	Reaping the Benefits of a GIS System
Palmer.B.D.	GIS or GIM?
Peacock.D. & Rutherford.I.	Concepts into reality: An Account of a GIS Implementation in the South Western Electricity Board
Purvis.J.	GIS - Moving Outside the Square
Reid.E.	Machiavelli and Data
Roberts.J.	The Hertsmere Experience
Sheath.N.	Geography - A Corporate Resource
Snoxell.J.D.	GIS - Satisfying business needs
Timms.T.	Relationships and Behaviour in GIS
Todd.P.	Cost-benefit - The Solution
Walker.J.C. & Hampson.E.	When the wind blows the GIS will be used
Waller.R.	GIS - A Cost Benefit study
Weatherill.D. & Keddie.J.	Building a Financial Case for GIS
Webb.G. & Todd.P.	Identifying GIS Benefits in the Regional Electricity Companies
Winter.P.	Selling a Corporate GIS
Winterkorn.E.	Local Authority GIS Usage in England, Scotland and Wales

Case Studies

Adnitt.N.C.	Number 1 in the Charts , Yorkshire Water GIS, currently available on CD
Alla.P. & Trow.S.W.	Implementation of a GIS - A way to optimising utilities operations

Andrews.D.E., Hartley.W.S.
& Walter.J.   GIS and ROMANSE - Integrating Traffic Systems
Annand.K.,
& Hookham.C.   A GIS Strategy for guarding the water environment
Arnott.D. & Keddie.A.R.   Data Capture - The Standards and Procedures Utilised within
                                    Northumbrian Water Group
Atkinson.R.   GIS: A catalyst for positive change
Bainbridge.P., Hinton.P.
& Curry.D.   GIS: Unlocking the Benefits by Implementing Applications
Boxer.A.   IS Government different?
Brown.N., Gerrard.F.,
& Parr.T.   Defining English Chalk and Limestone Grassland areas using GIS
Bundred.P., Hirschfield.A.,
& Marsden.J.   GIS in the Planning of Health Services in a District Health Authority
Burkmar.R.J. Petch. J.R. Basden.A.
& Yip.J.   Integrating GIS and Knowledge-based systems for Landscape Management
Campbell.H. & Masser.I.   Implementing GIS : The Organisational dimension
Capell.P.J. & Singh.G.   London Area Transport Survey; Automated Address Coding
Chell.M.   Patterns of energy use and air pollution
Clegg.P.   GIS Implementation - The Sheffield Experience
Dale.P.   GIS in property management - strengths, weaknesses, opportunities and threats
Dunn.R., Whitelegge.M., Barr.M.,
Swanwick.C., & Warnock.S.   The use of GIS in landscape classification : A case study of Southwest England
Elliott.L., McCallum.D.,
& Pretty.S.   Address-Point : Fusion of OS Mapping Precision and Royal Mail Postal Addresses
Ellis.S.   How not to use the 1991 Census
Farrell.B.   Corporate GIS in a Scottish Regional Council
Flowerdew.R. Green.M.
& Lucas.S.   Analysing Local House Price Variations with GIS
Gadd.S.   Training for Effective Implementation. Are we getting it right?
Godlement.H.J.R.   Managing the Geographic Information
Haines-Young.R.   GIS for Environmental Planning: Problems and Prospects
Hall.J., Ullyett.J.,
Hornung.M.,
& Hancock.S.   Using GIS for mapping the sensitivity of fresh waters to acidification
Hartley.J.C. Homer.I.R.
Trow.S. Hinton.P.H.
& Everett.C.   Inter Utility Exchange of Electronic Map Based Records in the
                                    North East of England

Hawker.L.
& Goodwin.R.   The Computerised Street Works Register - Streets ahead?
Homer.A.R.
& Watson.M.   Integrated Data Saves Money!!
Howes.D.A., Green.M.,
& Kurtz.T.   Issues in the assessment of rural deprivation in England
Ives.M.   Reaping the Benefits
Kawalek.J.P.   Information Systems and the use of multi-media in the water industries
Keddie.J.G., & Docherty.E.   GIS and the 1991 Census as a cohesive force in Local Government
Lang.N.   Why reinvent the Wheel? Solutions using off-the-shelf GIS
Leggett.D. & Dowie.P.J.   Avoiding deep water - getting the job done with GIS?
Lloyd.J.   Is Data Modelling Enough?
Longley.P. Batty.M.
Shepherd.J.
& Sadler.G.   On the use of Digitized Boundaries of Urban Areas in the Spatial Analysis of
                                    Settlements: The Case of South East England

Lukes.D.
& Vann.P.   Practical Implementation issues in a multi-function organisation
Manthorpe.J.
& Beardsall.T   Land Information and the Land Register
Matthews.S.A.   GIS Methodology and Spatial Statistics: The potential for Epidemiological Study
McMillan.R. & Williams.D.S.   Reaping the Benefits of a GIS System
Meaden.G.J.   Fishing for a marine GIS or how the world's first marine fisheries GIS might be
                                   implemented

Miller. D.R. Aspinall. R.J. Morrice. J.G.
& Ferrier.R.   Data Models for Environmental Applications
Miller.D.R., Morrice J.G.,
Horne.P.L., & Aspinall.R.J.   Analysis of Landscape views for visual impact
Mitchell.I.   Corporate GIS - Do the Real Users get the systems they want?
Moore.A.   GIS Implementation in Local Government - the central approach
Murray.D.,
& Dixon.P.,   GIS in Archaeological Survey
Newman.B.   Developing a Major GIS Project
Parsons.E   Virtual Worlds Technology: The Ultimate GIS Visualisation Tool?
Peacock.D.
& Rutherford.I.   Concepts into reality: An Account of a GIS Implementation in the South Western
                                    Electricity Board
Pitticas.N.   Land Value Information Unit. The Scottish NLIS?
Pritchard.M.
& Benson.L.   Geographical Information Systems and the NHS Internal Market
Rickman.D.   Pleiades - Evaluating the use of GIS for a planning traffic corridor between London
                                    and Paris
Robbins.R.   The Automobile Association - European Relationships and Activities
Smith.P.J.   Instant Access to Land Registry Data - Now a Reality
Snoxell.J.   Portable GIS in the Water Industry - Dealing with HAUC
Stringer.P. & Haslett.J.   Exploratory, Interactive analysis of spatial data: an illustration in the area of health
                                    inequalities
Swainston.R.   Integration of data within GIS - A case study in the Metropolitan Borough of Dudley,
                                    West Midlands
Trevor.I., & Russell.D.J.   Data Capture - in weeks not years
Vipond.D.L. & Harvey.R.A.   Gluing GIS into the System Integration environment
Walker.J.C., & Hampson.E.,   Integrating GIS with the Network using ISDN
Waller.R.   GIS - A Cost Benefit study
Wallwork.P. & Goodwin.R.   The Computerised Street and Road Works Register - Sharing Information
Watts.C.D.   Implementation of a 4-D GIS at a Utility Company
Weatherill.D. & Keddie.J.   Building a Financial Case for GIS
Wiseman.P.G.   Suffolk County Council's corporate GIS pilot study
Witten.A.P., Peacock.B.,
& Jacobs.R.   The Role of GIS in managing corporate data within Northeast Water's System
                                    architecture
Wood.T.R.   Experience of an operational GIS in the Water Industry
Wren.G., & MacKenzie.R.,   Deadline Oriented GIS - Task Specific

Data
Abbott.G., & Dickason.P.   Digital Chart of the World - A Global Perspective
Adnitt.N.   Data Capture - The Technology, The Issues
Adnitt.N.   Of Course we could always scan it
Afors.J., Chiu.M.,
& Martin.J.   Designing for corporate information
Albaredes.G.   Data Sharing - a hurdle for GIS
Archibald.I. & Jeffries-Harris.T.   A Vectorised Spaghetti-Hooped Boundary 10000 kms long
Arnott.D. & Keddie.A.R.   Data Capture - The Standards and Procedures Utilised within Northumbrian Water
                                    Group
Asabere. R.K. Durucan. S.
& Owen. D.B.   An Expert System for Application of Geographical Information Systems in Mineral
                                    Resources Management
Aspinall. R.J.   A Landscape Ecological Approach to Mapping Distribution and Abundance of Birds
                                    with GIS
Aybet.J.   Spatial data modelling for Geographic Information System design
Aybet.J., Martin.J.,
& Taskis.D.   Protecting investment by capturing asset data from existing systems
Aybet.J., & Walpole.R.   An Integrated GIS Approach to the Use of Earth Observation Data - Sugar Beet
                                    prediction and management system
Azizi.A. Clark.M.J.
& Davenport.J.   Air Photo or Video Inputs to Vector or Raster GIS
Barr.R.   Micro-analysis in GIS
Bird.A.C.   GIS Based Data on Land cover Change in the National Parks of England and Wales

Blakemore.M.	Sharing Data - Whose why how?
Blakemore.M.	Management issues in a Nationally Networked Geographic Information System
Booth.S.	GIS - are we missing the point?
Bourke.A.	How we Integrated GIS with Existing Central RDBMS over a wide area network
Boxer.A.	IS Government different?
Boyd.G., & Schaap.D.	Integrating Marine Geographical Information
Brand.M.J.D. & Gray.S.	Experience with Structured Data -Millstone or Milestone?
Brand.M.J.D. & Gray.S.	From Concept to reality
Brand.M.J.D., & Mitchell.G.	CORINE Land Cover (Ireland) - The Northern Perspective
Bridger.I. & Fukushima.Y.	Towards Intelligent Data
Bridger.I. & Land.N.	Structured Data - An Ordnance Survey Perspective
Brown.N.	Mapping of Land Cover from space and its use in a Geographical Information System
Brown.N., Gerrard.F., & Parr.T.	Defining English Chalk and Limestone Grassland areas using GIS
Brown.P.J.B., Batey.P.W.J., Hirschfield.A., & Marsden.J.	Poisson Chi Square Mapping, GIS & Geodemographic Analysis: The Spatial & Aspatial Analysis of Relatively Rare Conditions
Burrough.P.A.	Accuracy issues for future GIS
Calvert.C., & Holland.D.	GPS and Ordnance Survey map data: Issues and solutions
Cane.S.	Adding Corporate Value to Data - a sugGIStion
Carig.W.J.	Data to the People : North American efforts to empower communities with data and information
Cassettari. S.	Using Aerial Photography in your GIS
Chapman.D. Dowman.I. & Muller.J.P.	Digital photogrammetry - interfaces with GIS
Chell.M.	Patterns of energy use and air pollution
Clarke.K.E., & Norris.M.T.	Data and information highways
Clegg.P.	GIS Implementation - The Sheffield Experience
Clegg.P.	Basic Issues in Data Capture
Cooke.I.C.	Field Data Capture with Pen Computers
Coote.A.M.	Managing a large Spatial Archive
Coote.A.M. & Rackham.L.J.	Handling Update: Issues in Spatial Data Maintenance
Corbin.C.E.H.	GIS: A Catalyst for Data Integration
Cory.M.J.	Portable Data Capture
Cross.A. & Openshaw.S	Crime Pattern Analysis: The development of ARC/CRIME
Cross.D.	How to populate your GIS Database, on budget and on time
Dale.P.F.	Domesday 2000: The Professional as a Politician
Daniels.D.J.	High Performance Ground Penetrating Radar System
Devereux.B. & Mayo.T.	Task Oriented Tools for Cartographic Data Capture
Devereux.B. & Mayo.T.	In the Foreground: Intelligent Techniques for Cartographic Data Capture
Dirdal.P.	A Vendor's Perspective of the Chorley report four Years On
Dixon.P., Smallwood.J., & Dixon.M.	Development of a mobile GIS: Field data capture using a pen based note pad computer system
Dodson.A.H. & Basker.G.A.	GPS: GIS Problems Solved?
Doig.J., Atkinson.R. & Taylor.M.	Turning Public Registers into Public Information
Drake.P.	The use of different Data Conversion options for a GIS application in the Water Industry
Dugmore.K.	1991 Census: Outputs and Opportunities
Dunn.R. & Harrison.A.	A Feasibility Study for a National Land Use Stock Survey
Dunn.R., Harrison.A.R., Brown.L., & Turton.P.J.	The Use of Geographic Information Systems in the Analysis of Countryside Data
Dyson.J.	Data Integration via the Gazetteer
Easterfield.M.E., Newell.R.G., & Theriault.D.G.	Management of Time and Space in GIS
Ellis.S.	How not to use the 1991 Census
Ellis.S., & Gower.C.	A Generic data model for GIS Local Government
Gadd.S.	GIS - Forging a Key Role in Data Integration
Glickman.V.	Common Responses to Common Challenges Towards a Charter for the Marketing of Geographic Information
Godlement.H.J.R.	Managing the Geographic Information

Miller.D.R., Aspinall.R.J.,
Finch.P., & Fon.T.C.   A model of DEM and orthophotography quality using aerial photography
Miller.D.R. Aspinall.R.J. Morrice.J.G.
& Ferrier.R.   Data Models for Environmental Applications
Miller.D.R. Gauld.J.H. Bell.J.S.
& Towers.W.   Land Cover Changes in the Cairngorms
Miller.D.R., Morrice.J.G.,
Horne.P.L., & Aspinall.R.J.   Analysis of Landscape views for visual impact
Mills.D.   The Role of GIS in Safeguarding the Water Environment
Monckton.C.   Meta-Information and Future GIS
Morgan.R., & McKay.I.   Corporate Mapping Libraries a Local Authority experience
Morris.C.   Lifestyle Data in GIS - Treating Consumers as Individuals
Mounsey.H.   From Research to Reality - the Diffusion of Innovation
Mullins.S.   Business Strategy vs Data Management - No Contest!
Murray.K.J.,
& Watts.P.A.G.   Ordnance Survey national height models - mapping the third dimension
Newell.D.   Strategies for managing large AM/FM/GIS systems
Ovadia.D.C.,
& Loudon.T.V.   GIS in a geological survey's migration strategy
Parker.D. Taylor.G.   The 'S' in GIS
Parker.D. Taylor.G.
& Dauncey.D.   The Structured Concept: Realising the World
Platt.P.G.   Developments in Scanning, Application
& Compression of Colour Raster Data
Porter.L., & McNamee.C.   GIS and the review of parliamentary constituencies in Northern Ireland
Rackham.L.J., Coote.A.M.,
& Gower.R.J.   An operational requirement for spatial data
Raper.J.F., & Bundock.M.S.   GIS User Interfaces: A Window on the Future
Raper.J., McCarthy.T.,
& Livingstone.D.   Interfacing GIS with Virtual Reality Technology
Rasmussen.K. & Robson.P.   A Data Architecture for Geoscientific Data
Rayner.R.S.   Geographical Information Systems for the Marine Environment The Professional as a
                                Consultant
Reid.E.   Machiavelli and Data
Rhind.D.   Policy on the Supply and Availability of Ordnance Survey Information over the next
                                five years
Robbins.R.   The Automobile Association - European Relationships and Activities
Robinson.G.J.   Application of Expert Systems to Topographic Map Generalisation
Robinson.K.W., & Gill.S.J.   Property Management and O.S. Structured Data
Rowland.J., & Lilley.C.   The wide area connection
Shapiro.J.   GIS Data Conversion for Complex Documents
Sharman.D. & Haywood.P.   The National Transfer Format in 1991
Sheehan.R.   The UK Digital Marine Atlas Project and Issues in Data Capture
Short.M.   Time is Money - Vehicle Routing and Scheduling can save both
Sian.A.J. & Xavier.R.L.   Integrating Data Derived from within the British Local Government Institutional
                                Framework
Sinclair.N.   Selecting Locational Referencing Methods for GIS
Smith.A.B.   Small Scale Data - A Large Scale Problem
Smith.W.,   Here now, where next
State.R   A practical application of data capture techniques
Stuttard.M.J., Hayball.J.B.,
Narciso.G., Oroda.A.
& Suppo.M.   Use of a GIS to Assist Hydrological Modelling of Lake Basins in the Kenyan Rift
                                Valley
Swainston.R.   Integration of data within GIS - A case study in the Metropolitan Borough of Dudley,
                                West Midlands
Taylor.N.   The Third Dimension - Who Needs it?
Thomas.B.D.   Off-line Calibration and Validation Facilities at the Earth Observation Data Centre
Thomas.J.   Picking the Perfect Peripheral
Timms.T.   Relationships and Behaviour in GIS
Trevor.I., & Russell.D.J.   Data Capture - in weeks not years
Vipond.D.L. & Harvey.R.A.   Gluing GIS into the System Integration environment
Walker.J.C.,
& Hampson.E.,   Integrating GIS with the Network using ISDN

Walker.P.	The Digital Geographic Information Working Group
Walker.R.S.	Standards for GIS Data
Waters.R.	Air Photos to GIS, do we need the map?
Watts.C.D.	Implementation of a 4-D GIS at a Utility Company
Webb.H.	The Digitising Table - Is there a viable alternative?
Windsor.P. & Aybet.J.	What Object-Orientation Offers to Geographical Information Systems
Wise.S. & Haining.R.	The Role of Spatial Analysis in Geographical Information Systems
Witten.A.P., Peacock.B., & Jacobs.R.	The Role of GIS in managing corporate data within Northeast Water's System architecture
Wood.T.R.	Experience of an operational GIS in the Water Industry
Woodcock.M.E. & Clennell.R.	Can I Use a Digitising Bureau?
Woodsford.P.	The AGI Today
Woodsford.P.A.	Using Geographic Data - Practicalities and Politics
Woodsford.P.A. & Meader.D.	The State of the Art in Raster-to-Vector Conversion
Worboys.M.	Putting Time into GIS
Young.G.M. & Duley.C.	Spatial Marketing Information and its Analysis: A Pan European Perspective

Demographic Applications

Dugmore.K.	Paper 1.24	1991
Ellis.S.	Paper 3.20	1993
Green.P.	Paper 1.17	1991
Howes.D.A., Green.M., & Kurtz.T.	Paper 3.21	1993
Keddie.J.G., & Docherty.E.	Paper 3.23	1993
Langford.M., & Fisher.P.	Paper 3.22	1993
Morris.C.	Paper 1.11	1992
Todd.P., Bundred.P., & Brown.P.	Paper 17.1	1994
Tyler.J.	Paper 1.3	1989

Education And Training

Browne.T.	Teaching through Training?
Buchanan.H.	A Young Professional's View of the GIS Industry
Buxton.R. & Ball.C.E.	Geographic Information and the Expert System in Local Authority Development Control
Cassettari.S.	Introducing GIS into the National Curriculum for Geography
Green.D.R. & McEwen.L.J.	GIS as a component of Information Technology courses in Higher Education: Meeting the requirements of employers
Hosken.E. & Parker.D.	Civil Engineers need Geographic Information
Linsey.T., & Woods.A.	GIS Education - The Kingston Way
Maguire.D.J.	GIS Employment : Matching supply and Demand
Raper.J., & Green.N.	GIST: A new approach to a Geographical Information System Tutor
Unwin.D. & Dale.P.	An Educationalist's view of GIS

Environmental Applications

Allen.M., Schumacher.C., & Kutz.A.	The Use of GIS in the Environmental Assessment of the WW2 Mustard Gas Factory at Ergethan in the former East Germany
Atkinson.R.	GIS: A catalyst for positive change
Brand.M.J.D., & Mitchell.G.	CORINE Land Cover (Ireland) - The Northern Perspective
Brown.N., Gerrard.F., & Parr.T.	Defining English Chalk and Limestone Grassland areas using GIS
Brown.R., Slater.J., & Askew.D.	Environmental Monitoring of Protected Landscapes - monitoring environmentally sensitive areas using satellite imagery, within GIS
Chell.M.	Patterns of energy use and air pollution
Clark.M.J. Ball.J.H. & Gatward.J.	Costing the Impact of Sea-level rise: A GIS solution for the South Coast of England
Doig.J., Atkinson.R. & Taylor.M.	Turning Public Registers into Public Information
Gannon.P.J.	Managing Infrastructure Impacts
Green.D.R., & Morton.D.	Acquiring Environmental Remotely Sensed Data from Model Aircraft for Input to Geographic Information Systems

Haines-Young.R.	GIS for Environmental Planning: Problems and Prospects
Haines-Young.R.H., Bunce.R.G.H., & Parr.T.W.	Countryside Information System
Hall.J., Ullyett.J., Hornung.M., & Hancock.S.	Using GIS for mapping the sensitivity of fresh waters to acidification
Harrison.A. Dunn.R. & Turton.P.	Environmental GIS: Technology, Data and Policy
Jones.R.J.A., Bradley.R.I., & Siddons.P.A.	A Land Information System for environmental risk assessment
Leggett.D. & Dowie.P.J.	Avoiding deep water - getting the job done with GIS?
Meaden.G.J.	Fishing for a marine GIS or how the world's first marine fisheries GIS might be implemented
Miller.D.R. Aspinall.R.J. Morrice.J.G. & Ferrier.R.	Data Models for Environmental Applications
Miller.D.R., Morrice.J.G., Horne.P.L., & Aspinall.R.J.	Analysis of Landscape views for visual impact
Mills.D.	The Role of GIS in Safeguarding the Water Environment
Nathanail.P. & Rosenbaum.E.I.M.	Spatial Interpolation in GIS for environmental studies
Ovadia.D.C., & Loudon.T.V.	GIS in a geological survey's migration strategy
Petrie.G.	Photogrammetric input to an environmental GIS
Rayner.R.S.	Geographical Information Systems for the Marine Environment The Professional as a Consultant
Sheehan.R.	The UK Digital Marine Atlas Project and Issues in Data Capture
Thomas.B.D.	Off-line Calibration and Validation Facilities at the Earth Observation Data Centre
Vaughan.D.	Project-based Environmental GIS
Wachowicz.M., & Broadgate.M.L.	A Significant Challenge : Prediction of Environmental Changes using a temporal GIS
Wren.G., & MacKenzie.R.,	Deadline Oriented GIS - Task Specific

Health

Hirschfield.A. Brown.P.J.B. Marsden.J. & Bundred.P.	A GIS-linked Database for Analysing the Deployment of Community-based Health and Social Services on the Wirral
Pritchard.M. & Benson.L.	Geographical Information Systems and the NHS Internal Market
Stringer.P. & Haslett.J.	Exploratory, Interactive analysis of spatial data: an illustration in the area of health inequalities
Todd.P., Bundred.P., & Brown.P.	The Demography of Demand for Oncology Services: A Health Care Planning GIS Application
Wise.S. Haining.R. & Blake.M.	Analysing intra-urban variations in colorectal cancer - the role of GIS

Integration

Adrian.G. & Hardy.K.	Facilities & Estates Management with GIS
Albaredes.G.	Corporate GIS and access to information
Alla.P. & Trow.S.W.	Implementation of a GIS - A way to optimising utilities operations
Aybet.J., & Walpole.R.	An Integrated GIS Approach to the Use of Earth Observation Data - Sugar Beet prediction and management system
Bingham.P. & Thorneycroft.R.	GIS in a Small Shire County
Blakemore.M.	Sharing Data - Whose why how?
Bourke.A.	How we Integrated GIS with Existing Central RDBMS over a wide area network
Boyd.G., & Schaap.D.	Integrating Marine Geographical Information
Brand.M.J.D. & Gray.S.	From Concept to reality
Buxton.R. & Ball.C.E.	Geographic Information and the Expert System in Local Authority Development Control
Dale.P.F.	Domesday 2000: The Professional as a Politician
Gadd.S.	GIS - Forging a Key Role in Data Integration
Grimshaw.D.J.	The Transformation of Customer Databases
Gugan.D.J. & Gliddon.D.J.	User Requirements for an Integrated GIS
Hardy.G.A.K.	From GIS to Graphical MIS
Hobson.S.A.	Bridging Islands to Increase Return on Investment
Homer.A.R. & Watson.M.	Integrated Data Saves Money!!
Ives.M.	Reaping the Benefits
Ives.M.J., & Hartley.J.C.	Management of the Implementation of a Digital Records Systems

Keddie.J.G., & Docherty.E.   GIS and the 1991 Census as a cohesive force in Local Government
Langford.M., & Fisher.P.   Measuring the accuracy of cross areal population estimates
Lloyd.J.   Is Data Modelling Enough?
McLaren.R.A. & Healey.R.   Corporate Harmony - A Review of GIS Integration Tools
Mitchell.I.   Corporate GIS - Do the Real Users get the systems they want?
Moore.A.   GIS Implementation in Local Government - the central approach
Morgan.R., & McKay.I.   Corporate Mapping Libraries a Local Authority experience
Mounsey.H.   From Research to Reality - the Diffusion of Innovation
Newell.R.G., & Batty.P.M.   GIS Databases are different
Sian.A.J. & Xavier.R.L.   Integrating Data Derived from within the British Local Government Institutional
                                     Framework
Smith.A.B.   Small Scale Data - A Large Scale Problem
Swainston.R.   Integration of data within GIS - A case study in the Metropolitan Borough of Dudley,
                                     West Midlands
Thomas.B.D.   Off-line Calibration and Validation Facilities at the Earth Observation Data Centre
Vipond.D.L. & Harvey.R.A.   Gluing GIS into the System Integration environment
Waters.R. & Ternouth.P.   Kill the 'G' in GIS!
Weatherill.D. & Keddie.J.   Building a Financial Case for GIS

Local Government Applications

Bingham.P. & Thorneycroft.R.   GIS in a Small Shire County
Bosworth.M.   The Proposed National Street Gazetteer to a Standard Specification
Buxton.R. & Ball.C.E.   Geographic Information and the Expert System in Local Authority Development
                                     Control
Campbell.H. & Masser.I.   The Impact of GIS on Local Government in Great Britain
Cane.S.   Adding Corporate Value to Data - a sugGIStion
Clegg.P.   GIS Implementation - The Sheffield Experience
Clegg.P. & Keddie.A.R.   The Chorley Report 4 Years on - A User Perception
Ellis.S.   How not to use the 1991 Census
Ellis.S., & Gower.C.   A Generic data model for GIS Local Government
Ellis.S.P.   GIS 2000BC - AD2000
Farrell.B.   Corporate GIS in a Scottish Regional Council
Gower.C.J.   GIS Implementation Standards
Jenkins.J.C. & Yardy.A.S.   GIS : A strategic review for Northamptonshire
Keddie.J.G., & Docherty.E.   GIS and the 1991 Census as a cohesive force in Local Government
Litton.D.   GIS in action in Local Government - The benefits of partnerships
Lukes.D. & Vann.P.   Practical Implementation issues in a multi-function organisation
Mahoney.R.P. & McLaren.R.A.   Best practice guidelines for GIS Implementation in GB Local Government
Masser.I., & Campbell.H.   The take-up of GIS in Local Government: The LGMB/University of Sheffield Project
McAllister.C.S.   A Scottish Approach - The Scottish Geographic Information Systems Forum
Miller.P., & Oxley.J.,   Archaeology and Planning: GIS to the Rescue?
Moore.A.   GIS Implementation in Local Government - the central approach
Morgan.R., & McKay.I.   Corporate Mapping Libraries a Local Authority experience
Pugh.D. Black.T. & Mounsey.H.   A National Land and Property Gazetteer? Opportunities for development
Pugh.D. & Cushnie.J.   The Land and Property Gazetteer
Reid.J.A.   Getting Started in GIS - The Cornish Experience
Rietman.J.   The Utility base map of Rotterdam
Roberts.J.   The Hertsmere Experience
Robinson.K.W., & Gill.S.J.   Property Management and O.S. Structured Data
Rowland.J., & Lilley.C.   The wide area connection
Sian.A.J. & Xavier.R.L.   Integrating Data Derived from within the British Local Government Institutional
                                     Framework
Swainston.R.   Integration of data within GIS - A case study in the Metropolitan Borough of Dudley,
                                     West Midlands
Winter.P.   Selling a Corporate GIS
Wiseman.P.G.   Suffolk County Council's corporate GIS pilot study
Worrall.L.   Developing a Corporate Information Systems Strategy
Wren.G., & MacKenzie.R.,   Deadline Oriented GIS - Task Specific

Photogrammetry

Chapman.D. Dowman.I.
& Muller.J.P.   Digital photogrammetry - interfaces with GIS
Davies.A., & Pittard.S.   Photogrammetry : The first step to better process industry management

| Gugan.D.J. | Photogrammetry and GIS |
| Petrie.G. | Photogrammetric input to an environmental GIS |

Remote Sensing

Aspinall.R.J.	A Landscape Ecological Approach to Mapping Distribution and Abundance of Birds with GIS
Aybet.J., & Walpole.R.	An Integrated GIS Approach to the Use of Earth Observation Data - Sugar Beet prediction and management system
Bird.A.C.	GIS Based Data on Land cover Change in the National Parks of England and Wales
Brown.N.	Mapping of Land Cover from space and its use in a Geographical Information System
Brown.R., Slater.J., & Askew.D.	Environmental Monitoring of Protected Landscapes - monitoring environmentally sensitive areas using satellite imagery, within GIS
Daniels.D.J.	High Performance Ground Penetrating Radar System
Green.D.R., & Morton.D.	Acquiring Environmental Remotely Sensed Data from Model Aircraft for Input to Geographic Information Systems
Heard.M., Higgins.C., & Mather.P.,	Prototype Expert Systems for the Geometric Correction of Remotely-sensed Images
McDonald.A.J.W.	Remote mapping of UK geology
Thomas.B.D.	Off-line Calibration and Validation Facilities at the Earth Observation Data Centre

Standards

Adnitt.N.	The Relational Database, Ideal for a Corporate GIS
Arnott.D. & Keddie.A.R.	Data Capture - The Standards and Procedures Utilised within Northumbrian Water Group
Barr.R. & Campbell.K.	The Postal Address as a fundamental Geo-Reference
Bosworth.M.	The Proposed National Street Gazetteer to a Standard Specification
Dhillon.P.	Towards an Integrated Future
Dowers.S.	SQL - The Way Forward
Gower.C.J.	GIS Implementation Standards
Gradwell.D.J.L.	Can SQL Handle Geographic Data?
Haywood.P.	AGI and NTF (National Transfer Format): The Way Forward
Jenkins.J.C. & Yardy.A.S.	GIS : A strategic review for Northamptonshire
Land.N.	The Classification of Spatial Data - A proposal for a National Standard
Monckton.C.	Meta-Information and Future GIS
Pugh.D. Black.T. & Mounsey.H.	A National Land and Property Gazetteer? Opportunities for development
Rickman.D.	Open Systems GIS - Do We Need Them?
Rowley.J.	Summary of Standards Papers
Rowley.J.R.	A Strategy for Geographic Information Standards
Rowley.J.R.	Geographic Information - The Standards Viewpoint
Rowley.J.R.	Standards - A key building block of Geographic Information Strategy?
Sharman.D. & Haywood.P.	The National Transfer Format in 1991
Sian.A.J. & Xavier.R.L.	Integrating Data Derived from within the British Local Government Institutional Framework
Sowton.M.	European Standards - Where are they going?
Walker.P.	The Digital Geographic Information Working Group
Walker.R.S.	Standards for GIS Data
Walker.R.S.	Data quality standards
Woodsford.P.	The AGI Today
Woodsford.P.A.	Using Geographic Data - Practicalities and Politics

Strategy

Afors.J., Chiu.M., & Martin.J.	Designing for corporate information
Albaredes.G.	Data Sharing - a hurdle for GIS
Annand.K., & Hookham.C.	A GIS Strategy for guarding the water environment
Bainbridge.P., Hinton.P. & Curry.D.	GIS: Unlocking the Benefits by Implementing Applications
Callaghan.J.G.	GIS - Increasing business efficiency and profit
Cane.S.	Adding Corporate Value to Data - a sugGIStion

Chapman.D. Dowman.I.
& Muller.J.P.   Digital photogrammetry - interfaces with GIS
Clark.M.J.   GIS Awareness: The Technical and Educational Challenge
Clegg.P.   Basic Issues in Data Capture
Daniels.D.J.   High Performance Ground Penetrating Radar System
Davies.J.   Raster - The Forgotten Frontier
Dhillon.P.   Towards an Integrated Future
Dixon.B.   Transferring Data into Information - Tools and Technologies
Dunn.R. & Harrison.A.   Assessing User Requirements for GIS - the critical stage in implementation: how
                                       prototyping can help you get it right
Easterfield.M.E., Newell.R.G.,
& Theriault.D.G.   Management of Time and Space in GIS
Flowerdew.R. Green.M.
& Lucas.S.   Analysing Local House Price Variations with GIS
Gallagher.S.   The Personal GIS Revolution
Gittings.B.M. Dowers.S. Sloan.T.M. Healey.R.G.
& Waugh.T.C.   Turbo-Charging your GIS: Dealing with Performance Issues
Gradwell.D.J.L.   Can SQL Handle Geographic Data?
Gray.T.   Advanced Methods of Capturing and Using GIS Data in the Field
Green.D.R. & McEwen.L.J.   Turning Data into Information: Assessing GIS User Interfaces
Green.D.R. & McEwen.L.J.   GIS as a component of Information Technology courses in Higher Education:
                                       Meeting the requirements of employers
Grimshaw.D.J.   Broadening Your horizons: Shifting your thinking
Halls.P.J.   Video for Capture of Paper Material - Worthless for Some but Useful for Others?
Harrison. A. Dunn.R.
& Turton.P.   Environmental GIS: Technology, Data and Policy
Hendley.D.J.   Parallel Computing in GIS
Hobson.S.A.   Bridging Islands to Increase Return on Investment
Knott.J. & Shiers.D.   Mobile Maps - Sending vector map based maps across radio networks for use in
                                       unplanned locations by GIS Users
Longley.P. Batty.M. Shepherd.J.
& Sadler.G.   On the use of Digitized Boundaries of Urban Areas in the Spatial Analysis of
                                       Settlements: The Case of South East England
Maguire.D.J.   Spatial Data Management : Multi-User Access to Continuous Spatial Databases
Maguire.D.J., Kimber.B.,
& Laming.R.   A Corporate Map Server!
Masser.I., & Campbell.H.   The take-up of GIS in Local Government: The LGMB/University of Sheffield Project
McLaren.R.A.   Casual User Interfaces for GIS Through the Use of Hypertext Technology
McLaren.R.A. & Healey.R.   Corporate Harmony - A Review of GIS Integration Tools
Mitchell.I.   Corporate GIS - Do the Real Users get the systems they want?
Nathanail.P. & Rosenbaum.E.I.M.   Spatial Interpolation in GIS for environmental studies
Newell.D.   Strategies for managing large AM/FM/GIS systems
Newell.R.G., & Batty.P.M.   GIS Databases are different
Openshaw.S. Cross.A.
& Charlton.M.   Using a super computer to improve GIS analysis
Parsons.E.   The Development of a Multimedia Hypermap
Platt.P.G.   Developments in Scanning, Application & Compression of Colour Raster Data
Raper.J. & Bundock.M.   Extending GIS user interface concepts in the UGIX project
Raper.J.F., & Bundock.M.S.   GIS User Interfaces: A Window on the Future
Rickman.D.   Open Systems GIS - Do We Need Them?
Sheehan.R.   Coastal Zone Management is okay but it's not as good as the real thing!
Shepherd.I.D.H.   Meeting the design challenge of Multi-Sensory GIS
Stuart.N.   Spatial Analysis in GIS: Value Added or Vapour?
Thomas.J.   Picking the Perfect Peripheral
Walker.J.C., & Hampson.E.,   Integrating GIS with the Network using ISDN
Waters.R.   Air Photos to GIS, do we need the map?
Windsor.P. & Aybet.J.   What Object-Orientation Offers to Geographical Information Systems
Winter.B.   Using GIS - The Next Generation
Wood.A.J.   Geographic Information Systems: An Executive Decision Perspective

Utility Applications
Adnitt.N.C.   Number 1 in the Charts , Yorkshire Water GIS, currently available on CD
Alla.P. & Trow.S.W.   Implementation of a GIS - A way to optimising utilities operations

Arnott.D. & Keddie.A.R.	Data Capture - The Standards and Procedures Utilised within Northumbrian Water Group
Bainbridge.P., Hinton.P. & Curry.D.	GIS: Unlocking the Benefits by Implementing Applications
Barker.P.	Geographic Information Systems in BT : BT Uses GIS 40 different ways to manage its business
Clegg.P. & Keddie.A.R.	The Chorley Report 4 Years on - A User Perception
Drake.P.	The use of different Data Conversion options for a GIS application in the Water Industry
Dyson.J.	Data Integration via the Gazetteer
Gadd.S.	GIS - Forging a Key Role in Data Integration
Gadd.S.	Training for Effective Implementation. Are we getting it right?
Gross.H.	Very Clever....but does it really mean anything?
Hartley.J.C. Homer.I.R. Trow.S. Hinton.P.H. & Everett.C.	Inter Utility Exchange of Electronic Map Based Records in the North East of England
Homer.A.R. & Watson.M.	Integrated Data Saves Money!!
Hooper.B.D.	Operational Management of Geographical Information in South West Water
Hooper.B.D.	Five Years on and Meeting new challenges
Ives.M.	Reaping the Benefits
Ives.M.J., & Hartley.J.C.	Management of the Implementation of a Digital Records Systems
Jackson.G.H., & Platt.P.	Machine transfer of Circuit Information from film to computer
McMillan.R. & Williams.D.S.	Reaping the Benefits of a GIS System
Mitchell.I.	Corporate GIS - Do the Real Users get the systems they want?
Newman.B.	Developing a Major GIS Project
Newman.B.	Setting up a major GIS Implementation
Peacock.D. & Rutherford.I.	Concepts into reality: An Account of a GIS Implementation in the South Western Electricity Board
Roberts.G.V. Park.G.W.A. & Cottle.E.J.T.	Manweb's Implementation of a Mapless GIS
Shore.D.G.	Resourcing a Corporate GIS Implementation - A Strategic Viewpoint
Snoxell.J.	Portable GIS in the Water Industry - Dealing with HAUC
Snoxell.J.D.	GIS - Satisfying business needs
Trevor.I., & Russell.D.J.	Data Capture - in weeks not years
Vipond.D.L. & Harvey.R.A.	Gluing GIS into the System Integration environment
Weatherill.D. & Keddie.J.	Building a Financial Case for GIS
Webb.R.A., & Elkins.P.G.	Guidelines for the Implementation and Management of a GIS
Witten.A.P., Peacock.B., & Jacobs.R.	The Role of GIS in managing corporate data within Northeast Water's System architecture
Wood.T.R.	Experience of an operational GIS in the Water Industry
Woodcock.M.E. & Clennell.R.	Can I Use a Digitising Bureau?

AGI awards 1990-1995

Year	Category	Recipient	Description
1990	Chairmans	Dorothy Pugh	Work with the membership committee
1991	Industry	Hoskyns	Implementation of PIMMS at Shropshire County Council
	Journalist of the year	Nick Nuttall	'Shaking up the Old Town Halls' appeared in the Times
	Student of the year	Sally Bishop	Weighted digital terrain models to create land suitability models
	Technology	Action Information Management	Integration of low cost technologies
1992	Chairmans	Walter Smith	Services to the GIS community
	GIS Users	English Nature	Implementing a GIS to monitor sites of special scientific interest
	Industry	ED-Line Partnership	Creation of the Boundaries of census enumeration districts data set
	Journalist of the year	Tim Wright	GIS Article in Which Computing
	Journalist of the year	David Dobson	GIS Article in Which Computing
	Student of the year	John Baxendall	The travelling salesman problem (MSc Thesis)
1993	Journalist of the year	Tim Wright	'Target Markets' appeared in Which Computer Magazine
	Past chairmans	Alistair Keddie	AGI92 Conference Chairman
	Student of the year	Jonathan Makin	An object-oriented simulation of a complex geographic system using GIS
	Student of the year	Joe Aguilar Manjarrez	Construction of a GIS for Tabasco State Mexico
	Technological Progress	Hutchison Paging (UK)	Implementing a GIS and geographic referencing system in the development of a real-time paging network.
1994	Journalist of the year	Claire Gooding	Software at Work supplement to the Financial Times
	Past chairmans	John Leonard	Liaison between AGI & EC in forming Eurogi in 1993
	Student of the year	David Yarwood	Cartographic modelling in network space (MSc thesis)
	Technological Progress	Southern Water Ltd & Procis Software Ltd	Mobile GIS within a Water Utility
1995	Journalist of the year	David Butcher	GIS Special supplement in Planning Week 30.03.95
	Past chairmans	David R Green	Significant and sustained contribution to the AGI
	Student of the year		
	GIS Application	Dr Victor Mesev	Urban land use modelling from classified satellite imagery
	Student of the year		
	Technical achievement	Terence P Dawson	Towards a spatial thesaurus - an expert system approach to fuzzy specification of areas in GIS. (MPhil)
	Technological Progress	Gloucester County Council	Large and Complex corporate GIS implementation

AGI council

AGI officers

Chairman
Alastair Macdonald
Oakbank
Church Road
Bishopstoke
Eastleigh
Hants SO50 6LR

Senior Vice-Chairman
David Mitchell
Coopers & Lybrand
1 Embankment Place
London WC2N 6NN

Junior Vice-Chairman
David Green
Department of Geography
University of Aberdeen
Elphinstone Road
Aberdeen AB9 2UF

Honorary Treasurer
Tom Waugh
GIMMS
30 Keir Street
Edinburgh EH3 9EW

Honorary Secretary
Dorothy Salathiel
Department of the
Environment
2 Marsham Street
London SW1P 3EB

Council members

Chairman
Alastair Macdonald
Oakbank
Church Road
Bishopstoke
Eastleigh
Hants SO50 6LR

Barry Capper
London Borough of Barking
and Dagenham
Civic Centre
Dagenham
Essex RM10 7BH

Chris Corbin
Corbins Consultancy
50 Stanford Road
Brighton
East Sussex BN1 5PR

Bruce Gittings
University of Edinburgh
Geography Department
Drummond Street
Edinburgh EH8 9XP

David Green
University of Aberdeen
Geography Department
Elphinstone Road
Aberdeen AB9 2UF

Colin Hookham
Cambridge Computer
Consultants UK Ltd
18 Oaklands
Fenstanton
Huntingdon
Cambs PE18 9LS

Geoff Kendall
Dataview Solutions
40-42 Parker Street
London WC2B 5PQ

Nick Land
Ordnance Survey
Romsey Road
Southampton SO16 4GU

Vanessa Lawrence
GeoInformation International
307 Cambridge Science Park
Milton Road
Cambridge CB4 4ZD

Richard Ley
Military Survey MOD
GIS Standards
Elmwood Avenue
Feltham
Middlesex TW13 7AE

David Maguire
ESRI UK Ltd
23 Woodford Road
Watford WD1 1PB

Jon Makin
ESRI UK Ltd
23 Woodford Road
Watford WD1 1PB

David Mitchell
Coopers & Lybrand
1 Embankment Place
London WC2N 6NN

Helen Mounsey
Coopers & Lybrand
1 Embankment Place
London WC2N 6NN

David Parker
Department of Surveying
The Unversity
Newcastle upon Tyne
NE1 7RU

Nick Pearce
Railtrack Information
Systems
Room 199, 3rd Floor
Eastside Offices
Kings Cross Station
London N1 9AB

Dorothy Pugh
Cambridgeshire County
Council
Shire Hall
Castle Hill
Cambridge CB3 0AP

John Rowley
Geobase Consultants
28 Church Road
Epsom
Surrey KT17 4DX

Dorothy Salathiel
Room P1/170
Department of the
Environment
2 Marsham Street
London SW1P 3EB

Peter Smith
HM Land Registry
Lincolns Inn Fields
London WC2A 3PH

Sam Sowton
Bourne Cottage
Twyford
Winchester SO21 1NX

David Towns
Perot Systems Europe Ltd
PO Box 4, North PDO
398 Coppice Road
Arnold
Nottingham NH5 7HX

Matthew Tuohy
37 Rectory Close
Carlton
Bedford MK4 7JT

David Unwin
Birkbeck College
Geography Department
7-14 Gresse Street
London W1P 1PA

Tom Waugh
GIMMS
30 Kier Street
Edinburgh EH3 9EW

Major General Mike
Wilson
Room 202, Old War Office
Building
Whitehall
London SW1A 2EU

Mike Worboys
Keele University
Department of Computer
Science
Keele
Staffordshire ST5 5BH

AGI committees

Direct Action Committee

Chairman

Colin Hookham
Cambridge Computer
Consultants
18 Oaklands
Fenstanton
Huntingdon
Cambs PE18 9LS

Bruce Gittings
University of Edinburgh
Geography Department
Drummond Street
Edinburgh EH8 9XP

Richard Ley
Military Survey
Elmwood Avenue
Feltham
Middlesex TW13 7AE

Jon Makin
ESRI UK Ltd
23 Woodford Road
Watford WD1 1PB

John Rowley
Geobase Consultants
28 Church Road
Epsom
Surrey KT17 4DX

**Major General Mike
Wilson**
Room 202, Old War Office
Building
Whitehall
London SW1A 2EU

Peter Woodsford
Laser-Scan Ltd
Cambridge Science Park
Milton Road
Cambridge CB4 4FY

Corporate Affairs Committee

Chairman
Peter Smith
HM Land Registry
Lincolns Inn Fields
London WC2A 3PH

Chris Corbin
Corbins Consultancy
50 Stanford Street
Brighton
East Sussex BN1 5PR

Roland Cunningham
ICL UK Ltd
Observatory House
Windsor Road
Slough
Berkshire SL1 2EY

David Green
University of Aberdeen
Geography Department
Elphinstone Road
Aberdeen AB9 2UF

John Leonard
Delbrook
Hubert Road
St Cross
Winchester
SO23 9RG

Sam Sowton
Bourne Cottage
Twyford
Winchester SO21 1NX

Graham Stickler
Exor Corporation Ltd
Saville Court
Saville Place
Clifton
Bristol BS8 4ET

Information and Education Committee

Chairman
Mike Worboys
Department of Computer Sciences
University of Keele
Keele
Staffordshire ST5 5BG

Frank Bennion
3 Rowney Close
Loggerheads
Market Drayton
Shropshire FF9 4DB

Barry Capper
London Borough of Barking & Dagenham
Civic Centre
Dagenham
Essex RM10 7BH

Trevor Foale
W S Atkins
Woodcote Grove
Ashley Road
Epsom
Surrey

Diana Freeman
AUCE
126 Great North Road
Hatfield
Herts AL9 5JZ

David Green
Geography Department
University of Aberdeen
Elphinstone Road
Aberdeen AB9 2UF

Vanessa Lawrence
Longman Geoinformation
307 Cambridge Science Park
Milton Road
Cambridge CB4 4ZD

David Mitchell
Coopers & Lybrand
1 Embankment Place
London WC2N 6NN

Roy Newell
21 Hartington Villas
Hove
East Sussex BN3 6HF

Anne Tait
Consultant
72 Chatsworth House
Westmoreland Road
Bromley BR2 0RE

David Unwin
Geography Department
Birkbeck College
7-15 Gresse Street
London W1P 1PA

special interest groups

business

Chairman
Geoff Kendall
Dataview Solutions
40-42 Parker Street
London WC2B 5PQ

environment

Chairman
Jonathan Budd
English Nature
Northminster House
Northminster
Peterborough
PE1 1UA

health

Chairman
Paul Smith
Active Software
Datum House
Roentgen Road
Basingstoke
RG24 8NG

marine and coastal zone management

Chairman
David Green
Department of Geography
University of Aberdeen
Elphinstone Road
Aberdeen AB9 2UF

survey and mapping

Chairman
Roy Wood
Pantiles House
22 London Road
Bagshot
GU19 5HN

young GISers

Chairman
Jon Makin
ESRI (UK) Ltd
23 Woodford Road
Watford WD1 1PB

list of AGI members

All members of the AGI are listed below (except individual members). The four main categories are listed separately (sponsor, corporate, local authority and education). The list was correct as at October 1995.

sponsor members

CCTA - Government Centre for Information Systems

Coopers & Lybrand

Department of the Environment

ESRI (UK) Ltd

Genasys II Ltd

HM Land Registry

Intergraph (UK) Ltd

The Local Government Management Board

Military Survey

National Joint Utilities Group

National Rivers Authority

Ordnance Survey

Royal Institution of Chartered Surveyors

SEEBOARD plc

Smallworld Systems

Sysdeco (UK) Ltd

corporate members

Action Information (Management) Ltd

Active Software Ltd

Anglian Water

APIC systems

Architects & Surveyors Institute

Ashtech Europe Ltd

Babtie, Shaw & Morton

Bartholomew Electronic

Beacon Dodsworth Ltd

Binnie & Partners

BKS Surveys Ltd

Blenheim Online

Bristol Water Company

British Cartographic Society

British Gas PLC

British Geological Survey

Brown & Root Civil

C A Design Services Ltd

CACI Ltd

CADAC Ltd

Cambridge Computer Consultants UK Ltd

Concept Systems Ltd

Corbins Consultancy

Cray Systems

Crown Estate Office

Dataview Solutions Ltd

Department of Trade & Industry

Dept. of the Environment (NI) - Water Service

East Midlands Electricity

English Nature

ERDAS UK Ltd

Fairbairn Services Ltd

Forestry Commission

Gardline InfoTech

GDS (UK) Ltd

General Register Office for Scotland

Geobase Consultants Ltd

GeoInformation International Ltd

GIS Services Ltd

Glen Computing Ltd

GMAP Ltd

Graphical Data Capture Ltd

Greater Manchester Research

Greig Fester Ltd.

Guild of Incorporated Surveyors

Home Office

Hunting Engineering

Hydrographic Office

Hydrographic Society

ICL (UK) Ltd

IMASS Ltd

Inland Revenue

IT Southern Ltd

JPB Surveys Ltd

Kingswood Ltd

KPMG Peat Marwick

Laing Technology Group

Land Aspects Parkman Ltd

Laser-Scan Ltd

Leica UK Ltd

London Docklands Development Corporation

London Fire Brigade

London Underground Ltd

Loy Surveys Ltd

MAFF

MANWEB Plc

Mason Land Surveys Ltd

Merseyside Information Service

Metropolitan Police Service

Ministry of Defence

Moss Systems Ltd

MR-Data Graphics

MVA Systematica

MVM Consultants plc

National Grid Company

National Remote Sensing Centre Ltd

Navstar Systems

NERC

NI Housing Executive

North East Water

Northern Electric plc

Northern Ireland Electricity

Northumbrian Water Group

Northumbrian Water Ltd

NORWEB Plc

OPCS

Ordnance Survey

Ordnance Survey N I

Oscar Faber Water

P A Consulting Group

Philip's

Photogrammetric Data Services

Photogrammetric Society

Plowman Craven & Associates

Procis Software

Property Intelligence Ltd

QAS Systems Ltd

RAC Motoring Services

Racal Radar Defence Systems Ltd

RCAHMS

Registers of Scotland

Royal Comm on the Historical Monuments of England

Royal Geographical Society

Royal Mail

Royal Town Planning Institute

RPS Clouston

SAS Institute

Scott Wilson Kirkpatrick

Scottish Homes

Scottish Hydro Electric Plc

Scottish Natural Heritage

Scottish Power

SEMA Group Systems Ltd

Smith System Engineering Ltd

Society of Surveying Technicians

South West Water Services Ltd

South Western Electricity Plc

Southern Water Services Ltd

Structural Technologies Ltd

Survey & Development Services

TEAMS (Taylor Woodrow)

Terraquest Ltd

Thames Water Utilities Ltd

Unisys Ltd

Water Research Centre

Welsh Office

Welsh Water

Welsh Water,SW

Wigwam Information Plc

Yorkshire Electricity

Yorkshire Water Services Ltd

education members

Birkbeck College

De Montford University

English Heritage

Friends of the Earth

Geographical Association

Institute of Hydrology

Intergis, Manchester Metropolitan University

Junta de Andalucia

Kingston University

Lancaster University

London Research Centre

London School of Economics

Luton College of Higher Education

Macaulay Land Use Research Institute

Midland Regional Research Laboratory

National Council for Ed Technology

Soil Survey & Land Research Centre

The Advisory Unit: Computers in Education

University College Cork

University of East London

University of Edinburgh

University of Newcastle

University of Nottingham

University of Paisley

University of Reading

University of Sheffield

University of Southampton

University of Sunderland

University of Ulster

University of Wales

local authority members

Bedfordshire County Council

Birmingham City Council

Buckinghamshire County Council

Cambridgeshire County Council

Central Regional Council

Cheltenham Borough Council

Cheshire County Council

City of Newcastle upon Tyne

City of Stoke on Trent

Cornwall County Council

Cotswold District Council

Devon County Council

Dudley Metropolitan District Council

Durham County Council

Essex County Council

Fife Regional Council

Glasgow District Council

Gloucestershire County Council

Gordon District Council

Grampian Regional Council

Guildford Borough Council

Hampshire County Council

Hertsmere District Council

Highland Regional Council

Horsham District Council

Island Development Committee

Isle of Man Government

Kilmarnock & Loudoun District Council

Knowsley Metropolitan Borough Council

Lancashire County Council

Land Registry - Isle of Man Government

London Borough of Barking & Dagenham

London Borough of Barnet

London Borough of Islington

London Borough of Waltham Forest

Luton Borough Council

Malvern Hills District Council

Mid Glamorgan County Council

Mid-Suffolk District Council

Motherwell District Council

N Hertfordshire District Council

New Forest District Council

Northamptonshire County Council

Nottinghamshire County Council

Orkney Islands Council

Oxford City Council

Powys County Council

Royal Borough of Kensington & Chelsea

Royal Borough of Kingston upon Thames

Royal Borough of Windsor & Maidenhead

Sandwell Metropolitan Borough Council

Sevenoaks District Council

Shetland Islands Council

Shropshire County Council

Somerset County Council

South Glamorgan County Council

South Yorks P.T.E.

Strathclyde Regional Council

Suffolk County Council

Swansea City Council

Tameside Metropolitan Borough Council

Tayside Regional Council

Warwickshire County Council

Waveney District Council

West Lothian District Council

Wiltshire County Council

Wolverhampton Metropolitan Borough Council

Wrekin Council

thanks to our sponsors

The AGI depends on the support of its members in order to serve the geographic information community. Support from our sponsor members is particularly important to provide the resources necessary to carry out our mission. Thanks to:

CCTA

Coopers & Lybrand

Department of Environment

ESRI

Genasys II

HM Land Registry

Intergraph

LGMB

Military Survey

NJUG

National Rivers Authority

Ordnance Survey

RICS

Seeboard

Smallworld Systems

Sysdeco

Get the gist on GIS
with AGI specialist publications

GIS Dictionary version 2.1

ISBN 1 874059 16 0   £10 to members/£20 to non-members

The latest version of the glossary of GIS terms providing a common set of definitions for general use.

Report by the Copyright Working Party
free

This report was produced as a result of a series of meetings held in 1992 to discuss the important issue of copyright and how it affects the geographic information community.

Crime, policing and GIS

ISBN 0 85406 568 7   free to members/£5 to non-members

First published in 1992, this note aims to describe the ways in which GIS has been used for crime and policing applications.

The use of DXF for geographic data
free

Compares the use of DXF for geographic data with NTF, National Transfer Format.

GIS and epidemiology

ISBN 1 874059 06 5   free to members/£5 to non-members

First published in 1992 this note examines the need for a multi-disciplinary approach when tackling epidemiological issues.

Cartography and GIS

ISBN 1 874059 03 9   free to members/£5 to non-members

This note, first published in 1991, looks at the relationship between cartography and the growth of GIS.

GIS in schools

ISBN 1 874059 11 X   **free**

First published in 1993 this note aims to demonstrate that geography and IT courses within the National Curriculum have ample opportunities for using GIS.

The AGI source book for GIS 1995

ISBN 1 874059 15 2   £25 to members/£65 to non-members

Available from John Wiley & Son Ltd, Baffins Lane, Chichester, West Sussex PO19 1UD

Ethics and professionalism

ISBN 1 8740059 12 8   **free**

This GIS note tackles one of the most important issues facing the geographic information industry in the UK, and makes a number of suggestions that have enormous implications for both the industry and the AGI itself.

GIS and coastal zone management

ISBN 1 874059 13 6   free to members/£5 to non-members

This paper discusses the role that GIS might have in better coastal management and examines in detail some of the key issues still requiring resolution in order to progress the use of GIS at the shore.

AGI conference proceedings from 1989 – 1993

Copies of individual papers are available @ £5 for two or £10 for five.

AGI '94 conference papers   £50

ISBN 1 874059 14 4

AGI '95 conference papers   £65

ISBN 1 874059 21 7

BS7567: Electronic transfer of geographic information (NTF)

Part 1. Specification for NTF structures

Part 2. Specification for implementing plain NTF

Part 3. Specification for implementing NTF using BS6690

All three parts available for £100

All these publications are available from

AGI

12 Great George Street

London

SW1P 3AD

Payment with order please (Visa and Access credit card payments accepted); all orders of £10 or over include postage and packing; under £10 add £1 for postage and packing.